Specialised Metabolites of Australia's Customary Medicinal Flora

To this Land, its Stories and Custodians

Specialised Metabolites of Australia's Customary Medicinal Flora

Edward Owen Norman, Melissa Serrurier and Sylvia Urban

CABI

CABI is a trading name of CAB International

CABI
Nosworthy Way
Wallingford
Oxfordshire OX10 8DE
UK

Tel: +44 (0)1491 832111
E-mail: info@cabi.org
Website: www.cabi.org

CABI
200 Portland Street
Boston
MA 02114
USA

T: +1 (617)682-9015
E-mail: cabi-nao@cabi.org

CSIRO acknowledges the Traditional Owners of the lands that we live and work on across Australia and pays its respect to Elders past and present. CSIRO recognises that Aboriginal and Torres Strait Islander Peoples have made and will continue to make extraordinary contributions to all aspects of Australian life including culture, economy and science. CSIRO is committed to reconciliation and demonstrating respect for Indigenous knowledge and science. The use of Western science in this publication should not be interpreted as diminishing the knowledge of plants, animals and environment from Indigenous ecological knowledge systems.

A catalogue record for this book is available from the British Library, London, UK.

ISBN-13: 9781836991557 (hardback)
9781836991564 (ePDF)
9781836991571 (ePub)

DOI: 10.1079/9781836991571.0000

Edited by Kerry Brown
Cover design by Cath Pirret
Typeset by Envisage Information Technology

Front cover: (left to right) *Eucalyptus leucoxylon* (yellow gum), *Hakea multilineata* (grass-leafed hakea), *Hakea scoparia* Meisn. subsp. *scoparia* (kangaroo bush subspecies) (photos by Sylvia Urban and Rick Tinker).
Front and back covers: background graphic by pro500/Shutterstock.com
Printed and bound by CPI Group (UK) Ltd, Croydon, CR0 4YY

Contents

Preface

INTRODUCTORY REMARKS

In the vibrant tapestry of Earth's biodiversity, the intricate dance between nature and spirituality finds its most profound expression in the sacred realm of Indigenous flora. As we embark on a journey through the pages of this exploration, we are invited to explore the rich plant life that has woven itself into the cultural and spiritual fabric of Indigenous communities around the world.

The wisdom of the ages is encoded in the leaves, blossoms and roots of these resilient plants, each one a living testament to the profound connection between humanity and the natural world. In this preface, we set forth on a voyage that transcends the boundaries of mere botanical study, delving into the heart of a spiritual symbiosis that has flourished for centuries.

Here, amidst the verdant landscapes in ancient traditions, we encounter a profound reciprocity, one that extends beyond the utilitarian relationship between humans and the plant kingdom. Indigenous flora, revered and cherished, becomes a conduit to the divine, a manifestation of the sacred that intertwines seamlessly with daily life and ritual.

As we turn the pages that follow, we will explore the tales of plant spirits that whisper through the rustling leaves, the essential secrets harboured within the bark of the ancient trees and the symbolic residence of vibrant blooms in sacred ceremonies. This is an odyssey into the heart of spiritual ecologies, where each petal and frond speaks volumes about the cultural narratives, ceremonies and cosmologies of Indigenous Peoples.

May this journey be an awakening, an invitation to recognise and celebrate the sanctity of the natural world and its profound significance in the spiritual realm of humanity. Through the stories within these pages, we hope to cultivate a deeper understanding and appreciation for the intricate relationship between Indigenous communities and the flora that sustains not only their bodies but their souls.

Indigenous Peoples around Australia share a profound and sacred bond with the plants and lands that cradle their communities. For these communities, the land is not merely a physical expanse but a living, breathing entity, a spiritual guide and custodian of their cultural identity. The flora, from towering ancient trees to delicate wildflowers, serve as more than botanical companions; they are revered teachers, healers and messengers from the spiritual realm. Through intricate rituals, ceremonies and oral traditions passed down through generations, Indigenous Peoples forge an unbreakable connection to the plants and land, recognising them as essential partners in a sacred dance of reciprocity and reverence. This spiritual communion reflects a holistic worldview, where every plant, rock and stream is animated by a spiritual essence, fostering a harmony that extends beyond the tangible into the realms of the metaphysical and the eternal.

Through my ancestral lineage and teachings over the years, I have embraced the holistic and natural approach to medicine to treat a variety of illnesses. This preface describes my lived experiences with Indigenous medicinal flora and fauna with community members, as well as the teachings they passed down through my ancestral lineage. I wish to make note that every community is different and may use flora and fauna for different holistic purposes; the experiences I have encountered are the past lived experiences with specific community members in different parts of Queensland.

My name is Melissa and my ancestral connection to Indigenous culture is through the community in far north Queensland, Guugu Yimidhirr. I have worked in the education sector for many years as a teacher and also in the higher education space. I educate people around Indigenous perspectives and ways of working, as well as cultural traditions and responsible practice. In addition to working in the education and Indigenous spaces, I am also undertaking my doctorate in Indigenous education. I am hoping to educate and enlighten the education space when it comes to Indigenous practises and ways of learning.

I grew up in parts of Brisbane (Yuggera) and Townsville (Nyawaygi) and had connections to Cairns (Yidinjdji) as well (https://aiatsis.gov.au/explore/map-indigenous-australia). In my time living in Townsville, I was introduced to the holistic approach of medicinal plants by community members to assist in managing my rheumatoid arthritis. Through my years living in Townsville and in Brisbane my connection to mob became stronger and their teachings of medicinal plants became a pertinent part of my life. They showed me how they prepared medicinal plants and what specific parts of the plants did for the body. These experiences with native flora and Indigenous practices provided me with medicinal plants to help with the extreme pain of my rheumatoid arthritis, for common cold remedies, stomach settling practices for Crohn's disease and common skin treatments. Even though I was a young girl at the time, I still remember the experiences with community members and to this day I still use some of these holistic practices to treat aches and pains, common colds and stomach ailments. In the pages that follow I will talk about my lived experiences with some of the traditional flora and teachings from Indigenous community members.

SNAKE VINE (*TINOSPORA SMILACINA* (BENTH. 1861))

In the heart of many Indigenous communities, the snake vine emerges as a botanical marvel, revered not only for its sinuous tendrils but also for its potent medicinal properties (see *Fig. 0.1*). This resilient vine, gracefully weaving through the verdant landscape, holds a sacred place in traditional healing practices. Indigenous Peoples have harnessed its therapeutic potential for generations, creating a profound synergy between nature and wellness.

The snake vine, with its lush foliage and slender stems, is a repository of medicinal compounds that Indigenous healers expertly extract to address a myriad of ailments. The leaves, roots and even the bark of this plant contain a treasure trove of bioactive compounds, each with its unique healing prowess. Traditionally, the vine has been used to treat conditions ranging from digestive disorders to respiratory ailments, showcasing the holistic approach of Indigenous medicine.

In the hands of skilled healers, snake vine becomes a versatile remedy, used in various forms such as infusions, poultices or tinctures. Its anti-inflammatory and analgesic properties make it particularly valuable for soothing joint pains and muscle aches. Moreover, its role in alleviating digestive discomfort and promoting overall well being has earned it a revered status in the pharmacopeia of Indigenous cultures.

Beyond its physical healing attributes, the snake vine is often infused with spiritual significance. Indigenous communities believe that the plant not only mends the body but also acts as a conduit for spiritual healing, aligning the energies within and fostering a sense of balance and harmony.

The plant is found in and along the boundaries of dry rainforest and dry eucalyptus forest in Queensland, the Northern Territory, Western Australia and northern New South Wales. Specimens of the snake vine were collected near the Endeavour River at Cooktown in 1770. The vine photographed was on the roadside near the Rocky Bay lookout. This plant has long been used in Aboriginal medicine, mainly for the treatment of headaches, rheumatoid arthritis and other inflammatory disorders and for snakebite. For headaches, mashed stems were wrapped around the head. For stings and bites, heated leaves were applied.

Medicinal plants were always common plants. Indigenous Peoples carried no medicine kits and had to have remedies that grew at hand when needed. If a preferred herb was not available, there was usually a local substitute. Plants were prepared as remedies in several ways. Leafy branches were often placed over a fire while the patient squatted as close as possible to the fire and inhaled the steam and vapour. Sprigs of aromatic leaves might be crushed and inhaled, inserted into the nasal cavity or prepared into a pillow on which the patient slept. To make an infusion, leaves or bark were crushed and soaked in water (sometimes for a very long time); the infusion was then drunk or washed over the body. Ointment was prepared by mixing crushed leaves with animal fat. Other external treatments included rubbing down the patient with crushed leaf paste, fruit pulp or animal oil, or dripping milky sap or a gummy solution over the person. Most plant medicines were externally applied.

As a child I remember community members preparing the snake vine and rubbing it across my joints. They also prepared a tea-like drink that I drank over a few days. The prepared snake vine that was rubbed externally on my joints helped to relieve inflammation and the pain caused by the swollen joints. Mob showed great care and precision when preparing the snake vine for my treatment. When applying the snake vine ointment, they told me healing stories and spiritual connection to these plants.

Fig. 0.1. *Tinospora smilacina* growing in natural habit (QLD).

TEA TREE (*MELALEUCA ALTERNIFOLIA* ((MAIDEN & BETCHE) CHEEL, 1924))

The Indigenous use of tea tree in Australia has deep historical and cultural roots. It is often found in and is native to Australia's Bungawalbin Valley, in northern New South Wales, with some species also found in the Atherton Tablelands of Queensland (see *Fig. 0.2*).

Tea tree oil, derived from the leaves of this remarkable plant, has been a staple in the Indigenous pharmacopeia. Beyond its tangible healing properties, the oil carries with it the essence of cultural heritage. In the embrace of cultural rituals and ceremonies, tea tree becomes more than a remedy; it is a conduit to the spiritual realm. The aromatic infusion of tea tree leaves in hot water, creating a medicinal tea, becomes a ritualistic practice believed to fortify the immune system and maintain balance within the body.

As we delve into the Indigenous use of tea tree, we must recognise the importance of preserving this rich cultural heritage. It is a legacy that transcends time, offering profound insights into the symbiotic relationship between people and the natural world. This exploration is not just a journey into the past; it is an invitation to appreciate and respect the Indigenous wisdom that continues to shape the present and guide the future.

Indigenous communities have long recognised the medicinal properties of tea tree, utilising its leaves for a variety of purposes. Traditionally, tea tree leaves were crushed and used to make poultices to treat wounds, cuts and skin infections. The antiseptic and antibacterial properties of tea tree oil, extracted from the leaves, were harnessed to cleanse and heal skin ailments. Additionally, the leaves were often infused in hot water to create a medicinal tea, believed to have immune-boosting properties. Indigenous Peoples valued tea tree not only for its therapeutic uses but also for its spiritual significance, often incorporating it into cultural rituals and ceremonies.

In the 1920s, scientific experiments proved that the tea tree oil's antiseptic potency was far stronger than the commonly used antiseptic of the time. Since then, the oil has been used to treat everything from fungal infections to acne.

Today, tea tree oil has gained global recognition for its versatile applications to skincare and healthcare, thanks in part to the traditional knowledge passed down through generations by Indigenous communities.

Growing up, and even today, the use of tea tree is a staple in my household. As a child we often used tea tree to treat cuts and grazes, which was something we often did from falls when playing outside. I also used tea tree oil diluted in warm water for my hair; it kept it healthy and shiny and as a kid it kept away the dreaded headlice at school. Today, as an adult I still often use tea tree to treat skin wounds and also for throat ailments. Tea drinks and lozenges are my preferred method of using tea tree when I have a sore throat.

In conclusion, the use of tea tree in Indigenous communities stands as a poignant testament to the intricate interplay between nature, culture and wellbeing. This ancient practice encapsulates a profound understanding of the land, passed down through generations, where the tea tree (*Melaleuca alternifolia*) emerges as a sacred healer and guardian. The therapeutic applications, from poultices for physical ailments to spiritual ceremonies invoking the essence of tea tree, showcase a holistic approach to health that seamlessly blends the practical and the metaphysical.

Fig. 0.2. *Melaleuca alternifolia* growing in natural habit (QLD).

EUCALYPTUS (*EUCALYPTUS* (MYRTACEAE))

In the vast expanse of Australia's diverse landscape, the eucalyptus tree (genus *Eucalyptus*) stands tall and proud, embodying a botanical cornerstone of Indigenous knowledge and cultural significance (see *Fig. 0.3*). This section explores the myriad of ways in which this iconic tree has shaped and sustained Indigenous communities for millennia.

As custodians of a land, with a rich knowledge and understanding of ecosystems, Indigenous Peoples have developed a deep understanding of the eucalyptus's medicinal, spiritual and practical attributes. This ancient relationship finds expression in traditional healing practices where the aromatic leaves of various eucalyptus species are used to address a spectrum of ailments. From the soothing vapours of eucalyptus-infused steam to the application of eucalyptus oil for its antiseptic properties, these practices have not only endured through time but have become integral to the cultural identity of Indigenous communities.

Beyond the realm of healing, the eucalyptus tree plays a pivotal role in Indigenous spirituality and connection to the land. Some eucalyptus species are considered sacred, with their presence in the landscape marking significant ceremonial sites. The leaves, bark and even the distinct aroma of eucalyptus are interwoven into cultural practices, symbolising a profound communion with the natural world. Eucalyptus is often used during smoking ceremonies, as it is a symbol of strength, protection and abundance. The practice of burning eucalyptus leaves is to purify and negate negative energy and Indigenous Peoples continue to treat it as a sacred plant. Through stories, ceremonies and art, Indigenous Peoples continue to celebrate the eucalyptus as a living embodiment of their ancestral connection to Country.

Various eucalypt species can be found in all states and territories across Australia, except for the continent's driest regions (see *Fig. 0.4*). Many are found in Queensland where I grew up, so it was easily and readily available for medicinal use. While growing up, we infused eucalyptus leaves to treat body aches and pains, coughs and severe fevers when we were sick. The young shoots of the eucalyptus plant were sometimes chewed for treatment of colds. As with most eucalypt species, it was

Fig. 0.3. *Eucalyptus curtisii* growing in natural habit (NT).

also used as an effective antiseptic. Many specieshave roots that could be tapped for water for survival. As an adult today, I commonly use eucalyptus lozenges and specific mouthwashes to treat throat ailments and coughs .

This exploration of the Indigenous use of eucalyptus serves as a tribute to the resilience and ingenuity of Australia's Indigenous Peoples. It is an acknowledgment of a sustainable and reciprocal relationship with the environment that transcends mere utilitarianism. As we delve into the pages that follow, may we gain a deeper appreciation of the intricate knowledge and wisdom embedded in the Indigenous Peoples' use of eucalyptus, recognising it not only as a botanical ally but as a vital thread in the vibrant tapestry of Australia's cultural heritage.

WATTLE (*ACACIA* (FABACEAE))

In the sunburnt landscapes of Australia, the wattle tree (*Acacia* spp.) unfurls its golden blossoms, casting a vibrant glow across the terrain and beckoning us into the heart of Indigenous wisdom (see *Fig. 0.5*). From ancient times to the present, the Indigenous use of wattle resonates as a testament to the resilience, resourcefulness and profound connection between Australia's original custodians and the land they call Country.

Wattle, with its distinctive golden blooms and feathery leaves, holds a special place in the traditional practices of Australia's Indigenous communities. The versatile nature of the wattle tree is reflected in its myriad uses, from medicinal remedies to tools, ceremonial artefact and even as a significant source of food. Indigenous Peoples have harnessed the healing properties of wattle, utilising various parts of the tree to create poultices for wounds, infusions for treating ailments and decoctions for ceremonial purposes. These practices, handed down through generations, embody a holistic approach to health that intertwines seamlessly with the surrounding natural environment.

Beyond its practical applications, the wattle holds a symbolic significance in Indigenous culture. Its blossoms mark the changing seasons, and different species are associated with specific times of the year. Wattle has found expression in art, storytelling and cultural ceremonies, symbolising renewal, connection to Country and the cyclical nature of life. The use of wattle becomes a manifestation of the spiritual resonance between Indigenous communities and the Australian landscape, weaving a narrative that extends far beyond the material realm.

In Australia, the wattles are the largest genus of flowering plants. Of the 1380 species of acacia in the world, Australia has ~985 and they are extremely diverse – found in habitats from rainforest to arid lands. The largest number of species is found in the semi-arid wheatbelt region of Western Australia, but high numbers also occur in the rocky tablelands of the Great Divide in eastern Australia.

Not only used for medicinal purposes, the wattle's sap, pollen and wood were transformed into tools, musical instruments, dyes and glues, and even perfumes. Back then, the bloom of the wattle flowers also heralded springtime, eels arriving in streams and whales along the coast. Some wattle species have been used to treat headaches, rheumatism and general aches and pains, colds and toothaches.

The plant is also valued by Koori People in Victoria for its medicinal properties and food source. The gum of several wattle varieties is edible, with a mild sugary

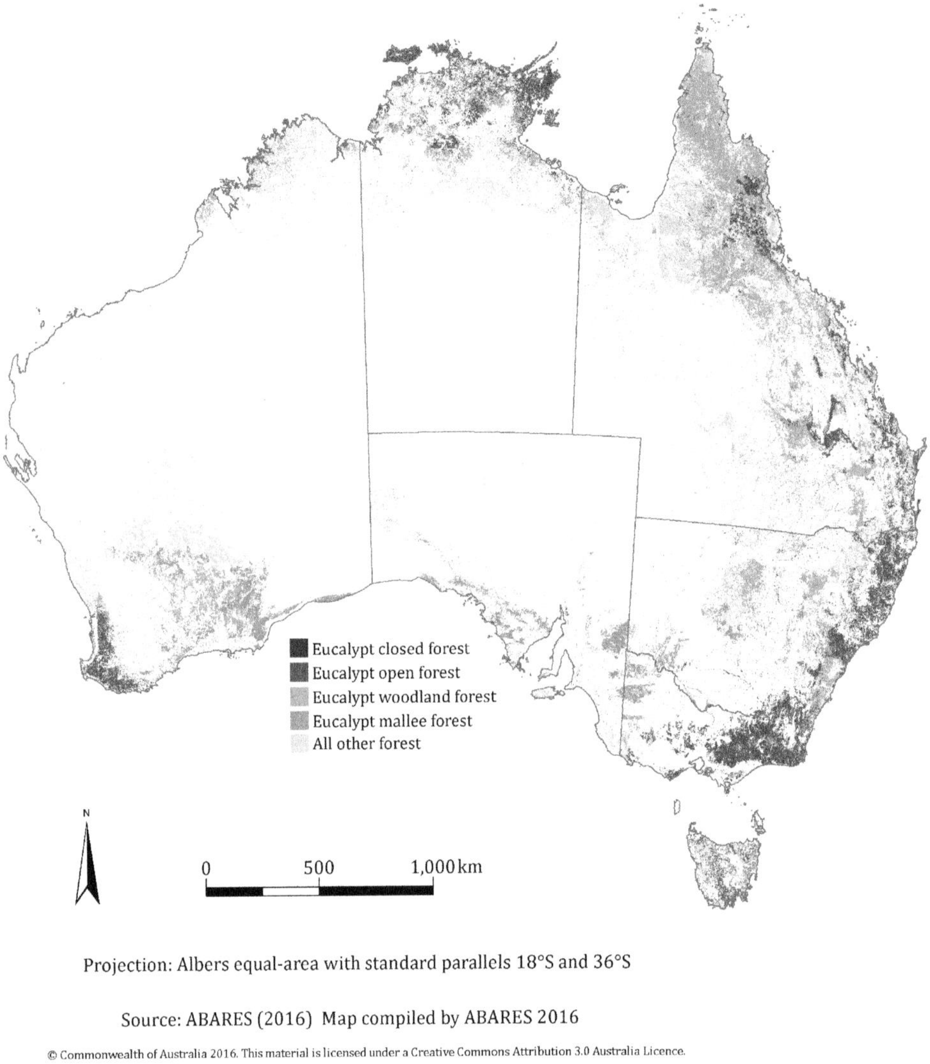

Fig. 0.4. Distribution of eucalypt native forest, 2013.

sweetness, but it can become unpalatable when offset by high amounts of the plant's tannin. The gum in some cases was utilised in the production of a drink, which served as a remedy for coughs, colds and chest infections. Gum could also be applied directly to cuts and abrasions as an antiseptic preparation, either in a solid or powdered form. Wattle bark has been used by Koori People as a traditional fish poison and pain killer, and the the smoke from the plant has been used in the treatment of diarrhoea.

From my personal experiences with community members, wattle can be used in a few different ways. First, the bark from some wattle species can be boiled and used to soothe sore joints and stomach problems and even help with sleep. I found this useful growing up when I was trying to find something to soothe my arthritic pain and to help me sleep at night. However, it is not the best tasting drink around, as you can probably guess from boiling bark. It did help soothe some of my mild aches and pains from my arthritis and it also was a nice natural way to assist with sleep when I was on medications that often prevented the ability to fall asleep.

The versatile and multifaceted nature of the wattle tree has shaped the daily lives, spiritual practices and cultural identity of Australia's Indigenous Peoples for centuries. As we reflect on the healing remedies, practical applications and symbolic significance embedded in the indigenous use of wattle, we come to appreciate a holistic approach to wellbeing that harmonises with the rhythms of the natural world. The golden blossoms of the

Fig. 0.5. *Acacia melanoxylon* growing in natural habit (QLD).

wattle tree not only signify the changing seasons but also serve as a radiant symbol of renewal and continuity.

WITCHETTY GRUB (*ENDOXYLA LEUCOMOCHLA* (TURNER 1915))

Yes, I am aware that a witchetty grub is not a plant species, but I couldn't resist its medicinal and survival properties. In the heart of Australia's vast and ancient landscapes, beneath the surface, lies a hidden treasure that has sustained Indigenous communities for generations – the witchetty grub (larvae of several moth species; see *Fig. 0.6*). The witchetty grub is often found in the roots of the witchetty bush plant *(Acacia kempeana* (F. Muell. 1882)). The larval stage of the large cossid wood moth, it grows up to 12 cm in length and then buries itself ~60 cm below the ground, feeding on the root sap of the witchetty bush. It is said in Indigenous communities that 10 witchetty grubs are sufficient to provide the daily protein needs of an adult.

For Indigenous communities, the witchetty grub is far more than a mere source of sustenance; it is a living embodiment of cultural practices and traditional knowledge. Revered for its nutritional richness, the witchetty grub has been a staple in the diet of Indigenous Peoples for centuries.

The harvesting and preparation of witchetty grubs are steeped in cultural rituals, passed down through generations. Knowledge about the seasonal availability of the grubs, their habitats and the methods for extracting and preparing them has been intricately woven into the fabric of Indigenous life. The significance of the witchetty grub extends beyond the realm of culinary practices; it is embedded in oral histories, art and Dreaming stories that illustrate the interconnectedness of humans, nature and the spiritual realm.

Eaten either raw or lightly cooked in hot ashes or barbecued, they are sought as a high-protein food by Indigenous Peoples. Having eaten witchetty grubs with some community elders, when they were teaching me about bush tucker and survival, the raw witchetty grub tastes similar to raw almonds or a bit like chicken. When cooked, the skin becomes crisp like roast chicken and the inside becomes chewy. Depending on your taste buds, these cooked grubs will either taste like chicken or prawns with peanut sauce. For me, it was definitely a chicken texture with the slight hint of peanut flavour. They're not all that bad once you get over the fact they are

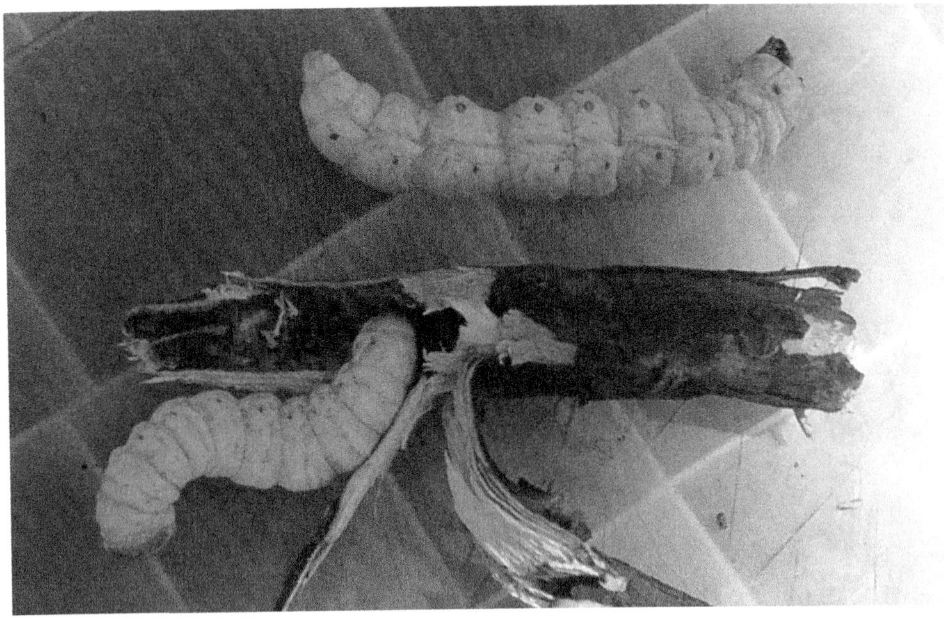

Fig. 0.6. *Endoxyla leucomochla* (witchetty grub).

an insect, quite tasty and nutritious. I highly recommend tasting them if you are ever given the opportunity.

Witchetty grubs are not only used as a food source, but also for medicinal purposes as well. Indigenous Peoples turned them into a paste and placed them on burns on the skin and covered it with a bandage to seal and soothe the skin.

The relationship between Australia's Indigenous Peoples and these natural resources transcends mere survival, weaving a narrative of resourcefulness, respect for the land and a deep understanding of the delicate balance between humans and their environment. The witchetty grub, a nutritional powerhouse, becomes more than a source of nourishment; it symbolises a profound connection to ancestral traditions, the cycles of nature and the sacred stories that echo through generations. Similarly, the witchetty bush, with its role in guiding the harvest of grubs and its presence in cultural practices, embodies the holistic wisdom that underscores Indigenous approaches to sustainable living.

The spiritual connection between Indigenous Peoples and the land, mediated through the profound relationship with Indigenous plants, is a testament to the depth of cultural and ecological wisdom embedded within these communities. The significance of native flora extends beyond its utilitarian value for sustenance and medicinal purposes; it encapsulates a spiritual interconnectedness that is integral to the cultural identity of Indigenous Peoples. The plants themselves become carriers of sacred stories, symbols, and rituals, fostering a sense of continuity between past, present and future generations. The spiritual significance of Indigenous plants is inseparable from the broader concept of Country, encapsulating not just physical landscapes but the spiritual essence of Place.

CONCLUDING REMARKS

This spiritual connection to the land through plants not only enriches Indigenous cultures but also offers valuable insights for contemporary society. It prompts reflection on sustainable and holistic approaches to environmental stewardship, challenging conventional notions of resource extraction and land use. As we navigate an era of ecological challenges and climate change, recognising and respecting the spiritual significance of Indigenous plants can inform more harmonious and reciprocal relationships between humanity and the natural world.

Ultimately, acknowledging and appreciating the spiritual connection between Indigenous Peoples and their native flora fosters a more profound understanding of the intricate web of life, culture and spirituality. It is a call to action, urging societies to embrace Indigenous knowledge, preserve biodiversity and cultivate a shared responsibility for the wellbeing of the land we all inhabit.

In doing so, we can collectively move towards a future that recognises the spiritual essence of the environment and honours the cultural heritage of Indigenous communities, ensuring a more sustainable and interconnected world for generations to come.

Over the years, I have been honoured to experience and hear the teachings of how these medicinal plants are used in everyday life for Indigenous communities, treating mild common conditions, all the way to inflammation treatment and anaesthetic properties. Indigenous cultural healings and teachings opened my eyes to holistic experiences and helped me understand the connection to Place and Land and how if we continue to look after the land, the land will continue to look after us.

Acknowledgements

The authors acknowledge the People of the Woi wurrung and Boon wurrung language groups of the eastern Kulin Nation (VIC), as well as the Gadigal and Garigal local clans of the Eora Nation (NSW), on whose unceded land they conduct their business. The authors respectfully acknowledge their Ancestors and Elders, past and present. The authors also acknowledge the Traditional Custodians and the Ancestors of the lands and waters across Australia. What follows is a short statement outlining some pertinent aspects regarding the terminology used throughout this book, as many of the sources and leads from which the research was drawn are predicated upon customary knowledge.

Historically, throughout the literature, terminology referring to Indigenous Peoples (a word choice which we consider below in the Australian context, and the context of this book) varies substantially, is conflicted and can verge on being imprecise, discriminatory and at times, offensive. For us as authors, it is important both to acknowledge variation and conflict and to be transparent with the reader on our thinking around certain terms. With this in mind, our use of the term 'Indigenous Peoples' here aims to be inclusive and takes into account the historical literature around customary medicine in Australia. Importantly, 'Indigenous Peoples' is often now used in collectively referring to both Aboriginal Australians (historically, the myriad original groups, nations and custodians of the Australian mainland and many of its islands, such as Tasmania, Fraser Island, Hinchinbrook Island, the Tiwi Islands and Groote Eylandt, but excluding the Torres Strait Islands that separate far northern continental Australia's Cape York Peninsula and the island of New Guinea) and Torres Strait Islanders. In turn, the historical literature pertaining to customary medicine in Australia does not explicitly differentiate between these two groups. In this way, 'Indigenous Peoples' in the text is a usage aiming to be as inclusive as possible, and we have used the term throughout. When referring to those Indigenous Peoples who have, over long periods of time including of course no less in the present, acquired and innovated vast amounts of evolving knowledge about the world, we have used the appropriate term: First Scientists.

The authors would also like to distinguish between 'traditional knowledge' and the related term, 'customary knowledge'. Indigenous medicinal plant knowledge is usually referred to as 'traditional knowledge'. This term refers to the knowledge held by respected Indigenous knowledge custodians and could bear the implication of a static body of knowledge, untouched by changing times. The knowledge of Indigenous peoples is constantly evolving. It is not static. Out of respect of this evolving knowledge, the term customary knowledge is used, in recognition that this knowledge incorporates traditional and contemporary knowledge, including innovations that may have occurred, such as using new resources (e.g. introduced plant species), technologies and practices. As such, customary knowledge connotes a modern, evolving knowledge system. The term 'customary medicine', used in the title and throughout this work, was adopted by the authors based on what was deemed to be the correct terminology from the literature (rather than any of the myriad 'unofficial' alternatives: Bush Medicine, Traditional Aboriginal Medicine, Australian Aboriginal Herbal Medicine, etc.). It is the intention of the authors to set a precedent that henceforth the most appropriate term for this field of research into the customary knowledge of Australian medicinal flora by Indigenous Peoples is 'customary medicine in Australia', with this book focusing specifically on the phytochemical constituents of the 'Australian customary medicinal flora'.

About the authors

Dr Edward Owen Norman is an applied scientist in chemistry and data science. Since completing his PhD in Natural Products at RMIT in 2022, he has been engaged in further postgraduate coursework in Health Data Science: Bioinformatics (UNSW School of Medicine and Health) while working as an academic in the Schools of CSE and Chemistry. Edward is working towards an understanding of ethnomedicine, through biostatistics, chemometrics and metabolomics, and is concerned about the sociocultural factors and ethical considerations that underlie work in this space.

Melissa Serrurier is an advocate and specialist in Indigenous and inclusive education, who brings a profound commitment to cultural inclusivity and equitable learning to her writing. A descendant of the Guugu Yimidhirr nation, Melissa draws upon her connection to her heritage and educational background in crafting her work. With over 10 years of experience in both traditional and academic settings, Melissa has been instrumental in advancing Indigenous education methodologies and fostering a sense of cultural pride among students. She holds a Bachelor's degree in education and is currently undertaking her Doctorate (PhD) in Education (Indigenous Perspectives) and has contributed to the development of curricula that authentically represent Indigenous perspectives. Beyond her writing, Melissa actively engages with Indigenous communities, participating in cultural and educational initiatives. She continues to pursue her passion for Indigenous and Inclusive Education, hoping that future generations of students are immersed in culture and history and feel accepted and successful in all types of educational journeys.

Professor Sylvia Urban is one of Australia's leading natural product chemists working in bioactive natural product discovery. This has included the discovery of anticancer natural products that have led to patents with industry. She has previously held research appointments at Griffith University and the University of Canterbury. She established and leads the Marine and Terrestrial Natural Product (MATNAP) research group at RMIT University and is a Fellow of the Royal Australian Chemical Institute (FRACI). Her interests are in the areas of separation science, NMR spectroscopy and mass spectrometry for small compound structural confirmation and the discovery of bioactive chemical compounds (principally antimicrobial and antiparasitic). As an expert in analytical organic chemistry, her research also extends into industry-based projects. Sylvia is also Program Manager of the Bachelor of Science degree at RMIT, is a passionate educator and leader in transforming education in STEM and is a Senior Fellow of the Higher Education Academcy (SFHEA) in the UK. In 2019 she was awarded an Australian Award for University Teaching Citation for Outstanding Contribution to Student Learning. Sylvia has served as the Chair and the Deputy Chair of the Women Researchers Network (WRN) at RMIT and is currently a commnittee member of the RACI Women in Chemistry group. Her focus is to promote better opportunities for women in STEM and is committed to Indigenous Science outreach with high school students and teachers. Sylvia is also the Reconciliation and Responsible Practice Facilitator for the School of Science (STEM College) at RMIT University, a role that she is passionate about.

Cultural sensitivity warning

Readers are warned that there may be words, descriptions and terms used in this book that are culturally sensitive, and which might not normally be used in certain public or community contexts. While this information may not reflect current understanding, it is provided by the authors in a historical context.

This publication may also contain quotations, terms and annotations that reflect the historical attitude of the original author or that of the period in which the item was written, and may be considered inappropriate today.

Aboriginal and Torres Strait Islander peoples are advised that this publication may contain the names and images of people who have passed away.

1 Introduction

Prior to embarking on this introductory chapter, it seems prudent to outline the origin and purpose of the text. The work presented here was compiled throughout the PhD candidature of the primary author, who is an Australian chemist working in the relatively complex and sociopolitically charged realm of ethnomedically based natural product (NP) chemistry in Australia. The literature that has been drawn upon was recorded over the past centuries by chemical researchers who were often prompted by academic interactions with Aboriginal and Torres Strait Islander Peoples (Indigenous Peoples) and Indigenous Cultural Intellectual Property (ICIP), but through an inescapably Western lens. As such, this work will also be biased in that direction, despite having every intention to explore, highlight and substantiate the deep understanding that Aboriginal and Torres Strait Islander people have of the land, its flora and their benefits. It is imperative that more Indigenous Peoples are involved in the direction the field takes, this being a task to be seriously addressed through a coordinated effort from tertiary institutions and the government. The authors hope that this resource will be utilised and developed by sympathetic entities over the coming decades.

The purpose of this book is to serve primarily as a comprehensive resource for researchers in the field of NPs, by bringing together a vast swathe of phytochemical research into the native flora reported to be in customary medicinal use among Indigenous Peoples. What emerges from this primary purpose is a less intentional benefit, which is the value that can be derived by all of the disparate intersecting fields that might be inspired to pursue research into the patterns and connections afforded by centralising this information. The rise of technologies that can provide insight into the concealed structures within complex systems, whether they be biological, ecological or cultural, has vastly expanded our capacity to reflect upon the applied knowledge derived from ICIP. An emphasis on the importance of respectful integration of customary knowledge into the scientific machine is perpetuated throughout this text; unfortunately, that does not necessarily extend to those who may seek to gain advantage from the valuable ideas detailed within. It is the belief of the authors that this work contributes meaningfully to the field and that it does so while honouring the knowledge systems from which it has resulted.

Integrating more disparate stakeholders into the development of this field is a critical aspect of this work. By highlighting the ethnomedical sophistication of First Scientists, public awareness can be raised regarding the value of customary knowledge and its relevance to contemporary healthcare. Integration of this knowledge into the educational curricula at the primary, secondary and tertiary levels can foster a deeper engagement with diverse worldviews and medical traditions, encouraging further exploration and respect. Healthcare professionals are key agents in this process, particularly in communities where customary medicine is widely practised. Cultural sensitivity and appreciation for effective treatments predicated upon the pharmacological potential of Australian native flora cultivates ongoing dialogues between evolving paradigms. Broader societal issues also emerge from work such as this, but lie outside the scope of the book itself, such as conservation of biodiversity, sustainable development and the preservation of cultural heritage. The plants addressed throughout this book, as well as Indigenous Peoples' customary knowledge about their botanical and medicinal properties, have their own trajectories that need to be championed. Each plant, each compound, each use, carries a story that holds significant meaning to many people. These plants are integral to their ecosystems and these stories are integral to our modern sociocutural system. The development of commercial endeavours relating to the compounds these plants bear can offer economic opportunity and self-sufficiency from which communities could really benefit. Yet this raises ethical issues implicit in the nature of bioprospecting, with sustainable development and the sharing of benefits to the source communities as key issues. Documentation of this knowledge helps in the process of cultural recognition and preservation, something that enriches our growing nation, ideally breaking down barriers that have historically separated Western science and Indigenous knowledge systems.

Growing up on a property bordered on three sides by Ku-Ring-Gai (Guringai) National Park, the lead author, Edward, has spent his adult life in an attempted comprehension of the what, how and why of the natural world. This process has instilled a profound respect for the seemingly inexplicable properties lying within and behind what we observe through a modern reductionist perspective. This respect is particularly pronounced in the field of ethnomedicine, where the convergence of Indigenous worldviews and modern science reveal a thriving organism of disease and remedy, diagnosis and cure. Where things get complex, they get chaotic, particularly when these things are interacting with their kin, and comprehension is not readily afforded. Advances in machine learning (ML) and artificial intelligence (AI) technologies, when applied within -informatic and -omic frameworks, can lower this entrance requirement to attainable levels. The author firmly believes the coming decades are going to be particularly fertile for substantial advances in our understanding of the operation of complex systems, something that will facilitiate the movement away from a reductionist one drug-one target-one disease model to a more holistic polypharmacological approach supported by the developing field of multi-omics; but more on this in the concluding chapter of this work. Echoes of this can be found in a more holistic approach to health and wellbeing, often ascribed to Traditional medical systems, which is increasingly being recognised by modern healthcare in the role of lifestyle, diet and environment in the prevention of disease. The incorporation of ICIP into the mainstream is going to be a continual source of ethical concern and the rights and autonomy of Indigenous communities must be respected through informed consent, research transparency and opportunities for collaboration.

In short, this introductory chapter will provide a three-part analysis of the complex field just described. It will begin with the historical underpinnings of ethnomedicine, focusing initially on the Western corpus due to its significant literary precedent (not by priority), before moving onto the Eastern tradition and finally the titular customary medicine in Australia. Anthropological, ethnobotanical and sociocultural speculation has been kept to a minimum, as this is addressed by parties that are far better equipped to explore these topics, with this section concerning itself quite specifically with the plants, their compounds and customary medicinal uses. The next introductory section will broadly outline chemical ecology: the biological basis of how and why plants produce such a diverse array of active molecules. As plants are ostensibly sessile, their evolution has resulted in an exquisite array of biosynthetic pathways that produce specialised metabolites to serve intensive and extensive functions for the benefit of their relatively immobile producer. The functions that these compounds play are largely an attempt to mediate interactions with the environment, defence from attack and offence to attract. This environmental mediation project, undertaken by the plant kingdom for the past ~3,000 Ma has had a serendipitous effect for thoughtful members of kingdom Animalia, in that these metabolites were precisely designed to interact in myriad ways with complex biosystems. This beneficial application, of plant chemistry as human drugs, is a natural segue to the final part of the introduction, which deals with the modern process of drug discovery. This section will go into further detail about the inner workings of NP chemistry and the drug development pipeline and will consider some perspectives on the future of the field - something that will be addressed in more detail in the reflective conclusion.

The body of this book will be dedicated to a presentation of the specialised metabolic constituents (i.e. NPs) that can be derived from the floral kingdom. A particular emphasis is placed upon the native plant orders/families/genera/species that have been recorded as being in use as customary medicines by Aboriginal and Torres Strait Islander peoples. The body text comprises four chapters based on a crude molecular ontology that is traditionally followed in chemistry textbooks. The vast diversity and dynamic interconnectivity of NPs makes their organisation problematic. More advanced ML/AI approaches to molecular ontology that have emerged recently are discussed in the conclusion, but they do not have the readability of this cruder classification into 'phenolic-type', 'terpenoid-type', 'alkaloid-type' and 'miscellaneous-type' classes of NPs. Each chapter will focus on a detailed breakdown of its class into subclasses and further into infraclasses and again into sub-infraclasses. This process should demonstrate to the uninitiated just how nuanced some of these NPs can get, particularly when one combines this with the knowledge that any changes in the presence or position of a single atom can yield drastically different biological effects. Each chapter will present the general representation of these groupings of NPs across the plant kingdom as a whole, any notable drugs that have been developed, specific biosynthetic pathways and detailed mechanisms of action if they have been elucidated. The end of each chapter will be a section titled 'Representation', which is a summation of any native floral genera that have been reported to bear compounds of each sub-/infraclass, also tabulated before each chapter's reference list, as well as speculative correlations between the compounds reported to be present and the customary uses of the plants reported to bear them.

A final caveat must be added here in that the convention stands where a specific chemical entity is reported from specific plant species (often even more specific than this), whereas this text outlines this information to a coarser-grain of chemical sub-/infraclass × plant genus. The reason for this is quite simple - complexity - as the engaged reader will soon experience the effort required to reconcile this field even at this level of detail. To complement this book, a website (www.cmfoa.info) has been developed to facilitate easy reference to the chemistry, botany and customary use outlined in the tables throughout the text. It is designed to be a dynamic resource, allowing for updates and additions as new research and discoveries emerge. For example, an expanded version incorporating a retrieval-augmentated generation (RAG) AI assistant has recently gone live and further iterations are anticipated over the coming years. By making this information accessible online, the authors all hope to reach a broader audience, including educators, healthcare professionals and policymakers, as well as fostering a collaborative environment for ongoing research and commercial development. By promoting the establishment of transparent and equitable partnerships, we can ideally create a framework that benefits both Indigenous communities and the broader scientific community as this field continues to evolve.

ETHNOMEDICINE

Documented evidence of early *Homo sapiens* exploring the natural world in the pursuit of better health outcomes begins with the discovery of the seeds of known medicinal plants being found in the possessions of a *Homo sapiens neanderthalensis*, frozen 60 millennia ago in a cave located in what is now Iraq.[1] This deceased humanoid was laid upon a bed of branches composed of the stimulating (ephedrine-containing) plant *Ephedra* sp. (Ephedraceae), along with scattered wildflowers of notable herbal genera such as *Achillea* (Asteraceae), *Centaurea* (Asteraceae), *Senecio* (Asteraceae), *Lupinus* (Fabaceae) and *Althaea* (Malvaceae).[1] Evidence of formalised herbal lore, in the form of 'Materia Medica', begins in the ancient Near East in the 'Ebers Papyrus' of Egypt (2900 BCE) and the cuneiform tablets of Mesopotamia (2600 BCE), before moving to and evolving independently within the Far East over the following two millennia.[2] These ancient origins not only show the significant conscious relationship forged between humans and their surroundings, but form the substrate from which the modern allopathic medical system has grown.

Traditional medicine (TM) is defined by the WHO as 'the sum of the total knowledge, skills and practices based on the theories, beliefs and experiences indigenous to different cultures, whether explicable or not, used in the maintenance of health, as well as the prevention, diagnosis, improvement or treatment of physical and mental illness'.[3] TM is an understandably huge and heterogeneous compendium of knowledge that contemporary scholars, particularly in the fields of anthropology, ethnobotany and ethnopharmacology, are continually working to piece together into a cohesive narrative. The field of TM is typically split into Western (WTM: based around the Mediterranean Sea in Europe, North Africa and Western Asia) and Eastern (ETM: based around southern and East Asia) traditions. Despite these two traditions taking up most of the contemporary TM narrative, it must be recognised that the lack of archaeological and literary evidence from Greater Africa, Australasia, Oceania and the Americas has unfortunately led to a relative dearth of information pertaining to the rich TM histories in these regions, a situation that is being rapidly and honestly rectified, both in this work and across many other fields.

Western tradition

Formal WTM ostensibly begins in the period known as Classical antiquity (800 BCE – 600 CE), comprising the interlocking civilisations surrounding the Mediterranean Sea (e.g. the Greek and Roman empires) in this period. Contemporaneous with and emerging out of the decline of the ancient civilisations of Egypt and Mesopotamia, the Classical West went on to form the basis of the European culture still present in this region today. Notably, the natural philosophers from Greece [Hippocrates (c. 460–c. 370 BCE), Theophrastus (c. 371–c. 287 BCE), Herophilos (c. 335–c. 280 BCE), Erasistratus (c. 304–c. 250 BCE)] and Rome [Pedanius Dioscorides (a Roman army physician of Greek descent) c. 40–c. 90 CE) and Claudius Galenus (c. 129–c. 216 CE)] were highly influential in the development of what would become the medical sciences: anatomy, physiology, pathology, pharmacology and neurology.[4–7] Plants commonly documented within this Classical herbal tradition can be inferred from the modern translation of Dioscorides' *De Materia Medica* (*Fig. 1.1*). This notable text is broken into sections titled 'aromatics' (e.g. cardamom (Zingiberaceae), nard (Caprifoliaceae), valerian (Caprifoliaceae), cassia (Lauraceae), hops (Cannabaceae), pine resin (Pinaceae), heather (Ericaceae), quince (Rosaceae), lemon (Rutaceae), etc.), 'herbs' (e.g. sea kale (Brassicaceae), beetroot (Amaranthaceae), asparagus (Asparagaceae), garlic (Amaryllidaceae), leek (Amaryllidaceae), onion (Amaryllidaceae), caper (Capparaceae), mustard (Brassicaceae) etc.), 'roots/seeds/herbs' (e.g. rhubarb (Polygonaceae), gentian (Gentianaceae), liquorice (Fabaceae), caraway (Apiaceae), cumin (Apiaceae), parsley (Apiaceae), lovage (Apiaceae), fennel (Apiaceae), clematis (Ranunculaceae), daffodil (Amaryllidaceae) etc.) and 'vines/wines/minerals' (e.g. wine made from grapes (Vitaceae) and concoctions prepared from mandrake (Solanaceae), hellebore (Ranunculaceae) and various other inorganic compounds).[8]

With the decline of the Classical world and the rise of Christianity throughout the so-called 'dark' and Middle ages, monastic scholasticism preserved a portion of the comparatively advanced Classical medical tradition, albeit in a much diminished, mainly palliative form.[9] Saint Isidore of Seville (c. 560–c. 636 CE) was a major bridging source for the medical ideas of the Classical world, as both the Hippocratic and Galenic corpuses were lost to the West in this period.[10] It was not until contact with Classical texts during the Italian Renaissance of the 15th century that WTM regained the level of medical sophistication attained during its Classical origin.[11–14] It is important to mention that these Classical texts were preserved by the Eastern Roman Empire in Byzantium until the Carolingian Renaissance (late 8th–early 9th centuries CE), as well as Islamic and Jewish cultures in the Near East until the 12th century Renaissance, at which point these texts gradually were returned to Western Europe and subsequently translated.[11–14] In this widely misunderstood period of relatively diminished healthcare, theological considerations (such as the belief that disease manifested in humanity as a result of its fallen state from God) dominated much of medical thought, but on the oft-ignored other hand, certain 'modern' medical notions were also widely accepted, such as illness spreading from person to person, lifestyle influencing health and certain individuals having a predisposition to poor health.[15]

WTM, 'folk medicine' or herbalism, as it was variously understood at the time, was for the greater part of society dependent upon the use of ancient, classical and 'pagan' herbal lore for the treatment of common ailments.[16] 'Physic gardens' based upon those of antiquity were maintained both by lay-people and the monastic cultures of post-Roman Europe.[17–19] These gardens featured areas for '*herbularis*' or '*hortus medicus*', medicinal plants that were used to remediate ill-health based upon what remained of the classical WTM culture. The Chelsea Physic Garden of England, Cowbridge Physic Garden of Wales and the Edinburgh Physic Garden of Scotland (now the Royal Botanic Garden, Edinburgh) continue to preserve the practice of maintaining medicinal herbs for the benefit of human health; indeed, the evolution of the modern botanic garden has its origins in this practice.[17–19] It is difficult to summarise the breadth of WTM

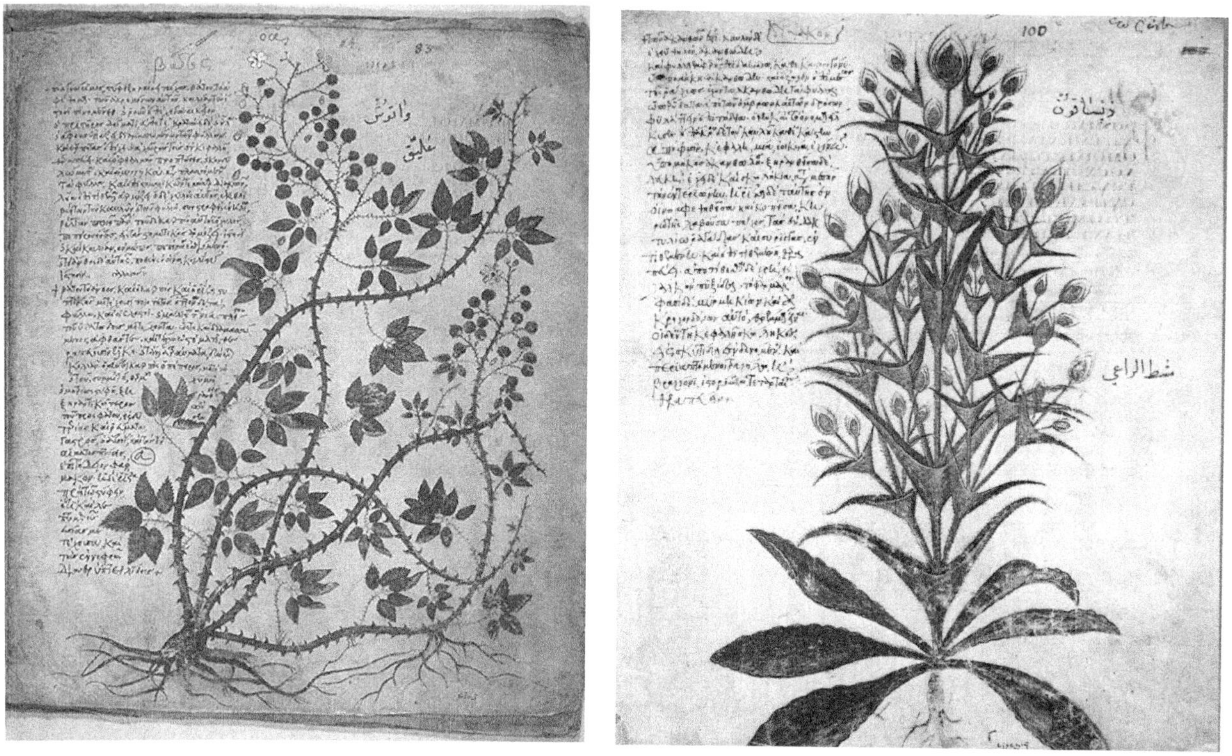

Fig. 1.1. Photographic representation of two illustrated pages of the 6th century 'Vienna Dioscorides' manuscript, an important early example of Dioscorides' De Materia Medica showing two medicinal plants of the genera *Rubus* [Rosaceae (Plate 83, left)] and *Dipsacus* [Caprifoliaceae (Plate 100, right)]. Wikimedia commons (public domain work of art).

lore at the time, but it is understood that essentially the whole gamut of the vegetal world was used, encompassing non-vascular 'flora' (bryophytes, lichens, algae, fungi), vascular plants (pteridophytes and conifers), flowering dicots (e.g. water lily (Nymphaceae), poppy (Papaveraceae), elm (Ulmaceae), dock (Polygonaceae), St John's-wort (Hypericaceae), primula (Primulaceae), currant (Grossulariaceae), rose (Rosaceae), legumes (Fabaceae), geraniums (Geraniaceae), umbellifers (Apiaceae), gentians (Gentianaceae), nightshades (Solanaceae), comfrey (Lamiaceae), foxgloves (Plantaginaceae), valerian (Caprifoliaceae) and daisies (Asteraceae) etc.) and flowering monocots (e.g. lilies (Liliaceae), orchids (Orchidaceae) etc.).[20]

The progress of WTM over the course of the medieval period, or early-, middle- and late- 'Middle Ages' (5th-15th centuries CE), is commensurate with the gradual resumption of a study of the natural sciences. As mentioned, increasing communication between Western Europe and the Near East, starting in the Carolingian period and accelerating throughout the crusades beginning in 1095 CE, led to a significant enrichment of the academic culture of the Western world due to knowledge and texts from the Classical world being preserved and greatly expanded upon by Arabic and Jewish scholars.[21] Renaissance humanism owes a lasting debt to the efforts of the Near East, beginning with Byzantine thinkers such as Paul of Aegina (c. 625–c. 690 CE) formalising textbooks, providing diagnostic explanations and technical drawings, and the development of hospitals, as well as the historically uncertain Jewish figure Asaph ben Berechiah (known as Asaph the Jew), credited with the authorship of the highly influential Sefer Refuot ('Book of Medicines').[22] The development continued in the Islamic world, beginning with the Umayyad Caliphate (661–750 CE) onwards through the 'Islamic Golden Age' of the Abbasid Caliphate (750–1258 CE), which saw an explosion of early scientific thought based upon the Greek tradition preserved in Alexandria and Egypt.[23] Hospitals and medical education flourished across the Islamic world during this period, with the ideas of Galen and Hippocrates being advanced by Persian physicians such as Abū Bakr Muḥammad ibn Zakariyyā' al-Rāzī (known as Al-Razi or Rhazes in the West: 965–1040 CE) and Ibn Sina (known as Avicenna in the West: c. 980–June 1037 CE).[24]

It was from this period, and well into the modern era, that the practice known as alchemy began to flourish.[25] Both Al-Razi and Avicenna were deeply interested in the alchemical arts, which, despite the somewhat unsavoury reputation today, were at the time a serious and fundamental aspect of what eventually became chemistry.[21] Al-Razi is credited with some of the earliest systematic classifications of chemical substances, reactions and equipment and Avicenna was instrumental in the development of distillation and his works on alchemy went on to form much of the basis for Renaissance alchemy in the West.[26] Another key Islamic figure in the history of medicinal chemistry is Abū Mūsā Jābir ibn Ḥayyān (known as 'the Father of Chemistry', who died c. 806–816), whose ~600 works, known collectively as the 'Jabirian Corpus' (some of which were subsequently published in the West under the name Geber and later pseudo-Geber) contained early descriptions of organic and inorganic chemical substances, as well as expanding upon many of the more philosophical underpinnings of this fledgling field, such as the Aristotelian elements (water, earth, fire, air), qualities (hotness, coldness, dryness, moistness) and his own 'mercury–sulphur theory' (to which Al-Razi later added the component 'salt').[27]

In the Renaissance, WTM (or 'Learned Medicine' as it was then known) benefitted greatly from the principle of '*ad fontes*' [(back) to the sources], which in essence meant the reincorporation of Classical period ideas preserved throughout the Middle Ages in the Near East back into the WTM framework, leading to what we would today call modern allopathic medicine.[28] This intense focus on scholarship and the translation of many Arabic, Greek and Byzantine scientific works into Latin led to monumental advances in the medical sciences.[29] Contrary to this more formalised work, the contentious Swiss figure Philippus Aureolus Theophrastus Bombastus von Hohenheim (known as Paracelsus: c. 1493–1541) rejected this scholarly approach and was instrumental in the development of experimental research.[30] Paracelsus insisted upon a foundational approach to medicine based on the natural sciences, particularly chemistry, which at the time was based upon the Islamic alchemical tradition. Paracelsus' *tria prima* ('three primes') of sulphur (soul/gas), mercury (spirit/liquid) and salt (body/solid) was based primarily on the alchemical work of Jabir ibn Hayyan and Al-Razi and was instrumental in the development of Renaissance theories of disease. By understanding the chemical nature of the *tria prima*, a physician could discover the means to curing a disease (viz. poisoning by excess of one of the three primes), by bringing the human system (microcosm) back into harmony with Nature (macrocosm).[31] Despite the intense and strange Hermetic mysticism that permeated much of Paracelsus' work, the focus on observing the extrapolated properties of material substances and their interaction with the homeostasis of the human body was a great conceptual leap in the history of medical chemistry, but as will be seen in the next section on ETM, this was not the only (or even the first) instance of thinking about human health in this way.[32]

Eastern tradition

As with the various aspects of WTM, such as alchemy or the Galenic theory of humours, ETM is often a complex, bewildering and alien idea for the typical 'modern' mind to grapple with. The most notable strains of ETM present in the modern world are Traditional Chinese Medicine (TCM), Traditional Indian medicine (Ayurvedic medicine), Traditional Perso-Arabic medicine (Unani medicine) and Traditional Japanese Medicine (TJM or Kampō).[33,34] It must be mentioned that due to constraints on length, the entire ETM sphere will not be detailed herein. TCM is arguably the most prevalent of these traditions and has thus benefited from the most detailed study in the contemporary academic sphere. As with WTM, alchemy, herbalism, etc., the literature expounding and analysing these fields is fraught with the issue that is now termed 'pseudoscience'; that is, the conventional stipulations for academic rigour according to the scientific method (e.g. causality, mechanism of action) are often absent. This issue is best left aside, with the current study limiting itself to the historical, botanical and molecular considerations that are pertinent to the field of NP chemistry.

Much work has been conducted since the mid-20th century, primarily in China beginning during the Cultural Revolution, to organise the disparate concepts within TCM that have accumulated over the three millennia since the Shang Dynasty (c. 1600–c. 1046 BCE) into a more cohesive, modern form of 'Chinese Medicine' (中医 – Zhongyi).[33] This typically involved a purging of aspects that were considered superstitious or unscientific to modern eyes but, as will be seen, many of the fundamental, esoteric components of this huge system of thought still retain aspects, such as qi/ki/ch'i (气), yin-yang (阴阳), the 'five phases'/wuxing (五行) etc. that are not amenable to qualification or quantification via the analytical instruments and techniques available today.[33,34] Additional complications arise because ETM often conflates the rigid categorical differences modern medicine has ascribed to biomedicine (body), psychology (mind) and theology (spirit), into one holistic system, which is a major point of contention between both advocates and detractors. This introductory review will now proceed through a brief history of this important portion of the Chinese herbal medical corpus.[35]

ETM has ancient roots going back into prehistory, sharing a worldview similar to that seen in WTM in which the individual, society and the universe are deeply interrelated ('as above, so below', etc.) and the patterns (zheng, 證), cycles and processes seen without are indisputably linked to those occurring within. Concepts such as these abound within TCM, in foundational texts such as the 'Inner Canon of the Yellow Emperor' (Huángdì Nèijīng, 黄帝内经) and the 'Treatise on Cold Damage Diseases' (Shānghán Lùn, 伤寒论), attributed to miscellaneous scholars of the Warring States Period (c. 260 BCE) and Zhang Zhongjing (c. 150–c. 219 CE).[33,36,37] Formalised herbal lore, of particular relevance to this work, bears significant representation across TCM and ETM, with various Chinese Materia Medica (CMM) being recorded and developed from the 1st millenium BCE, starting with a few dozen preparations, growing to over 12,800 herbal preparations as recorded today.[33] One of the first systematic summaries of TCM and CMM is 'Shennong's Herbal Classics' (Shénnóng Běncǎo Jīng, 神农本草经), compiled in the Eastern Han Dynasty (c. 200–c. 250 CE) and attributed to the deified mythological Chinese ruler 'Shennong' (神農).[38] This text comprises three volumes, devoted to the harmless noble or upper herbs (上品) (e.g. lingzhi (a fungus of genus *Ganoderma*), ginseng (*Panax*), jujube (*Ziziphus*), orange (*Citrus*), cinnamon (*Cinnamomum*), liquorice (*Glycyrrhiza*) etc.), the potentially toxic but also therapeutic human or middle herbs (中品) (e.g. ginger (*Zingiber*), peony (*Paeonia*), cucumber (*Cucuma*) etc.] and the physiologically violent, invariably poisonous lower herbs (下品) [e.g. rhubarb (*Rheum*), peaches (*Prunus*) etc.); the full English translation is now available for free, one click away (online), almost two thousand years later.[39]

As with WTM, much of the information available in the West pertaining to TCM has come through the lens of history and the mouthpieces of its dominant/victorious forces. In China's case, this is most apparent in the stance of the Chinese Communist Party, where, from its inception in 1950, modernity, cultural identity and socioeconomic reconstruction clashed with the traditions of a colonial and feudal past.[33,34] Modern TCM is in a perpetual dialogue with allopathic medicine, with the former often being reconstructed in response to the latter. Much of the content of this huge tradition lies far outside the scope of this book, but the CMM of Chinese herbal medicine is of obvious importance in this case. Clearly, the over 13,000 substances tabulated since the beginning of the Current Era are unwieldy to explicate at a glance, so a 'Handbook of Traditional Drugs' enumerating a more manageable 517 substances was produced in 1941.[40] Of these 517 drugs, 442 are of floral origin, with the remainder of more morally or physiologically dubious animal or mineral origins.[41] A more recent, commonly cited reference text (albeit in French) lists 50 fundamental herbs and species from genera in the following families: Achariaceae (*Hydonocarpus*), Apiaceae (*Ligusticum*), Apocynaceae (*Rauvolfia*), Aquafoliaceae (*Ilex*), Asteraceae (*Aster, Carthamus, Dolomiaea*), Campanulaceae (*Lobelia*), Cannabaceae (*Cannabis*), Cornaceae (*Alangium*), Cucurbitaceae (*Trichosanthes*), Cupressaceae (*Platycladus*), Ephedraceae (*Ephedra*), Ericaceae (*Rhododendron*), Eucommiaceae (*Eucommia*), Euphorbiaceae (*Croton, Euphorbia*), Fabaceae (*Astragalus, Gleditsia, Glycyrrhiza, Pueraria, Styphnolobium*), Gentianaceae (*Gentiana*), Hydrangeaceae (*Dichroa*), Lamiaceae (*Agastache, Leonurus, Pogostemon, Scutellaria*), Lauraceae (*Cinnamomum*), Menispermaceae (*Cissampelos, Stephania*), Oleaceae (*Forsythia*), Orchidaceae (*Dendrobium*), Orobanchaceae (*Rehmannia*), Pinaceae (*Pseudolarix*), Phyllanthaceae (*Flueggea*), Polygonaceae (*Rheum*), Primulaceae (*Ardisia*), Ranunculaceae (*Anemone, Coptis, Corydalis*), Rutaceae (*Phellodendron, Psilopeganum*), Schisandraceae (*Schisandra*), Solanaceae (*Anisodus, Datura*), Stemonaceae (*Stemona*), Theaceae (*Camellia*) and Thymelaceae (*Daphne, Wikstroemia*). However, universal agreement upon a species' rightful membership to this list is not forthcoming.[42]

The scope of this book will not permit even a cursory breakdown of the myriad other major and minor proponents of the ETM corpus, but the herbal aspects of a few notable members remain pertinent. The traditional medical system of the Indian subcontinent is primarily based upon the ancient text known as the Ayurveda (आयुर्वेद), loosely translated as 'the knowledge of life'. With roots going back to the Indus Valley civilisation (c. 3300–c. 1300 BCE), even beyond to the Proto-Indo-Europeans (c. ~5500–2500 BCE), Ayurvedic medicine ostensibly outdates any of the other forms of TM discussed thus far, yet problems involved with the archaeological understanding of the scant remaining texts from the period render the matter unresolved.[43] Ayurvedic medicine shares many of the problems experienced by other systems of TM integrating into the modern world. Many concepts such as life-force (prāṇa, प्राण), 'humoral biomaterials' (doṣa, दोषः) and the five elements (pañca-mahā-bhūta, पञ्चभूत, पञ्चमहाभूत) are akin to the non-quantificable aspects of TCM or WTM and accordingly attract negative press on the grounds of 'pseudoscience'.[44]

It is estimated that 70% of rural Indians use traditional Ayurvedic herbal remedies for their primary healthcare.[45] The herbal remedies used across the Indian subcontinent (as well as much of South-East Asia) are not only based upon Ayurveda, but also the Perso-Arabic TM known as Unani (tibb yūnānī, طب یونانی), among others. Notably, the term Unani (or Yunani) is based on the word for 'Greek' in Urdu, as this form of medicine was based upon early members of the WTM system such as Hippocrates and Galen and later Arab and Persian scholars such as Ibn Sina and Al-Razi, discussed previously.[46] It is estimated that ~2400 medicinal plants have a documented traditional use across the subcontinent, varying significantly with region (grouped primarily as Eastern Ghats, Western Ghats, North-East, West and Central Himalayas and Indo-Gangetic).[47–52] As with TCM, much of Ayurvedic and Unani medicine is aimed at restoring balance to the interlocking systems within and without the human body.[53] It is not possible to comprehensively enumerate these medicinal plants here, but some general trends can be outlined by mentioning some fundamental species belonging to genera in the following families: Acanthaceae (*Andrographis*, *Justicia*), Acoraceae (*Acorus*), Annonaceae (*Polyalthia*), Apiaceae (*Trachyspermum*), Apocynaceae (*Alstonia*), Asteraceae (*Ageratum*), Combretaceae (*Terminalia*), Cyperaceae (*Cyperus*), Ericaceae (*Rhododendron*), Fabaceae (*Abrus*, *Cassia*, *Glycyrrhiza*, *Mimosa*), Lamiaceae (*Callicarpa*, *Ocimum*), Malvaceae, Meliaceae (*Azadirachta*), Orchidaceae, Phyllanthaceae (*Phyllanthus*), Solanaceae (*Solanum*, *Withania*), Rubiaceae (*Piper*), Rutaceae (*Aegle*) and Zingiberaceae (*Curcuma*, *Zingiber*).[47–52]

The final space available for ETM must belong to the traditional medicinal system that has received the most significant effort at 'modernisation', namely 'Kampō' or Kanpō medicine (Kanpō igaku, 漢方医学) – *lit*. 'Chinese medicine' [in Japanese (from the Japanese translation of Han, *Kan*)]. After its introduction from China to Japan in the 7th century CE, TCM evolved over the next 10 centuries into its own form distinct from its mainland progenitor, now termed Traditional Japanese Medicine (TJM).[34,54] By the Edo Period (1603–1867 CE), Japanese medical practitioners were using elements of both TCM and TJM, while simultaneously beginning to incorporate aspects of Western medicine that were permeating Japanese society through recent and increasing contact with Dutch traders.[55] The Meiji Restoration of 1868 ushered in a drastic change to the medical paradigm, based upon the 'German medical system', where any attempt to preserve traditional medical practices had to do so within the Western paradigm.[55] From 1910, Japanese graduates who had trained in Western medical facilities began the gradual revival of TCM/TJM in Japan, where classical texts and techniques were interpreted through a modern, allopathic lens. These efforts led to what is known as Kampō today, a thoroughly modernised, state-controlled form of identified active-constituent, dose-quantified herbal medicines that are routinely tested for heavy metals, contaminants and microbial growth.[55]

The 14th edition of the Japanese Pharmacopoeia (Nihon yakkyokuhō) lists 165 herbal ingredients used in the preparation of the 148 approved Kampō medicines.[54] These 165 herbal species could be considered some of the most globally well studied and medically substantiated herbal remedies (in vitro, in vivo and clinically) and are distributed across the following families (and genera): Alismataceae (*Alisma*), Amaranthaceae (*Achyranthes*), Apiaceae (*Angelica*, *Bupleurum*, *Foeniculum*, *Glehnia*, *Ligusticum*, *Notopterygium*, *Peucedanum*, *Saposhnikovia*), Araceae (*Pinellia*), Araliaceae (*Aralia*, *Panax*), Arecaceae (*Areca*, *Arisaema*), Aristolochiaceae (*Asarum*), Asparagaceae (*Anemarrhena*, *Asparagus*, *Ophiopogon*), Asteraceae (*Arctium*, *Artemisia*, *Atractylodes*, *Carthamus*, *Chrysanthemum*, *Saussurea*), Boraginaceae (*Lithospermum*), Campanulaceae (*Platycodon*), Cannabaceae (*Cannabis*), Caprifoliaceae (*Lonicera*), Cornaceae (*Cornus*), Cucurbitaceae (*Benincasa*, *Trichosanthes*), Cyperaceae (*Cyperus*), Dioscoreaceae (*Dioscorea*), Ephedraceae (*Ephedra*), Eucommiaceae (*Eucommia*), Euphorbiaceae (*Croton*), Fabaceae (*Astragalus*, *Biancaea*, *Glycyrrhiza*, *Pueraria*, *Senna*, *Sophora*), Fagaceae (*Quercus*), Gentianaceae (*Gentian*, *Swertia*), Lamiaceae (*Leonurus*, *Mentha*, *Perilla*, *Schizonepeta*, *Scutellaria*), Lardizabalaceae (*Akebia*), Lauraceae (*Cinnamomum*, *Lindera*), Liliaceae (*Fritillaria*, *Lilium*), Magnoliaceae (*Magnolia*), Menispermaceae (*Sinomenium*), Moraceae (*Morus*), Myrtaceae (*Syzygium*), Nelumbonaceae (*Nelumbo*), Nymphaeaceae (*Nuphar*), Oleaceae (*Forsythia*), Orchidaceae (*Gastrodia*),

Orobanchaceae (*Rehmannia*), Paeoniaceae (*Paeonia*), Papaveraceae (*Corydalis*), Pedaliaceae (*Sesamum*), Poaceae (*Bambusa, Coix, Hordeum, Oryza, Triticum*), Plantaginaceae (*Plantago*), Polygalaceae (*Polygala*), Polygonaceae (*Reynoutria, Rheum*), Ranunculaceae (*Aconitum, Cimicifuga, Clematis, Coptis*), Rhamnaceae (*Ziziphus*), Rosaceae (*Crataegus, Eriobotrya, Prunus*), Rubiaceae (*Gardenia, Uncaria*), Rutaceae (*Citrus, Phellodendron, Tetradium, Zanthoxylum*), Sapindaceae (*Dimocarpus*), Saururaceae (*Houttuynia*), Schisandraceae (*Schisandra*), Solanaceae (*Lycium*), Theaceae (*Camellia*), Zingiberaceae (*Alpinia, Amomum, Zingiber*), Zygophyllaceae (*Tribulus*), as well as the Fungal families (and genera); Ganodermataceae (*Ganoderma*) and Polyporaceae (*Wolfiporia*).[54,56]

CUSTOMARY MEDICINE IN AUSTRALIA

The focus of the research project outlined in this book is the customary medicinal flora of Australia. Before embarking upon what will be an abridged summation of a subject that can (and has) filled many volumes, readers are reminded that the lead author is, for all intents and purposes, fundamentally a NP chemist. The rich history and sociopolitical plight of the original inhabitants of this country is of the utmost importance, but for the project being embarked upon an attempt is made to limit discussion as strictly as is possible to the phytochemical constituents of the customary medicinal flora of Australia. The author's awareness of the relative dearth of research into this topic led to the formulation of a plan to update the current state of the field with the significant, but fragmented, research that has been conducted on the phytochemical constituents of the native flora of Australia, since the publication of the primary text drawn from on this matter.[57]

Today, it is obvious that the utmost care and consideration must be taken when dealing with the cultural heritage of a marginalised group of people whose 'land and sovereignty was annexed without consent and without treaty with the country's rightful owners'.[58] The author sincerely wants to make it clear that, despite being consciously unprepared and ill-equipped to address the political, social, cultural and anthropological issues, this work has been conducted with a sense of proprietary bookkeeping, updating the information currently at hand regarding the chemical constituents of Australian plants, while leaving the greater humanitarian discussion to people who are better equipped to do so respectfully. The author attempts to approach this topic with the utmost sympathy for the plight of the original inhabitants of this land and truly wishes that the progenitors of the information contained within this work receive the respect, recognition and agency that they deserve for their expansive knowledge of Country, gained over at least 60,000 years of continuous occupation, the longest of any living culture.[58]

A poignant (albeit spurious) representation of the extremely fragmented state of the field of research pertaining to medicinal plants in Australia is that the Wikipedia page on the topic lists only a dozen plant species over seven genera: *Acacia, Backhousia, Centipeda, Cymbopogon, Erythrina, Eucalyptus* and *Ventilago*.[59] In contrast, the most notable book on the topic, Lassak and McCarthy's *Australian Medicinal Plants* lists 393 species over 264 genera (240 of these are chemotaxonomically extant).[57] Of the 240 extant genera, literature searching shows that 104 (43%) of them have not received a comprehensive phytochemical review in a top-ranked journal, if at all. Compared to the respect and detailed research that has been conducted on the traditional medicine systems that have been addressed in the preceding historical sections, 'customary medicine in Australia', as it is correctly termed, has been quite remarkably neglected.[60]

For an admittedly cursory example, at time of writing a search of SciFinder^n returns 579,796, 423,221 and 383,858 results for 'chinese medicine', 'herbal medicine' and 'ayurvedic medicine', respectively, but 'customary medicine in Australia' returns only 99 results, 'Aboriginal medicine' returns only 4,276 results, 'Australian Aboriginal medicine' returns only 3441 results and 'Aboriginal bush medicine' returns only 454 results (note: 'indigenous medicine' returns 184,936 results, but is not specific to Australia; when refined by the keyword 'Australia', the results drop to 4056, a similar situation is encountered with 'bush medicine' and 'First Nations medicine', etc.). It was this that led the author to compile, after some years of diligent bookkeeping, a reliable foundation for this field in the form of a database detailing 393 species of native Australian flora reported to be in use by First Scientists, as well as any chemical constituents of the 240 genera that have been reported in the literature, alongside tabulated (and networked) metadata pertaining to the customary use of these plants (recorded in Lassak and McCarthy), plus their botanical distribution and updated taxonomic phylogeny (up to publication, 2022, available at www.cmfoa.info).

The geography of Australia is one of the most unique on the planet due to an extended isolation (> 100 million years) after its separation from the Gondwanan supercontinent, leading to an abundance of endemic floral and faunal species.[61] In fact, Australia is considered to be one of the 17 'megadiverse' countries of the globe (defined as bearing at least 5,000 species of endemic plants), alongside Brazil, China, Colombia, the Democratic Republic of Congo, Ecuador, India, Indonesia, Madagascar, Malaysia, Mexico, Papua New Guinea, Peru, the Philippines, South Africa, the United States of America and Venezuela.[62] Of the estimated 1,900,000 species of living organisms worldwide, Australia has 147,759 accepted, described species and an estimated total of 566,398 (Australia contributes 29.8% of the estimated global total). Of these accepted species, 24,716 are plants, of 310,129 worldwide (Australia contributing 7.9%), an estimated 86% of which are considered endemic (21,255 plant spp.).[63] There are 14 ecoregions distributed across the globe, 8 of which are found in Australia: tropical and subtropical moist broadleaf forests, temperate broadleaf and mixed forests, tropical and subtropical grassland, savannas and shrublands, temperate grasslands, savannas and shrublands, montane grasslands and shrublands, Mediterranean forests, woodlands and shrubs, deserts and xeric shrublands and tundra (*Fig. 1.2*).[64]

The many distinct peoples who have inhabited Australia for over 60,000 years have only begun to define and self-identify as a single group in the past 200 years. Their origins, history and cultures are venerable and marvellously rich. Limited recorded historical documentation inhibits the accuracy of speculation upon how long Indigenous Peoples have been utilising the flora and fauna of this land to therapeutic ends, but it seems more than appropriate to say that these custodians have had a deep, unbroken and ongoing relationship with the medicinal properties of their surrounding environment for at least as long as any of the other cultures discussed in this introductory chapter, likely substantially longer.

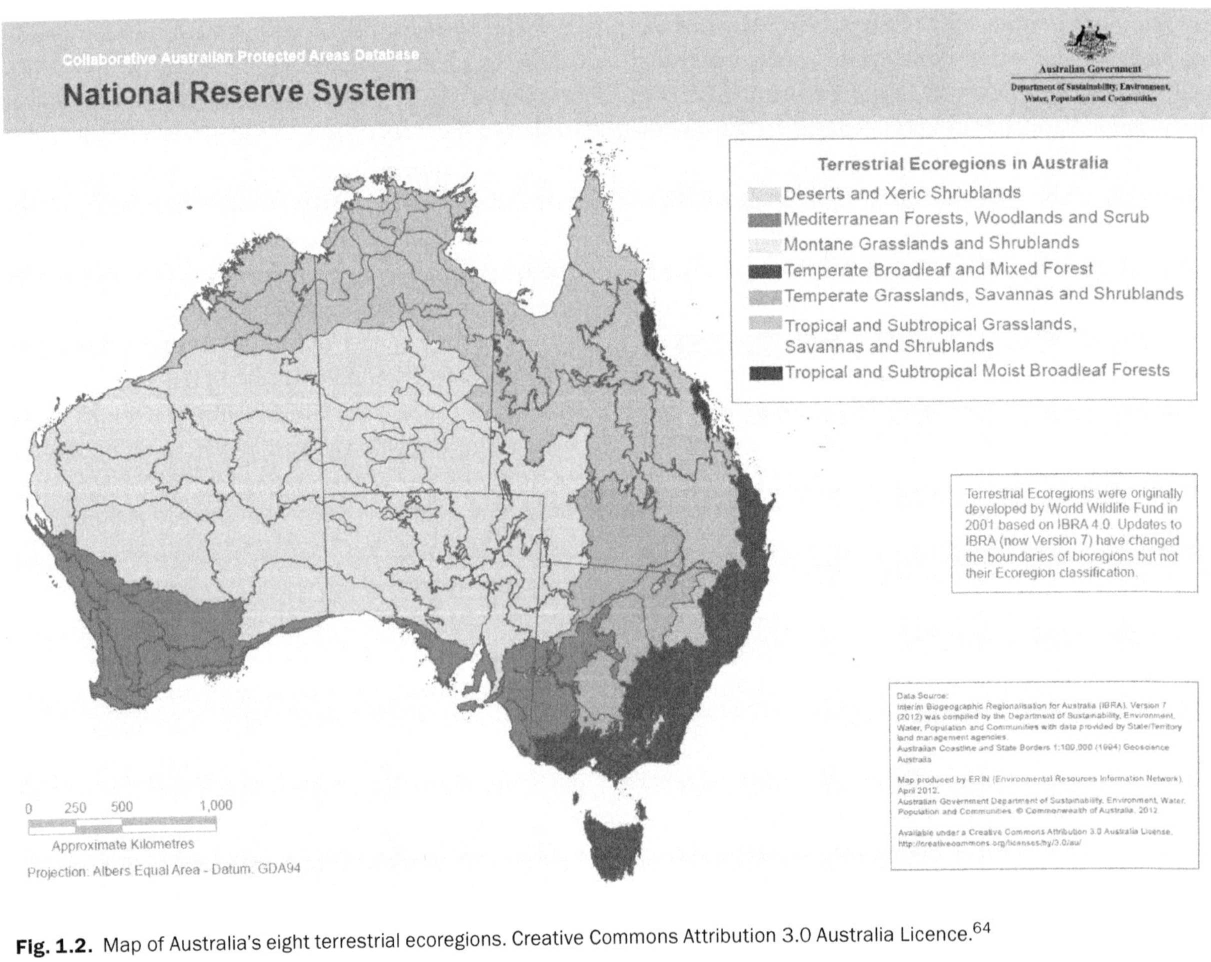

Fig. 1.2. Map of Australia's eight terrestrial ecoregions. Creative Commons Attribution 3.0 Australia Licence.[64]

A 1994 survey reported that 22% of the 15,000 Aboriginal and Torres Strait Islander Peoples consulted had practised customary medicine in the previous 6 months.[65] Customary medicinal traditions have largely been lost in southern and eastern Australia, but are still in common practice in the Northern Territory.[66] This decline is attributed largely to informational loss across the cultures of Indigenous Peoples, with tradition being preserved through Cultural Ceremonies, many of which have ceased to occur since the arrival of Western settlers. Customary medicine practitioners or healers, who go by various names in the over 250 languages and 800 dialects currently in use by Indigenous Peoples, such as ngangkari by the Anangu Peoples of the Western Desert or marrnggitj by the Yolngu of North-East Arnhem Land, have begun to be incorporated into health clinics, giving medical advice (diagnosing problems, advising on remedies, suggestion and performance of ritualised healing procedures, investigating sociocultural issues pertaining to the illness, etc.) and importantly giving emotional support (in a way akin to Western mental health services), particularly to Indigenous People who feel uncomfortable in the unfamiliar space of the Western clinic.[67–68]

As stated in the National Aboriginal Health Strategy report of 1989, Indigenous Peoples consider health as defined by 'not just the physical wellbeing of the individual, but the social, emotional and cultural wellbeing of the whole community. This is the whole-of-life approach and it also includes the cyclical concept of life-death-life'.[69] These fundamental differences in world-view from that of the modern West render the comprehension of the role of customary medicine in the lives of Indigenous Peoples, both presently and historically, dependent upon understanding the underlying basic structures and complex beliefs.[70] In lieu of this, an objective view can be taken by the NP chemist, in which the meaning of the system is artificially abstracted from its structure and as such the remainder of this section will be simply devoted to identifying and outlining the types of plants utilised by Indigenous Peoples in their system of customary medicine and to what ostensible therapeutic ends they were put to.

Lassak and McCarthy's *Australian Medicinal Plants* lists some significant recorded examples of customary medicine use among Indigenous Peoples: digestive issues treated with astringents [e.g. the exudates ('sap', an imprecise term) or 'kino' of gum trees (i.e. members of the Myrtaceae)], fevers treated with *Alstonia constricta* (among others), toothaches relieved by chewing 'wilga' [*Geijera parviflora* (Rutaceae)], cavities in teeth

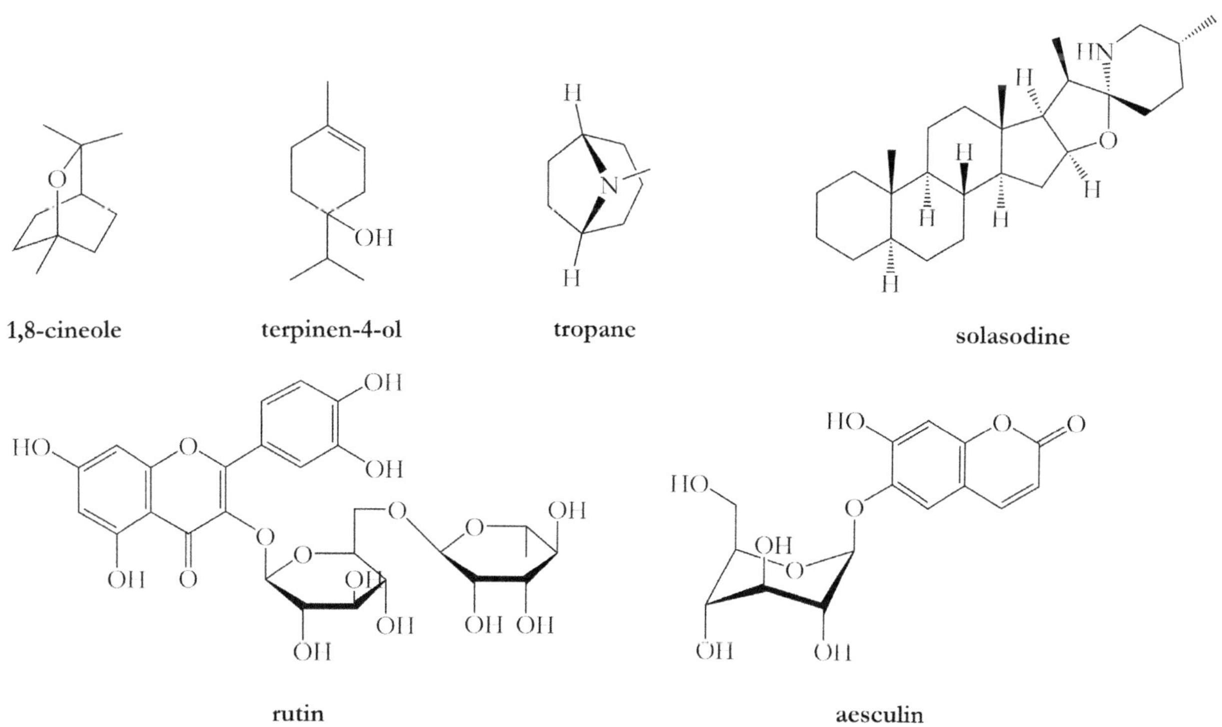

Fig. 1.3. Australian native phytochemicals under commercial production.

filled with resins (e.g. from *Melicope vitiflora* (Rutaceae) etc.).[57] A summarised list of the traditional uses of customary medicines among Indigenous Peoples, based on and updated from this excellent resource, is included in the Customary Medicinal Flora of Australia Database (www.cmfoa.info).[57] Chapters 2-5 of this book will go into substantial detail concerning the metabolic constituents of Australian flora with documented customary use by Indigenous Peoples. A few notable examples of the benefits that customary medicines had for their Western audience follow. British immigrants using 'Sydney peppermint' [*Eucalyptus piperita* (Myrtaceae)] in place of peppermint [*Mentha piperita* (Lamiaceae)] for the treatment of 'cholicky' complaints, also *Sebaea ovata* (Gentianaceae), *Alstonia constricta* (Apocynaceae), *Doryphora sassfras* (Atherospermataceae) and *Smilax glyciphylla* (Smilacaceae), which have been subsequently discovered to provide similar benefit.[57] Commercial production of Australian native phytochemicals has been underdeveloped, but some key examples are the essential oils bearing the cyclohexane monoterpenoids of the 1,8-cineole type ['eucalyptus oil' from the Myrtaceae (*Eucalyptus*)] and the terpinen-4-ol type ['tea tree oil' also from the Myrtaceae (*Melaleuca*)]; the tropane-type alkaloid hyoscine used for seasickness [from the Solanaceae (*Duboisia*)]; the steroidal alkaloid solasodine [also from the Solanaceae (*Solanum*)]; the phenolic flavonoid glycoside rutin (from many sources, largely the Rutaceae and Myrtaceae); and the phenolic coumarin glycoside aesculin [from the Pittosporaceae (*Bursaria*)] (*Fig. 1.3*).[57]

A trademark application regarding the customary medicine 'gumby gumby' [from the Pittosporaceae (*Pittosporum angustifolium*)], which was subsequently, graciously, denied, highlights the significant problems that can arise at the intersection of IP rights and Indigenous Peoples' culture.[71] The Gangulu of the Mount Morgan area of Queensland use this plant as part of their customary medicine culture, used topically for eczema and pruritus, to induce lactation in new mothers and used internally for coughs, colds and cramps. Academic research has identified biologically active, cytotoxic taraxastane-type triterpene saponins, which could be the active ingredient responsible for the traditional use of this plant, with further research currently underway at Central Queensland University.[72,73] The issue here is that an application was submitted (by non-Indigenous parties) in April 2017, which if approved, would have restricted the Gangulu people from using their own linguistic term, gumby gumby, for commercial purposes when marketing their own customary medicine. Due to this and other similar issues, IP Australia has released a report regarding the protection of Indigenous Knowledge (IK) in the IP system, similar to that which is in place for the Māori people of New Zealand.[74] This brings to light an aspect of this research, which is currently being pursued, that the customary knowledge documented in this work and the database that will shortly be hosted online could function as a foothold upon which Indigenous Peoples' IP in the field of NP chemistry could be formally recognised.

CHEMICAL ECOLOGY

The organic components of the higher plants discussed in the preceding historical section are grouped under the term 'phytochemicals', the study of which is termed 'phytochemistry'.[75] Traditionally, the diversity and function of these phytochemicals within the systems that produce them have resulted in a broad division between 'primary metabolites', which are chemical compounds that are essential to the growth, development and reproduction of the organism (standard biological macromolecules; carbohydrates, lipids, proteins, nucleic acids) and the more species-specific 'secondary metabolites', which arise due to the dynamic ecological and evolutionary circumstances with which the organism interacts and responds.[75,76] Contemporary understanding of the overlap between these two categories of phytochemicals has become increasingly convoluted and the sharp division between these two categories has subsequently been called into question, an issue that arises frequently throughout this document and as such, the term 'specialised metabolite' will be used for the bioactive NPs discussed herein.[77]

Specialised metabolites perform a selective role in the process of speciation, enabling their progenitors to survive and thrive under the adverse, biome-specific circumstances encountered in nature.[75,76] The scientific field concerned with this chemical relationship between organism and environment is termed 'chemical ecology', which studies the chemically mediated antagonistic (competition, predation) and mutualistic (attraction, pollination etc.) interactions of the natural world.[75,76] As plants are quite literally rooted to their space, one of the most important aspects of specialised metabolites in the plant kingdom is in the defence of the organism from pathogenic (microbial) attack. Specialised metabolites that function in this way are categorised more specifically as either 'phytoalexins' or 'phytoanticipins', which are produced in response to microbial hostility (i.e. invasion through a wound) or are constitutive to the metabolic functioning of the organism (anticipating pathogenic attack), respectively, a relevant distinction, as many of these specialised metabolites are energetically expensive to produce.[75,76]

Due to the ancient, evolutionary relationship between the specialised metabolites produced by organisms and the ecological vectors of antagonism-mutualism in their environment, NPs have been rather serendipitously tailored (for humanity) to be biologically active and so tend to have an effect on the biological systems to which they are directed.[75,76] This phenomenon is the key to the whole field of NP chemistry, which seeks to enable humans to benefit from this evolutionary relationship, particularly in the field of medicine. Specialised metabolites have evolved to be biologically active and so it is the role of the NP chemist to engage with the natural world to extract, purify and characterise the compounds contained therein and to identify what these metabolites are and how they contribute to the life of the system. Pragmatically, there is an ever present 'commercial' application to how these NPs can be of benefit to humans, but there is also a fundamental, philosophical 'blue skies' drive in the field to understand the complexity of the natural world in and of itself.

Biosynthesis

The highly interrelated, dynamic biochemical network of metabolites flowing within an organism is known as that organism's metabolism. When coupled with specific productive biochemical pathways, the catabolic breakdown of complex metabolites, either produced, ingested or otherwise, produces the chemical energy necessary for the maintenance and growth of cells. The reverse process (i.e. the anabolic production of complex metabolites from simple chemical precursors within a biological system) requires the input of energy and is termed 'biosynthesis'.[78] The study of biosynthesis mirrors the traditional biochemical separation of primary and secondary metabolism. The astonishing chemical diversity found across the kingdoms of life is maintained by myriad interconnecting metabolic pathways, with each step catalysed and facilitated by extremely specific and sophisticated enzymes (functionalised proteins).[79]

Primary metabolism involves the construction and degradation of complex molecules essential to the functioning of life; it functions in a very similar way across all the living kingdoms and is one of the key pieces of evidence for a universal common ancestor.[79] The compartmentalised, cellular construction of these essential biological macromolecules relies first upon the acquisition of each macromolecule's respective basic monomeric units: monosaccharides for carbohydrates, acetate for lipids, amino acids for proteins and nucleotides for ribonucleic acid (RNA) and deoxyribonucleic acid (DNA) (*Fig. 1.4*).[78] Energy is required for this constructive process, usually conveyed in the form of adenosine triphosphate (ATP), the vessel for coupling reaction pathways through phosphorylation and enthalpic heat generation. This energy can have differing sources depending on the organism, with autotrophs (e.g. plants) being able to generate their own energy from the fixation of environmental energy (i.e. sunlight via photosynthesis) and heterotrophs (e.g. animals) requiring the consumption of other organisms (i.e. herbivory, predation).[78] The key primary biosynthetic pathways are autotrophic carbon fixation (photosynthesis or chemosynthesis), carbohydrate anabolism (gluconeogenesis), lipid anabolism (fatty acid synthesis, sterol biosynthesis), protein biosynthesis (via DNA/RNA-mediated transcription and RNA/ribosomal-mediated translation) and nucleotide synthesis/salvage.[78] Some of these pathways are the most complex, most diverse and best understood concepts in biochemistry. Nevertheless, a detailed analysis of these fundamental processes of life lies outside the scope of this document, but any introductory textbook on biology or biochemistry will address them, should the reader be interested.[78]

Secondary (or specialised) metabolism is of critical significance to the field of NP chemistry. The specialised metabolites are conventionally divided into three major categories (phenolic-type, terpenoid-type, alkaloid-type), with a fourth miscellaneous-type category that contains all of the disparate metabolites that do not neatly fit into the preceding three major categories (often these metabolites combine two of the other major categories or diverge in their biosynthesis) (*Fig. 1.5*). The representation of these four major categories across the plant kingdom will be addressed in detail in chapters 2–5, respectively. These divisions of the complex metabolic space of the natural world are by their nature incomplete and imprecise. The clustering of such disparate classes as polyketides, RiPPs (ribosomally synthesised and post-translationally modified peptides) and NRPs (non-ribosomal peptides) into an artificial class labelled 'miscellaneous-type' NPs is far from ideal, but a method of grouping these classes loosely together so that this field could be analysed was necessary. Furthermore, RiPPs and NRPs belong to a huge group of peptide-based NPs that lie outside the scope of this work, as will be discussed later on in the text.

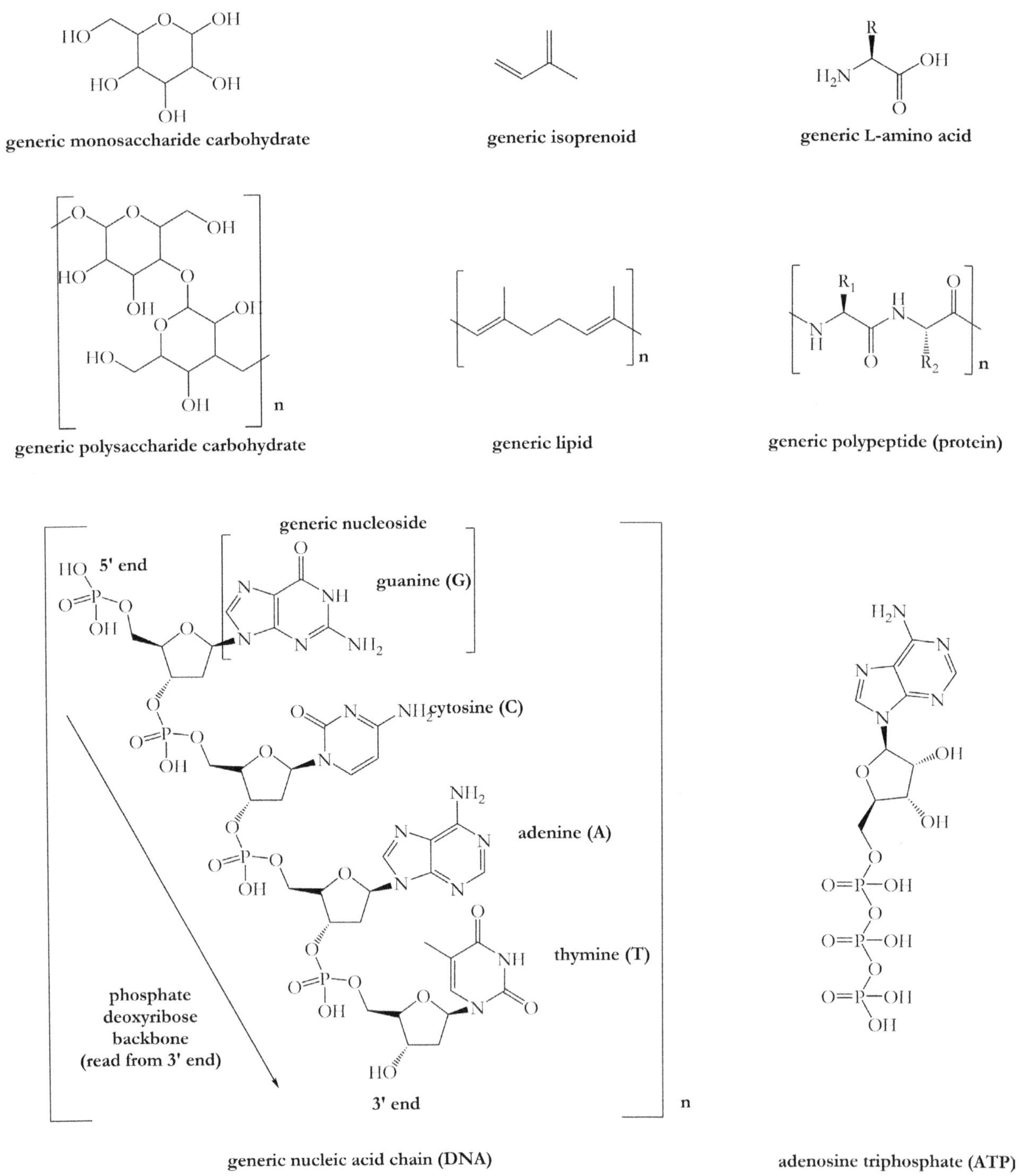

Fig. 1.4. Generic representations of key primary metabolites.

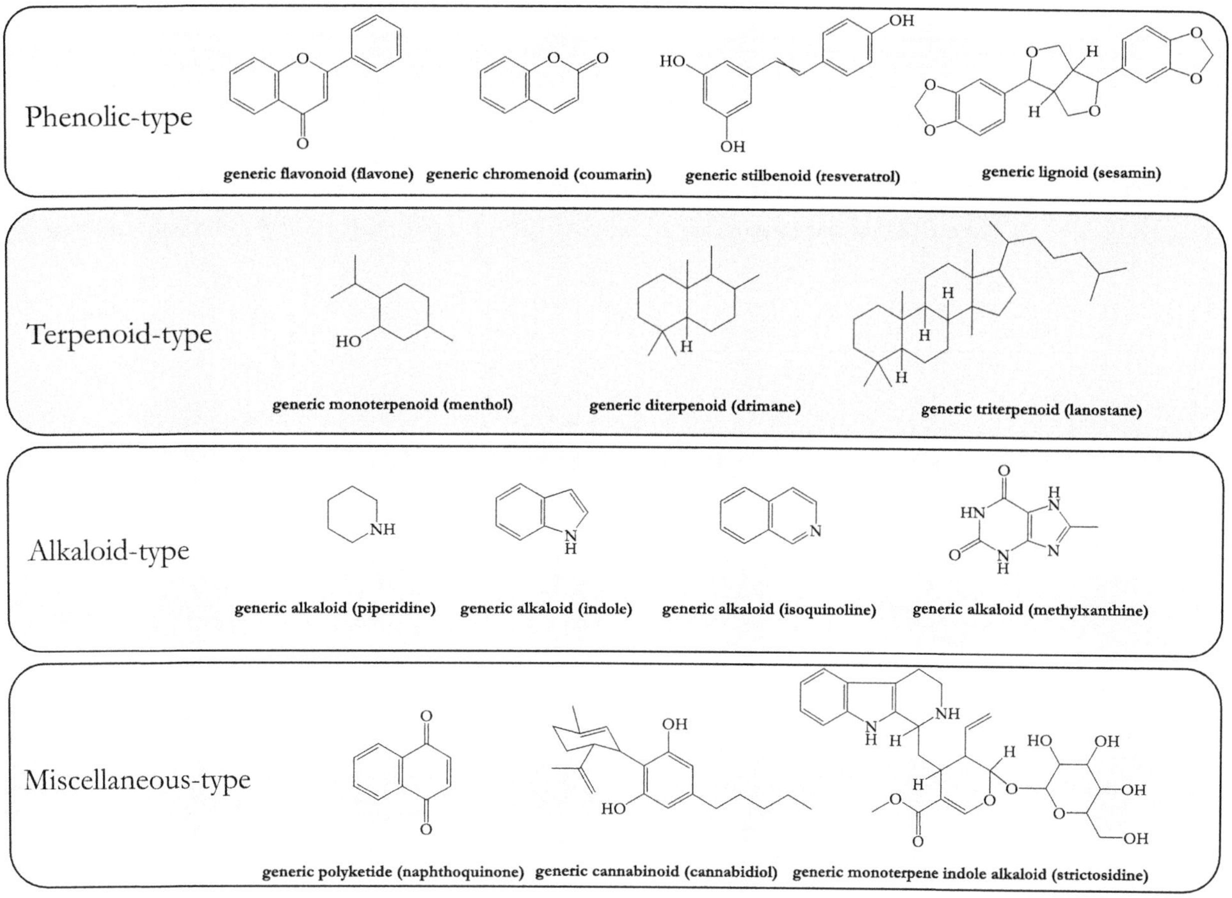

Fig. 1.5. Generic representations of the four key classes of specialised metabolites.

Function

Specialised metabolites perform a sophisticated array of functions *in planta*, which can be addressed generally herein, with more specific examples given in chapters 2–5. The biochemical processes occurring within a plant's specialised metabolic response to ecological signals are often facilitated by a suite of both primary and secondary metabolites that operate in synergy with each other and the various cellular machinery they encounter in their immediate and surrounding environment.[80] Negative feedback can be communicated to herbivores through the incorporation of indigestible polysaccharides (gum, mucilage, fibre) and toxic metabolites (alkaloids, proteins, carbohydrates, phenolics, terpenoids, cyanogens, organic acids, hydrocarbons).[80] Negative feedback can also be expressed to neighbouring plants, a process known as 'allelopathy', through the expression of fatty acids, terpenoids, phenolics, alkaloids, steroids and polyketides into the surrounding subterranean environment, known as the 'rhizosphere'.[80] Positive feedback can be communicated to prospective pollinators through visual cues [terpenoids (carotenoids), phenolics (flavonoids, anthocyanins), alkaloids (betalains)], olfactory cues (terpenoids, alkaloids, phenolics) and gustatory cues (carbohydrates, phenolics, terpenoids).[80] Chemical defence against pathogenic attack, or in response to physical damage, can be facilitated by the exudation of physical substances (complex mixtures known as oleoresins, gums etc.) that contain strong bioactive phytoalexins and phytoanticipins such as polyketides, alkaloids, phenolics, proteins, carbohydrates, terpenoids, cyanogens, organic acids and hydrocarbons.[80] Finally, and perhaps most surprisingly, chemical communication (intra-organism, intra-kingdom) can be facilitated by a myriad of factors into which researchers are only now starting to gain insight. A key example would be the mediation of the symbiotic nitrogen-fixing bacteria of the rhizosphere by *nod* factors [e.g. certain

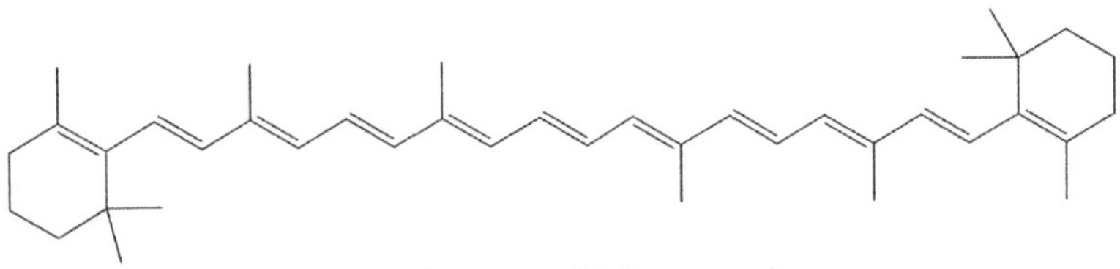

generic carotenoid (β-carotene)

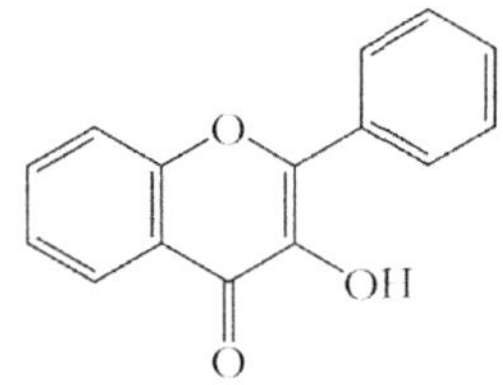

generic flavonoid (flavonol)

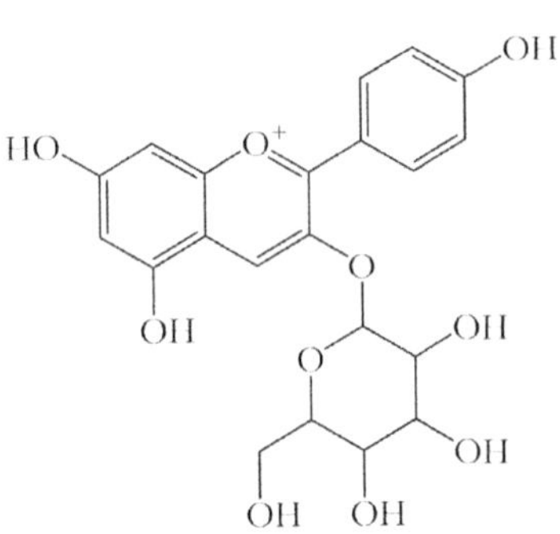

generic anthocyanin

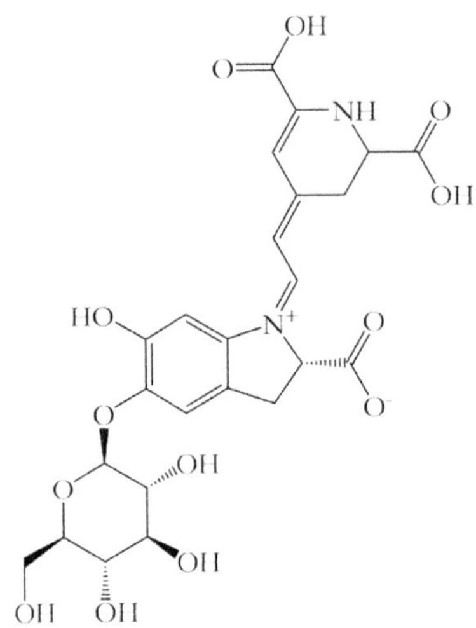

generic betalain (betanin)

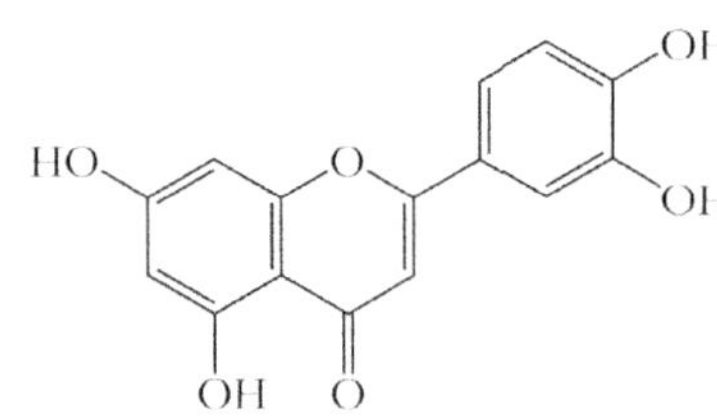

generic flavone (luteolin)

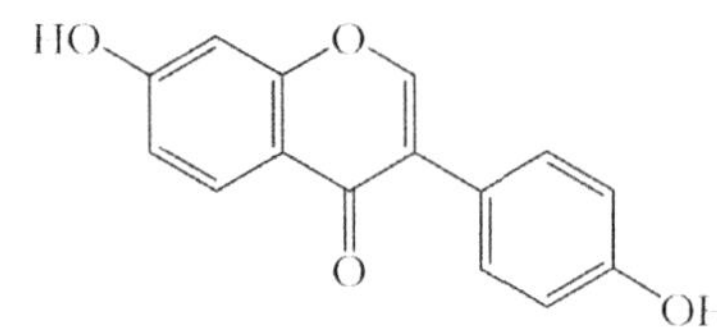

generic isoflavonoid (daidzein)

Fig. 1.6. Generic representation of key functional specialised metabolites.

flavonoids (luteolin) and isoflavonoids (daidzein)], with similar interactions purported to occur in the relationship between plants and mycorrhizal fungi (*Fig. 1.6*).[80]

These notable ecological activities are affected by a complex set of specific biochemical mechanisms of action, many of which have been revealed through dedicated researchers in the life sciences. Where relevant throughout this book, specific biochemical mechanisms will be discussed in detail, but the reader is encouraged to consider that many of the finer details of these effects are still undetermined. It is the pursuit of these details that provides the impetus for much of the research being conducted in the field of molecular biology, today and into the future. The complete set of metabolites produced within the plant kingdom is known as the plant metabolome and is a major focus of the modern field of metabolomics. The structural diversity represented by the breadth of the current plant metabolome is truly formidable, with (at the very least) ~40,000 terpenoids, 40,000 alkaloids, 10,000 phenolics and 2,500 miscellaneous compounds reported in the literature, although this number is continually increasing and estimates of the total number distributed throughout Nature go over 1,000,000.[81] The metabolome is created through the dynamic action of the proteins (enzymes) produced by the cells of an organism. The complete set of proteins produced by an organism is known as its proteome and is the focus of the modern field of proteomics. Similarly, the proteome is encoded from the genetic information contained in the DNA of the organism. The complete set of genes contained in an organisms DNA is known as its genome and is the focus of the modern field of genomics.[82,83] Despite their significant interest and relevance to NP chemistry, a detailed analysis of these three fields (metabolomics, proteomics, genomics) lies outside the scope of this work, although they will be touched upon again later in this chapter when discussing the relevance of computational intelligence and network science to the future scope of the life sciences.

Application

The primary application of NPs in society is the potential benefits that they can bring to the health sciences. Over the past 40 years of drug development, a total of

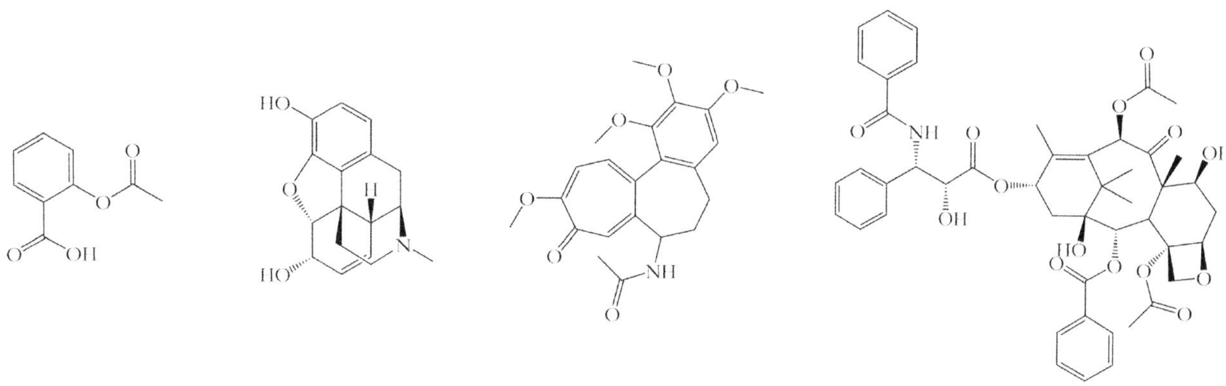

acetylsalicylic acid morphine colchicine taxol

Fig. 1.7. Important natural product-derived drugs in clinical use today.

67.9% (1276) of the 1881 newly approved drugs from 1981 to 2019 were derived in some way from a NP, with the following breakdown: synthetic drug either NP mimic or with NP pharmacophore [25.7% (489)], semi-synthetic NP derivative [18.9% (356)], biological molecule [18.4% (346)], NP – 'botanical' or otherwise [4.6% (85)].[84] Notable individual compounds of the four general classes (phenolic-type, terpenoid-type, alkaloid-type and miscellaneous-type) will be discussed in chapters 2–5, alongside their biological activity and biochemical mechanism of action, where applicable. Except for the application of NPs in the health sciences, a general discussion of which will be included in this section, NPs have also had significant applications in fields that could be loosely gathered under the broad umbrella of biotechnology (e.g. fertilisers, plastics, oils, biofuels, bioremediation, industrial catalysis, food production, environmental applications, etc., even bioterrorism would be a factor here). Unfortunately, these notable applications lie outside the scope of this book.[85]

As can be gathered from the preceding paragraph, the major role of NPs in the health sciences is as a scaffold (or pharmacophore) upon which new drugs can be modelled. This application as a drug-lead (or lead-like compound) within the pharmaceutical industry is significant and many NP researchers occupy positions in this field. Some of the most important drugs in medicine today have their roots set deeply in the NP field; the anti-inflammatory Aspirin or acetylsalicylic acid [from the Salicaceae (*Salix*)], the analgesic morphine [from the Papaveraceae (*Papaver*)], the gout-treatment colchicine [from the Colchicaceae (*Colchicum*)] and the chemotherapeutic paclitaxel or taxol [from the Taxaceae (*Taxus*)] (*Fig. 1.7*).[82,83] A final remark about the specific applications of NPs in the pharmaceutical sciences is their role in the eternal fight against pathogenic multidrug-resistant (MDR) bacteria or 'superbugs'. Superbugs can be both Gram-positive (e.g. *Clostridioides difficile, Staphylococcus aureus, Streptococcus pneumoniae* etc.) or Gram-negative (e.g. *Escherichia coli, Klebsiella pneumoniae, Pseudomonas aeruginosa* etc.) and pose a serious risk to human health. The previous iterations of 'front-line' antibiotics (e.g. penicillin, tetracycline, methicillin, erythromycin, gentamicin, ceftazidime, vancomycin, levofloxacin, iminipem, linezolid, ceftriaxome etc.) have become ineffective against MDR bacteria and thus new antimicrobial agents must be continually developed.[86]

Besides their pharmaceutical applications, many plants are still used in their raw 'whole-food' form as herbal remedies. These herbal applications can cause some consternation among modern medical practitioners, due to their dubious efficacy, inconsistency, bioavailability and potential toxicity, yet many of these plants (and the NPs contained therein) have shown considerable efficacy in vitro, in vivo and under clinical trial. Some of the directions in which this form of herbal remediation has been extensively studied are their direct antioxidant activity and their varying effects on cell signalling pathways and neoplastic (cancerous), cardiovascular, neurodegenerative, diabetic, gastrointestinal, optical and microbial diseases.[87] Specifically, some critical chronic, inflammatory or autoimmune diseases to which some promising efficacy has been demonstrated are atherosclerosis, stroke, myocardial infarction, certain cancers, diabetes mellitus, allergies, asthma, arthritis, Crohn's disease, multiple sclerosis, Alzheimer's

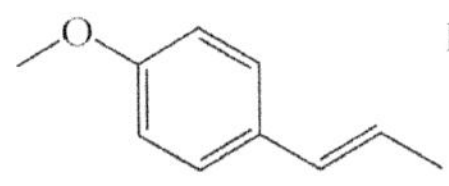

representative Apiaceae
monoterpenoid (*trans*-anethole)

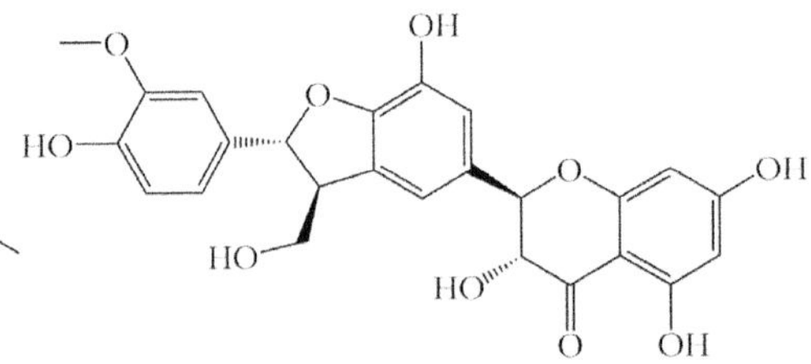

representative Asteraceae
flavanolignoid (silychristin)

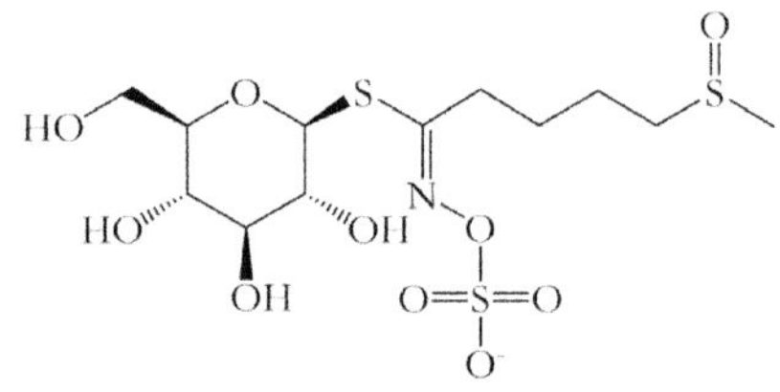

representative Brassicaceae
sulforaphane (glucoraphanin)

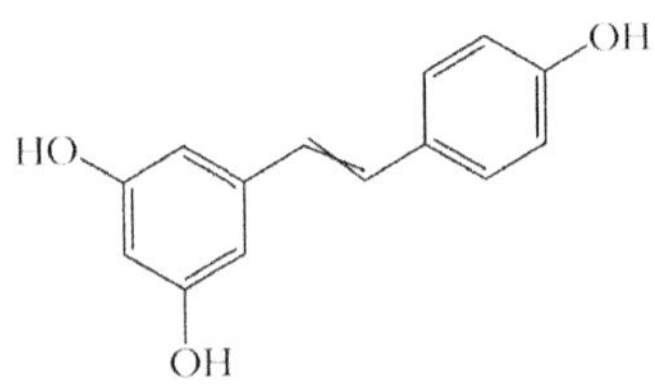

representative Ericaceae and Vitraceae
stilbenoid (resveratrol)

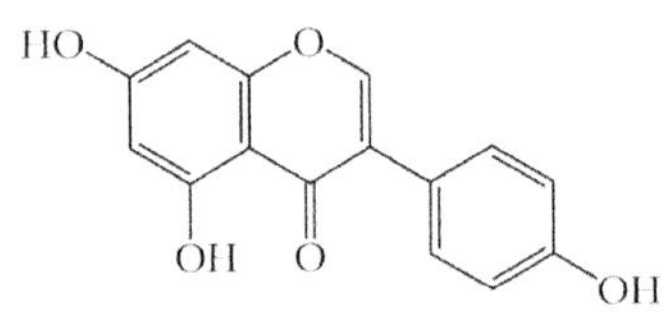

representative Fabaceae
isoflavonoid (genistein)

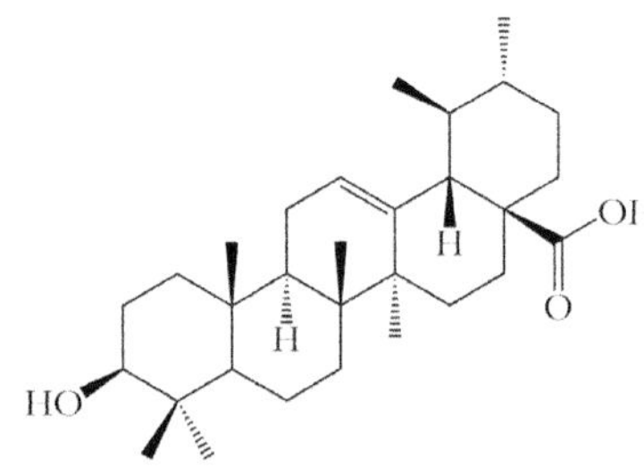

representative Lamiaceae
triterpenoid (ursolic acid)

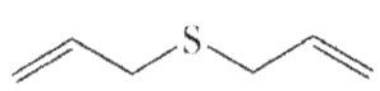

representative Liliaceae
organosulphur compound (diallyl sulfide)

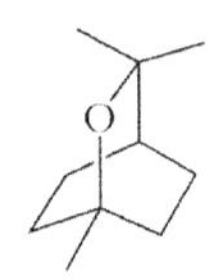

representative Myrtaceae
monoterpenoid (eucalyptol)

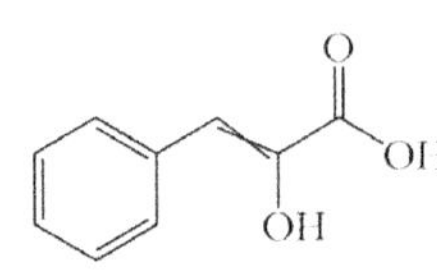

representative Rosaceae
phenylpropanoid (hydroxycinnamic acid)

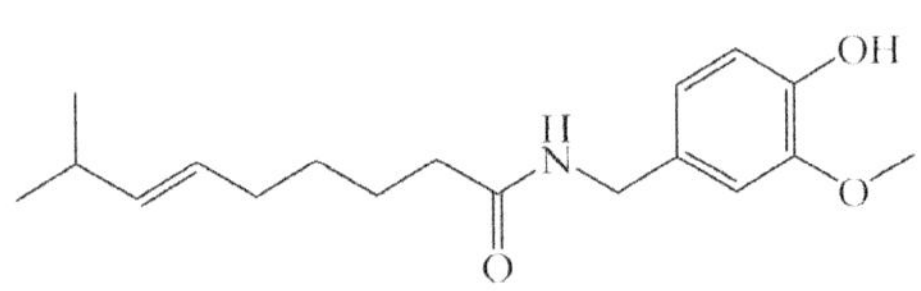

representative Solanaceae
capsaicinoid (capsaicin)

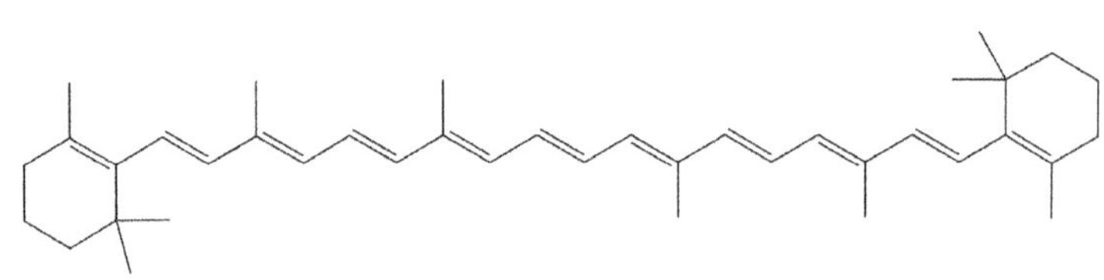

representative Solanaceae
carotenoid (beta carotene)

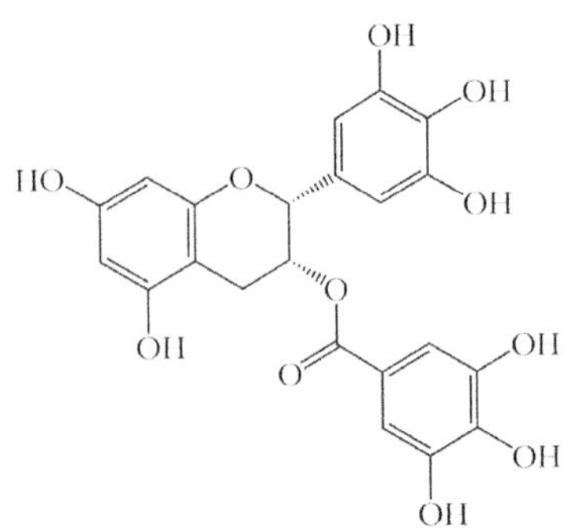

representative Theaceae
flavonoid (epigallocatechin gallate [EGCG])

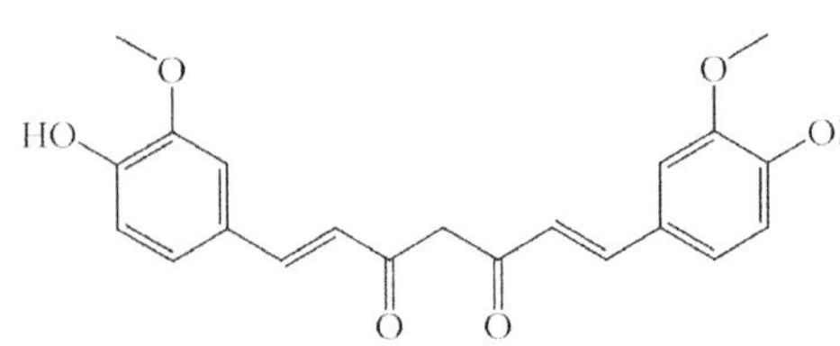

representative Zingiberaceae
curcuminoid (curcumin)

Fig. 1.8. Key nutritional metabolites from high-NP food plants.

disease, osteoporosis, psoriasis, septic shock, AIDS, menopausal symptoms and neurodegeneration.[87] Recent approaches to understanding the mode of action of traditional medicines through the unification of functional genomics, metabolomics, analytical biochemistry, immunology and traditional knowledge, rather than the reductionist approach of first purifying the bioactive principles traditionally employed by the general NP field, are forthcoming, but lie outside the scope of this book.[88]

Although certain claims about efficacy in the field of phytomedicine remain under heavy scrutiny, preventable and chronic 'lifestyle diseases' associated with ageing and poor diet are (undeniably) remarkably negatively correlated with the consumption of NP-rich foods (fruits, vegetables, herbs and spices). Some of the most well-studied plant families [and genera (and pertinent NP subclasses)] that have significant whole-food application in the health sciences include the Apiaceae [*Foeniculum* (monoterpenoids)], Asteraceae [*Silybum* (flavanolignoids), *Cynara* (flavanolignoids)], Brassicaceae [*Brassica* (sulforaphanes)], Ericaceae [*Vaccinium* (stilbenoids)], Fabaceae [*Glycine* (isoflavonoids)], Lamiaceae [*Ociumum* (triterpenoids), *Rosmarinus* (triterpenoids)], Liliaceae [*Allium* (organosulphur compounds)], Myrtaceae [*Syzygium* (monoterpenoid)], Rosaceae [*Fragaria* (phenylpropanoids), *Rubus* (phenylpropanoids)], Solanaceae [*Capsicum* (capsaicinoids), *Solanum* (carotenoids)], Theaceae [*Thea* (flavonoids)], Vitaceae [*Vitus* (stilbenoids)] and Zingiberaceae [*Curcuma* (curcuminoids), *Zingiber* (curcuminoids)] (*Fig. 1.8*).[85,87]

DRUG DISCOVERY

Natural product chemistry

NP chemistry is an interdisciplinary field primarily situated within organic and analytical chemistry. The general process followed by a NP chemist when conducting research, in order to determine the NP constituents contained within a given sample, is simplified under the following seven steps: collection, preparation, extraction, separation, purification, elucidation and characterisation (*Fig. 1.9*). It should be prefaced at the beginning of this section that the approach outlined herein is just one way in which the process of NP chemistry is undertaken, but other equally valid approaches also exist.

Collection involves the acquisition of specimens to be studied, which can be selected based upon myriad factors depending upon the direction of the research. Samples and collection details (e.g. taxonomy, collector, location) are catalogued and type-samples are preserved.[85] Samples are prepared by pulverisation and homogenisation before extraction, which involves maceration to abstract the solutes from the sample matrix. Typically, this is done via sequential solvent extraction (trituration), starting with a non-polar solvent [e.g. dicholoromethane (DCM - CH_2Cl_2)] which is macerated for a period of time, then the non-polar extract is filtered, dried (via rotary evaporation) and weighed, before the sample remnants are extracted again using the same approach with a more polar solvent [e.g. methanol (MeOH - CH_3OH)].[89] Once the various extracts are prepared and qualified (via preliminary analytical techniques (e.g. high-pressure liquid chromatography (HPLC), liquid chromatography mass spectrometry (LC-MS), nuclear magnetic resonance (NMR) spectroscopy), purification begins, which is reliant upon separations science, primarily chromatography.[90] Benchtop chromatography is performed initially, either via normal-phase silica column chromatography [for non-polar extracts (e.g. DCM)] or via reversed-phase C_{18} vacuum liquid chromatography for polar extracts (e.g. methanol).[90] Fractions generated by benchtop chromatography are again dried, weighed and qualified as before, before further chromatographic fractionation utilising a suite of sophisticated stationary phases (e.g. size-exclusion chromatography, solid-phase extraction, etc.) or purification using a more robust semipreparative HPLC.[91,90] Repeated fractionation yields increasingly purified samples, with fewer and fewer compounds contained in each fraction until individually isolated compounds are obtained.[91,90]

Once the compounds are isolated/purified, the process of elucidation can begin. Before embarking on the time-consuming elucidation stage, an important initial phase known as 'dereplication' should take place, wherein preliminary metrics obtained from analytical screening [e.g. HPLC-PDA (for indicative UV chromophores), LC-MS (for individual ionised molecular weights), or (LC-)NMR (for indicative chemical shifts)] can be compared with literature and database searches in order to either rapidly determine the identity of the compound, if known, or to gain a general understanding of the types of compounds present, a process that can be useful for prioritising fractions/compounds of interest.[83] Prioritised, purified compounds can then be submitted to more in-depth analyses to determine, or elucidate, their chemical structure. Elucidation is a

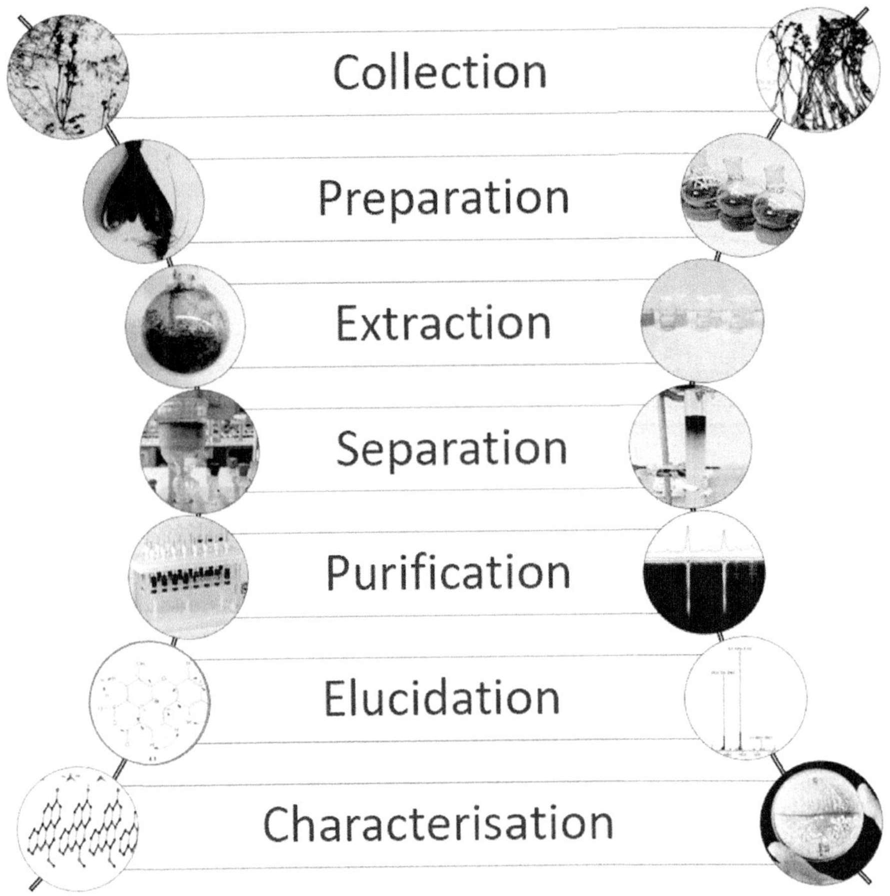

Fig. 1.9. Infographic depicting the generalised natural product research methodology.

complex and rigorous problem-solving process that relies on the physicochemical features of the isolated compound.[83] The most valuable tool for an NP chemist in the elucidation process of an unknown compound is NMR spectroscopy, which provides either one-dimensional information about the chemical environments of the atoms present in the molecule (typically ^{1}H or ^{13}C) or two-dimensional information (e.g. gCOSY, gHSQCAD, gHMBCAD etc.) about the relationship between the chemical environments present in the molecule.[83] This elucidation process is supported by a variety of other analytical spectroscopic techniques as the NP researcher accumulates information on how the compound being studied is arranged in space.

Once a preliminary structure has been determined, the assessment can be confirmed using high-resolution mass spectrometry (HRMS), which can determine the molecular weight of a pure compound to four decimal places (amu) and it can then be compared to the calculated molecular mass of the proposed molecular formula.[90] If these two figures are in agreement, the structures' absolute configuration in space must be determined, a process that is complicated in NP chemistry because isolated NPs rarely yield crystalline structures amenable to X-ray crystallographic analysis.[92] In lieu of this, various techniques can be used to determine a compound's 'relative' configuration, such as through-space nuclear Overhauser experiments performed using a suitably prepared NMR instrument.[93]

The final step in the NP journey is open-ended, because the final characterisation of a compound is never really 'finished'. Various physicochemical properties (e.g. IR and UV spectra, optical rotation of polarised light (OR) etc.) can be obtained relatively simply, yet the biological and pharmacological properties of isolated compounds are innumerable.[94] Standard biological assessments that can be conducted on isolated NPs (e.g. anti-microbial and antiparasitic assays, cytotoxicity assessment etc.), are routinely performed, but there is no technical limit to the types of assessment to which a NP can be submitted, apart from the physical amount of pure compound available. If an isolated NP shows significant biological or pharmacological activity in the assays to which it has been submitted, then this compound can be prioritised for further development as a potential drug candidate.[95] As a final note, it must be kept in mind that the process just detailed is the generalised, standardised strategy used by the research group to which the author belongs and that a great deal of variation is possible depending on the facilities and experience of other groups conducting similar research, although the general seven-step process remains the same.

A brief mention must be made regarding artefacts in NP chemistry. A NP is defined as a chemical that originates from, or can be detected in, a fresh extract of a source organism, provided the process of extraction and/or detection does not initiate a chemical transformation that is solely responsible for producing the chemical.[96] An artefact is a NP that has undergone chemical transformation during extraction, handling, storage and/or analysis. NPs can be chemically reactive, so artefacts can form under the mildest of conditions and just because a chemical compound is isolated from an extract does not mean it is a NP.[96] This information has been acknowledged in the NP literature for some time and most high-quality NP research institutions follow a tight protocol in an attempt to minimise any artefact formation, but it becomes increasingly apparent that this is easier said than done. Methylation, methoxylation and degradation are all phenomena that the lead author has experienced throughout the PhD research candidature, despite keeping these compounds well away from heat, light, air, acids, bases and harsh solvents. Artefact formation is a significant topic and in the coming years the author believes this aspect of NP chemistry will receive increasing attention.

Drug development

Although what follows is outside the scope of this book, because the role of the NP researcher typically concludes with the discovery of a lead compound, the end goal of the process is worth a short summation and readers are recommended to follow up with any of the numerous, substantial reference works on the topic if they are interested. To begin with, another term used in the pharmaceutical industry for what NP chemists would consider a promising drug candidate is a 'lead compound'. Lead

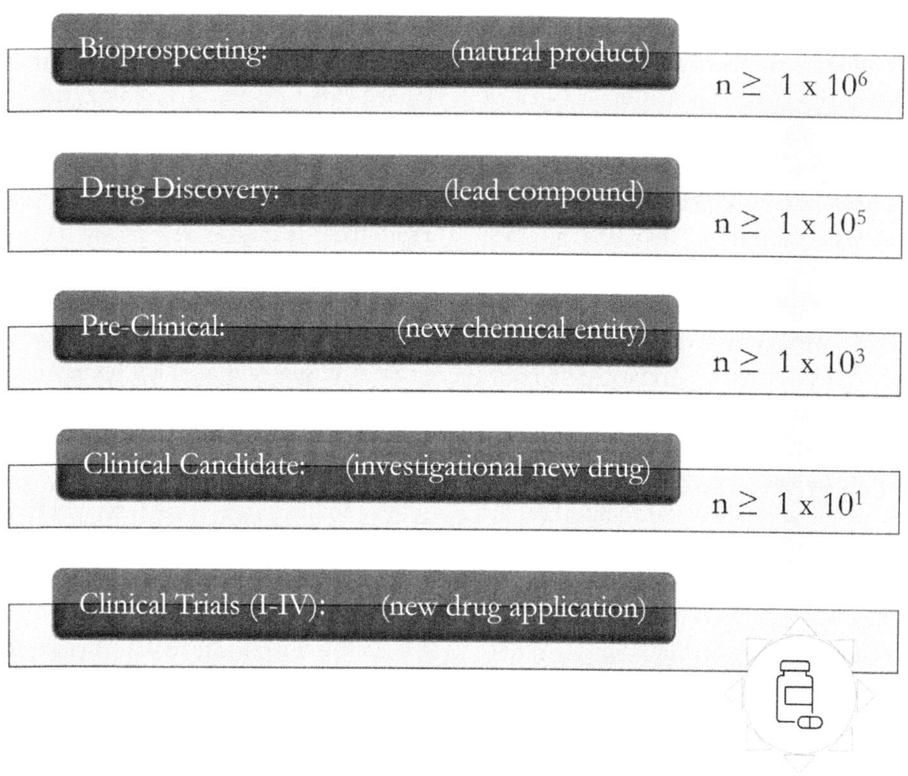

Fig. 1.10. Infographic depicting the generalised flow through the drug development process.

compounds generally have therapeutically useful biological or pharmacological activities, as determined by any of the numerous assays to which the compound has been submitted.[95] The sheer magnitude of combinations of compounds and the assays to which they can be submitted has led to a field of science in its own right, known as combinatorial chemistry.[83] The structure of a lead compound may well be suboptimal and it is possible that some structural relative or derivative may yield a more significant response in relevant assays.[83] Chemical modification (by varying the shape, size, polarity, functional groups etc.) can drastically improve the potency, selectivity or pharmacokinetic properties of a lead compound. Combinatorial chemistry involves the preparation of compound libraries in which lead compounds are iteratively modified upon set parameters in order to generate huge repositories of structurally similar molecules that, via high-throughput screening, can be analysed sequentially to determine the structure–activity relationship between a library and any number of assays or drug targets.[83]

Drug development, from lead compound to newly approved pharmaceutical drug, is an extremely lengthy (typically longer than 10 years), expensive and complex process that begins with the primary assays detailed at the end of the NP process (*Fig. 1.10*). In fact, an immense number of compounds are produced synthetically on the basis of predictive computational modelling, as well as by blind combinatorial iteration.[82] Promising primary screen candidates or 'new chemical entities' (NCEs) are submitted to secondary assays, which assess more pharmacologically relevant parameters than the primary screens, such as counter screens, bioavailability, toxicity, metabolism [absorption, distribution, metabolism, excretion (ADME)], structure–activity relationships (SAR/QSAR), etc.[82,97] This process is cyclical, with NCEs brought closer and closer to becoming a 'clinical candidate' through repeated iteration and analysis, typically in microbial, computational, physicochemical (e.g. stability, solubility etc.) and animal models. Clinical candidates are outstanding compounds that have performed well in pertinent preclinical assays and can then move into the clinical phase of drug development by satisfying the regulatory requirements for an 'investigational new drug application'.[97]

Clinical trials involving humans typically comprise phases I–IV, which become increasingly involved over time, in which the behaviour of the NCE is continually monitored using a variety of parameters to determine whether it has an acceptable toxicity and safety profile.[97,98] Clinical trials not only scrutinise safety, but efficacy, with significant failure rates between each phase to be expected, alongside commensurate financial losses to the developer. Phase I trials determine safety and dosing in a limited number of willing participants (up to a few dozen). Phase I trials carry the most potential risk and often occur with participants experiencing life-threatening illnesses or when all other treatment options have already been tried. Phase II trials begin to explore efficacy in patients afflicted with the targeted disease and further scrutinise safety. Phase III trials expands the sample size to be monitored and if successful at this point can be submitted for a 'new drug application'.[98,99] Phase IV trials are post-approval and monitor the population for long-term and chronic toxicities that were not apparent in the earlier phases. A group of 21,143 NCEs from a library of ~185,994 unique trials that made it to this initial point in the clinical drug development process showed the following breakdown of each phase transition: phase I to II (66.4%), phase II to III (48.6%) and phase III to IV (13.8%).[100] Over the whole clinical trial process, the specific therapeutic group to which the NCEs applies (e.g. oncology, cardiovascular, infectious disease), can vary widely in ultimate success rate (min: 3.4% – oncology, max: 33.4% – vaccine). These data demonstrate the absolute necessity of careful analytics and strategy that go in to avoiding costly 'attrition rate' problems.[101,102]

Future perspectives

From this broad overview of the 'how-to' of NP isolation and the drug discovery process, it is worth looking at future perspectives of the field. Clearly, the amount of time and effort that goes into the drug discovery process is substantial. This labour cost is one of the most significant hurdles to be overcome in the field and the solution (as it has in almost all research fields) appears to arise from the brute force of computational intelligence. Nature is so complex, biological systems are so dynamic and the parameters of the drug discovery process are so sophisticated that without the assistance of computing power, the Herculean task of drug development begins to look insurmountable. So, in the opinion of the author at least, it is to this end that the future of the field is directed. As mentioned in the previous two subsections, avoiding the 'attrition rate' of pursuing the relative 'dead-ends' of either previously characterised compounds (for the NP chemist) or unsafe/ineffective NCEs (for the pharmaceutical chemist) is of paramount importance. For the NP chemist, the most powerful tool available to this end is the process of dereplication.[82,83] For the pharmaceutical chemist, the most powerful tool available is computational molecular modelling.[103] Both strategies are directed at expediting the whole process, by determining whether a lead is worth pursuing before any significant work is done. Molecular modelling, the *in silico* determination of physicochemical properties of compounds (alone or in groups) and ligand-target binding of molecular docking systems, is a significant field in its own right, with applications across the life and materials sciences, but unfortunately a detailed analysis lies outside the scope of this book.

The concept of dereplication, introduced previously, involves the comparison of preliminary analytical results to those contained in (typically online) databases available to the researcher, in order to prioritise and expedite the isolation of new compounds only.[83] This process has special significance to the NP researchers of the future, because it exposes them to the fascinating and powerful field of data science. An individual researcher may compare a compound's UV chromophore (from HPLC-PDA), MS molecular ion (from LC-MS) or indicative chemical environments (from LC-NMR), even while this compound is still in a mixture/fraction/crude extract with numerous others.[82,83] Although this approach is excellent in theory, because it allows the superficial identification of compounds without the laborious process of purification, in practice the matrix effects of the mixture of compounds often convolutes this process, yielding unreliable results. Nevertheless, it is useful as a quick insight into the properties of the NPs in the mixture, giving valuable information on prioritisation, but the mixture will inevitably need to be purified further for any conclusive results to be obtained. That said, with the advance in raw computational and instrumentational power, the concept behind dereplication is becoming increasingly efficacious and is expanding in new and exciting directions. Several examples are the rising importance of molecular networking in NP research (e.g. liquid-chromatography tandem mass spectrometry (LC-MS/MS) coupled with the Global Natural Product Social (GNPS) molecular networking) and the exponential growth of the -omics sciences (e.g. genomics, proteomics, metabolomics, interactomics etc.), all of which are combined in the burgeoning field now known as informatics (both bioinformatics and cheminformatics are relevant here), which utilises computer science to help comprehend, study and represent large biological and chemical datasets.[104,105] Behind all of these emerging avenues is the mathematically grounded fields of AI/ML (deep learning), and computational network science.

Molecular networking is a powerful tool in the visualisation and comprehension of complex mixtures, such as a crude plant extract. Spectral data are obtained from an LC-MS/MS instrument, which uses liquid chromatography [typically ultraperformance (UPLC)] to first separate mixtures, then tandem mass spectrometry to not only fragment ionised compounds based on their mass-to-charge ratio (m/z), but to fragment these fragmented ions further, splitting the compounds up into their basic structural features.[106] The value of LC-MS/MS spectral data is that although similar compounds, which bear similar basic structural features, may bear significantly different retention times (often based on polarity), they have fragmentation patterns that are resolvable through comparison with one another. In the terminology of network science, each ionic spectra is treated as a 'node' and nodes that are similar to one another are given an 'edge' between them, with the resulting network clustering similar compounds based on this fragmentation pattern.[106] This process would be impossible for a human researcher to conduct, but with computational assistance, networks of this type can be created and disparate chromatographic peaks (viz. semi-pure or pure compounds) can be identified and then selectively targeted for purification.[106] Not only is this a powerful new tool in the dereplication toolkit of the NP chemist, but with addition of metadata to the network, or feature-based molecular networking, the space can be investigated in a way that was not possible before this technology, opening up entirely new areas for researchers to explore.[107] High-throughput screening of NP libraries, coupled with a technique such as LC-MS/MS GNPS molecular networking, is able to drastically increase the rate at which the overwhelming complexity of the natural world can be classified, with promising leads and new/novel metabolites being tagged for prioritised purification.[106,107]

The rapidly advancing fields with the (neo)suffix -omics, which is defined by the Oxford English Dictionary as the study of assemblages (with suffix -ome) where 'all constituents are considered collectively' ([i.e. metabolomics studies the metabolome (the totality of the metabolites that the organism's metabolism produces) of an

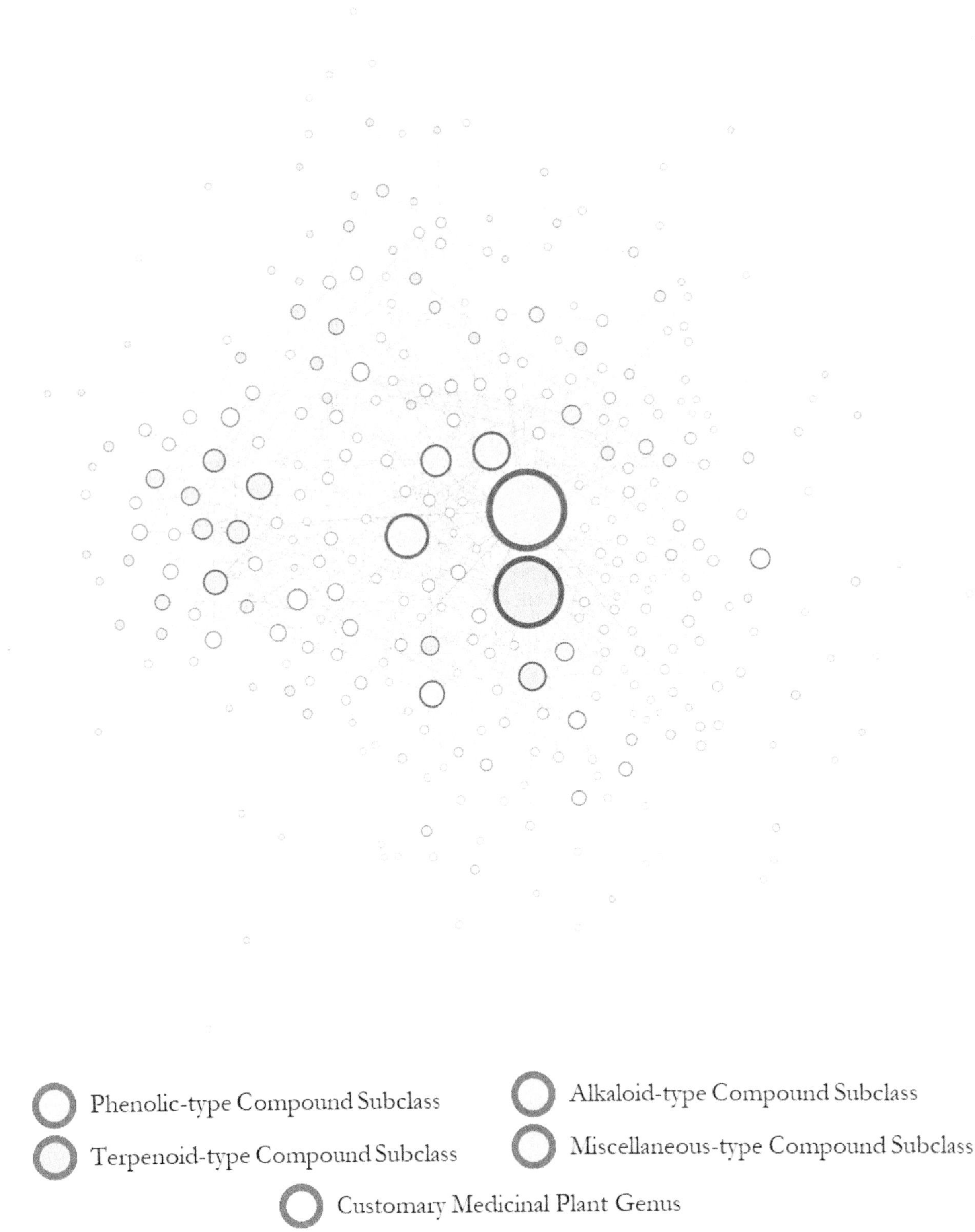

Fig. 1.11. Overall specialised metabolic network of customary medicinal flora of Australia. Node size indicates degree (number of edges).

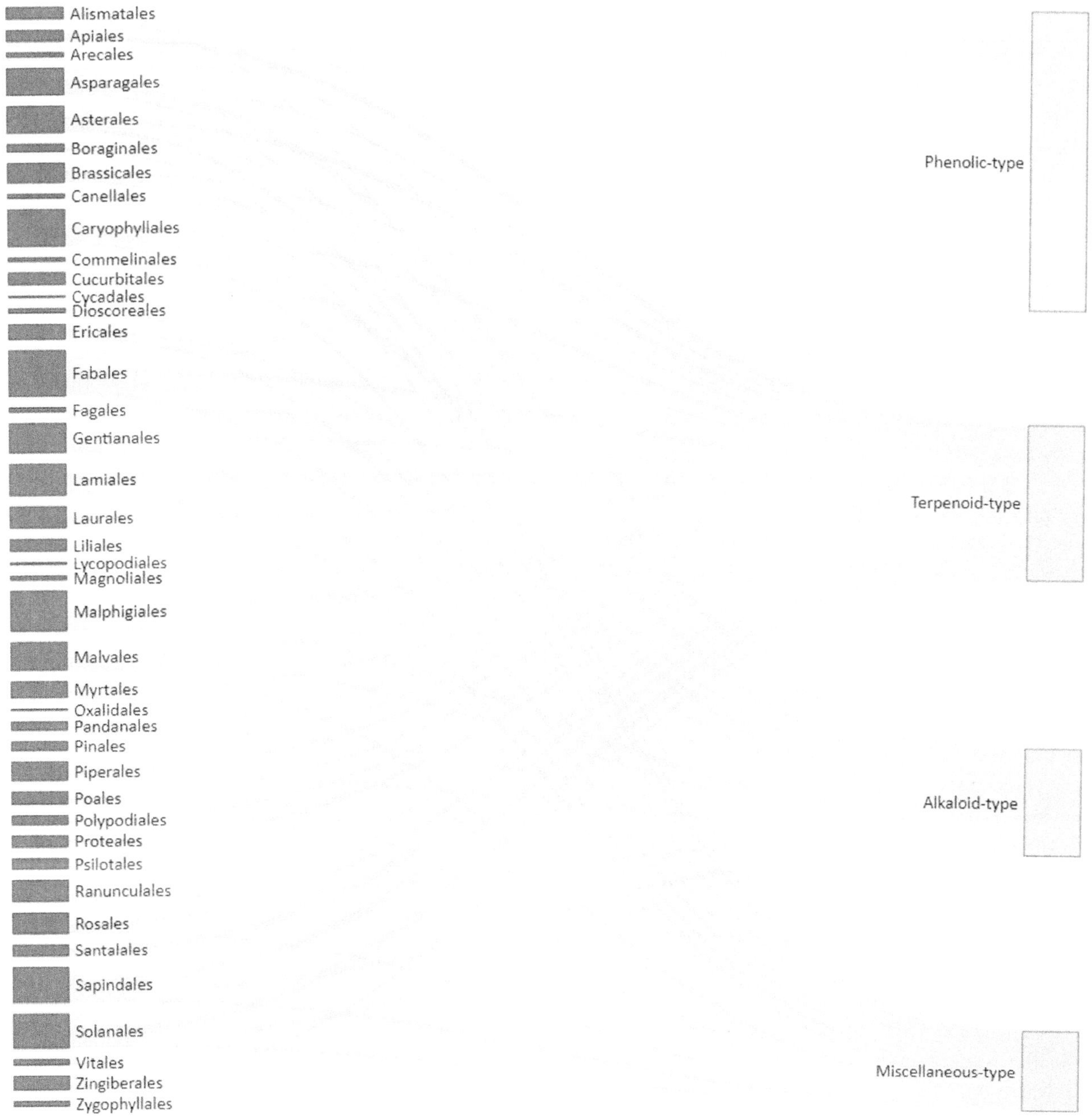

Fig. 1.12. Sankey diagram showing the distribution of the four major classes of specialised metabolites over the representative orders of the genera recorded in the customary medicinal flora of Australia database.

organism), is another example of the unprecedented directions in which the life sciences can head when collaborating with the computer sciences. Each of the -omics alone would be difficult to do justice to as individual introductions of their own, so only a cursory summation can be provided herein. The original -omes, the biome and the genome, have since expanded into a substantial array of differing projects applying the same concept of collective totality within a particular system. Some of the most relevant -omics to NP chemistry are of course metabolomics (the study of the metabolic fingerprint of an organism) and also proteomics (how these metabolites are produced by proteins), genomics [the genetic and evolutionary basis for how these proteins come to produce these metabolites, often related to taxonomy (or chemotaxonomy)] and interactomics (formally the study of protein interaction networks, but also any sort of chemical interaction occurring in a system, particularly relevant in medicine and pharmacology).[104,105] As can be seen, there is a significant overlap between many of these -omics fields, with each member furthering our collective understanding of how complex, dynamic biological systems exist and function in their environment.

The interdisciplinary nature of all these overlapping-omics is unified in the emerging field known as informatics. Informaticians conduct research in the life sciences with computer programming as part of their methodology, in the pursuit of understanding the organisational principles that drive the process of life.[104,105] The amount of data being produced by the ever-increasing instrumental power available to contemporary scientists necessitates the use of computational approaches to mine this data for meaningful insights. The computational tools available to the informatician can help compare, analyse and interpret biochemical data in order to catalogue the pathways and networks that form the basis of a living system.[104,105] Not only is the sheer size and complexity of even the simplest biological system impossible for a pre-informatic researcher to contend with, but these complex networks are in a state of constant processual change, with many of the meaningful insights not simply emerging from what is contained within the system, but how each of the parts of the system relate to each other over time.

It is in the interest of all scientists to familiarise themselves with these technological and computational developments. Finally, before moving on to a detailed breakdown of the major types of NPs distributed across the plant kingdom, the author wishes to state that it was only due to the difficult circumstances of a global pandemic occurring midway through the PhD candidature that the time and space was provided to consider the way in which that NP research progressed. The uncertainty surrounding the conduct of further research on campus led to a reconsideration of what could be done with the resources at hand. As a result, the evaluation of informatic ideas to systematise NP data prompted the author to embark upon what has resulted in this literary output and the creation of a database summarising the literature concerning the NP chemistry of those Australian native plants with documented customary use among Indigenous Peoples, presented in chapters 2–5 and hosted at www.cmfoa.info.

Snapshots of the network relations included in this database are shown in *Fig. 1.11* and *Fig. 1.12*, covering genus:infraclass and order:class, respectively. For the uninitiated, a network is a graphical space in which objects (nodes) are related to one another (edges). In the networks presented in *Fig. 1.11* and *Fig. 1.12*, nodes are plant groupings and chemical classes (of varying levels of depth; genus:sub/infraclass and order:class, respectively). An edge is created when one node (i.e. a plant genus or order) has been reported to produce the chemistry represented by another node (i.e. an NP class or sub/infraclass). The selected networking approach then uses various layout algorithms to project these relationships in the graphical space shown in the figures. It must be noted here that it has been a considerable task to represent a functionalised network space of this size (which can be zoomed in on and manipulated in a real-time capacity in the software), as a static image that is not overwrought with information. To address this, an interactive network is also hosted at www.cmfoa.info. Researchers are now able to utilise this resource to help direct their own work and focus on particular NPs. Importantly, the original producers of this IP, First Scientists, will benefit from the scientific substantiation of their immensely valuable knowledge of this land that they have so generously shared.

REFERENCES

1. Lietava J (1992) Medicinal plants in a Middle Paleolithic grave Shanidar IV? *Journal of Ethnopharmacology* **35**, 263–266. doi:10.1016/0378-8741(92)90023-K
2. Cragg GM, Newman DJ (2005) Biodiversity: a continuing source of novel drug leads. *Pure and Applied Chemistry* **77**, 7–24. doi:10.1351/pac200577010007
3. World Health Organization (2013) WHO Traditional Medicine Strategy: 2014–2023. https://www.who.int/publications/i/item/9789241506096 (accessed September 2024)
4. Garrison FH (1966) *History of Medicine*. W. B. Saunders Co., Philadelphia, PA.
5. Scarborough J, Nutton V (1982) The Preface of Dioscorides' Materia Medica: introduction, translation, and commentary. *Transactions & Studies of the College of Physicians of Philadelphia* **4**, 187–227.
6. Tiziano D (1999) Chronology. In *The Cambridge History of Hellenistic Philosophy*. (Ed. A Keimpe) pp. 52–53. Cambridge University Press, Cambridge, UK.
7. Wills A (1999) Herophilus, Erasistratus, and the birth of neuroscience. *Lancet* **354**, 1719–1720. doi:10.1016/S0140-6736(99)02081-4
8. Dioscorides P, Goodyer J (trans.) (2000) *De Materia Medica: Being an Herbal with Many Other Medicinal Matters. Written in Greek in the First Century of the Common Era* [c. 70]. Ibidis, Johannesburg, South Africa.
9. Porter R (1997) *The Greatest Benefit to Mankind: Medical History of Humanity*. Harper Collins, London, UK.

10. Sharpe WD (1964) Isidore of Seville: the medical writings. An English translation with an introduction and commentary. *Transactions of the American Philosophical Society* **54**, 1–75. doi:10.2307/1005938
11. Mckitterick R (2005) The Carolingian renaissance of culture and learning. In *Charlemagne: Empire and Society*. (Ed. J Story) pp. 151–167. Manchester University Press, Manchester.
12. De Vos P (2010) European Materia Medica in historical texts: longevity of a tradition and implications for future use. *Journal of Ethnopharmacology* **132**, 28–47. doi:10.1016/j.jep.2010.05.035
13. Rooney A (2012) *The History of Medicine*. The Rosen Publishing Group, New York, NY.
14. Syros V (2013) Galenic medicine and social stability in early modern Florence and the Islamic empires. *Journal of Early Modern History* **17**, 161–213. doi:10.1163/15700658-12342361
15. Conrad LI, Neve M, Nutton V, Porter R, Wear A (1995) *The Western Medical Tradition: 800 BC to AD 1800*. Cambridge University Press, Cambridge, UK.
16. Voigts L (1979) Anglo-Saxon plant remedies and the Anglo-Saxons. *Isis* **70**, 250–268. doi:10.1086/352199
17. Holmes EM (1906) Horticulture in relation to medicine. *Journal of the Royal Horticultural Society* **31**, 50–54.
18. Hill AW (1915) The history and function of botanic gardens. *Annals of the Missouri Botanical Garden* **2**, 185–240. doi:10.2307/2990033
19. O'Donnell K, Sharrock S (2018) Botanic gardens complement agricultural gene bank in collecting and preserving plant genetic diversity. *Biopreservation and Biobanking* **16**, 384–390. doi:10.1089/bio.2018.0028
20. Allen D, Hatfield G (2004) *Medicinal Plants in Folk Tradition: An Ethnobotany of Britain and Ireland*. Timber Press, Cambridge, UK.
21. Halleux R (1996) The reception of Arabic alchemy in the West. In *Encyclopedia of the History of Arabic Science*. (Eds R Rashed and R Morelon) pp. 840–856. Routledge, New York, NY.
22. Horden P (2005) The earliest hospitals in Byzantium, Western Europe, and Islam. *The Journal of Interdisciplinary History* **35**, 361–389. doi:10.1162/0022195052564243
23. Hajar R (2013) The air of history: part III. *Heart Views* **14**, 43–46[History of Medicine]. doi:10.4103/1995-705X.107125
24. Tschanz DW (2003) Arab roots of European medicine. *Heart Views* **4**, 9–10[History of Medicine].
25. Multhauf R (1954) John of Rupescissa and the origin of medical chemistry. *Isis* **45**, 359–367. doi:10.1086/348357
26. Essa A, Ali O (2010) *Studies in Islamic Civilization: The Muslim Contribution to the Renaissance*. Internation Institute of Islamic Thought (IIIT), Washington DC.
27. Norris J (2006) The mineral exhalation theory of metallogenesis in pre-modern mineral science. *Ambix* **53**, 43–65. doi:10.1179/174582306X93183
28. Nutton V (2022) The rediscovery of ancient medicine. In *Renaissance Medicine: A Short History of European Medicine in the Sixteenth Century*. Routledge, New York, NY.
29. Siraisi NG (2012) Medicine, 1450–1620, and the history of science. *Isis* **103**, 491–514. doi:10.1086/667970
30. Webster C (2008) *Paracelsus: Medicine, Magic and Mission at the End of Time*. Yale University Press, New Haven, CT.
31. Crone H (2004) *Paracelsus: The Man Who Defied Medicine – His Real Contribution to Medicine and Science*. Albarello Press, Melbourne.
32. Josephson-Storm J (2017) *The Myth of Disenchantment: Magic, Modernity, and the Birth of the Human Sciences*. University of Chicago Press, Chicago, Ill.
33. Chaudhury RR, Rafei UM (Eds) (2001) Traditional medicine in Asia. World Health Organisation, New Delhi, India.
34. Yuan H, Ma Q, Ye L, Piao G (2016) The traditional medicine and modern medicine from natural products. *Molecules (Basel, Switzerland)* **21**, 559. doi:10.3390/molecules21050559
35. Hinrichs TJ (2005) Healing and medicine in China. In *Healing and Medicine*. (Ed. L Jones) pp. 3859–3864. Macmillan Reference USA, Detroit, MI.
36. Sivin N (1988) Science and medicine in imperial China - the state of the field. *The Journal of Asian Studies* **47**, 41–90. doi:10.2307/2056359
37. Harper D (1998) *Early Chinese Medical Literature: The Mawangdui Medical Manuscripts*. Kegan Paul International, New York, NY.
38. Leung AY (2006) Traditional toxicity documentation of Chinese Materia Medica: an overview. *Toxicologic Pathology* **34**, 319–326. doi:10.1080/01926230600773958
39. Yang S-Z (2008). *A Translation of the Shen Nong Ben Cao Jing*. Blue Poppy Press, Boulder, CO. https://archive.org/stream/ShenNongBenCaoLingTheDivineFarmersMateriaMedica/ShenNong%20Ben%20Cao%20ling%28The%20Divine%20Farmers%20Materia%20Medica%29_djvu.txt (accessed 9 June 2022).
40. Chen K, Yu B (1999) Certain progress of clinical research on Chinese integrative medicine. *Chinese Medical Journal* **112**, 934–937.
41. Foster S, Yue C (1992) *Herbal Emissaries: Bringing Chinese Herbs to the West*. Healing Arts Press, Rochester, VT.
42. Wong M (1976) *La Médecine Chinoise par les Plantes*. Tchou, Paris, France.
43. Maas PA (2018) Indian medicine and Ayurveda. In *The Cambridge History of Science: Vol. 1, Ancient Science - Part IV, India*. (Eds A Jones and L Taub) pp. 532–550. Cambridge University Press, Cambridge, UK.
44. Mishra L-C, Singh BB, Dagenais S (2001) Ayurveda: a historical perspective and principles of the traditional healthcare system in India. *Alternative Therapies in Health and Medicine* **42**, 36–42.
45. Sen S, Chakraborty R (2017) Revival, modernization and integration of Indian traditional herbal medicine in clinical practice: Importance, challenges and future. *Journal of Traditional and Complementary Medicine* **7**, 234–244. doi:10.1016/j.jtcme.2016.05.006
46. Rahman HSZ (2001) Unani medicine in India: its origin and fundamental concepts. In *History of Science, Philosophy and Culture in Indian Civilisation, Vol. IV, Part*

2 (Medicine and Life Sciences in India). (Ed. Subbarayappa) Centre for Studies in Civilisations (Project of History of Indian Science, Philosophy and Culture), New Delhi, India.

47. Schipmann U, Leaman DJ, Cunningham AB, Walter S (2005) Impact of cultivation and collection on the conservation of medicinal plants: global trends and issues. *Acta Horticulturae* (676), 31–44. doi:10.17660/ActaHortic.2005.676.3
48. Choudhury S, Bahadur B, Pullaiah T (2017) Ethnomedicinal plants of north-east India. In *Ethnobotany of India. Vol. 3: North-East India and the Andaman and Nicobar Islands.* (Eds T Pullaiah, KV Krishnamurthy and B Bahadur) Apple Academic Press, Palm Bay, FL.
49. Prakash BN, Unnikrishnan PM, Hariramamurthi G (2017) Medicinal flora and related traditional knowledge of Western Ghats: a potential source for community-based malaria management through endogenous approach. In *Ethnobotany of India. Vol. 2: Western Ghats and West Coast of Peninsular India.* (Eds T Pullaiah, KV Krishnamurthy and B Bahadur) Apple Academic Press, Palm Bay FL.
50. Pullaiah T, Karuppusamy S (2017) Ethnomedicinal plants of Eastern Ghats and adjacent Deccan region. In *Ethnobotany of India. Vol. 1: Eastern Ghats and Deccan.* (Eds T Pullaiah, KV Krishnamurthy and B Bahadur) Apple Academic Press, Palm Bay, FL.
51. Pullaiah T, Krishnamurthy KV, Bahadur B (2017b) Introduction. In *Ethnobotany of India. Vol. 4: Western and Central Himalayas.* (Eds T Pullaiah, KV Krishnamurthy and B Bahadur) Apple Academic Press, Palm Bay, FL.
52. Rahaman CH, Mandal SK, Pullaiah T (2017) Ethnomedicinal plants of the Indo-Gangetic region and Central India. In *Ethnobotany of India. Vol. 5: The Indo-Gangetic Region and Central India.* (Eds T Pullaiah, KV Krishnamurthy and B Bahadur) Apple Academic Press, Palm Bay, FL.
53. Pole S (2006) *Ayurvedic Medicine: The Principles of Traditional Practice.* Churchill Livingstone, Edinburgh.
54. Matsumoto M, Inoue K, Kaiji E (1999) Integrating traditional medicine in Japan: the case of Kampo medicines. *Complementary Therapies in Medicine* **7**, 254–255. doi:10.1016/S0965-2299(99)80012-0
55. Yu F, Takahashi T, Moriya J, Kawaura K, Yamakawa J *et al.* (2006) Traditional Chinese medicine and Kampo: a review from the distant past for the future. *The Journal of International Medical Research* **34**, 231–239. doi:10.1177/147323000603400301
56. Watanabe K, Plotnikoff GA, Sakiyama T, Reissenweber-Hewel H (2014) Collaboration of Japanese Kampo medicine and Western biomedicine. *Evidence-Based Complementary and Alternative Medicine* **2014**, 646947. doi:10.1155/2014/646947
57. Lassak EV, Mccarthy T (1983) *Australian Medicinal Plants.* Methuen, Sydney.
58. Anderson P, Leibler M (2017) *Final Report of the Referendum Council: 30 June 2017.* Department of the Prime Minister and Cabinet, Canberra.
59. Wikipedia (2022). *Category: Medicinal plants of Australia* [Online]. https://en.wikipedia.org/wiki/Category:Medicinal_plants_of_Australia (accessed 9 June 2022).
60. Packer J, Gaikwad J, Harrington D, Ranganathan S, Jamie J *et al.* (2011) Medicinal plants of New South Wales, Australia. In *Genetic Resources, Chromosome Engineering, and Crop Improvement.* (Ed. RJ Singh) pp. 259–296. CRC Press, Boca Raton, FL.
61. Hill RS, Truswell EM, Mcloughlin S, Dettman ME (1999) The evolution of Australian flora: fossil evidence. *Flora of Australia* **1**, 251–320.
62. Cresswell I, Murphy H (2017) Cresswell I, Murphy H (2017) Australia State of the Environment 2016: biodiversity. Department of the Environment and Energy, Canberra.
63. Chapman AD (2009) Numbers of living species in Australia and the World: report for the Australian Biological Resources Study. Department of the Environment, Water, Heritage and the Arts, Canberra.
64. Department of Climate Change, Energy, the Environment and Water (2012) Collaborative Australian Protected Areas Database: National Reserve System. The Department, Canberra.
65. Mclennan W (1994) 1994 National Aboriginal and Torres Strait Islander Survey [archived]. Australian Bureau of Statistics, Canberra, Australia. https://www.abs.gov.au/AUSSTATS/abs@.nsf/Lookup/4190.0Main+Features11994?OpenDocument (accessed September 2024).
66. Oliver SJ (2013) The role of traditional medicine practice in primary healthcare within Aboriginal Australia: a review of the literature. *Journal of Ethnobiology and Ethnomedicine* **9**, 46. doi:10.1186/1746-4269-9-46
67. Clarke P (2008) Aboriginal healing practices and Australian bush medicine. *Journal of the Anthropological Society of South Australia* **33**, 3–38.
68. Dragon N (2019) Aboriginal healers treat patients in SA hospitals. *Australian Nursing & Midwifery Journal* 7 May. https://anmj.org.au/aboriginal-healers-treat-patients-in-sa-hospitals/ (accessed 8 June 6/2022).
69. National Aboriginal Health Strategy Working Party (1989) *A national Aboriginal health strategy.* NAHSWP, Canberra, Australia
70. Clarke PA (2011) The power of plants. In *Aboriginal People and Their Plants.* p. 96. Rosenburg Publishing, Sydney.
71. Lloyd O (2019) IP rights and Indigenous Knowledge: the case of gumby gumby. https://www.allens.com.au/insights-news/insights/2019/12/ip-rights-and-indigenous-knowledge-the-case-of-gumby-gumby/ [accessed 9 June 2022].
72. Baecker C, Jenett-Siems K, Siems K, Wurster M, Bodtke A *et al.* (2014) Cytotoxic saponins from the seeds of *Pittosporum angustifolium. Journal of Biosciences* **69**, 191–198.
73. Baecker C, Jenett-Siems K, Siems K, Niedermayer THJ, Wurster M *et al.* (2015) Taraxastane-type triterpenoid saponins isolated from *Pittosporum angustifolium* Lodd. *Journal of Chemical Sciences* **70**, 403–408.

74. IP Australia (2019) *Protection of Indigenous Knowledge in the Intellectual Property System*: consultation report. Australian Government, Canberra.
75. Brielmann HL, Setzer WN, Kaufman PB, Kirakosyan A, Cseke LJ (2006) Phytochemicals: the chemical components of plants. In *Natural Products from Plants.* (Eds LJ Cseke, A Kirakosyan, PB Kaufman, SL Warber, JA Duke *et al.*) pp. 1–50. CRC Press, Boca Raton, FL.
76. Gunatilaka LA (2012) Plant natural products. In *Natural Products in Chemical Biology.* (Ed. N Civjan) pp. 3–31. John Wiley & Sons, NJ.
77. Hartmann T (2007) From waste products to ecochemicals: 50 years research on plant secondary metabolism. *Phytochemistry* **68**, 2831–2846. doi:10.1016/j.phytochem.2007.09.017
78. Nelson DL (2005) *Lehninger Principles of Biochemistry*, 5th edin. W. H. Freeman, New York.
79. Wink M (2010) Introduction: biochemistry physiology and ecological functions of secondary metabolites. In *Biochemistry of Plant Secondary Metabolism.* (Ed. M Wink) pp. 1–19. Wiley-Blackwell, Sussex.
80. Cseke LJ, Lu CR, Kornfeld A, Kaufman PB, Kirakosyan A (2006) How and why these compounds are synthesised by plants. In *Natural Products from Plants.* (Eds LJ Cseke, A Kirakosyan, PB Kaufman, SL Warber, JA Duke *et al.*) pp. 51–100. CRC Press, Boca Raton, FL.
81. Kaufman PB, Kirakosyan A, Mckenzie M, Dayanandan P, Hoyt JE, Li C (2006) The uses of plant natural products by humans and risks associated with their use. In *Natural Products from Plants.* (Eds LJ Cseke, A Kirakosyan, PB Kaufman, SL Warber, JA Duke *et al.*) pp. 442–474. CRC Press, Boca Raton, FL.
82. Atanasov AG, Zotchev SB, Dirsch VM, International Natural Product Sciences Taskforce, Supuran CT (2021) Natural products in drug discovery: advances and opportunities. *Nature Reviews. Drug Discovery* **20**, 200–216. doi:10.1038/s41573-020-00114-z
83. Dias DA, Urban S, Roessner U (2012) A historical overview of natural products in drug discovery. *Metabolites* **2**, 303–336. doi:10.3390/metabo2020303
84. Newman DJ, Cragg GM (2020) Natural products as sources of new drugs over nearly four decades from 01/1981 to 09/2019. *Journal of Natural Products* **83**, 770–803. doi:10.1021/acs.jnatprod.9b01285
85. Cavalcanti RN, Forster-Carneiro T, Gomes MTMS, Rostagno MA, Prado JM, *et al.* (2013) Uses and applications of extracts from natural sources. In *Natural Product Extraction: Principles and Applications.* (Eds J Prado and M Rostagno) pp. 1–57. Royal Society of Chemistry, London.
86. Ventola CL (2015) The antibiotic resistance crisis. Part 1: causes and threats. *Pharmacology & Therapeutics* **40**, 277–283.
87. Reuben S, Cseke LJ, Bhinu VS, Narasimhan K, Jeyakumar M *et al.* (2006) Molecular biology of plant natural products. In *Natural Products from Plants.* (Eds LJ Cseke, A Kirakosyan, PB Kaufman, SL Warber, JA Duke *et al.*) pp. 166–202. CRC Press, Boca Raton, FL
88. Molimau-Samasoni S, Woolner VH, Foliga ST, Munkacsi AB (2021) Functional genomics and metabolomics advance the ethnobotany of the Samoan traditional medicine "matalafi". *Proceedings of the National Acadamy of Sciencesof the United States of America* **118**, e2100880118.
89. Palma M, Barbero GF, Pinero Z, Liazid A, Barroso CG *et al.* (2013) Extraction of natural products: principles and fundamental aspects. In *Natural Product Extraction: Principles and Applications.* (Eds J Prado and M Rostagno) pp. 58–88. Royal Society of Chemistry, London.
90. Xiao W, Lei F, Hengqiang Z, Xiaojing L (2013) Isolation and purification of natural products. In *Natural Product Extraction: Principles and Applications.* (Eds J Prado and M Rostagno) pp. 314–362. Royal Society of Chemistry, London.
91. Martinez J, Silva LPS (2013) Martinez J, Silva LPS (2013) Scale-up of extraction process. In *Natural Product Extraction: Principles and Applications.* (Eds J Prado and M Rostagno) pp. 363–398. Royal Society of Chemistry, London.
92. Wagner U, Kratky C (2015) Structure elucidation of natural compounds by X-ray crystallography. In *Progress in the Chemistry of Organic Natural Products.* (Eds AD Kinghorn, H Falk and J Kobayashi) pp. 1–75. Springer, Cham.
93. Wang Z-F, You Y-L, Li F-F, Kong W-R, Wang S-Q (2021) Research progress of NMR in natural product quantification. *Molecules (Basel, Switzerland)* **26**, 6308. doi:10.3390/molecules26206308
94. Sarker SD, Nahar L (2012) An introduction to natural products isolation. *Methods in Molecular Biology (Clifton, N.J.)* **864**, 1–25. doi:10.1007/978-1-61779-624-1_1
95. Vuorela P, Leinonenb M, Saikkuc P, Tammelaa P, Rauhad J-P *et al.* (2004) Natural products in the process of finding new drug candidates. *Current Medicinal Chemistry* **11**, 1375–1389. doi:10.2174/0929867043365116
96. Capon R (2022) Natural products: mainstream but not always natural. *Chemistry in Australia* **(Sept)**, 36–37.
97. Mohs RC, Greig NH (2017) Drug discovery and development: role of basic biological research. *Alzheimer's & Dementia* **3**, 651–657. doi:10.1016/j.trci.2017.10.005
98. Umscheid CA, Margolis DJ, Grossman CE (2011) Key concepts of clinical trials: a narrative review. *Postgraduate Medical Journal* **123**, 194–204. doi:10.3810/pgm.2011.09.2475
99. Ciociola AA, Cohen LB, Kulkarni P (2014) How drugs are developed and approved by the FDA: current process and future directions. *The American Journal of Gastroenterology* **109**, 620–623. doi:10.1038/ajg.2013.407
100. Wong CH, Siah KW (2019) Estimation of clinical trial success rates and related parameters. *Biostatistics (Oxford, England)* **20**, 273–286. doi:10.1093/biostatistics/kxx069
101. Wang Y (2012) Extracting knowledge from failed development programmes. *Pharmaceutical Medicine* **26**, 91–96. doi:10.1007/BF03256897
102. Fogel DB (2018) Factors associated with clinical trials that fail and opportunities for improving the likelihood of

success: a review. *Contemporary Clinical Trials* **11**, 156–164. doi:10.1016/j.conctc.2018.08.001

103. Aminpour M, Montemagno C, Tuszynski JA (2019) An overview of molecular modelling for drug discovery with specific illustrative examples of applications. *Molecules (Basel, Switzerland)* **24**, 1693. doi:10.3390/molecules24091693
104. Xia X (2017) Bioinformatics and drug discovery. *Current Topics in Medicinal Chemistry* **17**, 1709–1726. doi:10.2174/1568026617666161116143440
105. Romano JD, Tatonetti NP (2019) Informatics and computational methods in natural product drug discovery: a review and perspectives. *Frontiers in Genetics* **10**, 368. doi:10.3389/fgene.2019.00368
106. Quinn RA, Nothias L-F, Vining O, Meehan M, Esquenazi E *et al.* (2017) Molecular networking as a drug discovery, drug metabolism, and precision medicine strategy. *Trends in Pharmacological Sciences* **38**, 143–154. doi:10.1016/j.tips.2016.10.011
107. Nothias L-F, Petras D, Schmid R, Dührkop K, Rainer J *et al.* (2020) Feature-based molecular networking in the GNPS analysis environment. *Nature Methods* **17**, 905–908. doi:10.1038/s41592-020-0933-6

2
Phenolic-type specialised metabolites

INTRODUCTION

Phenolic-type compounds ('phenolics') are a large class of specialised metabolites naturally produced by plants.[1] These compounds are derived from a phenylalanine backbone (via the shikimate/chorismite pathway), which is subsequently channelled through the plant-specific phenylpropanoid metabolic pathway (*Fig. 2.1*).[2] Approximately 20% of the carbon fixed in photosynthesis is directed through this pathway, resulting in a chemical complexity that has been significant in the evolutionary success of the plant kingdom.[3] To date, more than 10,000 phenolic-type compounds have been isolated from the natural world, yet given the proportion of carbon channelled through the phenylpropanoid pathway, the total number could be reasonably estimated to lie between 20,000 and 200,000.[3-5] This chapter will proceed through an introduction to the class itself, followed by more detailed breakdowns of each of the phenolic-type metabolic subclasses it contains. The end of the chapter will present a summary of the representation of these metabolites reported to be present across the customary medicinal flora of Australia, detailed in *Table 2.1*, in an effort to uncover any patterns in the customary therapeutic uses of plants bearing these compound classes. This information can also be explored using the associated website (www.cmfoa.info).

Phenolic biosynthesis begins with the shikimate/chorismate (succinylbenzoate) pathway, common to bacteria, fungi and plants.[6] This pathway utilises simple products from two primary metabolic processes: glycolysis, which generates phosphoenolpyruvate (PEP), and the pentose phosphate pathway, which generates erythrose-4-phosphate. The shikimate/chorismate pathway comprises seven steps, beginning with the condensation of PEP and erythrose-4-phosphate to give chorismate, which can subsequently be converted to the aromatic amino acids: phenylalanine, tyrosine and tryptophan. The phenylpropanoid metabolic pathway is unique to plants, transforming phenylalanine (Phe) through hydroxylation and deamination into a key (C_6-C_3) intermediate, *p*-coumarate,[6] which forms a branching point for the phenolic-type specialised metabolic subclasses: phenolic acids/esters/glycosides, chalcones/flavonoids, stilbenes/phenanthrenes, lignans and chromenes, which in turn can be condensed (polymerised to form polyphenolics: hydrolysable tannins, condensed tannins, and lignins) and structurally reorganised to generate remarkable chemical diversity.[6-11] These regiospecific transformations are facilitated by a suite of enzymes and enzyme complexes that catalyse a variety of reactions: condensation, cyclisation, aromatisation, hydroxylation, glycosylation, acylation, prenylation, sulfation and methylation.[6] It should be mentioned here that some of these enzymes [i.e. chalcone synthase, stilbene synthase] are performing a malonyl-coenzyme A (CoA) extension, which is akin to those enzymatic

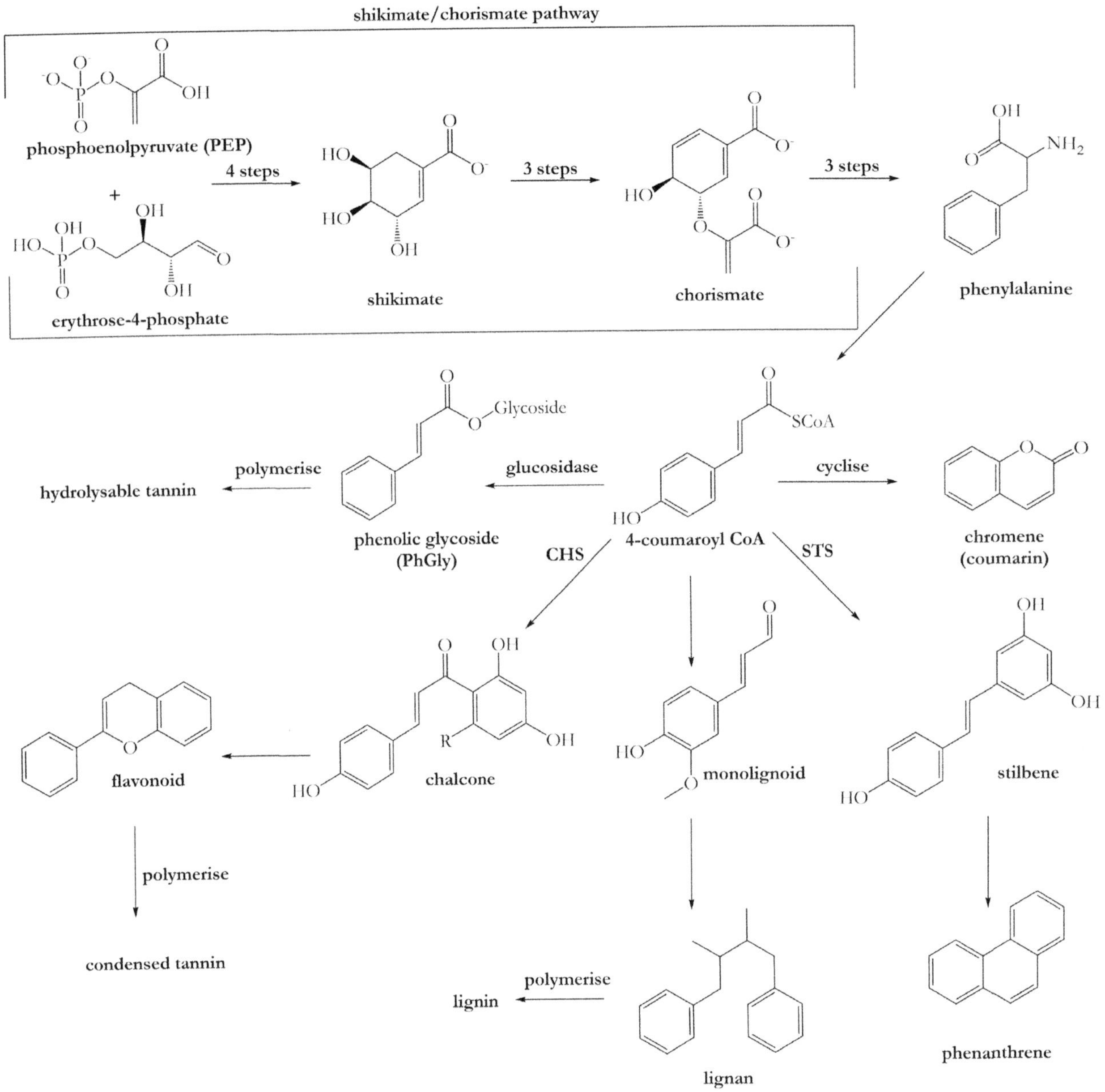

Fig. 2.1. The phenylpropanoid pathway. CHS = chalcone synthase; STS = stilbene synthase.

processes grouped under the polyketide synthase family, and as such could be considered a polyketide/phenolic hybrid (as the curcuminoids are); however, for the purposes of this text it is easier to treat these phenolic-type specialised metabolites as a group, as they share a common distribution and function *in planta*.[12,13]

The diverse results of this complex metabolic network are difficult to comprehensively and succinctly summarise; as was discussed in Chapter 1 (and expanded on in Chapter 6), chemical ontology is a burgeoning computational field that lies outside the scope of this book.[14,15] So, the following (somewhat simplified) structural breakdown into these will be followed: (1) phenylpropanoid (C_6-$C_{2/3}$; cinnamyl alcohols, cinnamic acids, cinnamamides, cinnamaldehydes and their glycosides), (2) flavonoid [C_6-C_3-C_6; major (2-phenyl), iso- (3-phenyl), neo (4-phenyl), minor (ring-opened; chalcone, aurone and auronol) and their glycosides)], (3) polycyclic (chromene, phenalene, stilbene, phenanthrene, xanthone and their glycosides), (4) lignoid (C_6-C_4-C_6; classic, norlignan, neolignan and their glycosides), (5) polyphenolic (condensed and hydrolysable tannins, and lignins),

Table 2.1. Phenolic-type specialised metabolite distribution across the customary medicinal flora of Australia

Order	Family	Genus	Phenolic glycoside	Flavonoid	Isoflavonoid	Neoflavonoid	Chalcone	Aurone	Chromene	Phenalene	Phenanthrene	Stilbene	Xanthone	Misc. polycyclic	Lignoid	Condensed tannin	Hydrolysable tannin
Alismatales	Araceae	*Alocasia*[56,57]													+		
Alismatales	Araceae	*Amorphophallus*[58,59]		+													
Alismatales	Araceae	*Rhaphidophora*[60,61]												+	+		
Alismatales	Hydrocharitaceae	*Ottelia*[62,63]												+			
Apiales	Apiaceae	*Centella*[64,65]	+	+													
Arecales	Arecaceae	*Calamus*[66]		+											+		
Asparagales	Amaryllidaceae	*Crinum*[67–70]	+	+			+		+								
Asparagales	Asparagaceae	*Asparagus*[71]			+									+			
Asparagales	Asphodelaceae	*Dianella*[72,73]		+										+			
Asparagales	Orchidaceae	*Dendrobium*[74]		+					+					+	+		
Asterales	Asteraceae	*Acmella*[75]		+					+								
Asterales	Asteraceae	*Ageratum*[76]	+	+	+				+						+		
Asterales	Asteraceae	*Centipeda*[77]															
Asterales	Asteraceae	*Chrysocephalum*[78,79]		+			+		+								
Asterales	Asteraceae	*Eclipta*[80,81]		+	+				+								
Asterales	Asteraceae	*Elephantopus*[82]		+													
Asterales	Asteraceae	*Gnaphalium*[83]		+			+		+								
Asterales	Asteraceae	*Helichrysum*[83]		+			+		+								
Asterales	Asteraceae	*Pterocaulon*[84,85]	+	+					+								
Asterales	Asteraceae	*Sphagneticolia*[86–89]		+	+		+		+								
Asterales	Campanulaceae	*Lobelia*[90]		+					+						+		
Asterales	Goodeniaceae	*Goodenia*[91]		+					+								
Asterales	Goodeniaceae	*Scaevola*[92,93]							+					I	I		
Boraginales	Boraginaceae	*Ehretia*[94,95]		+													
Boraginales	Boraginaceae	*Heliotropium*[96]		+										+			
Brassicales	Capparaceae	*Capparis*[97]		+													
Brassicales	Cleomaceae	*Cleome*[98]	+	+					+					+	+		
Canellales	Winteraceae	*Tasmannia*[99,100]	+	+													
Caryophyllales	Amaranthaceae	*Achyranthes*[101]	+	+													
Caryophyllales	Amaranthaceae	*Atriplex*[102,103]		+													
Caryophyllales	Amaranthaceae	*Dysphania*[104,105]	+	+	+				+						+		
Caryophyllales	Droseraceae	*Drosera*[106]	+	+										+			
Caryophyllales	Nyctaginaceae	*Boerhavia*[107]	+	+	+				+						+		
Caryophyllales	Plumbaginaceae	*Plumbago*[108]		+					+					+			
Caryophyllales	Polygonaceae	*Persicaria*[109–112]	+	+	+		+		+			+		+		+	+
Caryophyllales	Portulaceae	*Portulaca*[113]		+	+												
Commelinales	Haemodoraceae	*Haemodorum*[29]		+						+							

Table 2.1. continued

Order	Family	Genus	Phenolic glycoside	Flavonoid	Isoflavonoid	Neoflavonoid	Chalcone	Aurone	Chromene	Phenalene	Phenanthrene	Stilbene	Xanthone	Misc. polycyclic	Lignoid	Condensed tannin	Hydrolysable tannin
Cucurbitales	Cucurbitaceae	*Trichosanthes*[114–118]		+	+		+	+							+		
Dioscoreales	Dioscoreaceae	*Dioscorea*[10,119]												+			
Ericales	Lecythidaceae	*Planchonia*[120–124]		+													
Ericales	Primulaceae	*Aegiceras*[125]		+												+	
Ericales	Sapotaceae	*Planchonella*[126–132]	+	+													
Fabales	Fabaceae	*Abrus*[133]		+													
Fabales	Fabaceae	*Acacia*[134]		+												+	+
Fabales	Fabaceae	*Caesalpinia*[135]			+		+										
Fabales	Fabaceae	*Canavalia*[136–139]			+												
Fabales	Fabaceae	*Cassia*[140]		+									+				
Fabales	Fabaceae	*Chamaecrista*[140]		+									+				
Fabales	Fabaceae	*Crotalaria*[141,142]		+	+												
Fabales	Fabaceae	*Entada*[143,144]	+	+													
Fabales	Fabaceae	*Erythrophleum*[145]	+	+								+					
Fabales	Fabaceae	*Indigofera*[146]	+	+	+		+						+	+		+	+
Fabales	Fabaceae	*Lysiphyllum*[147]		+													
Fabales	Fabaceae	*Milettia*[148,149]	+	+	+		+										
Fabales	Fabaceae	*Rhynchosia*[150]	+	+	+								+	+		+	+
Fabales	Fabaceae	*Senna*[140]		+									+				
Fabales	Fabaceae	*Sesbania*[151,152]	+	+	+							+				+	+
Fabales	Fabaceae	*Sophora*[153–155]		+	+				+			+					
Fabales	Fabaceae	*Tephrosia*[156–158]		+	+	+	+		+					+	+		
Fabales	Fabaceae	*Vigna*[159–161]	+	+	+										+		
Fagales	Casuarinaceae	*Casuarina*[162]		+													
Gentianales	Apocynaceae	*Alstonia*[163]		+					+							+	+
Gentianales	Apocynaceae	*Alyxia*[164]							+						+		
Gentianales	Apocynaceae	*Carissa*[165]		+					+						+		
Gentianales	Apocynaceae	*Cerbera*[166]		+											+		
Gentianales	Apocynaceae	*Cynanchum*[167]	+	+											+		
Gentianales	Apocynaceae	*Marsdenia*[168]		+					+								
Gentianales	Apocynaceae	*Pentatropis*[169]		+													
Gentianales	Apocynaceae	*Tylophora*[170–173]		+													
Gentianales	Gentianaceae	*Schenkia*[174–177]		+					+				+				
Gentianales	Loganiaceae	*Strychnos*[178–182]	+	+											+		
Gentianales	Rubiaceae	*Morinda*[183–185]		+					+						+		
Gentianales	Rubiaceae	*Nauclea*[186,187]							+			+					
Lamiales	Lamiaceae	*Ajuga*[188]	+	+													
Lamiales	Lamiaceae	*Callicarpa*[189]	+	+											+		

Order	Family	Genus	Phenolic glycoside	Flavonoid	Isoflavonoid	Neoflavonoid	Chalcone	Aurone	Chromene	Phenalene	Phenanthrene	Stilbene	Xanthone	Misc. polycyclic	Lignoid	Condensed tannin	Hydrolysable tannin
Lamiales	Lamiaceae	*Clerodendrum*[90]	+	+													
Lamiales	Lamiaceae	*Mentha*[191]	+	+													
Lamiales	Lamiaceae	*Ocimum*[192]	+	+					+			+		+	+		
Lamiales	Lamiaceae	*Plectranthus*[193]	+	+										+			
Lamiales	Lamiaceae	*Prostanthera*[194–199]	+	+													
Lamiales	Lamiaceae	*Prunella*[200]	+	+					+								
Lamiales	Lamiaceae	*Teucrium*[201,202]	+	+													
Lamiales	Lamiaceae	*Vitex*[203–205]		+											+		
Lamiales	Linderniaceae	*Striga*[206–208]												+			
Lamiales	Phrymaceae	*Mimulus*[209,210]	+											+			
Lamiales	Plantaginaceae	*Bacopa*[211]	+														
Lamiales	Plantaginaceae	*Gratiola*[212,213]		+													
Lamiales	Plantaginaceae	*Scoparia*[214–217]	+	+													
Lamiales	Plantaginaceae	*Stemodia*[218–223]		+													
Lamiales	Scrophulariaceae	*Eremophila*[224]	+	+											+		
Lamiales	Scrophulariaceae	*Myoporum*[225–230]															
Lamiales	Verbenaceae	*Verbena*[205,231,232]	+														
Laurales	Atherospermataceae	*Atherosperma*[233]									+						
Laurales	Hernandiaceae	*Hernandia*[234,235]		+										+	+		
Laurales	Lauraceae	*Cassytha*[236,237]													+		
Laurales	Lauraceae	*Cinnamomum*[238]	+	+										+	+		
Laurales	Lauraceae	*Litsea*[239]	+	+					+		+			+	+		
Liliales	Smilacaceae	*Smilax*[240–245]	+	+		+	+	+	+						+		
Magnoliales	Myristicaceae	*Myristica*[246–248]		+										+			
Malphigiales	Calophyllaceae	*Calophyllum*[249,250]		+	+	+			+				+				
Malphigiales	Euphorbiaceae	*Acalypha*[251]															+
Malphigiales	Euphorbiaceae	*Aleurites*[252]		+					+						+		
Malphigiales	Euphorbiaceae	*Beyeria*[253]		+													
Malphigiales	Euphorbiaceae	*Chamaesyce*[254]		+					+								
Malphigiales	Euphorbiaceae	*Croton*[255,256]	+	+					+				+		+	+	+
Malphigiales	Euphorbiaceae	*Euphorbia*[254]		+					+								
Malphigiales	Euphorbiaceae	*Excoecaria*[257]		+			+										+
Malphigiales	Euphorbiaceae	*Mallotus*[258,259]	+	+			+		+					+	+		
Malphigiales	Phyllanthaceae	*Antidesma*		+												+	+
Malphigiales	Phyllanthaceae	*Breynia*[260,261]	+														
Malphigiales	Phyllanthaceae	*Flueggea*[262–264]	+	+					+								+
Malphigiales	Phyllanthaceae	*Phyllanthus*[265–269]	+	+					+					+	+	+	+
Malphigiales	Phyllanthaceae	*Securinega*[262–264]	+	+					+								+

Table 2.1. continued

Order	Family	Genus	Phenolic glycoside	Flavonoid	Isoflavonoid	Neoflavonoid	Chalcone	Aurone	Chromene	Phenalene	Phenanthrene	Stilbene	Xanthone	Misc. polycyclic	Lignoid	Condensed tannin	Hydrolysable tannin
Malphigiales	Rhizophoraceae	*Rhizophora*[270,271]	+	+									+	+	+	+	+
Malphigiales	Violaceae	*Hybanthus*[272,273]	+	+													+
Malvales	Malvaceae	*Grewia*[274–276]		+					+						+		
Malvales	Malvaceae	*Hibiscus*[277–280]	+	+	+				+					+	+		+
Malvales	Malvaceae	*Malva*[281,282]		+					+								
Malvales	Malvaceae	*Sida*[283]	+	+					+						+		
Malvales	Malvaceae	*Sterculia*[284,285]	+	+			+		+								
Malvales	Malvaceae	*Thespesia*[286–289]		+													
Malvales	Thymelaeaceae	*Wikstroemia*[290–294]	+	+					+			+			+		
Myrtales	Lythraceae	*Ammannia*[295]		+													+
Myrtales	Lythraceae	*Lythrum*[296]		+					+					+			+
Myrtales	Myrtaceae	*Angophora*[297–300]		+												+	+
Myrtales	Myrtaceae	*Corymbia*[297–300]		+												+	+
Myrtales	Myrtaceae	*Eucalyptus*[29–300]		+												+	+
Myrtales	Myrtaceae	*Syncarpia*[301–303]												+			
Myrtales	Myrtaceae	*Syzygium*[304–306]	+	+													+
Oxalidales	Cunoniaceae	*Eucryphia*[307–311]		+													
Pandanales	Pandanaceae	*Pandanus*[312,313]		+										+	+		
Pinales	Araucariaceae	*Araucaria*[314]	+	+	+										+		
Pinales	Cupressaceae	*Callitris*[315,316]													+		
Piperales	Piperaceae	*Piper*[317,318]	+	+			+							+	+		
Poales	Cyperaceae	*Cyperus*[319]	+	+					+								
Poales	Cyperaceae	*Eleocharis*															
Poales	Cyperaceae	*Remirea*[320–323]		+										+			
Poales	Cyperaceae	*Schoenoplectus*[324]												+			
Poales	Flagellariaceae	*Flagellaria*[325]		+													
Poales	Poaceae	*Cymbopogon*[326]	+	+													
Polypodiales	Dennstaedtiaceae	*Pteridium*[327–330]	+	+													
Polypodiales	Pteridaceae	*Adiantum*[331]	+	+					+								
Proteales	Nelumbonaceae	*Nelumbo*[332]		+													
Proteales	Proteaceae	*Grevillea*[333–338]	+	+										+			
Proteales	Proteaceae	*Hakea*[339,340]	+											+			
Proteales	Proteaceae	*Persoonia*[341–344]		+													
Psilotales	Psilotaceae	*Psilotum*[345–349]	+	+			+					+	+	+			
Ranunculales	Menispermaceae	Cissampelos[350]		+			+										
Ranunculales	Menispermaceae	*Tinospora*[351–353]		+										+	+		
Ranunculales	Ranunculaceae	*Clematis*[354]	+	+	+				+						+	+	+
Rosales	Moraceae	*Ficus*[355,356]		+					+								

Order	Family	Genus	Phenolic glycoside	Flavonoid	Isoflavonoid	Neoflavonoid	Chalcone	Aurone	Chromene	Phenalene	Phenanthrene	Stilbene	Xanthone	Misc. polycyclic	Lignoid	Condensed tannin	Hydrolysable tannin
Rosales	Rhamnaceae	*Alphitonia*[357,358]		+					+								
Rosales	Rhamnaceae	*Ventilago*[359–365]		+										+			
Rosales	Rhamnaceae	*Ziziphus*[366–369]	+	+	+				+								
Rosales	Rosaceae	*Rubus*[370,371]	+	+												+	+
Santalales	Olacaceae	*Ximenia*[372,373]	+	+													
Santalales	Santalaceae	*Santalum*[374,375]	+	+													
Sapindales	Anacardiaceae	*Semecarpus*[376–379]		+			+										
Sapindales	Burseraceae	*Canarium*[380]	+	+					+								+
Sapindales	Meliaceae	*Cedrela*[381–384]	+	+					+						+		
Sapindales	Meliaceae	*Melia*[385]		+					+				+		+		
Sapindales	Meliaceae	*Toona*[381–384]	+	+					+						+		
Sapindales	Rutaceae	*Bosistoa*[386]		+													
Sapindales	Rutaceae	*Melicope*[387–393]	+	+					+					+	+		
Sapindales	Rutaceae	*Philotheca*[394–399]	+	+					+						+		
Sapindales	Rutaceae	*Zieria*[400–404]	+														
Sapindales	Sapindaceae	*Allophylus*[405]		+													
Sapindales	Sapindaceae	*Dodonaea*[406]	+	+					+						+		
Sapindales	Simaroubaceae	*Ailanthus*[407]		+													
Solanales	Convolvulaceae	*Convolvulus*[408]	+	+	+				+			+				+	+
Solanales	Convolvulaceae	*Evolvulus*[409]		+													
Solanales	Convolvulaceae	*Ipomoea*[410]	+	+					+								
Solanales	Convolvulaceae	*Operculina*[411–414]							+					+			
Solanales	Solanaceae	*Nicotiana*[415–417]	+	+	+				+					+	+	+	+
Solanales	Solanaceae	*Solanum*[418]	+	+					+					+	+		
Vitales	Vitaceae	*Cayratia*[419]		+								+					
Vitales	Vitaceae	*Cissus*[420,421]	+	+								+					
Zingiberales	Musaceae	*Musa*[422,423]		+						+				+	+		
Zingiberales	Zingiberaceae	*Curcuma*[424]	+	+											+		
Zygophyllales	Zygophyllaceae	*Tribulus*[425–430]	+	+													

Note: grey rows have been taxonomically reclassified since the respective publications; genera with no reported phytochemistry have been omitted see (www.cmfoa.info)

and (6) a separate 'paraphyletic' group known as 'phytoestrogens', which are grouped on the basis of their important estradiol-mimicking binding affinity for human estrogen-receptors (certain lignans and flavonoids; isoflavone, coumestan and prenyl-flavonoid), rather than their structure.[1,16]

Phenolics function *in planta* as a diverse range of pigments and structural and defence compounds that directly influence a plant's growth, response to environment and productivity.[7] Chemical diversity and specific pathways vary widely across family, genus, species, variety and even ecotype and forms the basis of the long historical relationship humans have had with the global flora.[17] The biological properties of phenolics largely result from their phenolic hydroxyl moiety, which is relatively more acidic than other hydroxyl groups (such as those present in alcohols) due to the stability provided by the conjugated aromatic ring to which it is attached.[18] The resultant bioactivity has serendipitously given rise to their application in human health and physiology. As such, the phenolic-type specialised metabolites of the plant kingdom represent a large potential resource for drug discovery.

PHENYLPROPANOIDS

Phenylpropanoids, which contain one or several C_6-C_3 fragments, are the simplest form of biologically active phenolic-type specialised metabolites (*Fig. 2.2*).[1,2] The similarity of their structures, biosynthesis and biological function, together with their rarity in nature in an aglycone form, often results in varied monomeric and polymeric members of this subclass being grouped together as 'phenolic glycosides' (PhGlys). Due to their simplicity, PhGlys are widely distributed throughout the plant kingdom, often contributing to the bioactive efficacy of traditional medicinal plants.[19]

Phenylpropanoids have been divided into a variety of groups depending on the authors of the publication, but the most recent review to date treats the 48 reported phenylpropanoids as follows: cinnamyl alcohols, cinnamic acids, cinnamamides, cinnamaldehydes, phenylpropanes and phenylethanes.[20] Due to their simplicity and overlap into higher phenolic categories, it is difficult to estimate the total number of isolated natural PhGlys. The most recent comprehensive review specific to PhGly natural products (NPs), despite being almost 30 years old, concerns only the phenylethanoid glycosides (C_6-C_2), and lists 160 compounds over 26 families (Acanthaceae, Alangiaceae, Asteraceae, Berberidaceae, Bignoniaceae, Buddlejaceae, Caprifoliaceae, Cornaceae, Crassulaceae, Cucurbitaceae, Fagaceae, Gesneraceae, Lamiaceae, Lauraceae, Loganiaceae, Magnoliaceae, Oleaceae, Orobanchaeceae, Pedaliaceae, Plantaginaceae, Ranunculaceae, Rosaceae, Saliaceae, Scrophulariaceae, Theaceae and Verbenaceae).[21] Based on their simplicity, and centrality to the phenolic metabolic network, it is likely that PhGlys are ubiquitous across the plant kingdom and should be considered akin to primary metabolites. It should be noted that their small size (low molar mass) restricts their structural diversity and thus limits the total isolated population to a number far below that which is possible in the more diverse higher phenolics such as the flavonoids or lignoids.

FLAVONOIDS

The flavonoid (C_6-C_3-C_6) subclass can be divided into a variety of infraclasses: major (2-phenyl), iso- (3-phenyl), neo- (4-phenyl) and minor (ring-opened; chalcone, aurone and auronol) (summarised in *Fig. 2.3*).[5] Over 10,000 members of this class have been isolated, with their general structure being typified by an aromatic C_6 'A-ring', bound to a heterocyclic oxygenated C_3 'C-ring', which has a pendant aromatic C_6 'B-ring' located at either positions C-2, C-3 or C-4 to give the major,

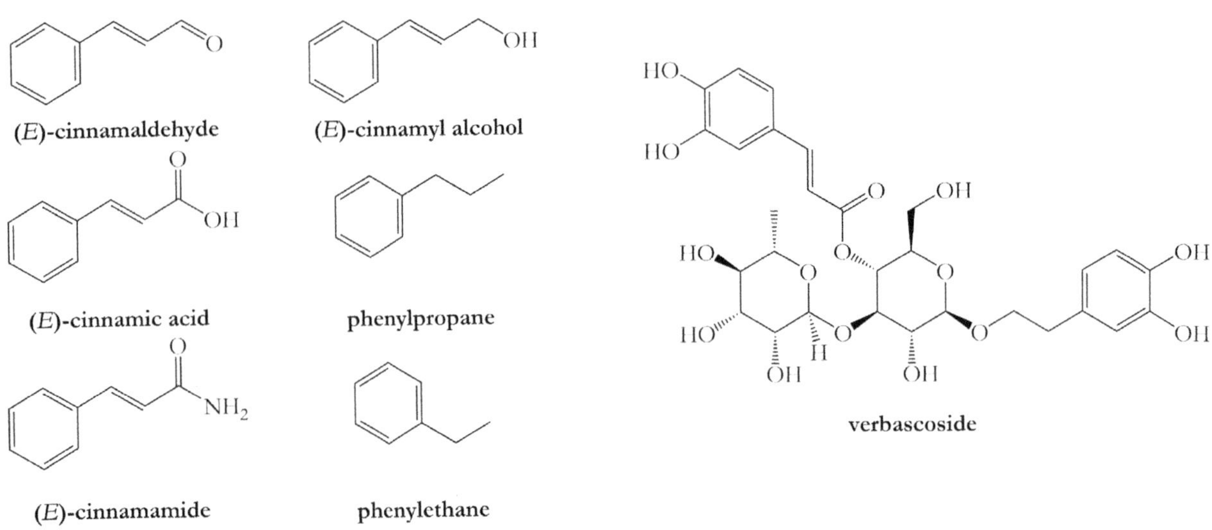

Fig. 2.2. Infraclasses of the phenolic-type phenylpropanoid subclass (left) and an example of a phenolic glycoside (PhGly): verbascoside (right).

iso- and neo- infraclasses, (*Fig. 2.3*).[5,22]Glycosylation is nearly ubiquitous in the natural flavonoids, which are predominantly isolated as *O*-glycosides, although some *C*-glycosylation is observed.[23] Within the plant, glycosylation is thought to improve the aqueous solubility of the flavanoids, which better facilitates their compartmentalisation in cell vacuoles, as well as mitigating some of their biological activity *in planta*, until it is required and subsequently activated by hydrolase enzymes at their site of action.[24]

Much of the driving force behind the 'return to nature' trend in pharmaceutical research since the turn of the 21st century is attributable to the considerable biological benefit of this large subclass of phenolic compounds.[23] Extensive epidemiological, nutritional and clinical research has been actively directed at a potential correlation between populations who consume a plant-rich diet and the generally healthier, longer lives of these populations.[25] Flavonoids, particularly in their glycosylated form, are the dominant subclass of phenolic compounds present in dietary fruits and vegetables. Current nutritional research in the field is concerned with the different physiological effects in humans depending on this glycosylation and the role played by the gut microbiome.[23] Considering their diversity, both in chemical structure and biological function, general or universally applicable remarks about this subclass are difficult to make. In short, modulation induced by flavonoids–microbiome interactions are mutual because flavonoids are altered through the biochemical machinery of the myriad gut microbiota and in turn these compounds affect microbial population dynamics and induce measurable physiological and biological effects to both the microbes and their host.[26]

As noted in the previous section on phenylpropanoids, the near-ubiquitous presence of flavonoids throughout the plant kingdom makes it difficult to detail their broad representation. The flavonoids are abundant in the spermatophytes, both gymnosperms and angiosperms, and have also been reported in more primitive taxa such as the bryophytes, pteridophytes and algae.[5] Flavones, isoflavones and flavonols have been detected in five different evolutionary lineages (Cyanobacteria, Rhodophyta, Chlorophyta, Haptophyta and Ochrophyta), utilising enzymes from the primary metabolic pathways of ancestral photosynthesisers, such as 2-oxoglutarate dependent dioxygenase, cytochrome P450 (CYP) and short-chain dehydrogenases/reductases, and thus are incredibly ancient and widespread.[5] Anthocyanins are significant because they are represented broadly across the plant kingdom, with the exception of the Caryophyllales in which the anthocyanins have been replaced with betalains/betacyanins (alkaloid-type pigmented specialised metabolites of this type that will be addressed separately in Chapter 4).[27]

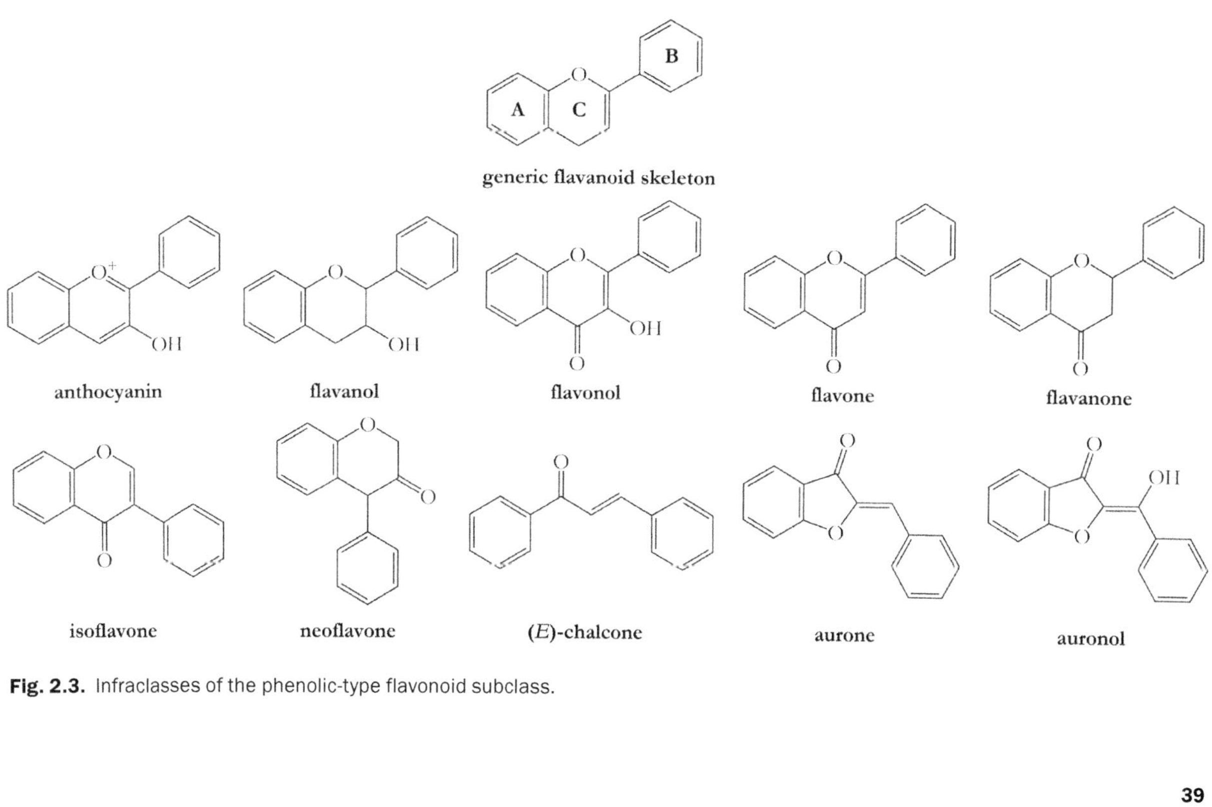

Fig. 2.3. Infraclasses of the phenolic-type flavonoid subclass.

POLYCYCLIC PHENOLICS

The polycyclic phenolic-type specialised metabolites are a paraphyletic grouping of compounds that have a common feature of multiple aromatic ring group moieties comprised of chromenes, phenalenes (phenalenones, phenylphenalenones), phenanthrenes, stilbenoids, xanthones and their glycosides.

Chromenes

Chromenes (benzopyrans) are heterocyclic C_6-C_3 compounds with an aromatic ring fused to a pyran nucleus and they are categorised into two major groups: the *2H*-chromenes [including the > 1,300 naturally occurring *2H*-chromen-2-ones (coumarins), the most notable being the anticoagulant drug warfarin] and the *4H*-chromenes (and 1,4-benzopyrone 'chromones', related to flavonoids) (*Fig. 2.4*).[28] Additionally, the fusion of benzene or naphthalene to differing locations around the general chromene structure, or varying degrees of dimerisation/polymerisation, gives rise to a diverse range of complex chromenes, furanochromenes (notably psoralen), pyranochromenes, benzochromenes (*c*/*f*/*g*/*h*), dibenzochromenes (c,*f*/c,*h*), naphthochromenes, bischromenes and trichromenes.[28]

Chromenes are biosynthetically simple, but exhibit great structural diversity and as such they are widely represented in the plant kingdom. *2H*-chromenes have been reported from the Asteraceae (*Ageratum, Encelia, Eupatorium, Helianthella, Myriactis, Tithonia*), Bromeliaceae (*Dyckia*), Ericaceae (*Rhododendron*), Euphorbiaceae (*Mallotus*), Fabaceae (*Crotalaria, Erythrina, Flemingia, Glycyrrhiza, Lonchocarpus*), Lamiaceae (*Orthosiphon*), Moraceae (*Artocarpus*), Myrtaceae (*Calyptranthes*), Piperaceae (*Peperomia, Piper*) and Rutaceae (*Euodia, Medicosma, Melicope, Neoraputia*).[28] Fused chromenes are variously represented in the plant kingdom and are particularly rich in the Rutaceae, Apiaceae, Fabaceae and Moraceae. Natural coumarins (*2H*-chromen-2-ones) are particularly rich in the Rutaceae (*Aegle, Amyris, Asterolasia, Chorilaena, Chloroxylon, Citrus, Clausena, Eriostemon, Geijera, Metrodorea, Murraya, Ruta, Skimmia, Zanthoxylum*) and Apiaceae (*Angelica, Bupleurum Deverra, Ferula, Peucedanum, Seseli*) families, but are also widely represented in the following higher plant orders: Asterales, Dipsacales, Ericales, Fabales, Geraniales, Gentianales, Lamiales, Laurales, Malphigiales, Malvales, Morales, Poales, Polytrichales, Rosales, Solanales and Tetraphidales.[28]

Phenalenes

Phenalenes are tricyclic C_6-C_4-C_3 compounds that form the central core of phenalenones (phenalenes that feature a ketone moiety). Phenalenes and phenalenones are not common in the plant kingdom, unlike phenylphenalenones, which feature a ketone moiety and a lateral phenyl ring, and are divided into four major groups: phenylphenalenones, oxabenzochrysenones, phenylbenzoisochromenones and phenylbenzoisoquinolindiones (*Fig. 2.5*).[29] The phenylphenalenones are encountered in the fungal kingdom, but arise through the acetate-polymalonate pathway, whereas the plant-derived phenylphenalenones arise from the phenylpropanoid pathway, resulting in differing structural motifs and substitution.[29] The ~140 phenylphenalenones derived from the plant kingdom function as chemotaxonomic markers for the Commelinid clade, comprising the orders Commelinales and Zingiberales.[29]

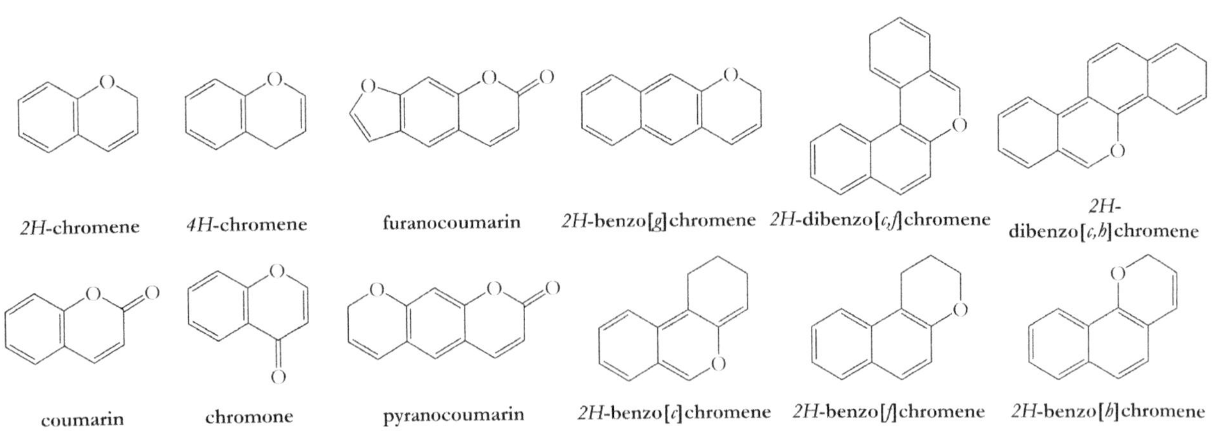

Fig. 2.4. Infraclasses of the phenolic-type chromene subclass.

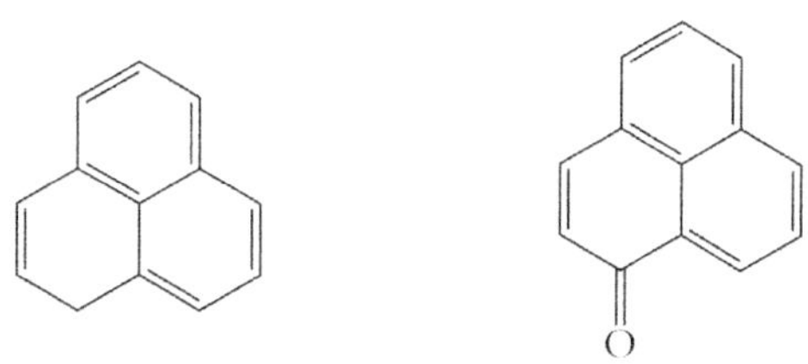

phenalene phenalenone

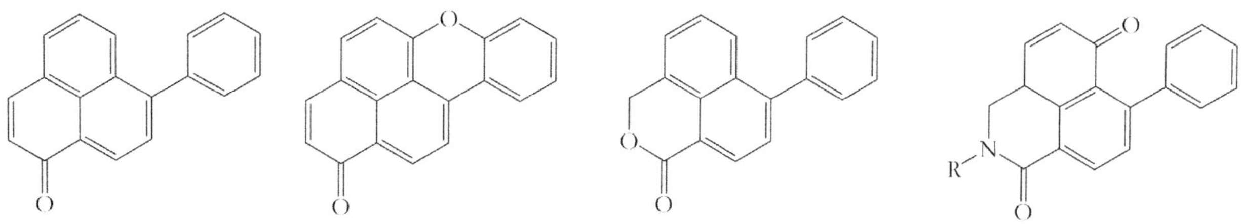

phenylphenalenone oxabenzochrysenone phenylbenzoisochromenone phenylbenzoisoquinolindione

Fig. 2.5. Infraclasses of the phenolic-type phenalene subclass.

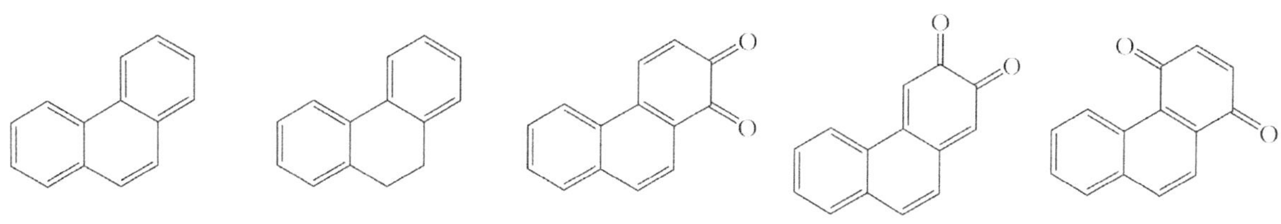

phenanthrene 9,10-dihydrophenanthrene phenanthrene-1,2-dione phenanthrene-2,3-dione phenanthrene-1,4-dione

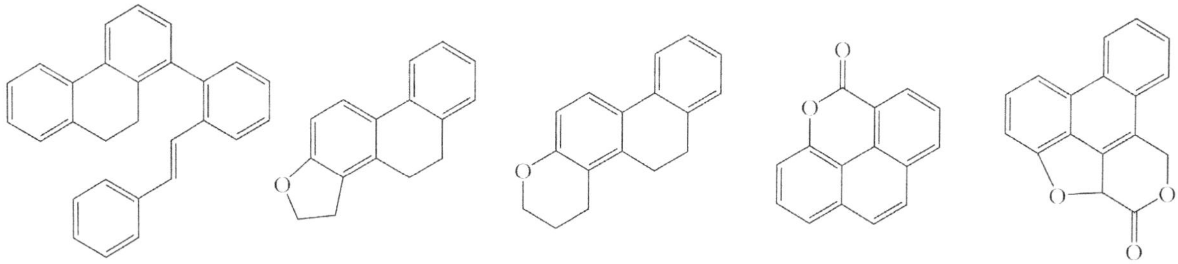

phenanthrostilbene phenanthrofuran phenanthropyran phenanthrenelactone phenanthrenelactone

Fig. 2.6. Infraclasses of the phenolic-type phenanthrene subclass.

Phenanthrenes

Phenanthrenes are C_6-C_2-C_6 compounds that are derived from the intramolecular oxidative coupling of stilbene precursors (*Fig. 2.6*). Approximately 450 natural phenanthrenes have been isolated and are typically treated as a singular group, but have been isolated as dimers and, in one instance, a triphenanthrene has also been recorded.[30] The phenanthrenes are a relatively small group of NPs, with the Orchidaceae as the richest source, followed by the Combretaceae, Dioscoreaceae and Juncaceae, but their presence has also been documented in the Annonaceae, Aristolochiaceae, Brassicaceae, Betulaceae, Cannabaceae, Euphorbiaceae, Lauraceae, Malphigiaceae and Stemonaceae families.[10,30]

Stilbenoids

Stilbenoids (stilbenes/diarylethenes) are C_6-C_2-C_6 compounds that are closely related to flavonoids, with more than 1,000 compounds of the stilbenoid subclass having been isolated to date (*Fig. 2.7*). Half of these recorded

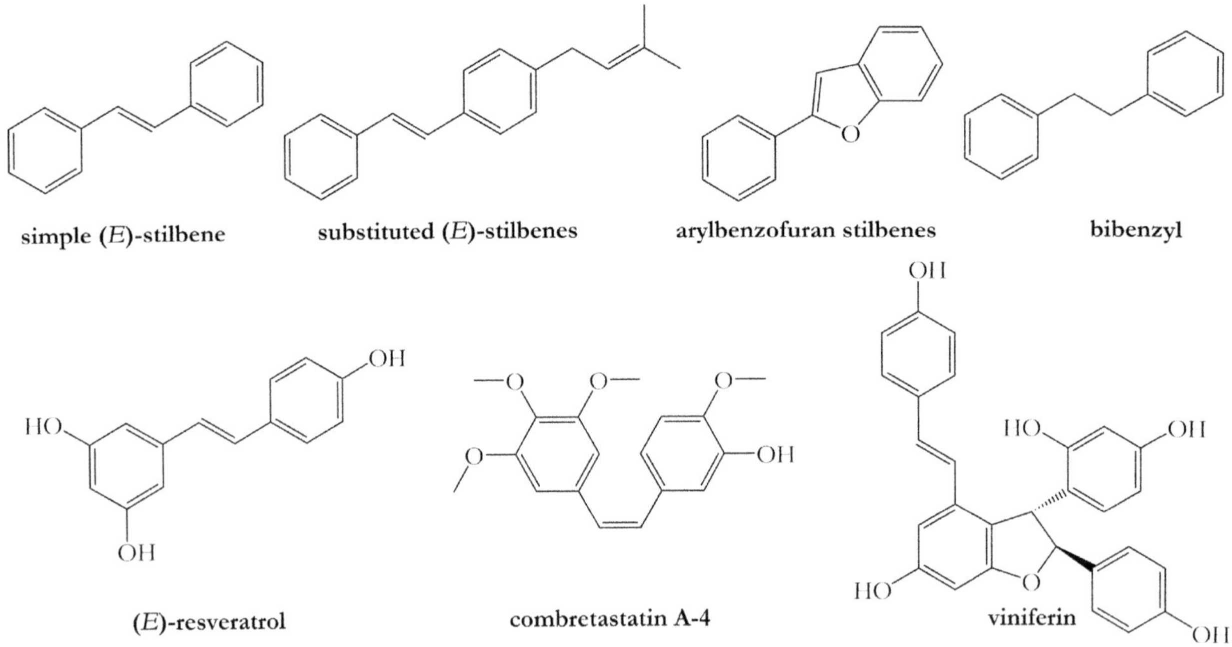

Fig. 2.7. Infraclasses of the phenolic-type stilbenoid subclass.

stilbenoids are 'true' stilbenes, which feature an ethene bridge and are categorised into three major groups (simple, substituted [prenylated/geranylated/hybrid], and arylbenzofuran) and their most notable representatives belong to the simple stilbene infraclass, resveratrol and combretastatin A-4, isolated from *Vitis vinifera* (L., 1753 - Vitaceae) and *Combretum leprosum* (Mart., 1841 - Combretaceae), respectively.[11] Bibenzyls, which bear a hydrated ethane bridge (both bisbibenzyl dimers and poly-bibenzyl polymers are common to the Bryophyte clade, particularly Lepidoziaceae, Lejeuneaceae and Marchantiaceae) and the stilbene-derived phenanthrenoids described above are often categorised within the stilbenoid family.[11] Polymerisation is extensive in the stilbenes, typically via regioisomeric 3-8', 8-10', 8-8' and 8-12' coupling, before cyclisation; the most notable example of this is the 3-8' dimeric arylbenzofuran viniferins (α/β/γ/δ/ε), isolated from *V. vinifera* (L., 1753 - Vitaceae).[31]

The stilbenes are widely distributed because they are formed biosynthetically using a pathway that shares many similarities with the ubiquitous flavonoids.[11] They have been reported in 45 higher plant families, as well as several bryophyte families. The families richest in stilbenes are the Orchidaceae (14 genera), Fabaceae (11 genera) and Dipterocarpaceae (8 genera), followed by the Asteraceae, Moraceae, Polygonaceae and Vitaceae. Stilbenes are also present in the Apiaceae, Arecaceae, Asparagaceae, Balantiopsaceae, Berberidaceae, Betulaceae, Blasiaceae, Burseraceae, Cannabaceae, Celastraceae, Combretaceae, Convulvulaceae, Corsiniaceae, Cyperaceae, Dioscoreaceae, Dryopteridaceae, Ericaceae, Euphorbiaceae, Frullaniaceae, Geocalycaceae, Gnetaceae, Haemodoraceae, Herbertaceae, Hydrangeaceae, Iridaceae, Juncaceae, Lauraceae, Lepidoziaceae, Lejeuneaceae, Marchantiaceae, Melanthiaceae, Meliaceae, Moraceae, Musaceae, Paeoniaceae, Plagiochilaceae, Rosaceae, Sapindaceae, Stemonaceae, Typhaceae, Welwitschiaceae and Zingiberaceae.[11]

Xanthones

Xanthones are symmetrical aromatic heterocyclic C_6-C_1-C_6 compounds, bearing a ketone moiety and a heterocyclic oxygen on the central ring. Approximately 500 naturally occurring xanthones have been isolated and are divided into six major groups: simple (148, and their glycosides, 61), substituted (285, typically prenylated, which variously condense into furoxanthones, pyranoxanthones, etc.), xanthonolignoids (2), bis-xanthones (12), and miscellaneous (8) (*Fig. 2.8*).[32] Xanthones are biosynthesised differently in fungi and higher plants, with higher plant xanthones arising through the convergent synthesis of the polyketide and shikimic acid pathways. In the process, 3-hydroxybenzoyl CoA (an

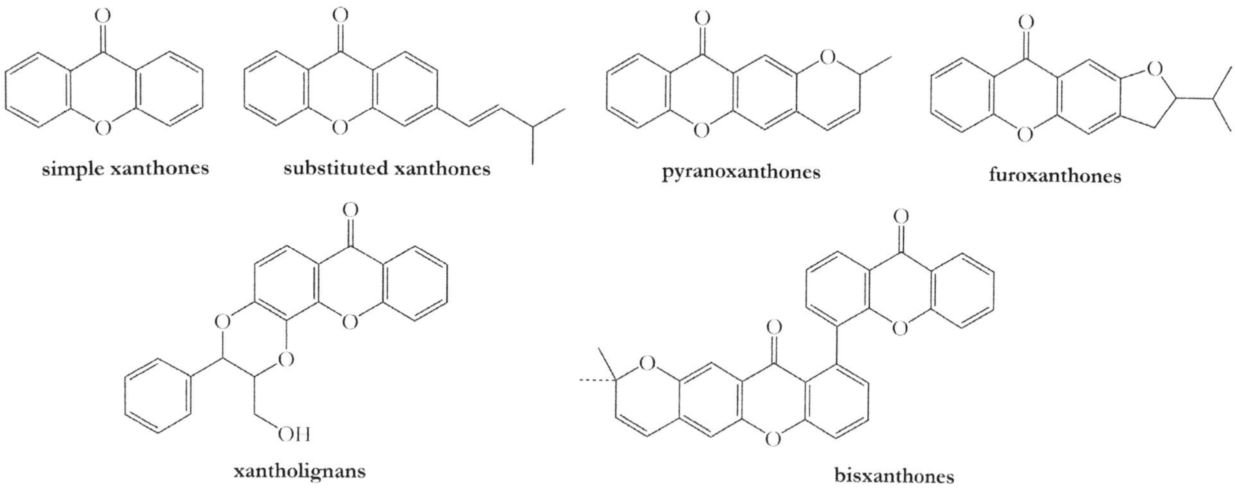

Fig. 2.8. Infraclasses of the phenolic-type xanthone subclass.

early shikimic acid pathway intermediate) is extended with three acetyl CoA units in a similar fashion to the chalcones and stilbenes to give a C_{14} backbone, which first cyclises to give a hydroxylated benzophenone, before further cyclisation to give the heterocyclic central ring of the completed tricyclic xanthone.[33,34]

The xanthones are present in 20 higher plant families, and are particularly abundant in the Clusiaceae (12 genera) and Gentianaceae (8 genera), but are also present in the Acanthaceae, Anacardiaceae, Annonaceae, Anthericaceae, Bignoniaceae, Bombacaceae, Euphorbiaceae, Fabaceae, Hippocrataceae, Hyacinthaceae, Lecythidaceae, Moraceae, Polygonaceae, Thymelaceae, Ulmaceae, Verbenaceae and Zingiberaceae.[32]

LIGNOIDS

Lignoids (C_6-C_3-C_3-C_6) are a large subclass of phenolic-type specialised metabolites, derived from the oxidative coupling of two C_6-C_3 phenylpropanoid (monolignol) units to form dimeric structures with a β, β'-linkage.[9] The lignans are divided into three major infraclasses: classic lignans (C-8 to C-8' bound phenylpropanoid dimers), norlignans (similar C_{16} or C_{17} compounds) and neolignans (other forms of dimerisation, including ether-bridges).[9] Classic lignans are divided into eight groups: arylnaphthalene, aryltetralin, dibenzylbutane, dibenzylbutyrolactol, dibenzylbutyrolactone, furan (2,5/2,4/3,4), furofuran and dibenzocyclooctadiene. The norlignans are less diverse and are treated as a singular group that varies in its bridging point (C7-C8'/C8-C8'/C9-C8') (*Fig. 2.9*).

Neolignans are further divided into 15 loose 'groups', distributed over the benzofuran, alkyl aryl ethers, biphenyls, benzodioxanes and 'other' structural groupings, with no special names assigned to them (*Fig. 2.10*).[35] Given this vast structural diversity, it is difficult to estimate the total number of lignans present in the plant kingdom, with more than 1,000 new examples having been published in the past 20 years.[35-37]

Lignoids are widely distributed in the plant kingdom, being isolated from over 70 families, with the various groupings of the infraclasses differing across genera, but some general trends can be observed.[36] It is thought that pinoresinol (furofuran), lariciresinol (furan), secoisolariciresinol (dibenzylbutane) and matairesinol (dibenzylbutyrolactone) form the core of the metabolic pathway for the lignans, and as such they are the most widely distributed forms, with the remainder of the groups being more specifically represented in particular genera (*Fig. 2.11*).[9] The furofurans are widely distributed in edible plant families such as Linaceae (flax), Pedaliaceae (sesame), Poaceae (cereal grains), Fabaceae (legume) and Brassicaceae (brassicas) and have been part of the human diet for aeons.[38]

Representatives of the Asterid clade, such as *Forsythia* (Oleaceae), *Carthamus* (Asteraceae) and *Anthriscus* (Apiaceae), metabolise matairesinol to arctigenin/isoarctigenin (dibenzylbutyrolactone) (*Fig. 2.11*). Representatives from the Ranunculales clade [*Podophyllum* (Berberidaceae)] and the Rosid clade [*Linum* (Linaceae)] metabolise matairesinol to the dibenzylbutyrolactones hinokinin, yatein and the aryltetralin podophyllotoxin (*Fig. 2.11*).[9] A unique branch of the lignan metabolic tree

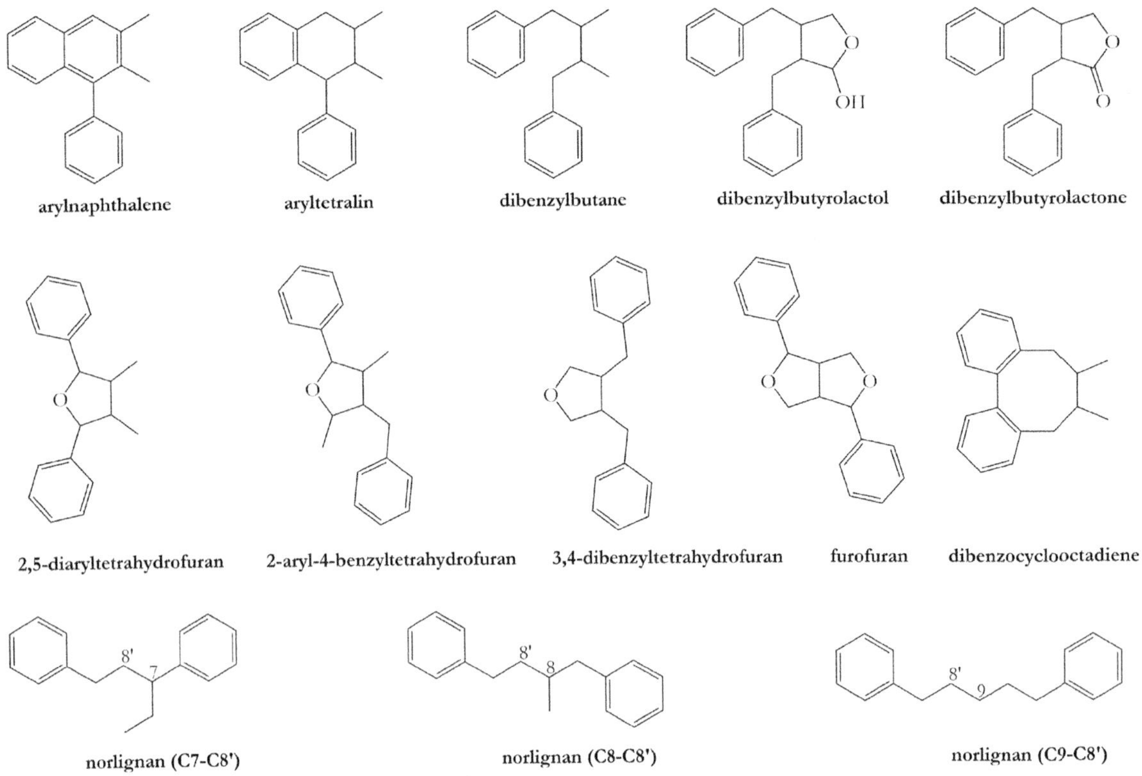

Fig. 2.9. Infraclasses of the phenolic-type simple lignan and norlignan groupings of the lignoid subclass.

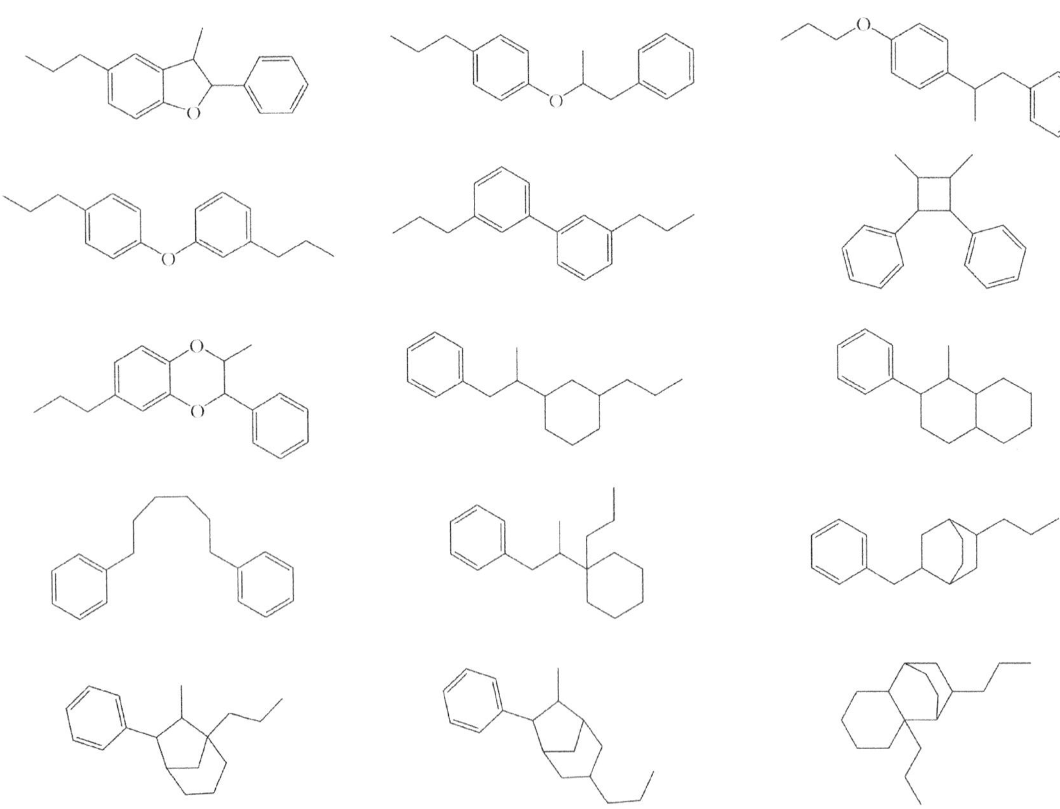

Fig. 2.10. Generic infraclasses of the phenolic-type neolignan grouping of the lignoid subclass.

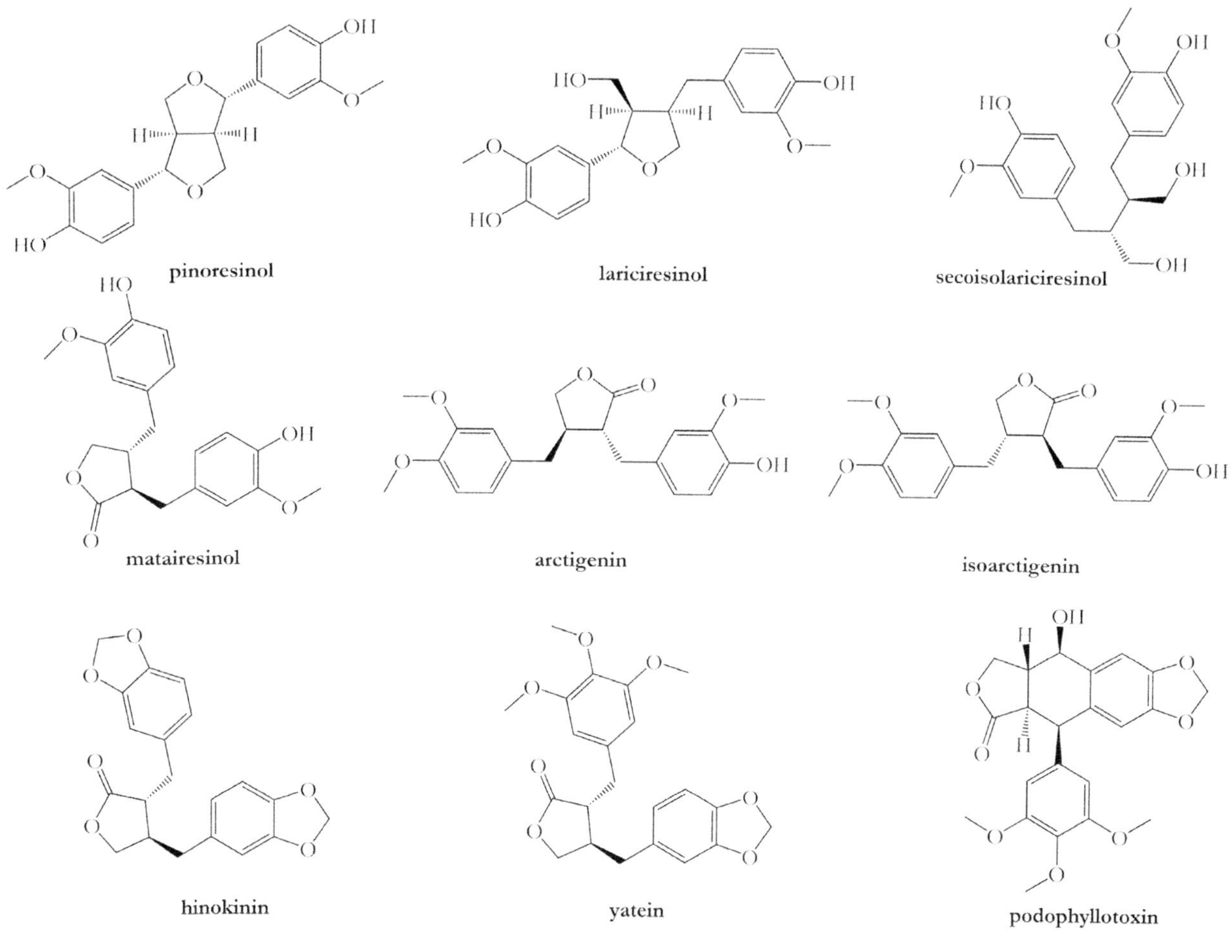

Fig. 2.11. Key examples of phenolic-type compounds of the lignoid subclass.

leads from pinoresinol to the sesamin subgroup of the furofurans, which get their name from the genus *Sesamum* (Pedaliaceae) of the Asterid clade.[9] The Schisandraceae family (*Kadsura, Illicium, Schisandra*) produce many lignans, but the dibenzocyclooctadiene group are thought to be unique to this family.[39] Norlignans are common in coniferous families such as Taxodiaceae, Cupressaceae and Araucariaceae. Norlignans are also found in some monocotyledonous families such as Liliaceae and Hypoxidaceae.[9,37] Neolignans, as the most structurally diverse lignan infraclass, are widely represented throughout the plant kingdom, with no general trends proposed.[35]

POLYPHENOLICS

Polyphenolics are a diverse, paraphyletic grouping of polymeric phenolic compounds of three major types: condensed tannins (proanthocyanidins/flavan-3-ols), hydrolysable tannins (carbohydrate: hydroxybenzoic acids) and lignins (lignans) (*Fig. 2.12*).[40] A huge variety of monomers can be incorporated into the polymeric structures of the polyphenolic subclass, so a detailed breakdown is impossible here, as it would take up most of this book. It is worth noting that the tannins are essentially absent from herbaceous plants.[40] Lignin can be characterised by the relative amounts of monolignoids (para-coumaroyl alcohol, coniferyl alcohol and sinapyl alcohol) that constitute it, with hardwoods and grasses being rich in coniferyl and sinapyl units and the softwoods predominantly composed of coniferyl units.[41] These compounds are often considered 'nuisance' compounds in NP chemistry, or more 'technically' as pan-assay interference compounds (PAINs).[42] PAINs are undesirable for several reasons: they are extremely difficult to work with (physically, they are sticky and do not filter well or solubilise well and are detrimental to small NP-oriented HPLC systems) and their biologically activity is often

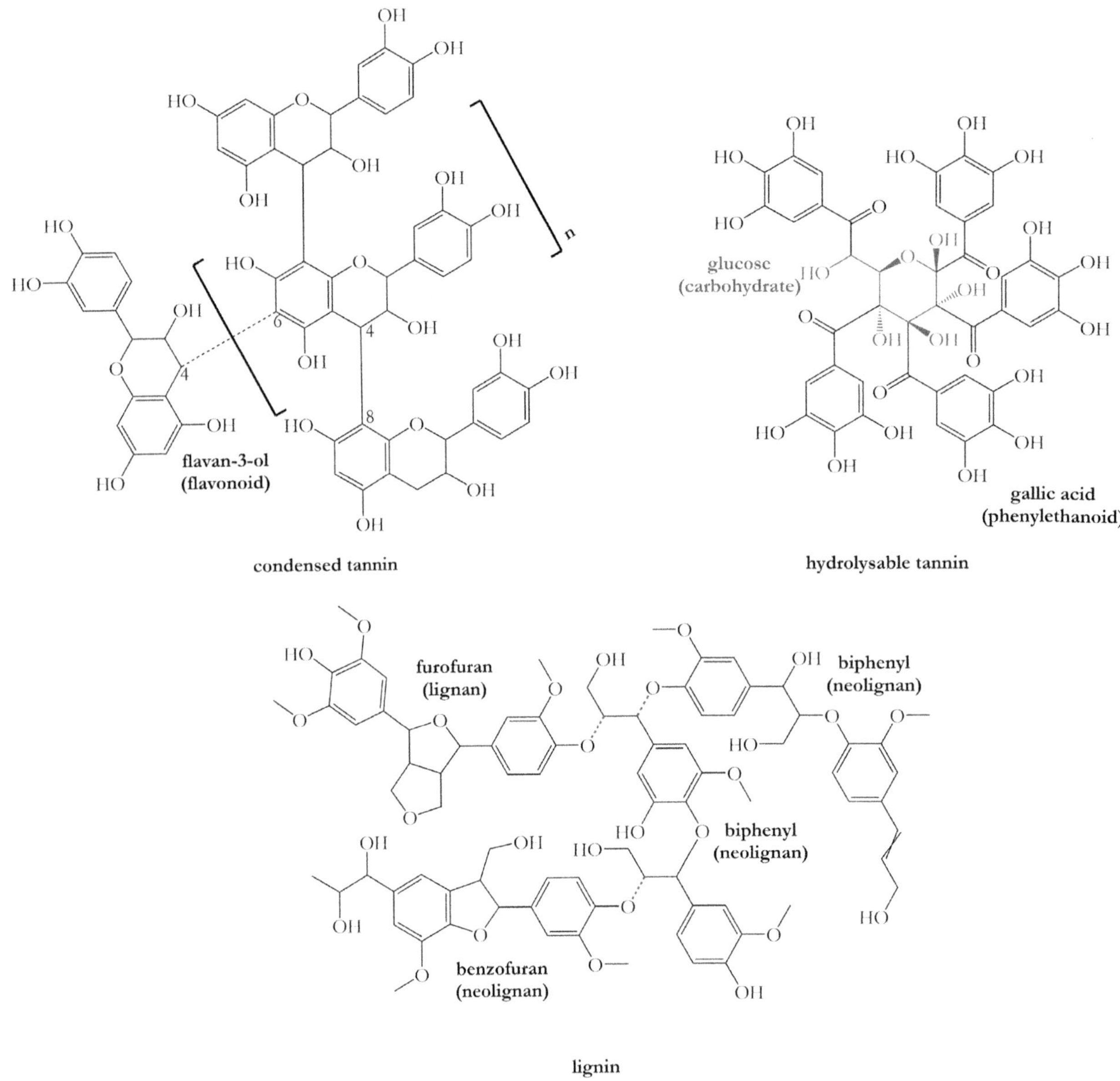

Fig. 2.12. Infraclasses of the phenolic-type polyphenolic subclass.

termed 'non-specific'; that is, they often show up as positive responses in bioassays, but are not amenable to further drug development protocols (which are generally searching for smaller molecules).

'PHYTOESTROGENS'

Phytoestrogens are included as a separate subsection because they are of particular significance given their medical applications as selective oestrogen receptor modulators (SERMs).[43] They are a paraphyletic grouping of non-steroidal plant compounds that resemble the structure of the steroid hormone estradiol and can come from the isoflavonoid, chromene (coumestan), lignan (neolignan) and prenylated flavonoid phenolic sub/infraclasses, among others (*Fig. 2.13*).[44] The oestrogenic properties of these plants have been recognised since the 1940s when a study demonstrated an infertility syndrome in sheep that grazed on clover, which is known to contain high concentrations of isoflavones. Conversely, modern epidemiological studies have demonstrated that phytoestrogen-rich diets have positive effects on menopausal symptoms, prostate and breast cancer, osteoporosis and cardiovascular disease.[43,44]

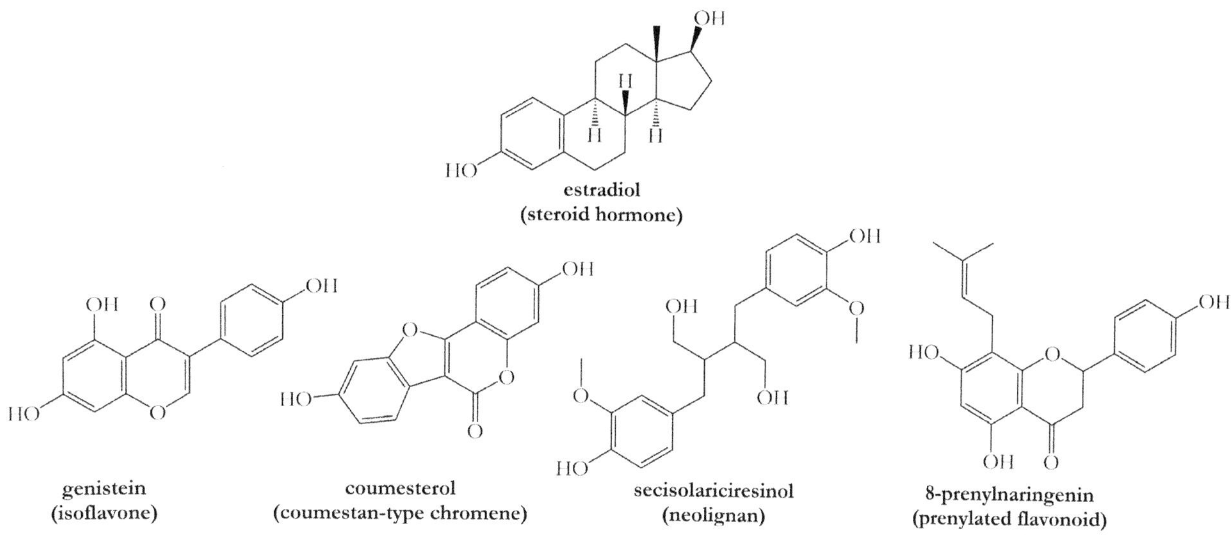

Fig. 2.13. Generic representatives of the paraphyletic phenolic-type phytoestrogen grouping.

Isoflavones are almost exclusively found in the Fabaceae (pea/legume) family, with the major dietary source being soybeans, which contain ~1,000–5,000 nMol/g.[45] The primary forms of isoflavone phytoestrogens are daidzein and genistein, which are converted metabolically by gut microflora to the more active form equol. Chromene phytoestrogens (coumestan/coumestrol) are derived from the isoflavonoid daidzein by a series of reactions, but are also present *in planta* throughout the Fabaceae, particularly in *Trifolium pratense* (L., 1753 - red clover).[46] Lignan phytoestrogens, such as secoisolariciresinol and matairesinol, are primarily sourced from the outer layers of cereals and grains such as flaxseed, which contains ~10,000 nMol/g.[45] Lignan phytoestrogens are metabolised by gut microflora to the significantly more potent 'enterolignans', enterolactone and enterodiol. Prenylated flavonoid phytoestrogens, such as 8-prenylnaringenin, are typically isolated from *Humulus lupulus* (L., 1753 - hops). These prenylated flavonoids show significant oestrogenicity compared with conventional flavonoids. It is thought that the lipophilic substitution of these flavonoid-type phenolics may help target these compounds to the oestrogen receptors or oestrogen metabolising enzymes in vivo.

REPRESENTATION

This section provides the reader with a higher-order hierarchical synopsis of the specialised metabolic trends contained within the flora listed in *Table 2.1*. By looking at this taxonomic data, while considering the customary uses of these plants, therapeutic patterns may (or may not) emerge that bely a common pharmacological basis to which these can be attributed. An obvious obstacle to this endeavour is the ubiquity of some of these structural sub/infraclasses across the floral kingdom, which will obfuscate any clear trends of therapeutic use because they essentially covers all bases. A better pattern emerges with chemical subclasses, which bear a narrower chemotaxonomic representation. A short summary of the reported biological activity for each sub/infraclass is included at the end of each grouping.

Phenylpropanoids ('phenolic glycosides')

The distribution of phenolic-type specialised metabolites of the phenolic glycoside subclass reported from the customary medicinal flora of Australia is as follows: Apiales [Apiaceae (*Centella*)], Asparagales [Amaryllidaceae (*Crinum*)], Asterales [Asteraceae (*Ageratum, Pterocaulon*)], Brassicales [Cleomaceae (*Cleome*)], Canellales [Winteraceae (*Tasmannia*)], Caryophyllales [Amaranthaceae (*Achyranthes, Dysphania*), Droseraceae (*Drosera*), Nyctaginaceae (*Boerhavia*), Polygonaceae (*Persicaria*)], Ericales [Sapotaceae (*Planchonella*)], Fabales [Fabaceae (*Entada, Erythrophleum, Indigofera, Miletta, Rhynchosia, Sesbania, Vigna*)], Gentianales [Apocynaceae (*Cynanchum*), Loganiaceae (*Strychnos*)], Lamiales [Lamiaceae (*Ajuga, Callicarpa, Clerodendrum, Mentha, Ocimum, Plectranthus, Prostanthera, Prunella, Teucrium*), Phrymaceae (*Mimulus*), Plantaginaceae (*Bacopa, Scoparia*), Scrophulariaceae (*Eremophila*), Verbenaceae (*Verbena*)], Laurales

[Lauraceae (*Cinnamomum*, *Litsea*)], Liliales (*Smilax*), Malphigiales [Euphorbiaceae (*Croton*, *Mallotus*), Phyllanthaceae (*Breynia*, *Flueggea*, *Phyllanthus*, *Securinega*), Rhizophoraceae (*Rhizophora*), Violaceae (*Hybanthus*)], Malvales [Malvaceae (*Hibiscus*, *Sida*, *Sterculia*), Thymelaeaceae (*Wikstroemia*)], Myrtales [Myrtaceae (*Syzygium*)], Pinales [Araucariaceae (*Araucaria*)], Piperales [Piperaceae (*Piper*)], Poales [Cyperaceae (*Cyperus*), Poaceae (*Cymbopogon*)], Polypodiales [Dennstaedtiaceae (*Pteridium*), Pteridaceae (*Adiantum*)], Proteales [Proteaceae (*Grevillea*, *Hakea*)], Psilotales (*Psilotum*), Ranunculales [Ranunculaceae (*Clematis*)], Rosales [Rhamnaceae (*Ziziphus*), Rosaceae (*Rubus*)], Santalales [Olacaceae (*Ximenia*), Santalaceae (*Santalum*)], Sapindales [Burseraceae (*Canarium*), Meliaceae (*Cedrela*, *Toona*), Rutaceae (*Melicope*, *Philotheca*, *Zieria*), Sapindaceae (*Dodonaea*)], Solanales [Convolvulaceae (*Convolvulus*, *Ipomoea*), Solanaceae (*Nicotiana*, *Solanum*)], Vitales [Vitaceae (*Cissus*)], Zingiberales [Zingiberaceae (*Curcuma*)] and Zygophyllales [Zygophyllaceae (*Tribulus*)].

No clear pattern of customary use emerges from this wide assortment of medicinal flora. Phenolic glycosides are often coupled with flavonoid glycosides when summarising their biological activity. The section on the flavonoid subclass below will outline this reported activity.

Flavonoids

The distribution of phenolic-type specialised metabolites of the flavonoid subclass reported from the customary medicinal flora of Australia is as follows: Alismatales [Araceae (*Amorphophallus*)], Apiales [Apiaceae (*Centella*)], Arecales [Arecaceae (*Calamus*)], Asparagales [Amaryllidaceae (*Crinum*), Asphodelaceae (*Dianella*), Orchidaceae (*Dendrobium*)], Asterales [Asteraceae (*Acmella*, *Ageratum*, *Chrysocephalum*, *Eclipta*, *Elephantopus*, *Gnaphalium*, *Helichrysum*, *Pterocaulon*, *Sphagneticolia*), Campanulaceae (*Lobelia*)], Goodeniaceae (*Goodenia*), Boraginaceae [Boraginales (*Ehretia*, *Heliotropium*)], Brassicales [Capparaceae (*Capparis*)], Cleomaceae (*Cleome*), Canellales [Winteraceae (*Tasmannia*)], Caryophyllales [Amaranthaceae (*Achyranthes*, *Atriplex*, *Dysphania*), Droseraceae (*Drosera*), Nyctaginaceae (*Boerhavia*), Plumbaginaceae (*Plumbago*), Polygonaceae (*Persicaria*), Portulacaceae (*Portulaca*)], Commelinales (*Haemodorum*), Cucurbitaceae (*Trichosanthes*), Ericales [Lecythidaceae (*Planchonia*), Primulaceae (*Aegiceras*), Sapotaceae (*Planchonella*)], Fabales [Fabaceae (*Abrus*, *Acacia*, *Cassia*, *Chamaecrista*, *Crotalaria*, *Entada*, *Erythrophleum*, *Indigofera*, *Lysiphyllum*, *Milettia*, *Rhynchosia*, *Senna*, *Sesbania*, *Sophora*, *Tephrosia*, *Vigna*)], Fagales [Casuarinaceae (*Casuarina*)], Gentianales [Apocynaceae (*Alstonia*, *Carissa*, *Cerbera*, *Cynanchum*, *Marsdenia*, *Pentatropis*, *Tylophora*), Gentianaceae (*Schenkia*), Loganiaceae (*Strychnos*), Rubiaceae (*Morinda*)], Lamiales [Lamiaceae (*Ajuga*, *Callicarpa*, *Clerodendrum*, *Mentha*, *Ocimum*, *Plectranthus*, *Prostanthera*, *Prunella*, *Teucrium*, *Vitex*), Plantaginaceae (*Gratiola*, *Scoparia*, *Stemodia*), Scrophulariaceae (*Eremophila*)], Laurales [Hernandiaceae (*Hernandia*), Lauraceae (*Cinnamomum*, *Litsea*)], Liliales (*Smilax*), Magnoliales (*Myristica*), Malphigiales [Calophyllaceae (*Calophyllum*), Euphorbiaceae (*Aleurites*, *Beyeria*, *Chamaesyce*, *Croton*, *Euphorbia*, *Excoecaria*, *Mallotus*), Phyllanthaceae (*Antidesma*, *Flueggea*, *Phyllanthus*, *Securinega*), Rhizophoraceae (*Rhizophora*), Violaceae (*Hybanthus*)], Malvales [Malvaceae (*Grewia*, *Hibiscus*, *Malva*, *Sida*, *Sterculia*, *Thespesia*), Thymelaeaceae (*Wikstroemia*)], Myrtales [Lythraceae (*Ammannia*, *Lythrum*), Myrtaceae (*Angophora*, *Corymbia*, *Eucalyptus*, *Syzygium*)], Oxalidales (*Eucryphia*), Pandanales (*Pandanus*), Pinales (*Araucaria*), Piperales (*Piper*), Poales [Cyperaceae (*Cyperus*, *Remirea*), Flagellariaceae (*Flagellaria*), Poaceae (*Cymbopogon*)], Polypodiales [Dennstaedtiaceae (*Pteridium*), Pteridaceae (*Adiantum*)], Proteales [Nelumbonaceae (*Nelumbo*), Proteaceae (*Grevillea*, *Persoonia*)], Psilotales (*Psilotum*), Ranunculales [Menispermaceae (*Cissampelos*, *Tinospora*), Ranunculaceae (*Clematis*)], Rosales [Moraceae (*Ficus*), Rhamnaceae (*Alphitonia*, *Ventilago*, *Ziziphus*), Rosaceae (*Rubus*)], Santalales [Olacaceae (*Ximenia*), Santalaceae (*Santalum*)], Sapindales [Anacardiaceae (*Semecarpus*), Burseraceae (*Canarium*), Meliaceae (*Cedrela*, *Melia*, *Toona*), Rutaceae (*Bosistoa*, *Melicope*, *Philotheca*), Sapindaceae (*Allophylus*, *Dodonaea*), Simaroubaceae (*Ailanthus*)], Solanales [Convolvulaceae (*Convolvulus*, *Evolvulus*, *Ipomoea*), Solanaceae (*Nicotiana*, *Solanum*)], Vitales [Vitaceae (*Cayratia*, *Cissus*)], Zingiberales [Musaceae (*Musa*), Zingiberaceae (*Curcuma*)] and Zygophyllales [Zygophyllaceae (*Tribulus*)].

No clear pattern of customary use emerges from this wide assortment of medicinal flora. A recent review of the therapeutic uses of medicinal plants containing phenolic glycosides and flavonoid glycosides lists the following biological activities: antineoplastic, anti-infective, anti-inflammatory, anti-obesity, use for skin ailments, protective effects for neurological and internal systems and some varying miscellaneous uses (reduced incidence of asthma, osteoporosis, the formation of blood vessels and altering drug availability).[47]

Chromenes

The distribution of phenolic-type specialised metabolites of the chromene subclass reported from the customary medicinal flora of Australia is as follows: Asparagales [Amaryllidaceae (*Crinum*), Orchidaceae (*Dendrobium*)], Asterales [Asteraceae (*Acmella*, *Ageratum*, *Chrysocephalum*, *Eclipta*, *Gnaphalium*, *Helichrysum*, *Pterocaulon*, *Sphagneticolia*), Campanulaceae (*Lobelia*), Goodeniaceae (*Goodenia*, *Scaevola*)], Brassicales [Cleomaceae (*Cleome*)], Caryophyllales [Amaranthaceae (*Dysphania*), Nyctaginaceae (*Boerhavia*), Plumbaginaceae (*Plumbago*), Polygonaceae (*Persicaria*)], Fabales [Fabaceae (*Sophora*, *Tephrosia*)], Gentianales [Apocynaceae (*Alstonia*, *Alyxia*, *Carissa*, *Marsdenia*), Gentianaceae (*Schenkia*), Rubiaceae (*Morinda*, *Nauclea*)], Lamiales [Lamiaceae (*Ocimum*, *Prunella*)], Laurales [Lauraceae (*Litsea*)], Liliales [Smilacaceae (*Smilax*)], Malphigiales [Calophyllaceae (*Calophyllum*), Euphorbiaceae (*Aleurites*, *Chamaesyce*, *Croton*, *Euphorbia*, *Mallotus*), Phyllanthaceae (*Flueggea*, *Phyllanthus*, *Securinega*)], Malvales [Malvaceae (*Grewia*, *Hibiscus*, *Malva*, *Sida*, *Sterculia*), Thymelaeaceae (*Wikstroemia*)], Myrtales [Lythraceae (*Lythrum*)], Poales [Cyperaceae (*Cyperus*)], Polypodiales [Pteridaceae (*Adiantum*)], Ranunculales [Ranunculaceae (*Clematis*)], Rosales [Moraceae (*Ficus*), Rhamnaceae (*Alphitonia*, *Ziziphus*)], Sapindales [Burseraceae (*Canarium*), Meliaceae (*Cedrela*, *Melia*, *Toona*), Rutaceae (*Geijera*, *Melicope*, *Philotheca*), Sapindaceae (*Dodonaea*)] and Solanales [Convolvulaceae (*Convolvulus*, *Ipomoea*, *Operculina*), Solanaceae (*Nicotiana*, *Solanum*)].

No clear pattern of customary use emerges from this wide assortment of medicinal flora. A review concerning the biological importance of chromenes as therapeutics highlights antineoplastic, antimicrobial, anti-inflammatory, antithrombotic and antipsychotic activities across the subclass.[48]

Phenalenes

The phenalene subclass is sparsely represented in the Australian medicinal flora, having only been reported from the Commelinales [Haemodoraceae (*Haemodorum*)] and the Zingiberales [Musaceae (*Musa*)]. Species from the family Haemodoraceae have been used as customary medicines by Aboriginal and Torres Strait Islander peoples for dysentery and bites/stings, which can be tentatively substantiated by the strong antifungal effect demonstrated by these metabolites.[29]

Phenanthrenes

The phenanthrene subclass is sparsely represented in the Australian medicinal flora, having only been reported from the Laurales [Atherospermataceae (*Atherosperma*), Lauraceae (*Litsea*)]. There is no overlap between the customary uses of these two medicinal genera. Species from the genus *Atherosperma* are used as a sedative/tonic, for inflammation, cough and fever or for laxative, diuretic or venereal disease applications.[43] Species from the genus *Litsea* were used as a general analgesic, as a bactericidal or for eye and skin ailments.[43] Plants containing phenanthrenes are known to be used customarily for pulmonary disorders, alongside antineoplastic, immunomodulatory and anthelmintic applications.[30]

Stilbenoids

The distribution of phenolic-type specialised metabolites of the stilbenoid subclass reported from the customary medicinal flora of Australia is as follows: Caryophyllales [Polygonaceae (*Persicaria*)], Fabales [Fabaceae (*Erythrophleum*, *Sesbania*, *Sophora*)], Gentianales [Rubiaceae (*Nauclea*)], Lamiales [Lamiaceae (*Ocimum*)], Malvales [Thymelaeaceae (*Wikstroemia*)], Psilotales [Psilotaceae (*Psilotum*)], Solanales [Convolvulaceae (*Convolvulus*)] and Vitales [Vitaceae (*Cayratia*, *Cissus*)].

There is some quite substantial overlap in the customary use of these genera, with *Erythrophleum*, *Ocimum*, *Persicaria*, *Sesbania* and *Wikstroemia* being used as bactericidal agents and *Convolvulus*, *Nauclea*, *Persicaria* and *Sophora* being used for stomach ailments.[43] *Nauclea*, *Ocimum* and *Persicaria* are all used as general tonics and *Cissus*, *Ocimum* and *Wikstroemia* are all used as a treatment for coughs/colds.[43] Plants containing stilbenoids and the stilbenoid-type metabolites isolated from them have been shown to have a range of biological activities: antioxidant, anti-inflammatory, antihyperglycemic, cardioprotective, neuroprotective and antineoplastic.[49]

Xanthones

The distribution of phenolic-type specialised metabolites of the xanthone subclass reported from the customary medicinal flora of Australia is as follows: Fabales [Fabaceae (*Cassia*, *Chamaecrista*, *Indigofera*, *Rhynchosia*, *Senna*)], Gentianales [Gentianaceae (*Schenkia*)], Malphigiales [Calophyllaceae (*Calophyllum*), Euphorbiaceae (*Croton*), Rhizophoraceae (*Rhizophora*)], Psilotales [Psilotaceae (*Psilotum*)] and Sapindales [Meliaceae (*Melia*)].

There is some significant overlap in the customary use of these floral genera, with *Calophyllum*, *Cassia*, *Chamaecrista*, *Melia* and *Senna* being used for both skin ailments and for their laxative effect, whereas *Calophyllum*, *Melia* and *Schenkia* were are used to reduce inflammation.[43] Xanthones isolated from the natural world, and the xanthone-derivatives generated from them, have a variety of biological activities ascribed to them: antineoplastic, anticoagulant, antiplatelet, anti-inflammatory, antimalarial, antimicrobial, hepatoprotective, antioxidant and multidrug resistance reversal effects.[50]

Lignoids

The distribution of phenolic-type specialised metabolites of the lignoid subclass reported from the customary medicinal flora of Australia is as follows: Alismatales [Araceae (*Alocasia*, *Rhaphidophora*)], Arecales [Arecaceae (*Calamus*)], Asparagales [Orchidaceae (*Dendrobium*)], Asterales [Asteraceae (*Ageratum*); Campanulaceae (*Lobelia*); Goodeniaceae (*Scaevola*)], Brassicales [Cleomaceae (*Cleome*)], Caryophyllales [Amaranthaceae (*Dysphania*), Nyctaginaceae (*Boerhavia*)], Cucurbitales [Cucurbitaceae (*Trichosanthes*)], Fabales [Fabaceae (*Tephrosia*, *Vigna*)], Gentianales [Apocynaceae (*Alyxia*, *Carissa*, *Cerbera*, *Cynanchum*), Loganiaceae (*Strychnos*), Rubiaceae (*Morinda*)], Lamiales [Lamiaceae (*Callicarpa*, *Ocimum*, *Vitex*), Scrophulariaceae (*Eremophila*)], Laurales [Hernandiaceae (*Hernandia*); Lauraceae (*Cassytha*, *Cinnamomum*, *Litsea*)], Liliales [Smilacaceae (*Smilax*)], Malphigiales [Euphorbiaceae (*Aleurites*, *Croton*, *Mallotus*); Phyllanthaceae (*Phyllanthus*); Rhizophoraceae (*Rhizophora*)], Malvales [Malvaceae (*Grewia*, *Hibiscus*, *Sida*), Thymelaeaceae (*Wikstroemia*)], Pandanales [Pandanaceae (*Pandanus*)], Pinales [Araucariaceae (*Araucaria*), Cupressaceae (*Callitris*)], Piperales [Piperaceae (*Piper*)], Ranunculales [Menispermaceae (*Tinospora*), Ranunculaceae (*Clematis*)], Sapindales [Meliaceae (*Cedrela*, *Melia*, *Toona*); Rutaceae (*Melicope*, *Philotheca*), Sapindaceae (*Dodonaea*)], Solanales [Solanaceae (*Nicotiana*, *Solanum*)] and Zingiberales [Musaceae (*Musa*); Zingiberaceae (*Curcuma*)].

No clear pattern of customary use emerges from this wide assortment of medicinal flora. Reviews of this broad subclass report a wide range of biological activities: antineoplastic, antioxidant, anticarcinogenic, antimutagenic and antioestrogenic (see 'Phytoestrogens', above).[51]

Polyphenolics

The distribution of phenolic-type specialised metabolites of the condensed tannin and hydrolysable tannin infraclasses of the polyphenolic subclass reported from the customary medicinal flora of Australia is as follows: Caryophyllales [Polygonaceae (*Persicaria*)], Ericales [Primulaceae (*Aegiceras*)], Fabales [Fabaceae (*Acacia*, *Indigofera*, *Rhynchosia*, *Sesbania*)], Gentianales [Apocynaceae (*Alstonia*)], Malphigiales [Euphorbiaceae (*Acalypha*, *Croton*, *Excoecaria*), Phyllanthaceae (*Antidesma*, *Flueggea*, *Phyllanthus*, *Securinega*), Rhizophoraceae (*Rhizophora*), Violaceae (*Hybanthus*)], Malvales [Malvaceae (*Hibiscus*)], Myrtales [Lythraceae (*Ammannia*, *Lythrum*), Myrtaceae (*Angophora*, *Corymbia*, *Eucalyptus*, *Syzygium*)], Ranunculales [Ranunculaceae (*Clematis*)], Rosales [Rosaceae (*Rubus*)], Sapindales [Burseraceae (*Canarium*)] and Solanales [Convolvulaceae (*Convolvulus*), Solanaceae (*Nicotiana*)].

Although the customary uses of this wide range of medicinal flora are difficult to summarise, and the substantial putative under-reporting in the literature due to these compounds being considered as PAINs, some general inferences can be drawn from the literature. A wide range of Aboriginal and Torres Strait Islander peoples' customary medicines were used for digestive issues, particularly the resinous (tannic) exudates from members of the Araucariaceae (*Araucaria*), Burseraceae (*Canarium*), Casuarinaceae (*Casuarina*), Fabaceae (*Acacia*, *Jacksonia*, *Vigna*), Lauraceae (*Cinnamomum*), Lythraceae (*Lythrum*), Myoporaceae (*Myoporum*), Myrtaceae (*Angophora*, *Backhousia*, *Eucalyptus*, *Syzygium*), Nelumbonaceae (*Nelumbo*), Rhizophoraceae (*Rhizophora*), Rosaceae (*Rubus*) and Rutaceae (*Flindersia*).[52] Although other more minor constituents may play a synergistic role in the efficacy of these compounds, the astringent properties of the tannic constituents of these resinous exudates are well known and substantiated in the literature, as tannin-type polyphenolic compounds have been shown to be antioxidant, anti-inflammatory, antimutagenic and antimicrobial, as well as having various other physiological effects such as modulation of immune responses.[53–55]

REFERENCES

1. Ververidis F, Trantas E, Douglas C, Vollmer G, Kretzschmar G *et al.* (2007) Biotechnology of flavonoids and other phenylpropanoid-derived natural products. Part I: Chemical diversity, impacts on plant biology and human health. *Biotechnology Journal* **2**, 1214–1234. doi:10.1002/biot.200700084
2. Yu O, Jez JM (2008) Nature's assembly line: biosynthesis of simple phenylpropanoids and polyketides. *The Plant Journal* **54**, 750–762. doi:10.1111/j.1365-313X.2008.03436.x

3. Ralston L, Subramanian S, Matsuno M, Yu O (2005) Partial reconstruction of flavonoid and isoflavonoid biosynthesis in yeast using soybean type I and II chalcone isomerases. *Plant Physiology* **137**, 1375–1388. doi:10.1104/pp.104.054502
4. Metcalf RL (1987) Plant volatiles as insect attractants. *Critical Reviews in Plant Sciences* **5**, 251–301. doi:10.1080/07352688709382242
5. Yonekura-Sakakibara K, Higashi Y, Nakabayashi R (2019) The origin and evolution of plant flavonoid metabolism. *Frontiers in Plant Science* **10**, 1–16. doi:10.3389/fpls.2019.00943
6. Vogt T (2010) Phenylpropanoid biosynthesis. *Molecular Plant* **3**, 2–20. doi:10.1093/mp/ssp106
7. Dixon RA, Achnine L, Kota P, Liu C-J, Srinivasa Reddy MS *et al.* (2002) The phenylpropanoid pathway and plant defence: a genomics perspective. *Molecular Plant Pathology* **3**, 371–390. doi:10.1046/j.1364-3703.2002.00131.x
8. Bourgaud F, Hehn A, Larbat R, Doerper S, Gontier E *et al.* (2006) Biosynthesis of coumarins in plants: a major pathway still to be unravelled for cytochrome P450 enzymes. *Phytochemistry Reviews* **5**, 293–308. doi:10.1007/s11101-006-9040-2
9. Suzuki S, Umezawa T (2007) Biosynthesis of lignans and norlignans. *Journal of Wood Science* **53**, 273–284. doi:10.1007/s10086-007-0892-x
10. Kovács A, Vasas A, Hohmann J (2008) Natural phenanthrenes and their biological activity. *Phytochemistry* **69**, 1084–1110. doi:10.1016/j.phytochem.2007.12.005
11. Xiao K, Zhang H-J, Xuan L-J, Zhang J, Xu Y-M *et al.* (2008) Stilbenoids: chemistry and bioactivities. *Studies in Natural Product Chemistry* **34**. doi:10.1016/S1572-5995(08)80032-4
12. Katsuyama Y, Kita T, Funa N, Horinouchi S (2009) Curcuminoid biosynthesis by two type III polyketide synthases in the herb *Curcuma longa*. *The Journal of Biological Chemistry* **284**, 11160–11170. doi:10.1074/jbc.M900070200
13. Davies KM, Schwinn KE (2006) Molecular biology and biotechnology of flavonoid biosynthesis. In *Flavonoids: Chemistry, Biochemistry and Applications.* (Eds OM Anderson and KR Markham) pp. 144–218. CRC Press, Boca Raton, FL.
14. Hastings J, Adams N, Ennis M, Hull D, Steinbeck C (2011) Chemical ontologies: what they are, what they are for and what are the challenges. *Journal of Cheminformatics* **3**, 04. doi:10.1186/1758-2946-3-S1-O4
15. Feunang YD, Eisner R, Knox C, Chepelev L, Hastings J *et al.* (2016) ClassyFire: automated chemical classification with a comprehensive, computable taxonomy. *Journal of Cheminformatics* **8**, 61.
16. Ververidis F, Trantas E, Douglas C, Vollmer G, Kretzschmar G *et al.* (2007) Biotechnology of flavonoids and other phenylpropanoid-derived natural products. Part II: reconstruction of multienzyme pathways in plants and microbes. *Biotechnology Journal* **2**, 1235–1249. doi:10.1002/biot.200700184
17. Douglas C, Ellard M, Hauffe KD, Molitor E, De Sa *et al.* (1992) General phenylpropanoid metabolism: regulation by environmental and developmental signals. In *Phenolic Metabolism in Plants.* (Eds HA Stafford and RK Ibrahim) pp. 63–89. Springer, New York, NY.
18. Denisov ET, Afanas'ev IB (2005) *Oxidation and Antioxidants in Organic Chemistry and Biology.* CRC Press, Boca Raton, FL.
19. Fu G, Pang H, Wong YH (2008) Naturally occuring phenylethanoid glycosides: potential leads for new therapeutics. *Current Medicinal Chemistry* **15**, 2592–2613. doi:10.2174/092986708785908996
20. Kurkin VA (2003) Phenylpropanoids from medicinal plants: distribution, classification, structural analysis, and biological activity. *Chemistry of Natural Compounds* **39**, 123–153. doi:10.1023/A:1024876810579
21. Jiménez C, Riguera R (1994) Phenylethanoid glycosides in plants: structure and biological activity. *Natural Product Reports* **11**, 591–606. doi:10.1039/NP9941100591
22. Veitch NC, Grayer RJ (2011) Flavonoids and their glycosides, including anthocyanins. *Natural Product Reports* **28**, 1626–1695. doi:10.1039/c1np00044f
23. Xiao J (2017) Dietary flavonoid aglycones and their glycosides: which show better biological significance? *Critical Reviews in Food Science and Nutrition* **57**, 1874–1905.
24. Gould KS, Lister C (2006) Flavonoid functions in plants. In *Flavonoids: Chemistry, Biochemistry and Applications.* (Eds OM Anderson and KR Markham) pp. 397–441. CRC Press, Boca Raton, FL.
25. Buettner D, Skemp S (2016) Blue Zones: lessons from the world's longest lived. *American Journal of Lifestyle Medicine* **10**, 318–321. doi:10.1177/1559827616637066
26. Baky MH, Elshahed M, Wessjohann J, Farag MA (2022) Interactions between dietary flavonoids and the gut microbiome: a comprehensive review. *British Journal of Nutrition* **128**, 577–591. doi:10.1017/S0007114521003627
27. Shimada S, Otsuki H, Sukuta M (2007) Transcriptional control of anthocyanin biosynthetic genes in the Caryophyllales. *Journal of Experimental Botany* **58**, 957–967. doi:10.1093/jxb/erl256
28. Pratap R, Ram VJ (2014) Natural and synthetic chromenes, fused chromenes, and versatility of dihydrobenzo[*h*]chromenes in organic synthesis. *Chemical Reviews* **114**, 10476–10526. doi:10.1021/cr500075s
29. Norman EO, Lever J, Brkljaca R, Urban S (2019) Distribution, biosynthesis, and biological activity of phenylphenalenone-type compounds derived from the family of plants, Haemodoraceae. *Natural Product Reports* **36**, 753–768. doi:10.1039/C8NP00067K
30. Tóth B, Hohmann J, Vasas A (2018) Phenanthrenes: a promising group of plant secondary metabolites. *Journal of Natural Products* **81**, 661–678. doi:10.1021/acs.jnatprod.7b00619
31. Keylor MH, Matsuura BS, Stephenson CR (2015) Chemistry and biology of resveratrol-derived natural products. *Chemical Reviews* **115**, 8976–9027. doi:10.1021/cr500689b

32. Vieira LMM, Kijjoa A (2005) Naturally-occuring xanthones: recent developments. *Current Medicinal Chemistry* **12**, 2413–2446. doi:10.2174/092986705774370682
33. Peres V, Nagem TJ, De Oliveira FF (2000) Tetraoxygenated naturally occurring xanthones. *Phytochemistry* **55**, 683–710. doi:10.1016/S0031-9422(00)00303-4
34. Wezeman T, Brase S, Masters KS (2015) Xanthone dimers: a compound family which is both common and privileged. *Natural Product Reports* **32**, 6–28. doi:10.1039/C4NP00050A
35. Teponno RB, Kusari S, Spiteller M (2016) Recent advances in research on lignans and neolignans. *Natural Product Reports* **33**, 1044–1092. doi:10.1039/C6NP00021E
36. Saleem M, Kim HJ, Ali MS, Lee YS (2005) An update on bioactive plant lignans. *Natural Product Reports* **22**, 696–716. doi:10.1039/b514045p
37. Pan JY, Chen SL, Yang MH, Wu J, Sinkkonen J *et al.* (2009) An update on lignans: natural products and synthesis. *Natural Product Reports* **26**, 1251–1292. doi:10.1039/b910940d
38. Rodríguez-García C, Sanchez-Quesada C, Toledo E, Delgado-Rodriguez M, Gaforio JJ (2019) Naturally lignan-rich foods: a dietary tool for health promotion? *Molecules (Basel, Switzerland)* **24**, 917. doi:10.3390/molecules24050917
39. Chang J, Song C (2013) Progress in the syntheses of dibenzocyclooctadiene lignans. In *Studies in Natural Product Chemistry*. (Ed. A-U Rahman) pp. 173–203. Elsevier, Oxford, UK.
40. Furlan CM, Motta LB, Cursino Dos Santos DYA (2010) Tannins: what do they represent in plant life. In *Tannins: Types, Foods Containing, and Nutrition*. (Ed. GK Petridis) pp. 1–13. Nova Science Publishers, Hauppauge, NY.
41. Boerjan W, Ralph J, Baucher M (2003) Lignin biosynthesis. *Annual Review of Plant Biology* **54**, 519–546. doi:10.1146/annurev.arplant.54.031902.134938
42. Bolz SN, Adasme MF, Schroeder M (2021) Toward an understanding of pan-assay interference compounds and promiscuity: a structural perspective on binding modes. *Journal of Chemical Information and Modeling* **61**, 2248–2262. doi:10.1021/acs.jcim.0c01227
43. Cos P, De Bruyne T, Apers S, Berghe DV, Pieters L *et al.* (2003) Phytoestrogens: recent developments. *Planta Medica* **69**, 589–599. doi:10.1055/s-2003-41122
44. Dixon RA (2004) Phytoestrogens. *Annual Review of Plant Biology* **55**, 225–261. doi:10.1146/annurev.arplant.55.031903.141729
45. Mazur W, Adlercreutz H (1998) Naturally occurring oestrogens in food. *Pure and Applied Chemistry* **70**, 1759–1776. doi:10.1351/pac199870091759
46. Wiseman H (2006) Isoflavonoids and human health. In *Flavonoids: Chemistry, Biochemistry and Applications*. (Eds OM Anderson and KR Markham) pp. 371–396. CRC PRess, Boca Raton, FL.
47. Johnson JB, Mani JS, Broszczak D, Prasad SS, Ekanayake CP *et al.* (2021) Hitting the sweet spot: asystematic review of the bioactivity and health benefits of phenolic glycosides from medicinally used plants. *Phytotherapy Research* **35**, 3484–3508. doi:10.1002/ptr.7042
48. Costa M, Dias TA, Brito A, Proença F (2016) Biological importance of structurally diversified chromenes. *European Journal of Medicinal Chemistry* **123**, 487–507. doi:10.1016/j.ejmech.2016.07.057
49. Duta-Bratu C-G, Nitulescu GM, Mihai DP, Olaru OT (2023) Resveratrol and other natural oligomeric stilbenoid compounds and their therapeutic applications. *Plants* **12**, 2935. doi:10.3390/plants12162935
50. Pinto MM, Palmeira A, Fernandes C, Resende DI, Sousa E *et al.* (2021) From natural products to new synthetic small molecules: A journey through the world of xanthones. *Molecules (Basel, Switzerland)* **26**, 431. doi:10.3390/molecules26020431
51. Yeung AWK, Tzvetkov NT, Balacheva AA, Georgieva MG, Gan R-Y *et al.* (2020) Lignans: quantitative analysis of the research literature. *Frontiers in Pharmacology* **11**, 37. doi:10.3389/fphar.2020.00037
52. Lassak EV, Mccarthy T (1983) *Australian Medicinal Plants*. Methuen, Sydney.
53. Guo Y, Sakulnarmrat K, Konczak I (2014) Anti-inflammatory potential of native Australian herbs polyphenols. *Toxicology Reports* **1**, 385–390. doi:10.1016/j.toxrep.2014.06.011
54. Chung KT, Wong TY, Wei CI, Huang YW, Lin Y (1998) Tannins and human health: a review. *Critical Reviews in Food Science and Nutrition* **38**, 421–464. doi:10.1080/10408699891274273
55. Ogawa S, Yazaki Y (2018) Tannins from *Acacia mearnsii* de Wild. bark: tannin determination and biological activities. *Molecules (Basel, Switzerland)* **23**, 837. doi:10.3390/molecules23040837
56. Huang W, Li C, Wang Y, Yi X, He X (2017) Anti-inflammatory lignanamides and monoindoles from Alocasia macrorrhiza. *Fitoterapia* **117**, 126–132. doi:10.1016/j.fitote.2017.01.014
57. Huang W, Yi X, Feng J, Wang Y, He X (2017) Piperidine alkaloids from Alocasia macrorrhiza. *Phytochemistry* **143**, 81–86. doi:10.1016/j.phytochem.2017.07.012
58. Mallik J, Das J, Banik RK (2018) Pharmacognostic profile and pharmacological activity of different parts of *Amorphophallus campanulatus* (Roxb.) Blume: a complete overview. *Asian Journal of Pharmaceutical Research & Development* **6**, 4–8. doi:10.22270/ajprd.v6i1.354
59. Padeepika C, Selvakumar R, Krishnakumar T, Sajad UNN, Sajeev MS (2018) Pharmacology and phytochemistry of underexploited tuber crops: a review. *Journal of Pharmacognosy and Phytochemistry* **7**, 1007–1019.
60. Zhang H-J, Tamez PA, Hoang VD, Tan GT, Hung NV *et al.* (2001) Antimalarial compounds from *Rhaphidophora decursiva*. *Journal of Natural Products* **64**, 772–777. doi:10.1021/np010037c
61. Zhang H, Qiu S, Tamez P, Tan GT, Aydogmus Z *et al.* (2002) Antimalarial agents from plants. II: decursivine, a new antimalarial indole alkaloid from *Rhaphidophora decursiva*. *Pharmaceutical Biology* **40**, 221–224. doi:10.1076/phbi.40.3.221.5832

62. Ayyad S-EN, Judd AS, Shier T, Hoye TR (1998) Otteliones A and B: potently cytotoxic 4-methylene-2-cyclohenones from *Ottelia alismoides*. *The Journal of Organic Chemistry* **63**, 8102–8106. doi:10.1021/jo971870a
63. Katoh T (2013) Enantioselective total synthesis of otteliones A and B, novel and powerful antitumor agents from the freshwater plant *Ottelia alismoides*. *Natural Product Communications* **8**, 973–980. doi:10.1177/1934578X1300800724
64. Brinkhaus B, Lindner M, Schuppan D, Hahn EG (2000) Chemical, pharmacological and clinical profile of the East Asian medical plant *Centella aslatica*. *Phytomedicine* **7**, 427–448. doi:10.1016/S0944-7113(00)80065-3
65. James JT, Dubery IA (2009) Pentacyclic triterpenoids from the medicinal herb, *Centella asiatica* (L.) Urban. *Molecules (Basel, Switzerland)* **14**, 3922–3941. doi:10.3390/molecules14103922
66. Agostini-Costa TDS (2018) Bioactive compounds and health benefits of some palm species traditionally used in Africa and the Americas: a review. *Journal of Ethnopharmacology* **224**, 202–229. doi:10.1016/j.jep.2018.05.035
67. Refaat J, Kamel MS, Ramadan MA, Ali AA (2012) *Crinum*: an endless source of bioactive principles: a review. Part I: *Crinum* alkaloids: lycorine-type alkaloids. *International Journal of Pharmaceutical Sciences and Research* **3**, 1883–1890.
68. Refaat J, Kamel MS, Ramadan MA, Ali AA (2012) *Crinum*: an endless source of bioactive principles: a review. Part II: *Crinum* alkaloids: crinine-type alkaloids. *International Journal of Pharmaceutical Sciences and Research* **3**, 3091–3100.
69. Refaat J, Kamel MS, Ramadan MA, Ali AA (2012) *Crinum*; an endless source of bioactive principles: a review. Part III: *Crinum* alkaloids: belladine-, galanthamine-, lycorenine-, tazettine-type alkaloids and other minor types. *International Journal of Pharmaceutical Sciences and Research* **3**, 3630–3638.
70. Refaat J, Kamel MS, Ramadan MA, Ali AA (2013) *Crinum*: an endless source of bioactive principles: a review. Part IV: non alkaloidal constituents. *International Journal of Pharmaceutical Sciences and Research* **4**, 941–948.
71. Negi JS, Singh P, Joshi GP, Rawat MS, Bisht VK (2010) Chemical constituents of *Asparagus*. *Pharmacognosy Reviews* **4**, 215–220. doi:10.4103/0973-7847.70921
72. Bloor SJ (2001) Deep blue anthocyanins from blue *Dianella berries*. *Phytochemistry* **58**, 923–927. doi:10.1016/S0031-9422(01)00343-0
73. Dias DA, Silva CA, Urban S (2009) Naphthalene aglycones and glycosides from the Australian medicinal plant, *Dianella callicarpa*. *Planta Medica* **75**, 1442–1447. doi:10.1055/s-0029-1185724
74. Xu J, Han Q-B, Li S-L, Chen X-J, Wang X-N *et al.* (2013) Chemistry, bioactivity and quality control of *Dendrobium*, a commonly used tonic herb in traditional Chinese medicine. *Phytochemistry Reviews* **12**, 341–367. doi:10.1007/s11101-013-9310-8
75. Paulraj J, Govindarajan R, Palpu P (2013) The genus spilanthes ethnopharmacology, phytochemistry, and pharmacological properties: a review. *Advances in Pharmacological and Pharmaceutical Sciences* **2013**, 510298.
76. Yadav N, Ganie SA, Singh B, Chhillar AK, Yadav SS (2019) Phytochemical constituents and ethnopharmacological properties of *Ageratum conyzoides* L. *Phytotherapy Research* **33**, 2163–2178. doi:10.1002/ptr.6405
77. Wu P, Su MX, Wang Y, Wang GC, Ye WC *et al.* (2012) Supercritical fluid extraction assisted isolation of sesquiterpene lactones with antiproliferative effects from *Centipeda minima*. *Phytochemistry* **76**, 133–140. doi:10.1016/j.phytochem.2012.01.003
78. Zdero C, Bohlmann F, King RM (1992) *Ent*-labdanes, manoyloxide and helipterol derivatives from *Chrysocephalum ambiguum*. *Phytochemistry* **31**, 1631–1638. doi:10.1016/0031-9422(92)83119-J
79. Lourens AC, Viljoen AM, Van Heerden FR (2008) South African *Helichrysum* species: a review of the traditional uses, biological activity and phytochemistry. *Journal of Ethnopharmacology* **119**, 630–652. doi:10.1016/j.jep.2008.06.011
80. Chung IM, Rajakumar G, Lee JH, Kim SH, Thiruvengadam M (2017) Ethnopharmacological uses, phytochemistry, biological activities, and biotechnological applications of *Eclipta prostrata*. *Applied Microbiology and Biotechnology* **101**, 5247–5257. doi:10.1007/s00253-017-8363-9
81. Feng L, Zhai YY, Xu J, Yao WF, Cao YD *et al.* (2019) A review on traditional uses, phytochemistry and pharmacology of *Eclipta prostrata* (L.) L. *Journal of Ethnopharmacology* **245**, 112109. doi:10.1016/j.jep.2019.112109
82. Hiradeve SM, Rangari VD (2014) A review on pharmacology and toxicology of *Elephantopus scaber* Linn. *Natural Product Research* **28**, 819–830. doi:10.1080/14786419.2014.883394
83. Zheng X, Wang W, Piao H, Xu W, Shi H *et al.* (2013) The genus *Gnaphalium* L. (Compositae): phytochemical and pharmacological characteristics. *Molecules (Basel, Switzerland)* **18**, 8298–8318. doi:10.3390/molecules18078298
84. Macleod JK, Rasmussen HB (1999) A hydroxy-B-caryophyllene from *Pterocaulon serrulatum*. *Phytochemistry* **50**, 105–108. doi:10.1016/S0031-9422(98)00460-9
85. Medeiros-Neves B, Teixeira HF, Von Poser GL (2018) The genus *Pterocaulon* (Asteraceae): a review on traditional medicinal uses, chemical constituents and biological properties. *Journal of Ethnopharmacology* **224**, 451–464. doi:10.1016/j.jep.2018.06.012
86. Li X, Dong M, Liu Y, Shi Q-W, Kiyota H (2007) Structures and biological properties of the chemical constituents from the genus *Wedelia*. *Chemistry & Biodiversity* **4**, 823–836. doi:10.1002/cbdv.200790070
87. That QT, Jossang J, Jossang A, Kim PPN, Jaureguiberry G (2007) Wedelolides A and B: novel sesquiterpene

sigma-lactones, (9R)-eudesman-9,12-olides, from *Wedelia trilobata*. *The Journal of Organic Chemistry* **72**, 7102–7105. doi:10.1021/jo070771m

88. Mizokami SS, Arakawa NS, Ambrosio SR, Zarpelon AC, Casagrande R *et al.* (2012) Kaurenoic acid from *Sphagneticola trilobata* inhibits inflammatory pain: effect on cytokine production and activation of the NO-cyclic GMP-protein kinase G-ATP-sensitive potassium channel signaling pathway. *Journal of Natural Products* **75**, 896–904. doi:10.1021/np200989t
89. Sun L, Wang Z, Wang Y, Xu J, He X (2020) Anti-proliferative and anti-neuroinflammatory eudesmanolides from *Wedelia* (*Sphagneticola trilobata* (L.) Pruski). *Fitoterapia* **142**, 104452. doi:10.1016/j.fitote.2019.104452
90. Folquitto DG, Swiech JND, Pereira CB, Bobek VB, Halila Possagno GC *et al.* (2019) Biological activity, phytochemistry and traditional uses of genus *Lobelia* (Campanulaceae): A systematic review. *Fitoterapia* **134**, 23–38. doi:10.1016/j.fitote.2018.12.021
91. Ghisalberti EL (2004) The Goodeniaceae. *Fitoterapia* **75**, 429–446. doi:10.1016/j.fitote.2004.01.018
92. Suthiwong J, Thongsri Y, Yenjai C (2017) A new furanocoumarin from the fruits of *Scaevola taccada* and antifungal activity against *Pythium insidiosum*. *Natural Product Research* **31**, 453–459. doi:10.1080/14786419.2016.1188100
93. Liao G, Wu P, Liu Z, Xue J, Li H, Wei X (2020) 2H-pyranone and isocoumarin derivatives from the endophytic fungus *Pestalotiopsis microspora* SC3082 derived from *Scaevola taccada* (Gaertn.) Roxb. *Natural Product Research* **35**, 3644–3651.
94. Li L, Peng Y, Yao X, Xu L-J, Wulan T-N, Liu Y *et al.* (2010) Chemical constituents and biological activities of plants from the genus *Ehretia* Linn. *Chinese Herbal Medicines* **2**, 106–111.
95. Shukla A, Kaur A (2018) A systematic review of traditional uses bioactive phytoconstituents of genus *Ehretia*. *Asian Journal of Pharmaceutical and Clinical Research* **11**, 88–100. doi:10.22159/ajpcr.2018.v11i6.25178
96. Goyal N, Sharma SK (2014) Bioactive phytoconstituents from genus Heliotropium. *International Journal of Green Pharmacy* **8**, 217–225.
97. Tlili N, Elfalleh W, Saadaoui E, Khaldi A, Triki S *et al.* (2011) The caper (*Capparis* L.): ethnopharmacology, phytochemical and pharmacological properties. *Fitoterapia* **82**, 93–101. doi:10.1016/j.fitote.2010.09.006
98. Singh H, Mishra A, Mishra AK (2018) The chemistry and pharmacology of *Cleome* genus: a review. *Biomedicine and Pharmacotherapy* **101**, 37–48. doi:10.1016/j.biopha.2018.02.053
99. Asakawa Y, Ludwiczuk A, Harinantenaina L, Toyota M, Nishiki M *et al.* (2012) Distribution of drimane sesquiterpenoids and tocopherols in liverworts, ferns and higher plants: Polygonaceae, Canellaceae, and Winteraceae species. *Natural Product Communications* **7**, 685–692. doi:10.1177/1934578X1200700601
100. Cock IE (2013) The phytochemistry and chemotherapeutic potential of *Tasmannia lanceolata* (Tasmanian pepper): a review. *Pharmacognosy Communications* **3**, 1–13. doi:10.5530/pc.2013.2.1
101. He X, Wang X, Fang J, Chang Y, Ning N *et al.* (2017) The genus Achyranthes: a review on traditional uses, phytochemistry, and pharmacological activities. *Journal of Ethnopharmacology* **203**, 260–278. doi:10.1016/j.jep.2017.03.035
102. Clauser M, Dall'acqua S, Loi MC, Innocenti G (2013) Phytochemical investigation on *Atriplex halimus* L. from Sardinia. *Natural Product Research* **27**, 1940–1944. doi:10.1080/14786419.2013.793684
103. Mroczek A (2015) Phytochemistry and bioactivity of triterpene saponins from Amaranthaceae family. *Phytochemistry Reviews* **14**, 577–605. doi:10.1007/s11101-015-9394-4
104. Kokanova-Nedialkova Z, Neialkov PT, Nikolov SD (2009) The genus *Chenopodium*: phytochemistry, ethnopharmacology and pharmacology. *Pharmacognosy Reviews* **3**, 280–306.
105. Koo KA, Lorent K, Gong W, Windsor P, Whittaker SJ *et al.* (2015) Biliatresone, a reactive natural toxin from *Dysphania glomulifera* and *D. littoralis*: discovery of the toxic moiety 1,2-diaryl-2-propenone. *Chemical Research in Toxicology* **28**, 1519–1521. doi:10.1021/acs.chemrestox.5b00227
106. Egan PA, van der Kooy F (2013) Phytochemistry of the carnivorous sundew genus *Drosera* (Droseraceae): future perspectives and ethnopharmacological relevance. *Chemistry & Biodiversity* **10**, 1774–1790. doi:10.1002/cbdv.201200359
107. Patil KS, Bhalsing SR (2016) Ethnomedicinal uses, phytochemistry and pharmacological properties of the genus *Boerhavia*. *Journal of Ethnopharmacology* **182**, 200–220. doi:10.1016/j.jep.2016.01.042
108. Garg V, Jyoti, Dutt R (2018) The genus *Plumbago*: an update. *International Journal of Green and Herbal Chemistry* **7**, 313–326.
109. Peng W, Qin R, Li X, Zhou H (2013) Botany, phytochemistry, pharmacology, and potential application of *Polygonum cuspidatum* Sieb.et Zucc.: a review. *Journal of Ethnopharmacology* **148**, 729–745. doi:10.1016/j.jep.2013.05.007
110. Lin L, Ni B, Lin H, Zhang M, Li X *et al.* (2015) Traditional usages, botany, phytochemistry, pharmacology and toxicology of *Polygonum multiflorum* Thunb.: a review. *Journal of Ethnopharmacology* **159**, 158–183. doi:10.1016/j.jep.2014.11.009
111. Ayaz M, Ahmad I, Sadiq A, Ullah F, Ovais M *et al.* (2020) *Persicaria hydropiper* (L.) Delarbre: a review on traditional uses, bioactive chemical constituents and pharmacological and toxicological activities. *Journal of Ethnopharmacology* **251**, 112516. doi:10.1016/j.jep.2019.112516
112. Gou KJ, Zeng R, Ma Y, Li AN, Yang K *et al.* (2020) Traditional uses, phytochemistry, and pharmacology of *Persicaria orientalis* (L.) Spach: a review. *Journal of Ethnopharmacology* **249**, 112407. doi:10.1016/j.jep.2019.112407

113. Iranshahy M, Javadi B, Iranshahi M, Jahanbakhsh SP, Mahyari S *et al.* (2017) A review of traditional uses, phytochemistry and pharmacology of *Portulaca oleracea* L. *Journal of Ethnopharmacology* **205**, 158–172. doi:10.1016/j.jep.2017.05.004
114. Kasai R, Sasaki A, Hashimoto T, Kaneko T, Ohtani K *et al.* (1999) Cycloartane glycosides from *Trichosanthes tricuspidata*. *Phytochemistry* **51**, 803–808. doi:10.1016/S0031-9422(99)00078-3
115. Kanchanapoom T, Kasai R, Yamasaki K (2002) Cucurbitane, hexanorcucurbitane and octanorcucurbitane glycosides from fruits of *Trichosanthes tricuspidata*. *Phytochemistry* **59**, 215–228. doi:10.1016/S0031-9422(01)00430-7
116. Mai LP, Guenard D, Franck M, Van TM, Gaspard C *et al.* (2002) New cytotoxic cucurbitacins from the pericarps of *Trichosanthes tricuspidata* fruits. *Natural Product Letters* **16**, 15–19. doi:10.1080/1057563029001/4755
117. Shaw PC, Lee KM, Wong KB (2005) Recent advances in trichosanthin, a ribosome-inactivating protein with multiple pharmacological properties. *Toxicon* **45**, 683–689. doi:10.1016/j.toxicon.2004.12.024
118. Yu X, Tang L, Wu H, Zhang X, Luo H *et al.* (2018) Trichosanthis Fructus: botany, traditional uses, phytochemistry and pharmacology. *Journal of Ethnopharmacology* **224**, 177–194. doi:10.1016/j.jep.2018.05.034
119. Lv H, She G (2010) Naturally occurring diarylheptanoids. *Natural Product Communications* **5**, 1687–1708. doi:10.1177/1934578X1000501035
120. Khong PW, Lewis KG (1977) New chemical constituents of *Planchonia careya*. *Australian Journal of Chemistry* **30**, 1311–1322. doi:10.1071/CH9771311
121. Khong PW, Lewis KG (1979) New chemical constituents of *Planchonia careya*. II: bark constituents soluble in light petroleum. *Australian Journal of Chemistry* **32**, 1621–1626. doi:10.1071/CH9791621
122. Crublet M-L, Long C, Sévenet T, Hadi HA, Lavaud C (2003) Acylated flavonol glycosides from leaves of *Planchonia grandis*. *Phytochemistry* **64**, 589–594. doi:10.1016/S0031-9422(03)00215-2
123. McRae JM, Yang Q, Crawford RJ, Palombo EA (2008) Antibacterial compounds from *Planchonia careya* leaf extracts. *Journal of Ethnopharmacology* **116**, 554–560. doi:10.1016/j.jep.2008.01.007
124. McRae JM, Yang Q, Crawford RJ, Palombo EA (2008) Acylated flavonoid tetraglycoside from *Planchonia careya* leaves. *Phytochemistry Letters* **1**, 99–102. doi:10.1016/j.phytol.2008.04.003
125. Rajeswari K, Bhaskara Rao T (2015) *Aegicera corniculatum* Linn (Myrsinaceae). *Journal of Chemical and Pharmaceutical Research* **7**, 305–316.
126. Hart NK, Lamberton JA (1966) Pyrrolizidine alkaloids from *Planchonella* species (family Sapotaceae). *Australian Journal of Chemistry* **19**, 1259–1264. doi:10.1071/CH9661259
127. Hart NK, Johns SR, Lamberton JA (1968) Alkaloids of the Sapotaceae: trans-B-methylthioacrylate and tiglate esters of (-)-isoretronecanol from a *Planchonella* species. *Australian Journal of Chemistry* **21**, 1393–1395. doi:10.1071/CH9681393
128. Eade RA, Hunt K, Simes JJH, Stern W (1969) Extractives of Australian timbers. IX: the saponin of *Planchonella pohlmanianum*. *Australian Journal of Chemistry* **22**, 2703–2707. doi:10.1071/CH9692703
129. Cambie RC, Ser NA, Kokubun T (1997) Heartwood constituents of *Planchonella vitiensis*. *Biochemical Systematics and Ecology* **25**, 677–678. doi:10.1016/S0305-1978(97)00063-X
130. Lee T-H, Juang S-H, Hsu F-L, Wu C-Y (2005) Triterpene acids from the leaves of *Planchonella duclitan* (Blanco) Bakhuizan. *Journal of the Chinese Chemical Society (Taipei)* **52**, 1275–1280. doi:10.1002/jccs.200500184
131. Lee S-S, Tseng C-C, Chen C-K (2010) Three new benzophenone glucosides from the leaves of *Planchonella obovata*. *Helvetica Chimica Acta* **93**, 522–529. doi:10.1002/hlca.200900264
132. Chen H-Y, Guh J-H, Chan S-H, Lee S-S (2015) Cytotoxic protobassic acid glycosides from *Planchonella obovata* leaf. *Phytochemistry Letters* **11**, 229–235. doi:10.1016/j.phytol.2015.01.005
133. Garaniya N, Bapodra A (2014) Ethno botanical and phytopharmacological potential of Abrus precatorius L.: a review. *Asian Pacific Journal of Tropical Biomedicine* **4**, S27–S34. doi:10.12980/APJTB.4.2014C1069
134. Seigler DS (2003) Phytochemistry of Acacia—*sensu lato*. *Biochemical Systematics and Ecology* **31**, 845–873. doi:10.1016/S0305-1978(03)00082-6
135. Zanin JL, De Carvalho BA, Martineli PS, Dos Santos MH, Lago JH *et al.* (2012) The genus *Caesalpinia* L. (Caesalpiniaceae): phytochemical and pharmacological characteristics. *Molecules (Basel, Switzerland)* **17**, 7887–7902. doi:10.3390/molecules17077887
136. Matsuzaki S, Hamana K, Okada M, Niitsu M, Samejima K (1990) Aliphatic pentaamines found in *Canavalia gladiata*. *Phytochemistry* **29**, 1311–1312. doi:10.1016/0031-9422(90)85449-P
137. Fujimura S, Terada S, Jayavardhanan KK, Panikkar KR, Kimoto E (1993) Primary structures of concanavalin A-like lectins from seeds of two species of *Canavalia*. *Phytochemistry* **33**, 985–987. doi:10.1016/0031-9422(93)85008-F
138. Pattamadilok D, Pengsuparp T, Phummiratch D, Ongpipattanakul B, Meksuriyen D *et al.* (2008) Canarosine: a new guanidine alkaloid from *Canavalia rosea* with inhibitory activity on dopamine D1 receptors. *Journal of Asian Natural Products Research* **10**, 915–918. doi:10.1080/10286020802181513
139. Huang X, Mu B, Lin W, Qiu Y (2012) Pterocarpin and isoflavan derivatives from *Canavalia maritima* (Aubl.) Thou. *Records of Natural Products* **6**, 166–170.
140. Yadav JP, Arya V, Yadav S, Panghal M, Kumar S *et al.* (2010) *Cassia occidentalis* L.: a review on its ethnobotany, phytochemical and pharmacological profile. *Fitoterapia* **81**, 223–230. doi:10.1016/j.fitote.2009.09.008
141. Pitanga BP, Nascimento RP, Silva VD, Costa SL (2012) The role of astrocytes in metabolism and neurotoxicity of the

pyrrolizidine alkaloid monocrotaline, the main toxin of *Crotalaria retusa*. *Frontiers in Pharmacology* **3**, 144. doi:10.3389/fphar.2012.00144

142. Al-Snafi AE (2016) The contents and pharmacology of *Crotalaria juncea*: a review. *IOSR Journal of Pharmacy* **6**, 77–86.
143. Singh O, Ali M, Akhtar N (2011) Phenolic acid glucosides from the seeds of *Entada phaseoloides* Merill. *Journal of Asian Natural Products Research* **13**, 682–687. doi:10.1080/10286020.2011.584444
144. Sugimoto S, Matsunami K, Otsuka H (2018) Biological activity of *Entada phaseoloides* and *Entada rheedei*. *Journal of Natural Medicines* **72**, 12–19. doi:10.1007/s11418-017-1146-x
145. Son NT (2019) Genus *Erythrophleum*: botanical description, traditional use, phytochemistry and pharmacology. *Phytochemistry Reviews* **18**, 571–599. doi:10.1007/s11101-019-09640-0
146. Gerometta E, Grondin I, Smadja J, Frederich M, Gauvin-Bialecki A (2020) A review of traditional uses, phytochemistry and pharmacology of the genus *Indigofera*. *Journal of Ethnopharmacology* **253**, 112608. doi:10.1016/j.jep.2020.112608
147. Farag MA, Sakna ST, El-Fiky NM, Shabana MM, Wessjohann LA (2015) Phytochemical, antioxidant and antidiabetic evaluation of eight *Bauhinia* L. species from Egypt using UHPLC-PDA-qTOF-MS and chemometrics. *Phytochemistry* **119**, 41–50. doi:10.1016/j.phytochem.2015.09.004
148. Harrison JJEK, Dankyi E, Kingsford-Adaboh R, Ishida H (2011) In search of new leads: a closer look at the therapeutic potential of the constituents of *Millettia thonningii*, *Millettia machycarpa* and their structural analogues. *International Journal of Pharmacy and Pharmaceutical Sciences* **3**, 71–81.
149. Al Muqarrabun LM, Ahmat N, Ruzaina SA, Ismail NH, Sahidin I (2013) Medicinal uses, phytochemistry and pharmacology of *Pongamia pinnata* (L.) Pierre: a review. *Journal of Ethnopharmacology* **150**, 395–420. doi:10.1016/j.jep.2013.08.041
150. Rammohan A, Reddy GM, Bhaskar BV, Gunasekar D, Zyryanov GV (2020) Phytochemistry and pharmacological activities of the genus *Rhynchosia*: a comprehensive review. *Planta* **251**, 9. doi:10.1007/s00425-019-03311-2
151. Hasan N, Osman H, Mohamad S, Chong WK, Awang K *et al.* (2012) The chemical components of *Sesbania grandiflora* root and their antituberculosis activity. *Pharmaceuticals* **5**, 882–889. doi:10.3390/ph5080882
152. Noviany N, Samadi A, Yuliyan N, Hadi S, Aziz M *et al.* (2020) Structure characterization and biological activity of 2-arylbenzofurans from an Indonesian plant, *Sesbania grandiflora* (L.) Pers. *Phytochemistry Letters* **35**, 211–215. doi:10.1016/j.phytol.2019.12.008
153. Krishna PM, Rao KNV, Sandhya S, Banji D (2012) A review on phytochemical, ethnomedical and pharmacological studies on genus *Sophora*, Fabaceae. *Revista Brasileira de Farmacognosia-Brazilian Journal of Pharmacognosy* **22**, 1145–1154. doi:10.1590/S0102-695X2012005000043
154. He X, Fang J, Huang L, Wang J, Huang X (2015) *Sophora flavescens* Ait.: traditional usage, phytochemistry and pharmacology of an important traditional Chinese medicine. *Journal of Ethnopharmacology* **172**, 10–29. doi:10.1016/j.jep.2015.06.010
155. He X, Bai Y, Zhao Z, Wang X, Fang J *et al.* (2016) Local and traditional uses, phytochemistry, and pharmacology of *Sophora japonica* L.: a review. *Journal of Ethnopharmacology* **187**, 160–182. doi:10.1016/j.jep.2016.04.014
156. Crombie L, Whiting DA (1998) Biosynthesis in the rotenoid group of natural products: applications of isotope methodology. *Phytochemistry* **49**, 1479–1507. doi:10.1016/S0031-9422(98)00178-2
157. Touqeer S, Saeed MA, Ajaib M (2013) A review on the phytochemistry and pharmacology of genus *Tephrosia*. *Phytopharmacology* **4**, 598–637.
158. Chen Y, Yan T, Gao C, Cao W, Huang R (2014) Natural products from the genus *Tephrosia*. *Molecules (Basel, Switzerland)* **19**, 1432–1458. doi:10.3390/molecules19021432
159. Onyilagha JC, Islam S, Ntamatungiro S (2009) Comparative phytochemistry of eleven species of *Vigna* (Fabaceae). *Biochemical Systematics and Ecology* **37**, 16–19. doi:10.1016/j.bse.2008.11.013
160. Leu YL, Hwang TL, Kuo PC, Liou KP, Huang BS *et al.* (2012) Constituents from *Vigna vexillata* and their anti-inflammatory activity. *International Journal of Molecular Sciences* **13**, 9754–9768. doi:10.3390/ijms13089754
161. Tang D, Dong Y, Ren H, Li L, He C (2014) A review of phytochemistry, metabolite changes, and medicinal uses of the common food mung bean and its sprouts (*Vigna radiata*). *Chemistry Central Journal* **8**, 4. doi:10.1186/1752-153X-8-4
162. Mutharaj S, Muthusamy P, Ilango K (2019) Pharmacognostical, phytochemical and pharmacological review of *Casuarina equisetifolia* Linn. *World Journal of Pharmaceutical Research* **8**, 328–339.
163. Khyade MS, Kasote DM, Vaikos NP (2014) *Alstonia scholaris* (L.) R. Br. and *Alstonia macrophylla* Wall. ex G. Don: a comparative review on traditional uses, phytochemistry and pharmacology. *Journal of Ethnopharmacology* **153**, 1–18. doi:10.1016/j.jep.2014.01.025
164. Sriphana U, Thongsri Y, Ardwichai P, Poopasit K, Prariyachatigul C *et al.* (2013) New lignan esters from *Alyxia schlechteri* and antifungal activity against *Pythium insidiosum*. *Fitoterapia* **91**, 39–43. doi:10.1016/j.fitote.2013.08.005
165. Kaunda JS, Zhang YJ (2017) The genus *Carissa*: an ethnopharmacological, phytochemical and pharmacological review. *Natural Products and Bioprospecting* **7**, 181–199. doi:10.1007/s13659-017-0123-0
166. Shen L-R, Jin S-M, Yin B-W, Du X-F, Wang Y-L *et al.* (2007) Chemical constituents of plants from the genus *Cerbera*. *Chemistry & Biodiversity* **4**, 1438–1449. doi:10.1002/cbdv.200790122

167. Han L, Zhou X, Yang M, Zhou L, Deng X *et al.* (2018) Ethnobotany, phytochemistry and pharmacological effects of plants in genus *Cynanchum* Linn. (Asclepiadaceae). *Molecules (Basel, Switzerland)* **23**, 1194. doi:10.3390/molecules23051194
168. Wang P, Yang J, Zhu Z, Zhang X (2018) *Marsdenia tenacissima*: a review of traditional uses, phytochemistry and pharmacology. *The American Journal of Chinese Medicine* **46**, 1449–1480. doi:10.1142/S0192415X18500751
169. Rasool N, Ahmad VU, Malik A (1991) Terpenoids from *Pentatropis spiralis. Phytochemistry* **30**, 1331–1332. doi:10.1016/S0031-9422(00)95230-0
170. Ali M, Bhutani KK (1989) Alkaloids from *Tylophora indica. Phytochemistry* **28**, 3513–3517. doi:10.1016/0031-9422(89)80376-0
171. Abe F, Iwase Y, Yamauchi T, Honda K, Hayashi N (1995) Phenathroindolizidine alkaloids from *Tylophora tanakae. Phytochemistry* **39**, 695–699. doi:10.1016/0031-9422(95)00034-5
172. Gurav S, Senthilkumar DGP, Tembare R, Mani T (2011) *Tylophora indica*: a review on its ethnobotany, phytochemical and pharmacological profile. *Asian Journal of Biochemical and Pharmaceutical Research* **3**, 405–414.
173. Zhao JY, Wang XJ, Liu Z, Meng FX, Sun SF *et al.* (2019) Nonadride and spirocyclic anhydride derivatives from the plant endophytic fungus *Talaromyces purpurogenus. Journal of Natural Products* **82**, 2953–2962. doi:10.1021/acs.jnatprod.9b00210
174. Shahat AA, Apers S, Van Miert S, Claeys M, Pieters L *et al.* (2001) Structure elucidation of three new acetylated flavonoid glycosides from *Centaurium spicatum. Magnetic Resonance in Chemistry* **39**, 625–629. doi:10.1002/mrc.900
175. Bibi H, Ali I, Sadozai SK, Atta Ur R (2006) Phytochemical studies and antibacterial activity of *Centaurium pulchellum* Druce. *Natural Product Research* **20**, 896–901. doi:10.1080/14786410500162047
176. Aberham A, Pieri V, Croom EM Jr, Ellmerer E, Stuppner H (2011) Analysis of iridoids, secoiridoids and xanthones in *Centaurium erythraea, Frasera caroliniensis* and *Gentiana lutea* using LC-MS and RP-HPLC. *Journal of Pharmaceutical and Biomedical Analysis* **54**, 517–525. doi:10.1016/j.jpba.2010.09.030
177. El-Shanawany MA, Mohamed GA, Nafady AM, Ibrahim SRM, Radwan MM *et al.* (2011) A new xanthone from the roots of *Centaurium spicatum. Phytochemistry Letters* **4**, 126–128. doi:10.1016/j.phytol.2010.12.008
178. Itoh A, Tanaka Y, Nagakura N, Nishi T, Tanahashi T (2006) A quinic acid ester from *Strychnos lucida. Journal of Natural Medicines* **60**, 146–148. doi:10.1007/s11418-005-0021-3
179. Galliford CV, Scheidt KA (2007) Pyrrolidinyl-spirooxindole natural products as inspirations for the development of potential therapeutic agents. *Angewandte Chemie International Edition* **46**, 8748–8758. doi:10.1002/anie.200701342
180. Tchinda AT, Tamze V, Ngono ARN, Ayimele GA, Cao M *et al.* (2012) Alkaloids from the stem bark of *Strychnos icaja. Phytochemistry Letters* **5**, 108–113. doi:10.1016/j.phytol.2011.11.004
181. Guo R, Wang T, Zhou G, Xu M, Yu X *et al.* (2018) Botany, phytochemistry, pharmacology and toxicity of *Strychnos nux-vomica* L.: a review. *The American Journal of Chinese Medicine* **46**, 1–23. doi:10.1142/S0192415X18500015
182. Li W, Tang GH, Chen L, Tang YQ, Xu YK *et al.* (2018) New pyridocarbazole alkaloids from *Strychnos nitida. Natural Product Research* **32**, 1532–1536. doi:10.1080/14786419.2017.1385016
183. Pawlus AD, Kinghorn AD (2007) Review of the ethnobotany, chemistry, biological activity and safety of the botanical dietary supplement *Morinda citrifolia* (noni)*. *The Journal of Pharmacy and Pharmacology* **59**, 1587–1609. doi:10.1211/jpp.59.12.0001
184. Potterat O, Hamburger M (2007) *Morinda citrifolia* (noni) fruit: phytochemistry, pharmacology, safety. *Planta Medica* **73**, 191–199. doi:10.1055/s-2007-967115
185. Zhang JH, Xin HL, Xu YM, Shen Y, He YQ *et al.* (2018) *Morinda officinalis* How.: a comprehensive review of traditional uses, phytochemistry and pharmacology. *Journal of Ethnopharmacology* **213**, 230–255. doi:10.1016/j.jep.2017.10.028
186. Boucherle B, Haudecoeur R, Queiroz EF, De Waard M, Wolfender JL *et al.* (2016) *Nauclea latifolia*: biological activity and alkaloid phytochemistry of a West African tree. *Natural Product Reports* **33**, 1034–1043. doi:10.1039/C6NP00039H
187. Haudecoeur R, Peuchmaur M, Peres B, Rome M, Taiwe GS *et al.* (2018) Traditional uses, phytochemistry and pharmacological properties of African *Nauclea* species: a review. *Journal of Ethnopharmacology* **212**, 106–136. doi:10.1016/j.jep.2017.10.011
188. Luan F, Han K, Li M, Zhang T, Liu D *et al.* (2019) Ethnomedicinal uses, phytochemistry, pharmacology, and toxicology of species from the genus *Ajuga* L.: a systematic review. *The American Journal of Chinese Medicine* **47**, 959–1003. doi:10.1142/S0192415X19500502
189. Tu Y, Sun L, Guo M, Chen W (2013) The medicinal uses of *Callicarpa* L. in traditional Chinese medicine: an ethnopharmacological, phytochemical and pharmacological review. *Journal of Ethnopharmacology* **146**, 465–481. doi:10.1016/j.jep.2012.12.051
190. Wang JH, Luan F, He XD, Wang Y, Li MX (2018) Traditional uses and pharmacological properties of *Clerodendrum* phytochemicals. *Journal of Traditional and Complementary Medicine* **8**, 24–38. doi:10.1016/j.jtcme.2017.04.001
191. Mimica-Dukic N, Bozin B (2008) *Mentha* L. species (Lamiaceae) as promising sources of bioactive secondary metabolites. *Current Pharmaceutical Design* **14**, 3141–3150. doi:10.2174/138161208786404245
192. Singh D, Chaudhuri PK (2018) A review on phytochemical and pharmacological properties of Holy basil (*Ocimum sanctum* L.). *Industrial Crops and Products* **118**, 367–382. doi:10.1016/j.indcrop.2018.03.048
193. Abdel-Mogib M, Albar HA, Batterjee SM (2002) Chemistry of the genus *Plectranthus. Molecules (Basel, Switzerland)* **7**, 271–301. doi:10.3390/70200271

194. Southwell IA, Tucker DJ (1993) Cis-dihydroagarofuran from *Prostanthera* sp. aff. *ovalifolia*. *Phytochemistry* **33**, 857–862. doi:10.1016/0031-9422(93)85290-8
195. Dellar JE, Cole MD, Gray AI, Gibbons S, Waterman PG (1994) Antimicrobial sesquiterpenes from *Prostanthera* aff. *melissifolia* and *P. rotundifolia*. *Phytochemistry* **36**, 957–960. doi:10.1016/S0031-9422(00)90471-0
196. Kisiel W, Piozzi F (1999) Phenylethanoid glycosides from *Prostanthera melissifolia*. *Phytochemistry* **51**, 1083–1085. doi:10.1016/S0031-9422(99)00167-3
197. Russell M, Southwell IA, Waterman PG (2001) α-Phellandren-8-ol, a rare monoterpenene alcohol from the oil of *Prostanthera staurophylla* F. Muell. *The Journal of Essential Oil Research* **13**, 446–447. doi:10.1080/10412905.2001.9699723
198. Collins TL, Jones GL, Sadgrove NJ (2014) Volatiles from the rare Australian desert plant *Prostanthera centralis* B.J.Conn (Lamiaceae): chemical composition and antimicrobial activity. *Agriculture* **4**, 308–316. doi:10.3390/agriculture4040308
199. Tang KSC, Konczak I, Zhao J (2017) Phenolic compounds of the Australian native herb *Prostanthera rotundifolia* and their biological activities. *Food Chemistry* **233**, 530–539. doi:10.1016/j.foodchem.2017.04.137
200. Bai Y, Xia B, Xie W, Zhou Y, Xie J *et al.* (2016) Phytochemistry and pharmacological activities of the genus *Prunella*. *Food Chemistry* **204**, 483–496. doi:10.1016/j.foodchem.2016.02.047
201. Ulubelen A, Topcu G, Sonmez U (2000) Chemical and biological evaluation of genus *Teucrium*. *Studies in Natural Products Chemistry* **23**, 591–648. doi:10.1016/S1572-5995(00)80139-8
202. Bahramikia S, Yazdanparast R (2012) Phytochemistry and medicinal properties of *Teucrium polium* L. (Lamiaceae). *Phytotherapy Research* **26**, 1581–1593. doi:10.1002/ptr.4617
203. Filho JGS, Duringer J, Maia GLA, Tavares JF, Xavier HS *et al.* (2008) Ecdysteroids from *Vitex* species: distribution and compilation of their 13C-NMR spectral data. *Chemistry & Biodiversity* **5**, 707–713. doi:10.1002/cbdv.200890067
204. Rani A, Sharma A (2013) The genus *Vitex*: a review. *Pharmacognosy Reviews* **7**, 188–198. doi:10.4103/0973-7847.120522
205. Yao JL, Fang SM, Liu R, Oppong MB, Liu EW *et al.* (2016) A review on the terpenes from genus *Vitex*. *Molecules (Basel, Switzerland)* **21**, 1179. doi:10.3390/molecules21091179
206. Xie X, Yoneyama K, Yoneyama K (2010) The strigolactone story. *Annual Review of Phytopathology* **48**, 93–117. doi:10.1146/annurev-phyto-073009-114453
207. Koua FHM (2011) *Striga hermonthica* (Del.) Benth: phytochemistry and pharmacological properties outline. *Journal of Applied Pharmaceutical Science* **1**, 1–5.
208. Koua FHM, Babiker HA, Halfawi A, Ibrahim RO, Abbas FM *et al.* (2011) Phytochemical and biological study of *Striga hermonthica* (Del.) Betnh callus and intact plant. *Research in Pharmaceutical Biotechnology* **3**, 85–92.
209. Piovano M, Garbarino J, Tomassini L, Nicoletti M (2009) Cyclohexanones from *Mimulus glabratus* and *M. luteus*. *Natural Product Communications* **4**, 1637–1638. doi:10.1177/1934578X0900401204
210. Holeski LM, Keefover-Ring K, Bowers MD, Harnenz ZT, Lindroth RL (2013) Patterns of phytochemical variation in *Mimulus guttatus* (yellow monkeyflower). *Journal of Chemical Ecology* **39**, 525–536. doi:10.1007/s10886-013-0270-7
211. Russo A, Borrelli F (2005) Bacopa monniera, a reputed nootropic plant: an overview. *Phytomedicine* **12**, 305–317. doi:10.1016/j.phymed.2003.12.008
212. Tomas-Barberan FA, Grayer-Barkmeijer RJ, Gil MI, Harborne JB (1988) Distribution of 6-hydroxy-, 6-methoxy- and 8-hydroxyflavone glycosides in the Labiatae, the Scrophulariaceae and related families. *Phytochemistry* **27**, 2631–2645. doi:10.1016/0031-9422(88)87034-1
213. Grayer-Barkmeijer RJ, Tomas-Barberan FA (1993) 8-hydroxylated flavone *O*-glycosides and other flavonoids in chemotypes of *Gratiola officinalis*. *Phytochemistry* **34**, 205–210. doi:10.1016/S0031-9422(00)90806-9
214. Hayashi T, Kishi MP, Kawasaki M, Arisawa M, Shimizu M *et al.* (1987) Scopadulcic acid -A and -B, new diterpenoids with a novel skeleton, from a Paraguayan crude drug "Typycha Kuratu" (*Scoparia dulcis* L.). *Tetrahedron Letters* **28**, 3693–3696. doi:10.1016/S0040-4039(00)96357-3
215. Kawasaki M, Hayashi T, Arisawa M, Morita N, Berganza LH (1988) 8-hydroxytricetin 7-glucuronide, a B-glucuronidase inhibitor from *Scoparia dulcis*. *Phytochemistry* **27**, 3709–3711. doi:10.1016/0031-9422(88)80811-2
216. Hayashi T, Kawasaki M, Miwa Y, Taga T, Morita N (1990) Antiviral agents of plant origin. III: scopadulin, a novel tetracyclic diterpene from *Scoparia dulcis* L. *Chemical & Pharmaceutical Bulletin* **38**, 945–947. doi:10.1248/cpb.38.945
217. Mishra MR, Behera RK, Jha S, Panda AK, Mishra A *et al.* (2011) A brief review on phytoconstituents and ethnopharmacology of *Scoparia dulcis* Linn. (Scrophulariaceae). *International Journal of Phytotherapy and Phytopharmacology* **3**, 422–438.
218. Manchand PS, White JD, Wright H, Clardy J (1973) Structures of stemodin and stemodinone. *Journal of the American Chemical Society* **95**, 2705–2706. doi:10.1021/ja00789a060
219. Chamy MC, Piovano M, Barbarino JA, Gambaro V (1991) Stemodane diterpenoids from *Stemodia chilensis*. *Phytochemistry* **30**, 1719–1721. doi:10.1016/0031-9422(91)84245-N
220. Hufford CD, Oguntimein BO, Muhammad I (1992) New stemodane diterpenes from *Stemodia maritima*. *Journal of Natural Products* **55**, 48–52. doi:10.1021/np50079a007
221. Allen JG, Colegate SM, Mitchell AA, Mulder RJ, Raisbeck MF (2006) The bioactivity-guided isolation and structural identification of toxic cucurbitacin steroidal glucosides from *Stemodia kingii*. *Phytochemical Analysis* **17**, 226–235. doi:10.1002/pca.914

222. Rodrigues FEA, Lima JQ, de Oliveir MCF, Vasconcelos JN, Santiago GMP *et al.* (2010) Diterpene and other constituents from *Stemodia maritima* (Scrophulariaceae). *Journal of the Brazilian Chemical Society* **21**, 1581–1586. doi:10.1590/S0103-50532010000800025
223. Leonelli F, Mostarda A, De Angelis L, Lamba D, Demitri N *et al.* (2016) Proof of the structure of the *Stemodia chilensis* tetracyclic diterpenoid (+)-19-acetoxystemodan-12-ol by synthesis from (+)-podocarpic acid: X-ray structure determination of a key intermediate. *Journal of Natural Products* **79**, 1155–1159. doi:10.1021/acs.jnatprod.5b00834
224. Singab AN, Youssef FS, Ashour ML, Wink M (2013) The genus *Eremophila* (Scrophulariaceae): an ethnobotanical, biological and phytochemical review. *The Journal of Pharmacy and Pharmacology* **65**, 1239–1279. doi:10.1111/jphp.12092
225. Blackburne ID, Park RJ, Sutherland MD (1971) Terpenoid chemistry XVIII: myodesmone and isomyodesmone, toxic furanoid ketones from *Myoporum deserti* and *M. acuminatum*. *Australian Journal of Chemistry* **24**, 995–1007. doi:10.1071/CH9710995
226. Blackburne ID, Park RJ, Sutherland MD (1972) Terpenoid chemistry XX: myoporone and dehydromyoporone, toxic furanoid ketones from *Myoporum* and *Eremophila* species. *Australian Journal of Chemistry* **25**, 1787–1796. doi:10.1071/CH9721787
227. Metra PL, Sutherland MD (1983) Further skeletal variety in the toxic furanosesquiterpene ketones in the *Myoporum* genus. *Tetrahedron Letters* **24**, 1749–1752. doi:10.1016/S0040-4039(00)81761-X
228. Sutherland MD, Rodwell JL (1989) Terpenoid chemistry XXVIII: furanosesquiterpene B-ketols from *Myoporum betcheanum*, *M. deserti*, *M. montanum* and other Myoporaceae. *Australian Journal of Chemistry* **42**, 1995–2019. doi:10.1071/CH9891995
229. Menut C, Cabalion P, Hnawia E, Agnaniet H, Waikedre J *et al.* (2005) Two new furanosesquiterpenes from *Myoporum crassifolium* from New Caledonia. *Flavour and Fragrance Journal* **20**, 621–625. doi:10.1002/ffj.1509
230. Kanemoto M, Matsunami K, Otsuka H, Shinzato T, Ishigaki C *et al.* (2008) Chlorine-containing iridoid and iridoid glucoside, and other glucosides from leaves of *Myoporum bontioides*. *Phytochemistry* **69**, 2517–2522. doi:10.1016/j.phytochem.2008.07.002
231. Deepak M, Handa SS (1998) 3a,24-dihydroxy-urs-12-en-28-oic acid from *Verbena officinalis*. *Phytochemistry* **49**, 269–271. doi:10.1016/S0031-9422(97)01004-2
232. Deepak M, Handa SS (2000) Antiinflammatory activity and chemical composition of extracts of *Verbena officinalis*. *Phytotherapy Research* **14**, 463–465. doi:10.1002/1099-1573(200009)14:6<463::AID-PTR611>3.0.CO;2-G
233. Urzua AM, Cassels BK (1978) Alkaloid chemosystematics, chemotaxonomy and biogenesis in the Atherospermataceae. *Lloydia* **41**, 98–113.
234. Lakshmi V, Pandey K, Mishra SK, Srivastava S, Mishra M *et al.* (2009) An overview of family Hernandiaceae. *Records of Natural Products* **3**, 1–22.
235. Lai YW, Wang SW, Hu YY, Hwang TL, Cheng MJ *et al.* (2020) Anti-inflammatory alkaloids from the root bark of *Hernandia nymphaeifolia*. *Phytochemistry* **173**, 112326. doi:10.1016/j.phytochem.2020.112326
236. Chang F-R, Chao Y-C, Teng C-M, Wu Y-C (1998) Chemical constituents from *Cassytha filiformis* II. *Journal of Natural Products* **61**, 863–866.
237. Tsai T-H, Wang G-J, Lin L-C (2008) Vasorelaxing alkaloids and flavonoids from *Cassytha filiformis*. *Journal of Natural Products* **71**, 289–291. doi:10.1021/np070564h
238. Zhao J, Ma J-S (2016) Phytochemicals and biological activities of the genus *Cinnamomum*. *Journal of Pharmacognosy and Phytochemistry* **4**, 27–34.
239. Wang YS, Wen ZQ, Li BT, Zhang HB, Yang JH (2016) Ethnobotany, phytochemistry, and pharmacology of the genus *Litsea*: an update. *Journal of Ethnopharmacology* **181**, 66–107. doi:10.1016/j.jep.2016.01.032
240. Cox SD, Jayasinghe KC, Markham JL (2005) Antioxidant activity in Australian native sarsaparilla (*Smilax glyciphylla*). *Journal of Ethnopharmacology* **101**, 162–168. doi:10.1016/j.jep.2005.04.005
241. Wang WX, Li TX, Ma H, Zhang JF, Jia AQ (2013) Tumoral cytotoxic and antioxidative phenylpropanoid glycosides in *Smilax riparia* A. DC. *Journal of Ethnopharmacology* **149**, 527–532. doi:10.1016/j.jep.2013.07.011
242. Chen W, Shou X-A, Chen Y, Qin N, Qiao W *et al.* (2014) A new aurone from *Smilax riparia*. *Chemistry of Natural Compounds* **50**, 989–993. doi:10.1007/s10600-014-1143-1
243. Shu J, Liang F, Zhu G, Liu X, Yu J *et al.* (2017) Lignan glycosides from the rhizomes of *Smilax trinervula* and their biological activities. *Phytochemistry Letters* **20**, 1–8. doi:10.1016/j.phytol.2017.03.002
244. Tian LW, Zhang Z, Long HL, Zhang YJ (2017) Steroidal saponins from the genus *Smilax* and their biological activities. *Natural Products and Bioprospecting* **7**, 283–298. doi:10.1007/s13659-017-0139-5
245. Hua S, Zhang Y, Liu J, Dong L, Huang J *et al.* (2018) Ethnomedicine, phytochemistry and pharmacology of *Smilax glabra*: an important traditional Chinese medicine. *The American Journal of Chinese Medicine* **46**, 261–297. doi:10.1142/S0192415X18500143
246. Kalbhen DA (1971) Nutmeg as a narcotic: a contribution to the chemistry and pharmacology of nutmeg (*Myristica fragrans*). *Angewandte Chemie International Edition* **10**, 370–374. doi:10.1002/anie.197103701
247. Valderrama JCM (2000) Distribution of flavonoids in the Myristicaceae. *Phytochemistry* **55**, 505–511. doi:10.1016/S0031-9422(00)00114-X
248. Nagja T, Vimal K, Sanjeev A (2016) *Myristica fragrans*: a comprehensive review. *International Journal of Pharmacy and Pharmaceutical Sciences* **8**, 27–30.
249. Su X-H, Zhang M-L, Li L-G, Huo C-H, Gu Y-C *et al.* (2008) Chemical constituents of the plants of the genus *Calophyllum*. *Chemistry & Biodiversity* **5**, 2579–2608. doi:10.1002/cbdv.200890215
250. Filho VC, Meyre-Silva C, Niero R (2009) Chemical and pharmacological aspects of the genus *Calophyllum*.

Chemistry & Biodiversity **6**, 313–327. doi:10.1002/cbdv.200800082

251. Seebaluck R, Gurib-Fakim A, Mahomoodally F (2015) Medicinal plants from the genus Acalypha (Euphorbiaceae): a review of their ethnopharmacology and phytochemistry. *Journal of Ethnopharmacology* **159**, 137–157. doi:10.1016/j.jep.2014.10.040
252. Dilpreet K, Amandeep K, Jaswinder K (2018) A review of *Aleurites moluccana. World Journal of Pharmacy and Pharmaceutical Sciences* **7**, 341–358.
253. Bakker HJ, Ghisalberti EL, Jefferies PR (1972) Biosynthesis of diterpenes in *Beyeria leschenaultii. Phytochemistry* **11**, 2221–2231. doi:10.1016/S0031-9422(00)88381-8
254. Shi Q-W, Su X-H, Kiyota H (2008) Chemical and pharmacological research of the plants in genus *Euphorbia. Chemical Reviews* **108**, 4295–4327. doi:10.1021/cr078350s
255. Salatino A, Faria Salatino ML, Negri G (2007) Traditional uses, chemistry and pharmacology of *Croton* species (Euphorbiaceae). *Journal of the Brazilian Chemical Society* **18**, 11–33. doi:10.1590/S0103-50532007000100002
256. Gupta D, Bleakley B, Gupta RK (2008) Dragon's blood: botany, chemistry and therapeutic uses. *Journal of Ethnopharmacology* **115**, 361–380. doi:10.1016/j.jep.2007.10.018
257. Yin B-W, Shen L-R, Zhang M-L, Zhao L, Wang Y-L *et al.* (2008) Chemical constituents of plants from the genus *Excoecaria. Chemistry & Biodiversity* **5**, 2356–2371. doi:10.1002/cbdv.200890201
258. Rivière C, Nguyen Thi Hong V, Tran Hong Q, Chataigné G, Nguyen Hoai N *et al.* (2010) *Mallotus* species from Vietnamese mountainous areas: phytochemistry and pharmacological activities. *Phytochemistry Reviews* **9**, 217–253. doi:10.1007/s11101-009-9152-6
259. Gangwar M, Goel RK, Nath G (2014) *Mallotus philippinensis* Muell. Arg (Euphorbiaceae): ethnopharmacology and phytochemistry review. *BioMed Research International* **2014**, 213973.
260. Morikawa H, Kasai R, Otsuka H, Hirata E, Shinzato T *et al.* (2004) Terpenic and phenolic glycosides from the leaves of *Breynia officinalis* Hemsl. *Chemical & Pharmaceutical Bulletin* **52**, 1086–1090. doi:10.1248/cpb.52.1086
261. Meng D, Chen W, Zhao W (2007) Sulfur-containing spiroketal glycosides from *Breynia fructicosa. Journal of Natural Products* **70**, 824–829. doi:10.1021/np0606300
262. Raj D, Luczkiewicz M (2008) Securinega suffruticosa. *Fitoterapia* **79**, 419–427. doi:10.1016/j.fitote.2008.02.011
263. Wehlauch R, Gademann K (2017) Securinega alkaloids: complex structures, potent bioactivities, and efficient total syntheses. *Asian Journal of Organic Chemistry* **6**, 1146–1159. doi:10.1002/ajoc.201700142
264. Wang XF, Liu FF, Zhu Z, Fang QQ, Qu SJ *et al.* (2019) Flueggenoids A-E, new dinorditerpenoids from *Flueggea virosa. Fitoterapia* **133**, 96–101. doi:10.1016/j.fitote.2018.12.025
265. Huang Y-L, Chen C-C, Hsu F-L, Chen C-F (1996) A new lignan from *Phyllanthus virgatus. Journal of Natural Products* **59**, 520–521. doi:10.1021/np9603380
266. Calixto JB, Santos ARS, Filho VC, Yunes RA (1998) A review of the plants of the genus *Phyllanthus:* their chemistry pharmacology and therapeutic potential. *Medicinal Research Reviews* **18**, 225–258. doi:10.1002/(SICI)1098-1128(199807)18:4<225::AID-MED2>3.0.CO;2-X
267. Huang Y-L, Chen C-C, Hsu F-L, Chen C-F (1998) Tannins, flavonol sulfonates, and a norlignan from *Phyllanthus virgatus. Journal of Natural Products* **61**, 1194–1197. doi:10.1021/np970336v
268. Bagalkotkar G, Sagineedu SR, Saad MS, Stanslas J (2006) Phytochemicals from *Phyllanthus niruri* Linn. and their pharmacological properties: a review. *The Journal of Pharmacy and Pharmacology* **58**, 1559–1570. doi:10.1211/jpp.58.12.0001
269. Patel JR, Tripathi P, Sharma V, Chauhan NS, Dixit VK (2011) *Phyllanthus amarus*: ethnomedicinal uses, phytochemistry and pharmacology: a review. *Journal of Ethnopharmacology* **138**, 286–313. doi:10.1016/j.jep.2011.09.040
270. Nebula M, Harisankar HS, Chandramohanakumar N (2013) Metabolites and bioactivities of Rhizophoraceae mangroves. *Natural Products and Bioprospecting* **3**, 207–232. doi:10.1007/s13659-013-0012-0
271. Bibi SN, Fawzi MM, Gokhan Z, Rajesh J, Nadeem N *et al.* (2019) Ethnopharmacology, phytochemistry, and global distribution of mangroves - A comprehensive review. *Marine Drugs* **17**, 231.
272. Patel DK, Kumar R, Sairam K, Hemalatha S (2013) Hybanthus enneaspermus (L.) F. Muell: a concise report on its phytopharmacological aspects. *Chinese Journal of Natural Medicines* **11**, 199–206. doi:10.1016/S1875-5364(13)60017-5
273. Rajsekhar PB, Arvind Bharani RS, Angel J, Ramachandran M, Rajsekhar SPV (2016) *Hybanthus enneaspermus* (L) F. Muell: a phytopharmacological review on herbal medicine. *Journal of Chemical and Pharmaceutical Research* **8**, 351–355.
274. Goyal PK (2012) Phytochemical and pharmacological properties of the genus *Grewia*: a review. *International Journal of Pharmacy and Pharmaceutical Sciences* **4**, 72–78.
275. Ullah W, Uddin G, Siddiqui BS (2012) Ethnic uses, pharmacological and phytochemical profile of genus *Grewia. Journal of Asian Natural Products Research* **14**, 186–195. doi:10.1080/10286020.2011.639764
276. Zia-Ul-Haq M, Stankovic MS, Rizwan K, Feo VD (2013) *Grewia asiatica* L., a food plant with multiple uses. *Molecules (Basel, Switzerland)* **18**, 2663–2682. doi:10.3390/molecules18032663
277. Wu P-L, Chuang T-H, He C-X, Wu T-S (2004) Cytotoxicity of phenylpropanoid esters from the stems of *Hibiscus taiwanensis. Bioorganic & Medicinal Chemistry* **12**, 2193–2197. doi:10.1016/j.bmc.2004.02.020
278. Wu P-L, Wu T-S, He C-X, Su C-H, Lee K-H (2005) Constituents from the stems of *Hibiscus taiwanensis. Chemical & Pharmaceutical Bulletin* **53**, 56–59. doi:10.1248/cpb.53.56

279. Vasudeva N, Sharma SK (2008) Biologically active compounds from the genus *Hibiscus*. *Pharmaceutical Biology* **46**, 145–153. doi:10.1080/13880200701575320
280. Maganha EG, Halmenschlager RD, Rosa RM, Henriques Ja P, Ramos ALLD *et al.* (2010) Pharmacological evidences for the extracts and secondary metabolites from plants of the genus *Hibiscus*. *Food Chemistry* **118**, 1-10. doi:10.1016/j.foodchem.2009.04.005
281. Gasparetto JC, Martins CA, Hayashi SS, Otuky MF, Pontarolo R (2012) Ethnobotanical and scientific aspects of *Malva sylvestris* L.: a millennial herbal medicine. *The Journal of Pharmacy and Pharmacology* **64**, 172–189. doi:10.1111/j.2042-7158.2011.01383.x
282. Sharifi-Rad J, Melgar-Lalanne G, Hernandez-Alvarez AJ, Taheri Y, Shaheen S *et al.* (2020) *Malva* species: insights on its chemical composition towards pharmacological applications. *Phytotherapy Research* **34**, 546–567. doi:10.1002/ptr.6550
283. Dinda B, Das N, Dinda S, Dinda M, Silsarma I (2015) The genus *Sida* L.: a traditional medicine: its ethnopharmacological, phytochemical and pharmacological data for commercial exploitation in herbal drugs industry. *Journal of Ethnopharmacology* **176**, 135–176. doi:10.1016/j.jep.2015.10.027
284. Al Muqarrabun LM, Ahmat N (2015) Medicinal uses, phytochemistry and pharmacology of family Sterculiaceae: a review. *European Journal of Medicinal Chemistry* **92**, 514–530. doi:10.1016/j.ejmech.2015.01.026
285. Thabet AA, Youssef FS, El-Shazly M, Singab ANB (2018) *Sterculia* and *Brachychiton*: a comprehensive overview on their ethnopharmacology, biological activities, phytochemistry and the role of their gummy exudates in drug delivery. *The Journal of Pharmacy and Pharmacology* **70**, 450–474. doi:10.1111/jphp.12876
286. Vadivel V, Sriram S, Brindha P (2016) Distribution of flavonoids among Malvaceae family members: a review. *International Journal of Green Pharmacy* **10**, s33–s45.
287. Hiteksha SP, Mamta BS (2017) *Thespesia populnea* Linn.: a review. *International Journal of Pharmaceutics* **4**, 1–5.
288. Reddy MVB (2017) Review on chemistry, biological activity and biosynthesis of naturally occurring compounds of cadalene type sesquiterpenoids from Malvaceae. *Der Pharma Chemica* **9**, 40–49.
289. Suvarna CHM, Sriya P, Pavan KA (2018) A review on phytochemical and pharmacological properties of *Thespesia populnea*. *Journal of Drug Delivery and Therapeutics* **8**, 1–4. doi:10.22270/jddt.v8i4.1718
290. Duh C-Y, Phoebe CHJ, Pezzuto JM, Kinghorn AD, Farnsworth NR (1986) Plant anticancer agents. XLII: cytotoxic constituents from *Wikstroemia elliptica*. *Journal of Natural Products* **49**, 706–709. doi:10.1021/np50046a031
291. Liao S-G, Wu Y, Yue J-M (2006) Lignans from *Wikstroemia hainanensis*. *Helvetica Chimica Acta* **89**, 73–80.
292. Wang G-C, Zhang X-L, Wang Y-F, Li G-Q, Ye W-C *et al.* (2012) Four new dilignans from the roots of *Wikstroemia indica*. *Chemical & Pharmaceutical Bulletin* **60**, 920–923. doi:10.1248/cpb.c12-00132
293. Fan Q, Jiang Y-P, Zhu D-Q, Xu W, Huang W-Q *et al.* (2018) Phenols from the rhizome of *Wikstroemia indica*. *Biochemical Systematics and Ecology* **78**, 59–62. doi:10.1016/j.bse.2018.04.004
294. Li SF, Jiao YY, Zhang ZQ, Chao JB, Jia J *et al.* (2018) Diterpenes from buds of *Wikstroemia chamaedaphne* showing anti-hepatitis B virus activities. *Phytochemistry* **151**, 17–25. doi:10.1016/j.phytochem.2018.01.021
295. Poornima V, Sharanya M, Jeyam M (2014) An ethnomedical, pharmacological and phytochemical review of *Ammannia baccifera* L. *World Journal of Pharmaceutical Research* **3**, 1771–1789.
296. Piwowarski JP, Granica S, Kiss AK (2015) *Lythrum salicaria* L.: underestimated medicinal plant from European traditional medicine: a review. *Journal of Ethnopharmacology* **170**, 226–250. doi:10.1016/j.jep.2015.05.017
297. Ghisalberti EL (1996) Bioactive acylphloroglucinol derivatives from *Eucalyptus* species. *Phytochemistry* **41**, 7–22. doi:10.1016/0031-9422(95)00484-X
298. Vuong QV, Chalmers AC, Bhuyan DJ, Bowyer MC, Scarlett CJ (2015) Botanical, phytochemical, and anticancer properties of the *Eucalyptus* species. *Chemistry & Biodiversity* **12**, 907–924. doi:10.1002/cbdv.201400327
299. Marsh KJ, Kulheim C, Blomberg SP, Thornhill AH *et al.* (2017) Genus-wide variation in foliar polyphenolics in eucalypts. *Phytochemistry* **144**, 197–207. doi:10.1016/j.phytochem.2017.09.014
300. Senadeera SPD, Duffy S, Avery VM, Carroll AR (2017) Antiplasmodial beta-triketones from the flowers of the Australian tree *Angophora woodsiana*. *Bioorganic & Medicinal Chemistry Letters* **27**, 2602–2607. doi:10.1016/j.bmcl.2017.03.065
301. Hodgson D, Ritchie E, Taylor WC (1960) Chemical studies of the Myrtaceae. II: the constituents of *Syncarpia laurifolia* Tenn. *Australian Journal of Chemistry* **13**, 385–389. doi:10.1071/CH9600385
302. Brophy JJ, Craig DC, Goldsack RJ, Fookes CJR (1994) Hillyl acetate, a keto-acetate from the leaf steam volatiles of *Syncarpia hillii*. *Phytochemistry* **37**, 1645–1647. doi:10.1016/S0031-9422(00)89584-9
303. Su Q, Dalal S, Goetz M, Cassera MB, Kingston DG (2016) Antiplasmodial phloroglucinol derivatives from *Syncarpia glomulifera*. *Bioorganic & Medicinal Chemistry* **24**, 2544–2548. doi:10.1016/j.bmc.2016.04.020
304. Baliga MS, Bhat HP, Baliga BRV, Wilson R, Palatty PL (2011) Phytochemistry, traditional uses and pharmacology of *Eugenia jambolana* Lam. (black plum): a review. *Food Research International* **44**, 1776–1789. doi:10.1016/j.foodres.2011.02.007
305. Ayyanar M, Subash-Babu P (2012) *Syzygium cumini* (L.) Skeels: a review of its phytochemical constituents and traditional uses. *Asian Pacific Journal of Tropical Biomedicine* **2**, 240–246. doi:10.1016/S2221-1691(12)60050-1
306. Cortés-Rojas DF, De Souza CRF, Oliveira WP (2014) Clove (*Syzygium aromaticum*): a precious spice. *Asian Pacific Journal of Tropical Biomedicine* **4**, 90–96. doi:10.1016/S2221-1691(14)60215-X

307. Bate-Smith EC, Harborne JB, Davenport SM (1966) Occurrence of azaleatin and caryatin in *Eucryphia*. *Nature* **212**, 1065–1066. doi:10.1038/2121065a0
308. Bate-Smith EC, Davenport SM, Harborne JB (1967) Comparative biochemistry of flavonoids. III: a correlation between chemistry and plant geography in the genus *Eucryphia*. *Phytochemistry* **6**, 1407–1413. doi:10.1016/S0031-9422(00)82883-6
309. Tschesche R, Delhvi S, Sepulveda S, Breitmaier E (1979) Eucryphin, a new chromone rhamnoside from the bark of *Eucryphia cordifolia*. *Phytochemistry* **18**, 867–869. doi:10.1016/0031-9422(79)80032-1
310. Sepulveda-Boza S, Delhvi S, Cassels BK (1993) Flavonoids from the twigs of *Eucryphia glutinosa*. *Phytochemistry* **32**, 1301–1303. doi:10.1016/S0031-9422(00)95110-0
311. Wollenweber E, Dorr M, Rozefelds AC, Minchin P, Forster PI (2000) Variation in flavonoid exudates in *Eucryphia* species from Australia and South America. *Biochemical Systematics and Ecology* **28**, 111–118. doi:10.1016/S0305-1978(99)00045-9
312. Nonato MG, Takayama H, Garson MJ (2008) *Pandanus* alkaloids: chemistry and biology. *The Alkaloids* **66**, 215–249.
313. Adkar PP, Bhaskar VH (2014) *Pandanus odoratissimus* (Kewda): a review on ethnopharmacology, phytochemistry, and nutritional aspects. *Advances in Pharmacological and Pharmaceutical Sciences* **2014**, 1–19. doi:10.1155/2014/120895
314. Aslam MS, Choudhary BA, Uzair M, Ijaz AS (2013) Phytochemical and ethno-pharmacological review of the genus *Araucaria* – review. *Tropical Journal of Pharmaceutical Research* **12**, 651–659. doi:10.4314/tjpr.v12i4.31
315. Simoneit BRT, Cox RE, Oros DR, Otto A (2018) Terpenoid compositions of resins from *Callitris* species (Cupressaceae). *Molecules (Basel, Switzerland)* **23**, 3384. doi:10.3390/molecules23123384
316. Sadgrove NJ, Jones GL (2014) Medicinal compounds, chemically and biologically characterised from extracts of Australian *Callitris endlicheri* and *C. glaucophylla* (Cupressaceae): used traditionally in Aboriginal and colonial pharmacopoeia. *Journal of Ethnopharmacology* **153**, 872–883. doi:10.1016/j.jep.2014.03.054
317. Parmar VS, Jain SC, Bisht KS, Jain R, Taneja P *et al.* (1997) Phytochemistry of the genus *Piper*. *Phytochemistry* **46**, 597–673. doi:10.1016/S0031-9422(97)00328-2
318. Xiang C-P, Shi Y-N, Liu F-F, Li H-Z, Zhang Y-J *et al.* (2016) A survey of the chemical compounds of *Piper* spp. (Piperaceae) and their biological activities. *Natural Product Communications* **11**, 1403–1408. doi:10.1177/1934578X1601100948
319. Peerzada AM, Ali HH, Naeem M, Latif M, Bukhari AH *et al.* (2015) *Cyperus rotundus* L.: traditional uses, phytochemistry, and pharmacological activities. *Journal of Ethnopharmacology* **174**, 540–560. doi:10.1016/j.jep.2015.08.012
320. Allan RD, Correll RL, Wells RJ (1969) Two new phenolic ketones from *Remirea maritima* (Cyperaceae). *Tetrahedron Letters* **10**, 4673–4674. doi:10.1016/S0040-4039(01)88779-7
321. Allan RD, Correll RL, Wells RJ (1969) A new class of quinones from certain members of the family Cyperaceae. *Tetrahedron Letters* **10**, 4669–4672. doi:10.1016/S0040-4039(01)88778-5
322. Allan RD, Wells RJ, Macleod JK (1970) Further phenolic ketones from *Remirea maritima* Aubl. *Tetrahedron Letters* **11**, 3945–3946. doi:10.1016/S0040-4039(01)98632-0
323. Rabelo AS, Oliveira ID, Guimaraes AG, Quintans JS, Prata AP *et al.* (2013) Antinociceptive, anti-inflammatory and antioxidant activities of aqueous extract from *Remirea maritima* (Cyperaceae). *Journal of Ethnopharmacology* **145**, 11–17. doi:10.1016/j.jep.2012.10.020
324. Peng D, Lin X, Jiang L, Huang J, Zeng G *et al.* (2019) Five macrocyclic glycosides from *Schoenoplectus tabernaemontani*. *Natural Product Research* **33**, 427–434. doi:10.1080/14786419.2018.1455195
325. Williams CA, Harborne JB, Clifford HT (1971) Flavonoid patterns in the monocotyledons: flavanols and flavones in some families associated with the Poaceae. *Phytochemistry* **10**, 1059–1063. doi:10.1016/S0031-9422(00)89939-2
326. Avoseh O, Oyedeji O, Rungqu P, Nkeh-Chungag B, Oyedeji A (2015) *Cymbopogon* species; ethnopharmacology, phytochemistry and the pharmacological importance. *Molecules (Basel, Switzerland)* **20**, 7438–7453. doi:10.3390/molecules20057438
327. Agnew MP, Lauren DR (1991) Determination of ptaquiloside in bracken fern (*Pteridium esculentum*). *Journal of Chromatography. A* **538**, 462–468. doi:10.1016/S0021-9673(01)88870-1
328. Tanaka N, Yuhara H, Wada H, Murakami T, Cambie RC *et al.* (1993) Phenolic constituents of *Pteridium esculentum*. *Phytochemistry* **32**, 1037–1039. doi:10.1016/0031-9422(93)85251-L
329. Yamada K, Ojika M, Kigoshi H (2007) Ptaquiloside, the major toxin of bracken, and related terpene glycosides: chemistry, biology and ecology. *Natural Product Reports* **24**, 798–813. doi:10.1039/b614160a
330. Fletcher MT, Hayes PY, Somerville MJ, De Voss JJ (2010) Ptesculentoside, a novel norsesquiterpene glucoside from the Australian bracken fern *Pteridium esculentum*. *Tetrahedron Letters* **51**, 1997–1999. doi:10.1016/j.tetlet.2010.02.032
331. Pan C, Chen YG, Ma XY, Jiang JH, He F *et al.* (2011) Phytochemical constituents and pharmacological activities of plants from the genus *Adiantum*: a review. *Tropical Journal of Pharmaceutical Research* **10**, 681–692. doi:10.4314/tjpr.v10i5.18
332. Sharma BR, Gautam LN, Adhikari D, Karki R (2017) A comprehensive review on chemical profiling of *Nelumbo nucifera*: potential for drug development. *Phytotherapy Research* **31**, 3–26. doi:10.1002/ptr.5732
333. Ridley DD, Ritchie E, Taylor WC (1970) Chemical studes of the Proteaceae. IV: the structures of the major phenols of *Grevillea striata*; a group of novel cyclophanes.

Australian Journal of Chemistry **23**, 147–183. doi:10.1071/CH9700147

334. Cannon JR, Chow PW, Fuller MW, Hamilton BH, Metcalf BW *et al.* (1973) Phenolic constituents of *Grevillea robusta* (Proteaceae): the structure of robustol, a novel macrocyclic phenol. *Australian Journal of Chemistry* **26**, 2257–2275. doi:10.1071/CH9732257
335. Ahmed AS, Nakamura N, Meselhy MR, Makhboul MA, El-Emary N *et al.* (2000) Phenolic constituents from *Grevillea robusta. Phytochemistry* **53**, 149–154. doi:10.1016/S0031-9422(99)00484-7
336. Castillo U, Harper JK, Strobel GA, Sears J, Alesi K *et al.* (2003) Kakadumycins, novel antibiotics from *Streptomyces* sp. NRRL 30566, an endophyte of *Grevillea pteridifolia. FEMS Microbiology Letters* **224**, 183–190. doi:10.1016/S0378-1097(03)00426-9
337. Chuang T-H, Wu P-L (2007) Cytotoxic 5-alkylresorcinol metabolites from the leaves of *Grevillea robusta. Journal of Natural Products* **70**, 319–323. doi:10.1021/np0605687
338. Ovenden SPB, Cobbe M, Kissell R, Birrell GW, Chavchich M *et al.* (2011) Phenolic glycosides with antimalarial activity from *Grevillea* "Poorinda Queen". *Journal of Natural Products* **74**, 74–78. doi:10.1021/np100737q
339. Manju M, Varma RS, Parthasarathy MR (1977) New arbutin derivatives from leaves of *Grevillea robusta* and *Hakea saligna. Phytochemistry* **16**, 793–794. doi:10.1016/S0031-9422(00)89268-7
340. Lytollis W, Scannell RT, An H, Murty VS, Sambi Reddy K *et al.* (1995) 5-Alkylresorcinols from *Hakea trifurcata* that cleave DNA. *Journal of the American Chemical Society* **117**, 12683–12690. doi:10.1021/ja00156a004
341. Cannon JR, Metcalf BW (1971) Phenolic constituents of *Persoonia elliptica* (Proteaceae). *Australian Journal of Chemistry* **24**, 1925–1931. doi:10.1071/CH9711925
342. MacLeod JK, Rasmussen HB, Willis AC (1997) A new glycoside antimicrobial agent from *Persoonia linearis* x *pinifolia. Journal of Natural Products* **60**, 620–622. doi:10.1021/np970006a
343. Bieleski RL, Briggs BC (2005) Taxonomic patterns in the distribution of polyols within the Proteaceae. *Australian Journal of Botany* **53**, 205–217. doi:10.1071/BT04098
344. Deans BJ, Kilah NL, Jordan GJ, Bissember AC, Smith JA (2018) Arbutin derivatives isolated from ancient Proteaceae: potential phytochemical markers present in *Bellendena, Cenarrhenes,* and *Persoonia* genera. *Journal of Natural Products* **81**, 1241–1251. doi:10.1021/acs.jnatprod.7b01038
345. Wallace JW, Markham KR (1978) Apigenin and amentoflavone glycosides in the Psilotaceae and their phylogenetic significance. *Phytochemistry* **17**, 1313–1317. doi:10.1016/S0031-9422(00)94580-1
346. Leete E, Muir A, Neil Towers GH (1982) Biosynthesis of psilotin from [2′,3′-13C2,1′-14C,4-3H]phenylalanine studied with 13C-NMR. *Tetrahedron Letters* **23**, 2635–2638. doi:10.1016/S0040-4039(00)87416-X
347. Markham KR (1984) The structures of amentoflavone glycosides isolated from *Psilotum nudum. Phytochemistry* **23**, 2053–2056. doi:10.1016/S0031-9422(00)84969-9
348. Yamazaki Y, Suh DY, Sitthithaworn W, Ishiguro K, Kobayashi Y *et al.* (2001) Diverse chalcone synthase superfamily enzymes from the most primitive vascular plant, *Psilotum nudum. Planta* **214**, 75–84. doi:10.1007/s004250100586
349. Samec D, Pierz V, Srividya N, Wust M, Lange BM (2019) Assessing chemical diversity in *Psilotum nudum* (L.) Beauv., a pantropical whisk fern that has lost many of its fern-like characters. *Frontiers in Plant Science* **10**, 868. doi:10.3389/fpls.2019.00868
350. Semwal DK, Semwal RB, Vermaak I, Viljoen A (2014) From arrow poison to herbal medicine: the ethnobotanical, phytochemical and pharmacological significance of *Cissampelos* (Menispermaceae). *Journal of Ethnopharmacology* **155**, 1011–1028. doi:10.1016/j.jep.2014.06.054
351. Hungerford NL, Sands DPA, Kitching W (1998) Isolation and structure of some constituents of the Australian medicinal plant *Tinospora smilacina* ('snakevine'). *Australian Journal of Chemistry* **51**, 1103–1111. doi:10.1071/C98034
352. Pathak AK, Jain DC, Sharma RP (2008) Chemistry and biological activities of the genera *Tinospora. International Journal of Pharmacology* **33**, 277–287.
353. Singh D, Chaudhuri PK (2017) Chemistry and pharmacology of *Tinospora cordifolia. Natural Product Communications* **12**, 299–308. doi:10.1177/1934578X1701200240
354. Chawla R, Kumar S, Sharma A (2012) The genus *Clematis* (Ranunculaceae): chemical and pharmacological perspectives. *Journal of Ethnopharmacology* **143**, 116–150. doi:10.1016/j.jep.2012.06.014
355. Ahmed F, Urooj A (2010) Traditional uses, medicinal properties, and phytopharmacology of *Ficus racemosa*: a review. *Pharmaceutical Biology* **48**, 672–681. doi:10.3109/13880200903241861
356. Singh D, Singh B, Goel RK (2011) Traditional uses, phytochemistry and pharmacology of *Ficus religiosa*: a review. *Journal of Ethnopharmacology* **134**, 565–583. doi:10.1016/j.jep.2011.01.046
357. Muhammad D, Lalun N, Bobichon H, Le Magrex Debar E, Gangloff SC *et al.* (2016) Triterpenoids from the leaves of Alphitonia xerocarpus Baill and their biological activity. *Phytochemistry* **129**, 45–57. doi:10.1016/j.phytochem.2016.07.005
358. Raju R, Gunawardena D, Ahktar MA, Low M, Reddell P *et al.* (2016) Anti-inflammatory chemical profiling of the Australian rainforest tree Alphitonia petriei (Rhamnaceae). *Molecules (Basel, Switzerland)* **21**, 1521. doi:10.3390/molecules21111521
359. Cooke RG, Johnson BL (1963) Colouring matters of Australian plants. XI: quinones from *Ventilago viminalis. Australian Journal of Chemistry* **16**, 695–702. doi:10.1071/CH9630695
360. Rao BK, Hanumaiah T, Rao CP, Rao GSR, Rao KVJ *et al.* (1983) Anthraquinones in *Ventilago* species. *Phytochemistry* **22**, 2583–2585. doi:10.1016/0031-9422(83)80170-8

361. Hanumaiah T, Marshall DS, Rao BK, Rao CP, Rao GSR *et al.* (1985) Benzoisochromanquinones in *Ventilago* species. *Phytochemistry* **24**, 2373–2378. doi:10.1016/S0031-9422(00)83044-7
362. Hanumaiah T, Rao BK, Rao CP, Rao GSR, Rao JUM *et al.* (1985) Naphthalenes and naphthoquinones from *Ventilago* species. *Phytochemistry* **24**, 1811–1815. doi:10.1016/S0031-9422(00)82557-1
363. Jammula SR, Pepalla SB, Hanumaiah T, Rao KVJ, Thomson RH (1991) Benzisochromanquinones from *Ventilago goughii*. *Phytochemistry* **30**, 3741–3744. doi:10.1016/0031-9422(91)80100-F
364. Lin L-C, Chou C-J, Kuo Y-C (2001) Cytotoxic principles from *Ventilago leiocarpa*. *Journal of Natural Products* **64**, 674–676. doi:10.1021/np000569d
365. Periyasamy K, Kaliyaperumal S (2016) Ethnobotanical, phytochemical and pharmaceutical studies of medicinal plant, *Ventilago maderaspatana* Gaertn. (red creeper): a review. *International Journal of Current Pharmaceutical Research* **8**, 16–18.
366. Cassels BK, Eckhardt G, Kaussmann EU, Tschesche R (1974) Cyclopeptide alkaloids of *Zizyphus oenoplia*. *Tetrahedron* **30**, 2461–2466. doi:10.1016/S0040-4020(01)97117-0
367. Mahajan RT, Chopda MZ (2009) Phytopharmacology of *Ziziphus jujuba* Mill.: a plant review. *Pharmacognosy Reviews* **3**, 320–329.
368. Gao QH, Wu CS, Wang M (2013) The jujube (*Ziziphus jujuba* Mill.) fruit: a review of current knowledge of fruit composition and health benefits. *Journal of Agricultural and Food Chemistry* **61**, 3351–3363. doi:10.1021/jf4007032
369. Prabhavathi S, Vajayalakshmi S (2015) New hydroxycarboxylic acid from the root bark of *Ziziphus oenoplia*. *Journal of Pharmacognosy and Phytochemistry* **3**, 150–152.
370. Rocabado GO, Bedoya LM, Abad MJ, Bermejo P (2008) *Rubus*: a review of its phytochemical and pharmacological profile. *Natural Product Communications* **3**, 423–436. doi:10.1177/1934578X0800300319
371. Lee J, Dossett M, Finn CE (2012) *Rubus* fruit phenolic research: the good, the bad, and the confusing. *Food Chemistry* **130**, 785–796. doi:10.1016/j.foodchem.2011.08.022
372. Le NH, Malterud KE, Diallo D, Paulsen BS, Nergard CS, *et al.* (2012) Bioactive polyphenols in *Ximenia americana* and the traditional use among Malian healers. *Journal of Ethnopharmacology* **139**, 858–862. doi:10.1016/j.jep.2011.12.031
373. Maroyi A (2016) *Ximenia caffra* Sond. (Ximeniaceae) in sub-Saharan Africa: a synthesis and review of its medicinal potential. *Journal of Ethnopharmacology* **184**, 81–100. doi:10.1016/j.jep.2016.02.052
374. Baldovini N, Delasalle C, Joulain D (2011) Phytochemistry of the heartwood from fragrant *Santalum* species: a review. *Flavour and Fragrance Journal* **26**, 7–26. doi:10.1002/ffj.2025
375. Kumar R, Anjum N, Tripathi YC (2015) Phytochemistry and pharmacology of *Santalum album* L.: a review. *World Journal of Pharmaceutical Research* **4**, 1842–1876.
376. Oelrichs PB, Macleod JK, Seawright AA, Ng JC (1997) Isolation and characterisation of urushiol components from the Australian native cashew (*Semecarpus australiensis*). *Nature Toxicology* **5**, 96–98.
377. Premalatha B (2000) *Semecarpus anacardium* Linn. Nuts: a boon in alternative medicine. *Indian Journal of Experimental Biology* **38**, 1177–1182.
378. Nagabhushana KS, Umamaheshwari S, Tocoli FE, Prabhu SK, Green IR *et al.* (2002) Inhibition of soybean and potato lipoxygenases by bhilawanols from bhilawan (*Semecarpus anacardium*) nut shell liquid and some synthetic salicylic acid analogues. *Journal of Enzyme Inhibition and Medicinal Chemistry* **17**, 255–259. doi:10.1080/1475636021000006243
379. Semalty M, Semalty A, Badola A, Joshi GP, Rawat MS (2010) *Semecarpus anacardium* Linn.: a review. *Pharmacognosy Reviews* **4**, 88–94. doi:10.4103/0973-7847.65328
380. Mogana R, Wiart C (2011) *Canarium* L.: a phytochemical and pharmacological review. *Journal of Pharmacy Research* **4**, 2482–2489.
381. Negi JS, Bisht VK, Bandari AK, Bharti MK, Sundriyal RC (2011) Chemical and pharmacological aspects of *Toona* (Meliaceae). *Research Journal of Phytochemistry* **5**, 14–21. doi:10.3923/rjphyto.2011.14.21
382. Divakar PR (2017) Phytopharmacology of *Toona ciliata*: a review. *International Journal of Pharmaceutics* **8**, 30–35.
383. Peng W, Liu Y, Hu M, Zhang M, Yang J *et al.* (2019) *Toona sinensis*: a comprehensive review on its traditional usages, phytochemisty, pharmacology and toxicology. *Revista Brasileira de Farmacognosia-Brazilian Journal of Pharmacognosy* **29**, 111–124. doi:10.1016/j.bjp.2018.07.009
384. Almubayedh H, Ahmad R (2020) Ethnopharmacology, phytochemistry, biological activities, and therapeutic applications of *Cedrela serrata* Royle: a mini review. *Journal of Ethnopharmacology* **246**, 112206. doi:10.1016/j.jep.2019.112206
385. Zhao L, Huo C-H, Shen L-R, Yang Y, Zhang Q *et al.* (2010) Chemical constituents of plants from the genus *Melia*. *Chemistry & Biodiversity* **7**, 839–859. doi:10.1002/cbdv.200900043
386. Auzi AA, Hartley TG, Waterman PG (1997) Distribution of flavonoids, alkaloids, acetophenones and phloroglucinols in the genus *Bosistoa* (Rutaceae). *Biochemical Systematics and Ecology* **25**, 611–617. doi:10.1016/S0305-1978(97)00062-8
387. Simonsen HT, Adsersen A, Bremner P, Heinrich M, Wagner Smitt U *et al.* (2004) Antifungal constituents of *Melicope borbonica*. *Phytotherapy Research* **18**, 542–545. doi:10.1002/ptr.1482
388. Yamamoto M, Akita T, Koyama Y, Sueyoshi E, Matsunami K *et al.* (2008) Euodionosides A-G: megastigmane glucosides from leaves of *Euodia meliaefolia*. *Phytochemistry* **69**, 1586–1596. doi:10.1016/j.phytochem.2008.01.020

389. Zhao N, Li ZL, Li DH, Sun YT, Shan DT *et al.* (2015) Quinolone and indole alkaloids from the fruits of *Euodia rutaecarpa* and their cytotoxicity against two human cancer cell lines. *Phytochemistry* **109**, 133–139. doi:10.1016/j.phytochem.2014.10.020
390. Eliaser EM, Ho JH, Hashim NM, Rukayadi Y, Ee GCL *et al.* (2018) Phytochemical constituents and biological activities of *Melicope lunu-ankenda. Molecules (Basel, Switzerland)* **23**, 2708.
391. Fan S, Zhang C, Luo T, Wang J, Tang Y *et al.* (2019) Limonin: a review of its pharmacology, toxicity, and pharmacokinetics. *Molecules (Basel, Switzerland)* **24**, 3679. doi:10.3390/molecules24203679
392. Vu VT, Chen XL, Kong LY, Luo JG (2020) Melipatulinones A-C, three lignan-phloroglucinol hybrids from *Melicope patulinervia. Organic Letters* **22**, 1380–1384. doi:10.1021/acs.orglett.9b04680
393. Yao Q, Gao Y, Lai C, Wu C, Zhao CL *et al.* (2020) The phytochemistry, pharmacology and applications of *Melicope pteleifolia*: a review. *Journal of Ethnopharmacology* **251**, 112546. doi:10.1016/j.jep.2020.112546
394. Rashid MA, Armstrong JA, Gray AI, Waterman PG (1991) Pyranocoumarins as chemotaxonomic markers in *Eriostemon coccineus* and *Philotheca citrina. Phytochemistry* **30**, 4033–4035. doi:10.1016/0031-9422(91)83459-X
395. Ghisalberti EL (1998) Phytochemistry of the Australian Rutaceae: *Boronia*, *Eriostemon* and *Phebalium* species. *Phytochemistry* **47**, 163–176. doi:10.1016/S0031-9422(97)00571-2
396. Sultana N, Sarker SD, Armstrong JA, Wilson PG, Waterman PG (2003) The coumarins of *Philotheca* sensu lato: distribution and systematic significance. *Biochemical Systematics and Ecology* **31**, 681–691. doi:10.1016/S0305-1978(02)00226-0
397. Chlouchi A, Muyard F, Girard C, Waterman PG, Bévalot F (2005) Coumarins from the twigs of *Diplolaena mollis* P. G. Wilson (Rutaceae). *Biochemical Systematics and Ecology* **33**, 967–969. doi:10.1016/j.bse.2005.03.002
398. Chlouchi A, Girard C, Tillequin F, Bévalot F, Waterman PG *et al.* (2006) Coumarins and furoquinoline alkaloids from *Philotheca deserti* var. *deserti* (Rutaceae). *Biochemical Systematics and Ecology* **34**, 71–74. doi:10.1016/j.bse.2005.07.002
399. Chlouchi A, Girard C, Bévalot F, Waterman PG, Muyard F (2007) Taxonomically significant coumarins from three *Philotheca* species (Rutaceae). *Biochemical Systematics and Ecology* **35**, 251–254. doi:10.1016/j.bse.2006.09.011
400. Southwell IA (1981) Methoxystyrenes from the genus *Zieria. Phytochemistry* **20**, 1448–1450. doi:10.1016/0031-9422(81)80068-4
401. Flynn TM, Southwell IA (1987) Cyanogenesis in the genus *Zieria. Phytochemistry* **26**, 1669–1673. doi:10.1016/S0031-9422(00)82267-0
402. Flynn TM, Southwell IA (1987) Essential oil constituents of the genus *Zieria. Phytochemistry* **26**, 1673–1686. doi:10.1016/S0031-9422(00)82268-2
403. Southwell IA, Armstrong JA (1987) Chemical variation within the genus *Zieria. Phytochemistry* **26**, 1687–1692. doi:10.1016/S0031-9422(00)82269-4
404. Griffin SG, Leach DN, Markham JL, Johnstone R (1998) Antimicrobial activity of essential oils from *Zieria. The Journal of Essential Oil Research* **10**, 165–174. doi:10.1080/10412905.1998.9700870
405. Chavan RB, Gaikwad DK (2016) The ethnobotany, phytochemistry and biological properties of *Allophylus* species used in traditional medicine: a review. *World Journal of Pharmacy and Pharmaceutical Sciences* **5**, 664–682.
406. Ghisalberti EL (1998) Ethnopharmacology and phytochemistry of *Dodonaea* species. *Fitoterapia* **69**, 99–113.
407. kundu P, Laskar S (2010) A brief resume on the genus *Ailanthus*: chemical and pharmacological aspects. *Phytochemistry Reviews* **9**, 379–412. doi:10.1007/s11101-009-9157-1
408. Salehi B, Krochmal-Marczak B, Skiba D, Patra JK, Das SK *et al.* (2020) *Convolvulus* plant: a comprehensive review from phytochemical composition to pharmacy. *Phytotherapy Research* **34**, 315–328. doi:10.1002/ptr.6540
409. Austin DF (2008) *Evolvulus alsinoides* (Convolvulaceae): an American herb in the Old World. *Journal of Ethnopharmacology* **117**, 185–198. doi:10.1016/j.jep.2008.01.038
410. Meira M, Da Silva EP, David JM, David JP (2012) Review of the genus *Ipomoea*: traditional uses, chemistry and biological activities. *Revista Brasileira de Farmacognosia-Brazilian Journal of Pharmacognosy* **22**, 682–713. doi:10.1590/S0102-695X2012005000025
411. Ding W, Zeng F, Xu L, Chen Y, Wang Y *et al.* (2011) Bioactive dammarane-type saponins from *Operculina turpethum. Journal of Natural Products* **74**, 1868–1874. doi:10.1021/np200274m
412. Ding W, Jiang ZH, Wu P, Xu L, Wei X (2012) Resin glycosides from the aerial parts of *Operculina turpethum. Phytochemistry* **81**, 165 174. doi:10.1016/j.phytochem.2012.05.010
413. Gupta S, Ved A (2017) *Operculina turpethum* (Linn.) Silva Manso as a medicinal plant species: a review on bioactive components and pharmacological properties. *Pharmacognosy Reviews* **11**, 158–166.
414. Lira-Ricardez J, Pereda-Miranda R, Castaneda-Gomez J, Fragoso-Serrano M, Simas RC *et al.* (2019) Resin glycosides from the roots of *Operculina macrocarpa* (Brazilian Jalap) with purgative activity. *Journal of Natural Products* **82**, 1664–1677. doi:10.1021/acs.jnatprod.9b00222
415. Nugroho LH, Verpoorte R (2002) Secondary metabolism in tobacco. *Plant Cell, Tissue and Organ Culture* **68**, 105–125. doi:10.1023/A:1013853909494
416. Dewey RE, Xie J (2013) Molecular genetics of alkaloid biosynthesis in *Nicotiana* tabacum. *Phytochemistry* **94**, 10–27. doi:10.1016/j.phytochem.2013.06.002
417. Jassbi AR, Zare S, Asadollahi M, Schuman MC (2017) Ecological roles and biological activities of secialized

metabolites from the genus *Nicotiana*. *Chemical Reviews* **117**, 12227–12280. doi:10.1021/acs.chemrev.7b00001

418. Kaunda JS, Zhang YJ (2019) The genus *Solanum*: an ethnopharmacological, phytochemical and biological properties review. *Natural Products and Bioprospecting* **9**, 77–137. doi:10.1007/s13659-019-0201-6
419. Kumar D, Kumar S, Gupta J, Arya R, Gupta A (2011) A review on chemical and biological properties of *Cayratia trifolia* Linn. (Vitaceae). *Pharmacognosy Reviews* **5**, 184–188. doi:10.4103/0973-7847.91117
420. Stohs SJ, Ray SD (2013) A review and evaluation of the efficacy and safety of *Cissus quadrangularis* extracts. *Phytotherapy Research* **27**, 1107–1114. doi:10.1002/ptr.4846
421. Siddiqua A, Mittapally S (2017) A review of *Cissus quadrangularis*. *Journal of Pharmaceutical Innovation* **6**, 329–334.
422. Pereira A, Maraschin M (2015) Banana (*Musa* spp) from peel to pulp: ethnopharmacology, source of bioactive compounds and its relevance for human health. *Journal of Ethnopharmacology* **160**, 149–163. doi:10.1016/j.jep.2014.11.008
423. Mathew NS, Negi PS (2017) Traditional uses, phytochemistry and pharmacology of wild banana (*Musa acuminata* Colla): a review. *Journal of Ethnopharmacology* **196**, 124–140. doi:10.1016/j.jep.2016.12.009
424. Nahar L, Sarker SD (2007) Phytochemistry of the genus *Curcuma*. In *Turmeric: The Genus Curcuma*. (Eds PN Ravindran, K Nirmal Babu and K Sivaraman) pp. 71–106. CRC Press, Boca Raton, FL.
425. Achenbach H, Hubner H, Brandt W, Reiter M (1994) Cardioactive steroid saponins and other constituents from the aerial parts of *Tribulus cistoides*. *Phytochemistry* **35**, 1527–1543. doi:10.1016/S0031-9422(00)86890-9
426. Achenbach H, Hubner H, Reiter M (1996) Cholestane- and pregnane-type glycosides from the roots of *Tribulus cistoides*. *Phytochemistry* **41**, 907–917. doi:10.1016/0031-9422(95)00622-2
427. Kostova I, Dinchev D (2005) Saponins in *Tribulus terrestris*: chemistry and bioactivity. *Phytochemistry Reviews* **4**, 111–137. doi:10.1007/s11101-005-2833-x
428. Chhatre S, Nesari T, Somani G, Kanchan D, Sathaye S (2014) Phytopharmacological overview of *Tribulus terrestris*. *Pharmacognosy Reviews* **8**, 45–51. doi:10.4103/0973-7847.125530
429. Neychev V, Mitev V (2016) Pro-sexual and androgen enhancing effects of *Tribulus terrestris* L.: fact or fiction. *Journal of Ethnopharmacology* **179**, 345–355. doi:10.1016/j.jep.2015.12.055
430. Santos HO, Howell S, Teixeira FJ (2019) Beyond tribulus (*Tribulus terrestris* L.): the effects of phytotherapics on testosterone, sperm and prostate parameters. *Journal of Ethnopharmacology* **235**, 392–405. doi:10.1016/j.jep.2019.02.033

3
Terpenoid-type specialised metabolites

INTRODUCTION

Terpenoid-type compounds are comparable with the alkaloids as the largest class of specialised metabolites naturally produced by plants.[1] The term 'terpene' [$(C_5H_8)_n$] applies specifically to natural hydrocarbons derived from the polymerisation of isopentene (isoprene, C_5H_8) units derived from either of two biosynthetic pathways present within higher plants: (1) the *2C*-methyl-D-erythritol-4-phosphate (MEP) pathway [also known as the 1-deoxy-D-xylulose-5-phosphate (DXP) pathway] and (2) the mevalonic acid (MVA) pathway.[2] Formally, any structural modification to these hydrocarbon terpenes, such as oxidation or rearrangement, results in the compound being reclassified as a terpenoid, although both terms are used interchangeably throughout the literature. The terpenoids are categorised based on the number of isoprene units they are derived from, with most falling within one of five subclasses: (1) hemiterpenes (1 × isoprene unit, C_5), (2) monoterpenes (2 × isoprene units, C_{10}), (3) sesquiterpenes (3 × isoprene units, C_{15}), (4) diterpene (4 × isoprene units, C_{20}) and (5) triterpene (6 × isoprene units, C_{30}) (*Fig. 3.1*). The tetraterpene (8 × isoprene units, C_{40}) subclass will not be addressed here because the tetrapenes are constitutive primary metabolites (as pigments and light harvesters/energy dissipators in photosynthesis [the carotenoids and xanthophylls]), although it is worth mentioning that their role in human health is vast, particularly their retinoid degradation products [e.g. retinol (vitamin A)].[3,4] To date, over 40,000 plant-derived terpenoids have been reported and, given the significant conformational diversity facilitated by the repeating isoprene units, the total number of natural terpenoids is likely to be significantly higher, somewhere in the order of 10^5.[1,2,5] This chapter will proceed through an introduction to the class itself, followed by more detailed breakdowns of each of the terpenoid-type metabolic subclasses. The end of the chapter will present a summary of the representation of these metabolites reported to be present across the customary medicinal flora of Australia, detailed in *Tables 3.1–3.4*. This information can be also explored using the associated website (www.cmfoa.info).

The two pathways (MEP and MVA) involved in terpenoid biosynthesis both produce the central intermediates of the terpenoid metabolic network: isopentenyl pyrophosphate (IPP) and its structural isomer dimethylallyl pyrophosphate (DMAPP) (*Fig. 3.2*).[2] The MEP and MVP pathways have different evolutionary origins and are constitutively expressed in different regions of the plant cell, with limited evidence of exchange between their respective IPP/DMAPP pools.[6] The MEP pathway is of prokaryotic (eubacterial) origin and confined to the plastids of higher plants, providing further support to the already well established notion of cyanobacterial endosymbiosis.[7,8] The MEP pathway combines one unit of IPP and one unit of DMAPP to produce geranyl

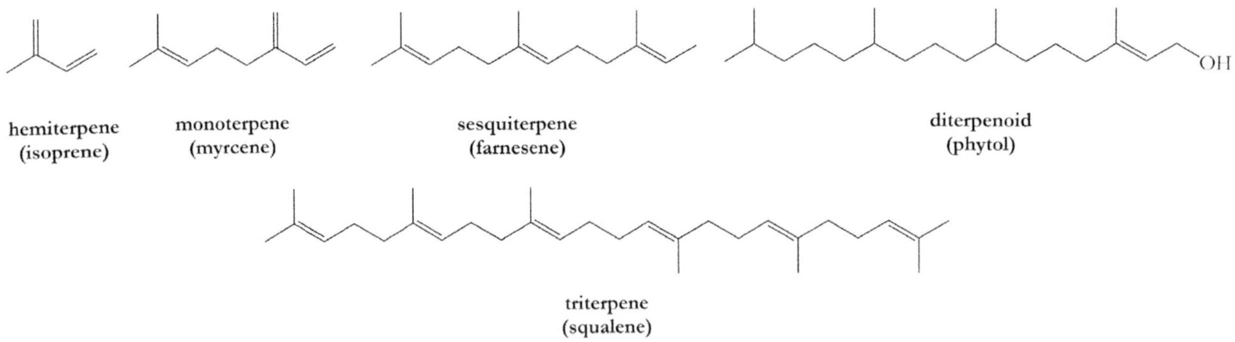

Fig. 3.1. The five major subclasses of natural terpenoid-type specialised metabolites.

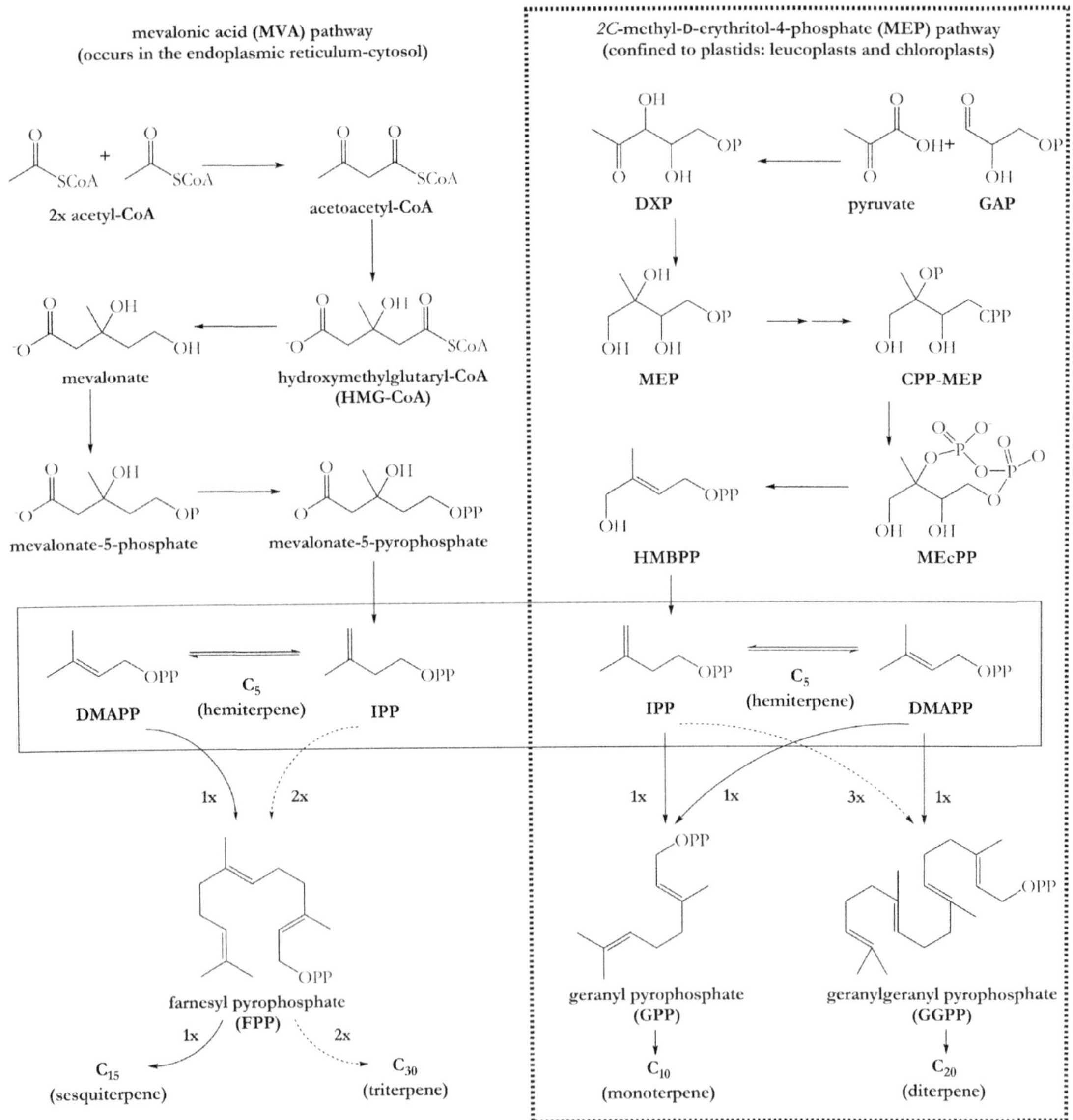

Fig. 3.2. Comparison of the mevalonic acid (MVA) and 2C-methyl-D-erythritol-4-phosphate (MEP) terpenoid biosynthetic pathways. IPP = isopentenyl pyrophosphate, DMAPP = dimethylallyl pyrophosphate, GAP = glyceraldehyde-3-phosphate, DXP = 1-deoxy-D-xylulose-5-phosphate, MEP = 2C-methyl-D-erythritol 4-phosphate, CPP-MEP = citydyl pyrophosphate 2C-methyl-D-erythritol 2-phosphate, MEcPP = 2C-methyl-D-erythritol-2,4-cyclopyrophosphate, HMBPP = 1-hydroxy-2-methyl-2-(E)-butenyl-4-pyrophosphate.

pyrophosphate (GPP, C_{10}), from which the monoterpenes (C_{10}) are derived. Additionally, the MEP pathway can combine three units of IPP and one unit of DMAPP to produce geranylgeranyl pyrophosphate (C_{20}), from which the diterpenes (C_{20}) and tetraterpenes (C_{40}) are derived (*Fig. 3.2*). The MVA pathway is of eukaryotic origin and occurs within the endoplasmic reticulum-cytosol of the cells of higher plants.[8] The MVA pathway combines two units of IPP and one unit of DMAPP to produce farnesyl pyrophosphate (FPP, C_{15}) from which the sesquiterpenes (C_{15}) and triterpenes (C_{30}) are derived, as well as the structurally rearranged sterols (C_{27-29}) and glycosylated saponin derivatives (*Fig. 3.2*). It is worth noting that as with the phenolic-type specialised metabolites described in Chapter 2, the terpenoid biosynthetic network traverses the line between primary and secondary metabolism, with triterpenes (C_{30}) and tetraterpenes (C_{40}) being considered primary metabolites and many of the genetic origins of these pathways being modified from the primary genome.[2]

Isoprene-derived metabolites are integral to the physiological function of all organisms, including electron transfer (plastoquinone), respiration (ubiquinone), membrane structure/permeability (sterols) and hormones (steroids).[1,6] Within the plant kingdom specifically, isoprene derivatives play essential roles in light harvesting (the phytol tail of chlorophyll) and photoprotection (carotenoids), as well as specialised roles in ecological plant-animal, plant-pathogen and plant-plant interactions such as defence (phytoalexin/phytoanticipin) and attraction (pigments/aromas).[9-11] The consistency in the structure and function of the terpenoids across the kingdoms of the natural world has facilitated a natural affinity between plant-derived metabolites and animal-based proteins. In fact, the evolution of these compounds, and the organisms that produce and interact with them, has formed the basis for a biological preselection that has proved felicitous in the application of these compounds in medicine.[12-14] Each of the terpenoid subclasses will be addressed concisely below, but some general and well-established examples of the bioactivity of terpenoids include their antimicrobial activity (e.g. monoterpenoids (*p*-menthanes) via membrane disruption),[15] antimalarial activity (e.g. sesquiterpenoids (artemisinin) via endoperoxyl carbon-centred radical heme/protein alkylation),[16] antineoplastic activity (e.g. sesquiterpenoids (thapsigargin) via disruption of Ca^{2+} homeostasis and resultant apoptosis, diterpenoids [paclitaxel] via microtubule stabilisation and prevention of mitosis),[17,18] insecticidal properties [e.g. monoterpenoids (pinene-type, limonene-type), monoterpene esters (pyrethroid), triterpenoid (limonoid)],[19,20] and even psychoactive/anxiolytic/anaesthetic effects (e.g. meroterpenoids (cannabinoids) via specific receptor ligation, addressed separately in Chapter 5).[21]

HEMITERPENOIDS

Apart from the simple building block isoprene, which is the most abundant hydrocarbon released into the canopy of pine forests, and despite being fundamental to terpenoid biosynthesis, the constituents of the small, simple subclass of C_5 hemiterpenoids are rarely encountered in nature, with only 90 hemiterpenoid glycosides recorded.[5] This includes the common aglycone forms comprising isoprene itself, prenol (Annonaceae, Cannabaceae, Rutaceae) and 4-methoxy-2-methyl-2-butanthiol (Saxifragaceae) (*Fig. 3.3*).[5] Their scarcity is likely due to the efficiency of the MEP/MVA pathways, with their more substantially bioactive descendants being synthesised as fast as these hemiterpene precursors are produced.[6] Hemiterpenoids can be incorporated in other natural products (NPs) as angelic, tiglic and isovaleric acid moieties, forming mixed biosynthetic NPs known as the meroterpenoids, which are detailed in Chapter 5. Hemiterpenoids *sensu stricto* are both ubiquitous and widely overlooked in NP journal articles, so for the purposes of this book, they were omitted from this general review, as well as from the website (www.cmfoa.info).

MONOTERPENOIDS

Approximately 1,500 C_{10} monoterpenoid type NPs have been isolated to date. They are generally colourless,

isoprene   prenol   4-methoxy-2-methyl-2-butanthiol   angelic acid   tiglic acid   isovaleric acid

Fig. 3.3. Infraclasses of the terpenoid-type hemiterpenoid subclass.

lipophilic and volatile specialised metabolites, notable for being the constituents of essential oils and defensive oleoresins of the plant kingdom, imparting the characteristic flavours and fragrances of many plant genera.[5,22,23] They are often produced and exuded from specialised plant structures, such as peltate glandular trichomes (e.g. in *Mentha*), secretory cells (e.g. in *Citrus*) or resin ducts (e.g. in *Eucalyptus*), in a mixture with sesquiterpenoids (C_{15}) and diterpenoids (C_{20}) and their function is largely to mediate ecological plant-environment interactions.[1,5,24] Monoterpenoids arise from GPP produced from the MEP pathway, which is confined to the plastids of plant cells. GPP can remain acyclic (as 2,6-dimethyloctane derivatives), but often cyclises to form the important monocyclic *p*-menthane infraclass, as well as the numerous bicyclic monoterpenoids detailed below. The distribution of terpenoid-type specialised metabolites of the monoterpenoid subclass reported from the customary medicinal flora of Australia is presented in *Table 3.1* (see Representation, below).

Acyclic monoterpenoids

There are at least seven loosely defined infraclasses within the regular acyclic monoterpenoids (2,6-dimethyloctane derivatives), which display significant isomerism (*Fig. 3.4*).[1,22] β-Myrcene and β-ocimene are characteristic of the Lamiaceae (*Ocimum* - basil), Myrtaceae (*Pimenta* - bay leaf), Rutaceae (*Citrus*) and Cannabaceae (*Humulus* - hops), among others.[1,5,22,24,25] In terms of aroma, linalool is characteristic of the Apiaceae (*Coriandrum* - coriander), geraniol is characteristic of the Geraniaceae (*Geranium*), nerol is characteristic of the Poaceae (*Cymbopogon* - lemongrass) and citronellol is characteristic of the Rosaceae (Rosa); citronellal gives the characteristic aroma of citronella oil, which is extracted from *Cymbopogon* and is present as a mixture with several other monoterpenoids.[1,5,22,24,25] Several other irregular 'acyclic' monoterpenoids are recognised, although some cyclisation arises within these groups, which makes clear delineation of infraclasses problematic, a common issue that will be seen throughout this chapter (*Fig. 3.4*). The pyrethrins/pyrethroids are notable insecticidal meroterpenoids derived from the genus *Chrysanthemum* (Asteraceae) that feature a chrysanthemyl (chrysanthemic acid) core. The pyrethrins have potent insecticidal activity through interference with voltage-gated Na^+ channels in insect nerve-cells, causing paralysis and eventually death. The natural origin and ultraviolet-induced breakdown of the pyrethrins has attracted interest as an organic alternative to conventional organophosphate/organochloride pesticides.[20]

Monocyclic monoterpenoids

Iridoid (cyclopentane)

The iridoids are a large group of cyclopentane monoterpenoids based on the cis-2-oxabicyclo(4.3.0)nonane skeleton and are present in many traditional medicines used throughout the world where they function as bitter tonics, sedatives, antipyretics, cough medicines, hypotensives and general remedies for wounds and skin disorders.[26-28] Over 600 iridoids (including secoiridoids, which have undergone pentacyclic ring-cleavage;

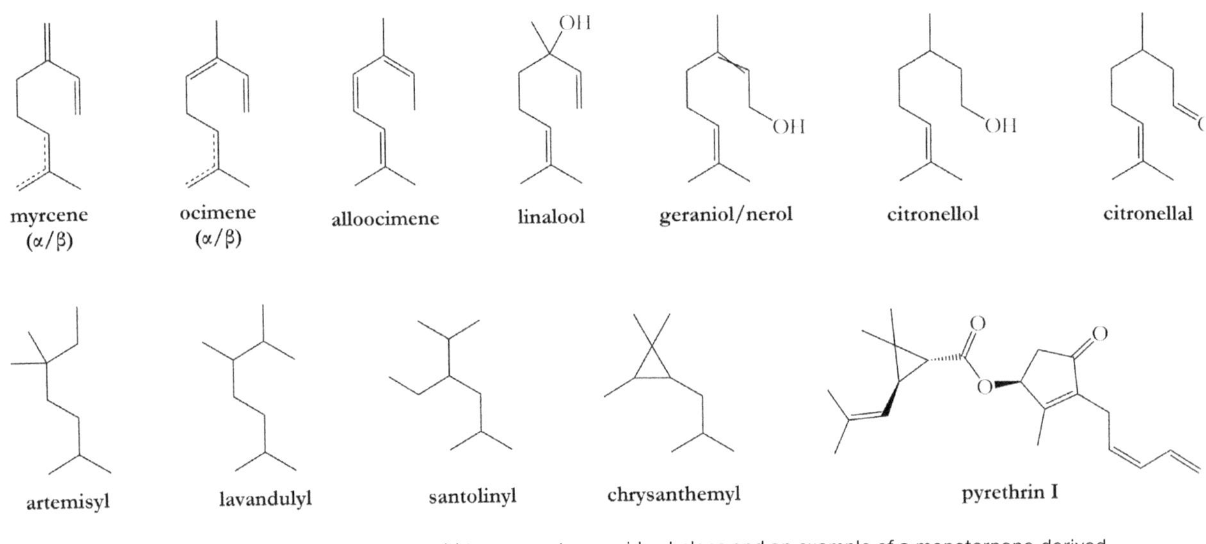

Fig. 3.4. Acyclic-type infraclasses of the terpenoid-type monoterpenoid subclass and an example of a monoterpene-derived meroterpenoid (pyrethrin I).

Fig. 3.5) have been recorded in the literature and they are present across the Cornanae, Ericanae, Gentiannae, Lamianae and Loasanae superorders, with specific members functioning as the chemotaxonomic markers of the Plantaginaceae (aucubin for *Plantago*), Rubiaceae (asperuloside for *Galium*) and Scrophulariaceae (harpagide for *Scrophularia*), but are also present among the Ericaceae, Loganiaceae, Gentianaceae, Verbenaceae, Lamiaceae, Oleaceae, Valerianaceae (Caprifoliaceae) and Menyanthaceae.[26–28] Iridoids are biosynthetically derived *in planta* from the acyclic monoterpenoid 9-hydroxy nerol, via sequential phosphorylation, cyclisation, oxidation and glycosylation.[27,28] The iridoids have demonstrated a wide range of bioactivities, including anti-inflammatory, antispasmodic, antineoplastic, antimicrobial, antiviral, cardiovascular, antihepatotoxic, choleretic, immunomodulatory, purgative, hypoglycaemic and hypolipidaemic.[27,28]

Cyclohexane

The monocyclic cyclohexane monoterpenoids, which also include the important *p*-menthane (1-methyl-4-isopropylcyclohexane) infraclass are well represented across the plant kingdom, but are particularly concentrated in the essential oils of the Lamiaceae (menthol, menthone and (-)-(4*R*)-*trans*-carvone in *Mentha* - mint), Apiaceae ((+)-(4*S*)-*trans*-carvone in *Carum* - caraway), Rutaceae (limonene in *Citrus*), Myrtaceae (1,8-cineole, β-phellandrene in *Eucalyptus*, terpinen-4-ol in *Melaleuca*) and Pinaceae (terpinene, terpinolene in *Pinus*) (*Fig. 3.6*).[5] The aromatic cyclohexane monoterpenoids are more specifically distributed with *p*-cymene concentrated in the essential oil of the Lauraceae (*Cinnamomum* - cinnamon), Cupressaceae (*Cupressus* - cyprus), Myrtaceae (*Eucalyptus*) and *m*-cymene in the Saxifragaceae (*Ribes* - blackcurrant), while the phenolic

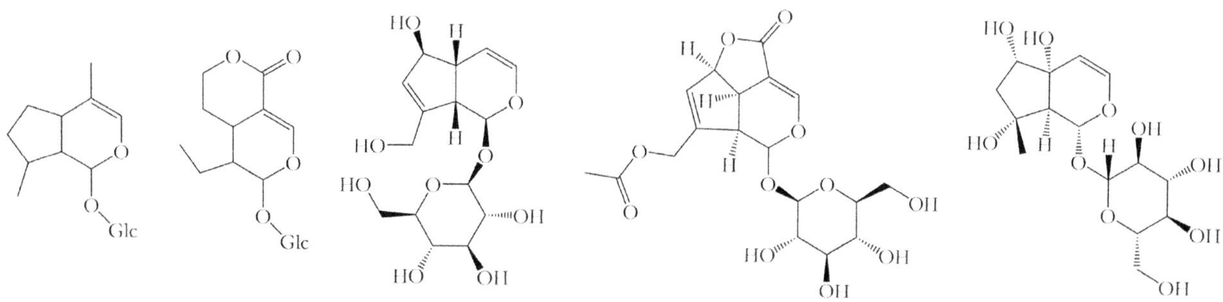

iridoid | secoiridoid | aucubin (*Plantago*) | asperuloside (*Galium*) | harpagide (*Scrophularia*)

Fig. 3.5. Iridoid-type infraclasses of the terpenoid-type monoterpenoid subclass and some key representative compounds (aucubin, asperuloside, harpagide).

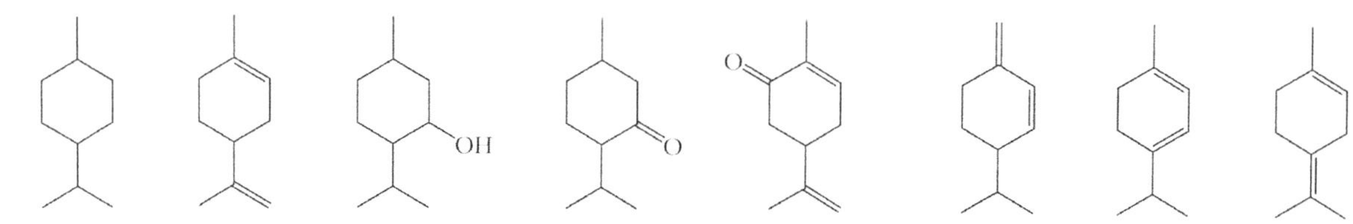

generic p-menthane | limonene | menthol | menthone | carvone | β-phellandrene | terpinene | terpinolene

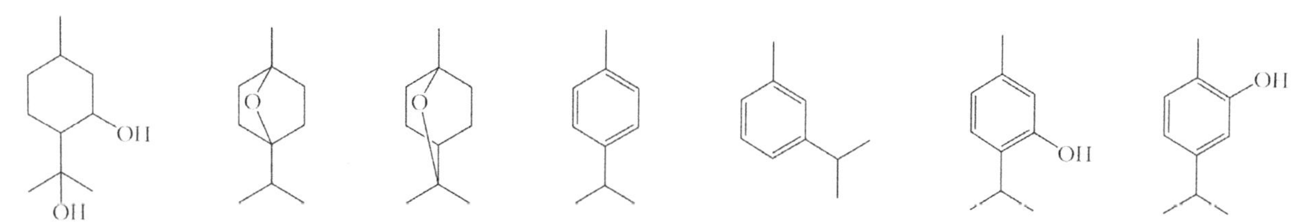

p-menthane 3,8-diol | 1,4-cineole | 1,8-cineole | *p*-cymene | *m*-cymene | thymol | carvacrol

Fig. 3.6. Cyclohexane-type infraclasses of the terpenoid-type monoterpenoid subclass.

monoterpenoids thymol and carvacrol are present in the Lamiaceae (*Thymus* - thyme, etc.).[5] It is worth mentioning that the *R/S* stereoisomers of carvone give rise to the drastically different flavours/aromas of spearmint and caraway seeds, a fine example of the supremely sophisticated vertebrate olfactory system.[29] Biosynthetically, (-)-limonene is the common precursor to the *Mentha* (-)-*p*-menthanes and is generated from guanosine diphosphate (GDP) by (-)-limonene synthase, while (+)-limonene is the common precursor to the carvone (+)-*p*-menthanes and is generated from GDP by (+)-limonene synthase.[24] (-)-Menthol is notable as a full agonist of the homotetrameric Na^+/Ca^{2+} ion channel cold and menthol receptor 1, which when activated by (-)-menthol, confers mild pain relief through indirect activation of the κ opioid receptor, useful in the treatment of muscle pain and irritable bowel syndrome.[2]

Megastigmane

The C_{13} isonorterpenoid compounds bearing a cyclohexane moiety and termed 'megastigmane' are often grouped with the sesquiterpenoids but will be discussed here. Biosynthetically the megastigmanes are thought to be apocarotenoids, arising from the oxidative degradation of tetraterpenoids (carotenoids), ultimately arising from the MEP pathway.[30] Plants containing megastigmanes have been used by customary medical practitioners globally for the treatment of chronic viral hepatitis, dysentery, inflammation, fever and haemonchosis.[30] Megastigmanes have been isolated from the Acanthaceae (*Acanthus*), Aquafoliaceae (*Ilex*), Asclepiadaceae (*Caralumma*), Asteraceae (*Lagerra*), Caprifoliaceae (*Lonicera*), Celastraceae (*Salacia*), Cornaceae (*Alangium*), Crassulaceae (*Sedum*), Cucurbitaceae (*Cucumis, Gynostemma*), Cupressaceae (*Juniperus*), Equisetaceae (*Equisetum*), Euphorbiaceae (*Breynia, Bridelia, Croton, Euphorbia, Glochidon, Macaranga*), Fabaceae (*Crotalaria*), Lamiaceae (*Phlomis, Salvia*), Lauraceae (*Cinnamomum, Laurus*), Myrsinaceae (*Myrsine*), Orobanchaceae (*Rehmannia*), Piperaceae (*Piper*), Phyllanthaceae (*Glochidon*), Plantaginaceae (*Veronica*), Rubiaceae (*Lasianthus, Tricalysia*), Rutaceae (*Boronia, Euodia*), Staphylaceae (*Turpinia*), Verbenaceae (*Clerodendrum*) and Vitaceae (*Ciccus*).[30] Although isolated from such a wide variety of plants, their biological and pharmacological properties have not yet been studied in much depth.

Bicyclic monoterpenoids

The bicyclic monoterpenoids (more correctly 'carbobicyclic' to exclude heterocyclic monocycles such as the cineoles) are diverse and can be divided into four major groups: the bicyclopentanes (camphanes/bornanes/fenchanes), pinanes, caranes and thujanes (*Fig. 3.7*).[5] Their representation across the plant kingdom is more specific than the other monoterpenoids and they are generally found in the camphanes (Lauraceae, Lamiaceae, Asteraceae), fenchanes (Cupressaceae, Apiaceae), pinanes (Pinaceae, Rutaceae, Myrtaceae), caranes (Pinaceae, Cupressaceae) and thujanes (Cupressaceae, Asteraceae).

SESQUITERPENOIDS

More than 10,000 sesquiterpenoids are known, but the diversity of the structures observed in this subclass makes them very difficult to reconcile. Since the 1970s this has often been attempted on biosynthetic grounds and an effort will be made to follow this strategy here.[5,31,32] The structural variations encountered within the sesquiterpenoids are not amenable to simple categorisation, as any additional bond/inversion/cyclisation can drastically change the overall shape and perspective of the compound, altering the infraclass into which it would naturally fall. The major biosynthetic groupings (and general cyclic structure - #carbons per ring) are as follows, with ':' denoting directly bound to and '-' denoting a spiro structure; '—' denoting separation along a C-C bond: farnesane (acyclic), cyclofarnesane (mono/bi/polycyclic - 6:6), bisabolane (monocyclic - 6), sesquipinane/

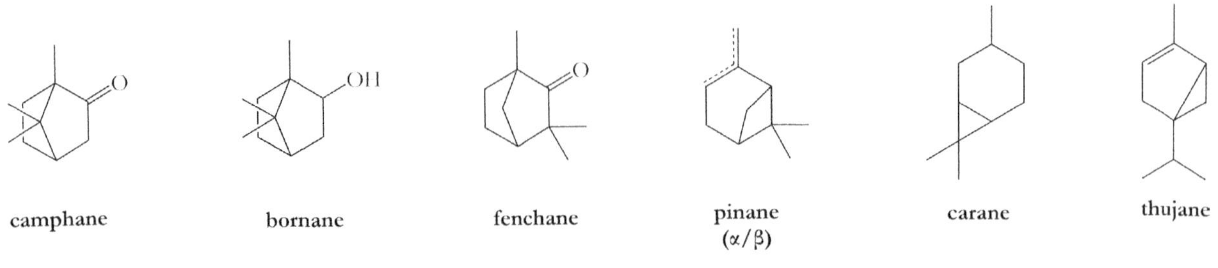

Fig. 3.7. Bicyclic-type infraclasses of the terpenoid-type monoterpenoid subclass.

sesquicamphane (bicyclic - 6:3), germacrane (originally monocyclic - 10, cyclise in various ways, e.g. 6/6-5/7:5/8:5), humulane (originally monocyclic - 11, cyclise in various ways e.g. 9:4/3-6:5/3:6:5/5:5:5), drimane/cadinane/eudesmane (bicyclic - 6:6), himachalane (bicyclic - 6:7, often with intramolecular bonding), guaiane (bicyclic - 5:7), pterosin (bicyclic - 5:6), cuparane/laurane/trichothecane (bi/polycyclic - 6-5), chamigrane/omphalane (bicyclic - 6-6) and vetivane (bicyclic - 6-5/6-5/6-4);[5,31,32] as well there is the vast category of disesquiterpenoids (C_{30} sesquiterpene dimers).[33] These general categories are far from exhaustive and it is doubtful a perfect breakdown of this (or any chemical class for that matter) will ever emerge.

As with the volatile monoterpenoids, the sesquiterpenoids are a common and characteristic feature of plant essential oils. It is not possible to quickly summarise and enumerate their representation across the plant kingdom, but some notable examples are provided: the acyclic farnesanes in the Rutaceae (*Citrus*), bisabolanes in the Zingiberaceae (*Zingiber* - ginger), sesquisabinanes in the Piperaceae (*Piper* - pepper), germacranes in the Rutaceae (*Citrus*), eremophilanes in the Asteraceae (*Ligularia*) and Scrophulariaceae (*Eremophila*), elemanes in the Lamiaceae (*Mentha*), eudesmane dihydro-β-agarofurans in the Celastraceae, humulanes in the Lauraceae (*Lindera*), Myrtaceae (*Syzygyium* - clove), Cannabaceae (*Humulus* - hops) and Zingiberaceae (*Zingiber* - ginger), as well as

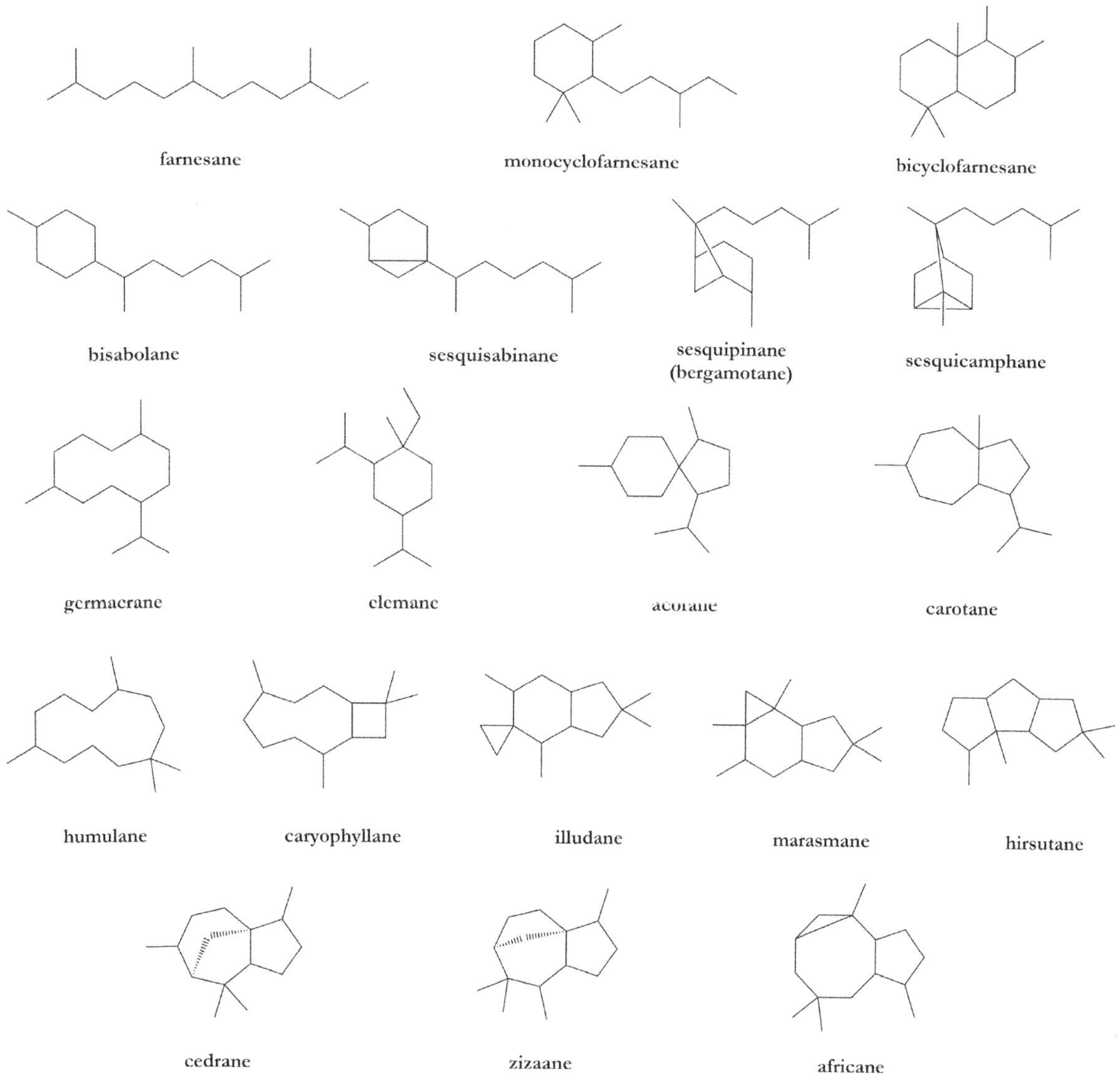

Fig. 3.8. Acyclic-, mono/macrocyclic-, bicyclic- and polycyclic-type infraclasses of the terpenoid-type sesquiterpenoid subclass, categorised on biosynthetic grounds.

guaiane in the Asteraceae (*Artemisia* – wormwood).[5,34,33] Bioactivity varies across the sesquiterpenoid subclass as much as their structures; infraclasses would have to be addressed on a case-by-case basis, much of which is detailed at length in the literature but lies outside the scope of this book. Grouping these various infraclasses is difficult, so the strategy used here is based on biosynthetic grounds, with the acyclic and monocyclic farnesanes, bisabolanes, germacranes, humulanes and their respective bi- and polycyclic derivatives presented in one figure and the remaining bi- and polycyclic drimane-, himachalane-, guaiane-, pterosin-, cuparane-, trichothecane-, chamigrane- and vetivane-type infraclasses are presented separately (*Fig. 3.8* and *Fig. 3.9*).

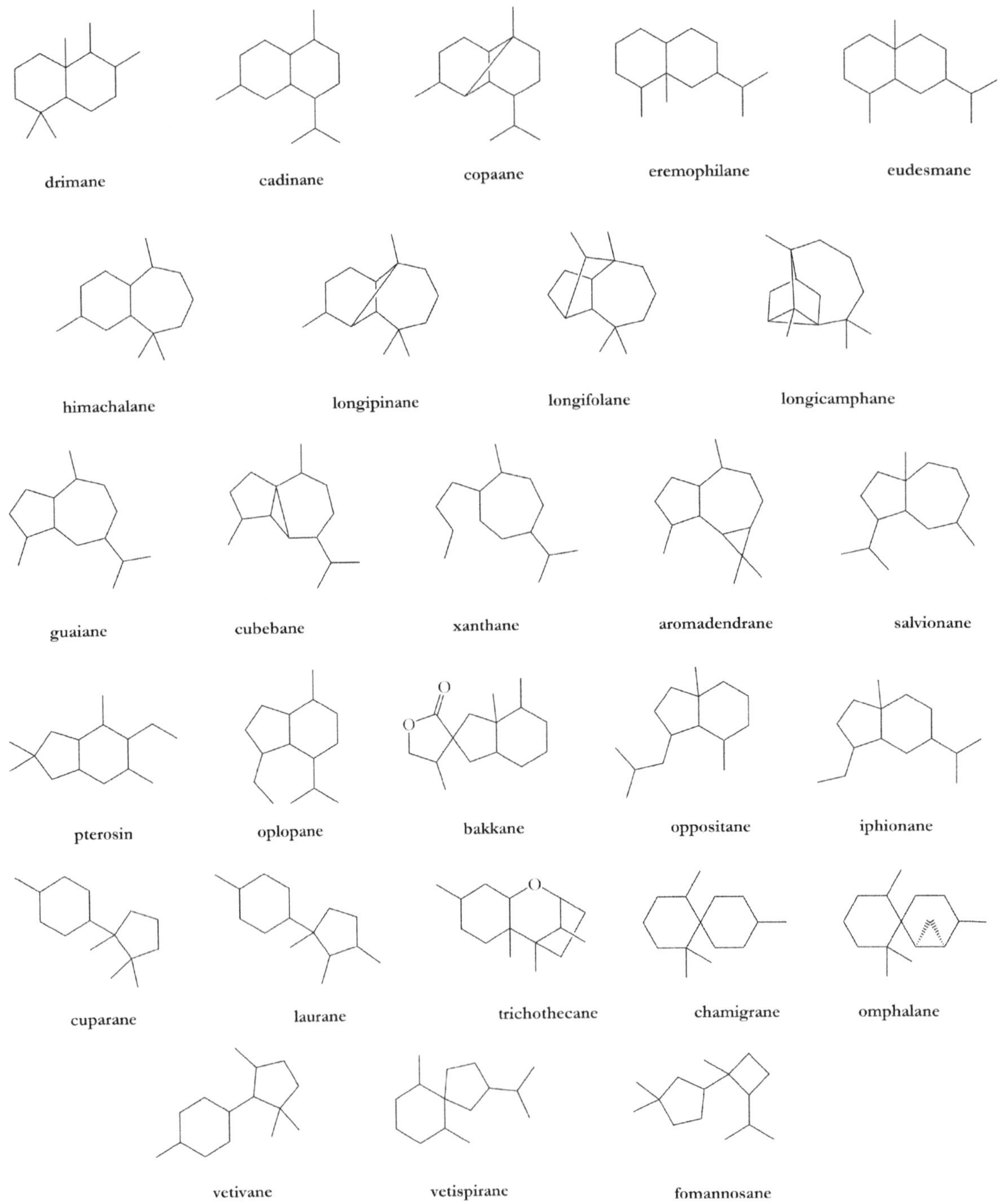

Fig. 3.9. Bicyclic- and polycyclic-type infraclasses of the terpenoid-type sesquiterpenoid subclass, categorised on biosynthetic grounds.

DITERPENOIDS

Over 12,000 natural diterpenoids (C_{20}) have been reported, primarily from the Araliaceae, Asteraceae, Cistaceae, Cupressaceae, Euphorbiaceae, Fabaceae, Lamiaceae and Pinaceae.[5,35] The diterpenoid subclass is divided largely upon biosynthetic grounds into acyclic, mono/macrocyclic (cembrane and its polycyclic condensates [casbane, dolabellane, jatrophane, daphnane, taxane]), bicyclic (labdane, halimane, clerodane), tricyclic (abietane, pimarane, rosane, cassane), tetracyclic (kaurane, trachylobane, gibberellin, grayanotoxin) and miscellaneous (briarane, xenicane, pseudopterane, amphilectane) infraclasses, with the mono- and bicyclic-type infraclasses presented separately to the tricyclic-, tetracyclic- and miscellaneous-type infraclasses (*Fig. 3.10* and *Fig. 3.11*).[36,37] As with the monoterpenoids, the diterpenoid subclass is the result of the MEP pathway localised in the plastids (leucoplasts and chloroplasts). Of the 12,000 natural diterpenoids, ~7,000 belong to the labdane infraclass and are basal to the formation of the tetracyclic gibberellins (gibberellic acids – GAs). GAs are significant diterpenoid phytohormones, the development of which is so integral to the evolution of the terrestrial plant kingdom that their requisite genetic reservoir likely gave rise to the entire terpenoid-type natural product class (450–400 million years ago).[35]

GAs are directly implicated in stem elongation, flowering and seed germination (GA-deficient plants are sterile). The primary metabolic status of GA phytohormones and their *ent*-kaurenoic acid ancestors makes the representation of diterpenoids in the plant kingdom essentially ubiquitous, with their genetic pool present in all higher plant genomes (Gymnospermata, Angiospermata [monocotyledonous and dicotyledonous]).[35] Curiously, although GAs appear to be quite structurally sophisticated, they almost certainly evolved before the other diterpenoid infraclasses (putatively before all terpenoids). The function of the diterpenoid subclass *in planta* is vast (even beyond the primary metabolic function of GAs), including antifeedant, allelochemical and

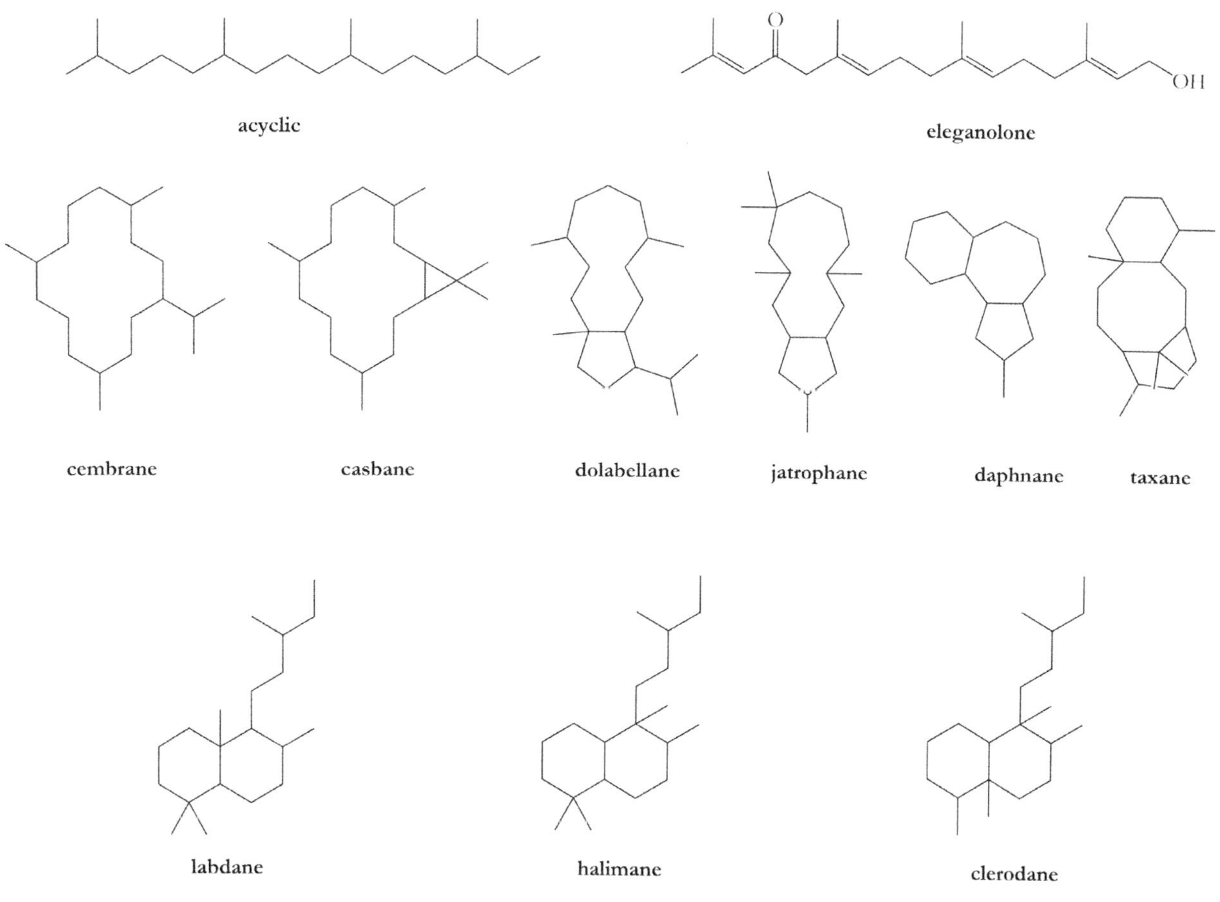

Fig. 3.10. Mono- and bicyclic-type infraclasses of the terpenoid-type diterpenoid subclass, including some polycyclic derivatives categorised on biosynthetic grounds.

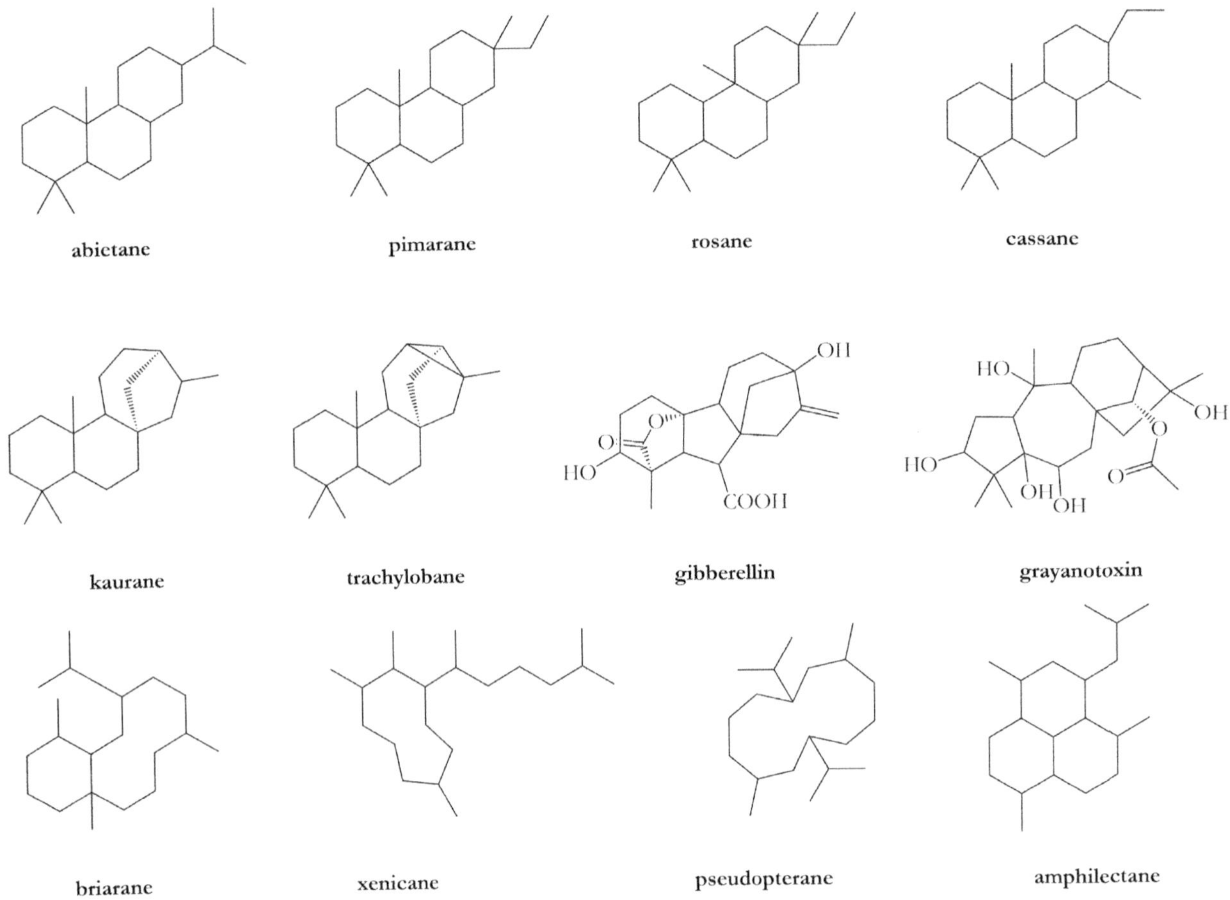

Fig. 3.11. Tricyclic-, tetracyclic- and miscellaneous-type infraclasses of the terpenoid-type diterpenoid subclass.

antibiotic (phytoalexin). In terms of pharmacological applications, the diterpenoids have significant potential; a key example is Taxol (paclitaxel, of the taxane-infraclass), a multi-billion-dollar chemotherapy drug, and a wide variety of other diterpenoids display general antimicrobial, antituberculosis, antineoplastic, antiviral, antiprotozoal, abortifacient, antihypertensive, hepatoprotective and anti-inflammatory activities.[38] Many of these reported bioactivities are due to direct interaction with enzyme systems (adenosine transferase, aldose reductase, adenylated cyclase, protein kinases, phospholipase A_2, β-glucuronidase, etc.) and physiological systems (e.g. immune, endocrine).[38]

TRITERPENOIDS

There are more than 20,000 C_{30} triterpenoids that originate from the cyclisation of the acyclic 2,3-oxidosqualene unit, which arises from the condensation of two FPP subunits produced from the MVA pathway in the cytosol of plant cells.[39] The classification of the triterpenoid-type subclass can be based on biosynthetic grounds into acyclic (squalane/squalene), tetracyclic (6/6/6/5: lanostane/fusidane (including the cucurbitanes), dammarane/euphane (including the tetranortriterpenoid limonoids and the quassinoids), cycloartanes and brassinosteroids), pentacyclic (6/6/6/6/5: lupane, hopane, and 6/6/6/6/6: ursane, oleanane), those that are 'unusually cyclised' and the glycosylated saponin (including the cardiac glycosides) infraclasses, with the general triterpenoid infraclasses presented separately to the steroid/sterol-type and saponin-type infraclasses (*Fig. 3.12* and *Fig. 3.13*).[39,40]

Once again, the triterpenoids, particularly the constitutive sterols (fundamental to membranes and cell signalling), tread a fine line between primary and secondary metabolism.[39] As such they are present across the entire plant kingdom in some shape or form and this ubiquity, combined with their wide-reaching biological activity discussed below, has resulted in an overwhelming amount of literature on the topic. The past 10 years has seen published reviews of the triterpenoid content of genera within the Amaranthaceae, Apocynaceae (*Alstonia*, *Gymnema*), Aquafoliaceae (*Ilex*), Araliaceae (*Panax*, *Schefflera*), Asteraceae (*Calendula*), Burseraceae (*Boswellia*) Campanulaceae (*Codonopsis*, *Platycodon*), Caprifoliaceae (*Lonicera*), Combretaceae (*Terminalia*), Cucurbitaceae (*Siraitia*), Euphorbiaceae (*Euphorbia*), Fabaceae (*Albizia*, *Astragalus*, *Glycyrrhiza*, *Medicago*), Lamiaceae (*Clinopodium*, *Leucas*) Meliaceae (*Aphanamixis*, *Toona*) Oleaceae (*Olea*), Rhizophoraceae (*Ceriops*), Sapindaceae (*Aesculus*, *Sapindus*), Sapotaceae (*Argania*), Schisandraceae and Simaroubaceae (*Ailanthus*) families.[41-46] The distribution of terpenoid-type specialised metabolites of the triterpenoid subclass reported from the customary medicinal flora of Australia is presented in *Table 3.4* (see Representation, below).

The constitutive effects of sterols (divided into the five groups: sitosterol, campesterol, cholesterol, stigmasterol, stigmastanol; *Fig. 3.13*) and phytosteroids are

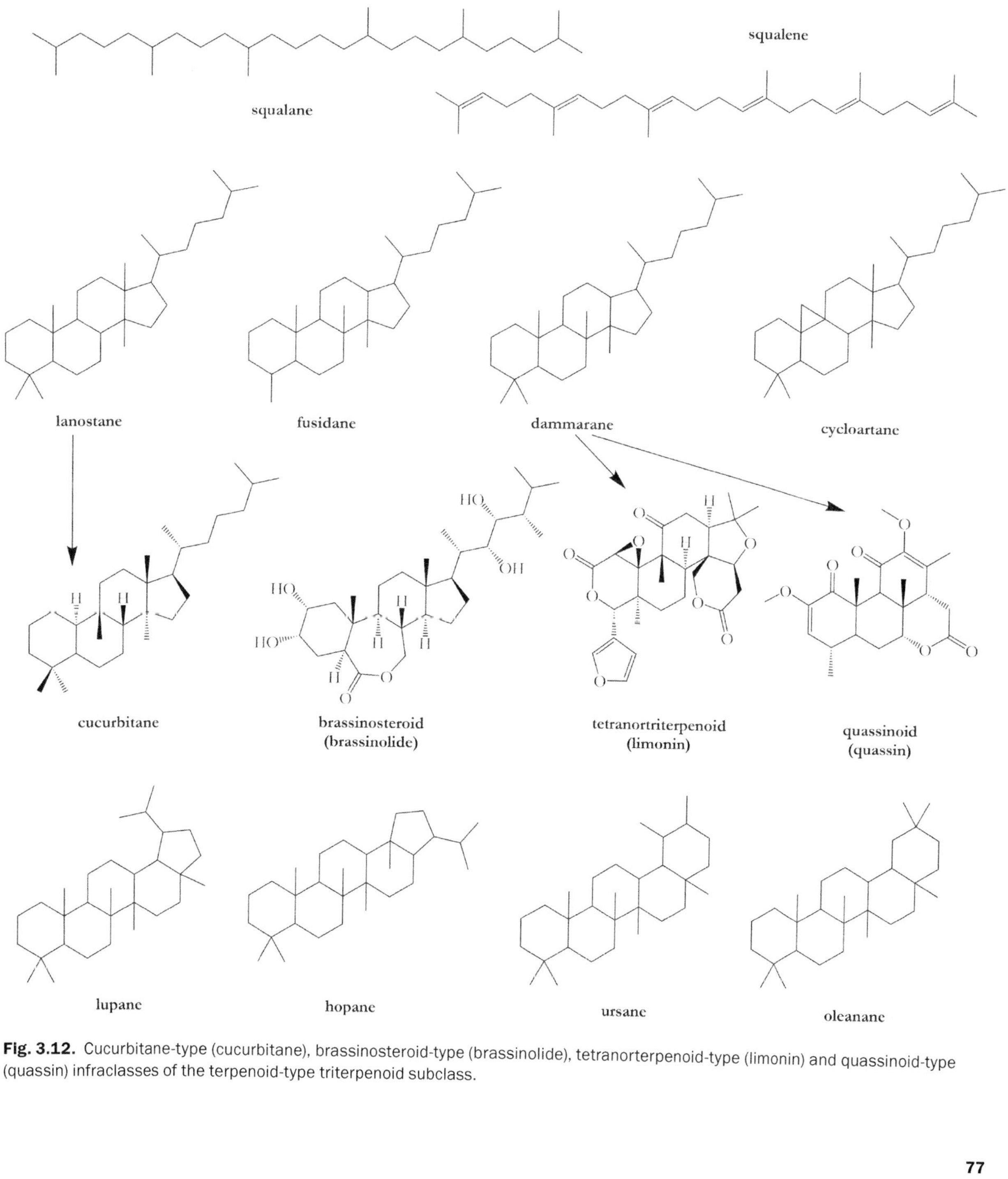

Fig. 3.12. Cucurbitane-type (cucurbitane), brassinosteroid-type (brassinolide), tetranorterpenoid-type (limonin) and quassinoid-type (quassin) infraclasses of the terpenoid-type triterpenoid subclass.

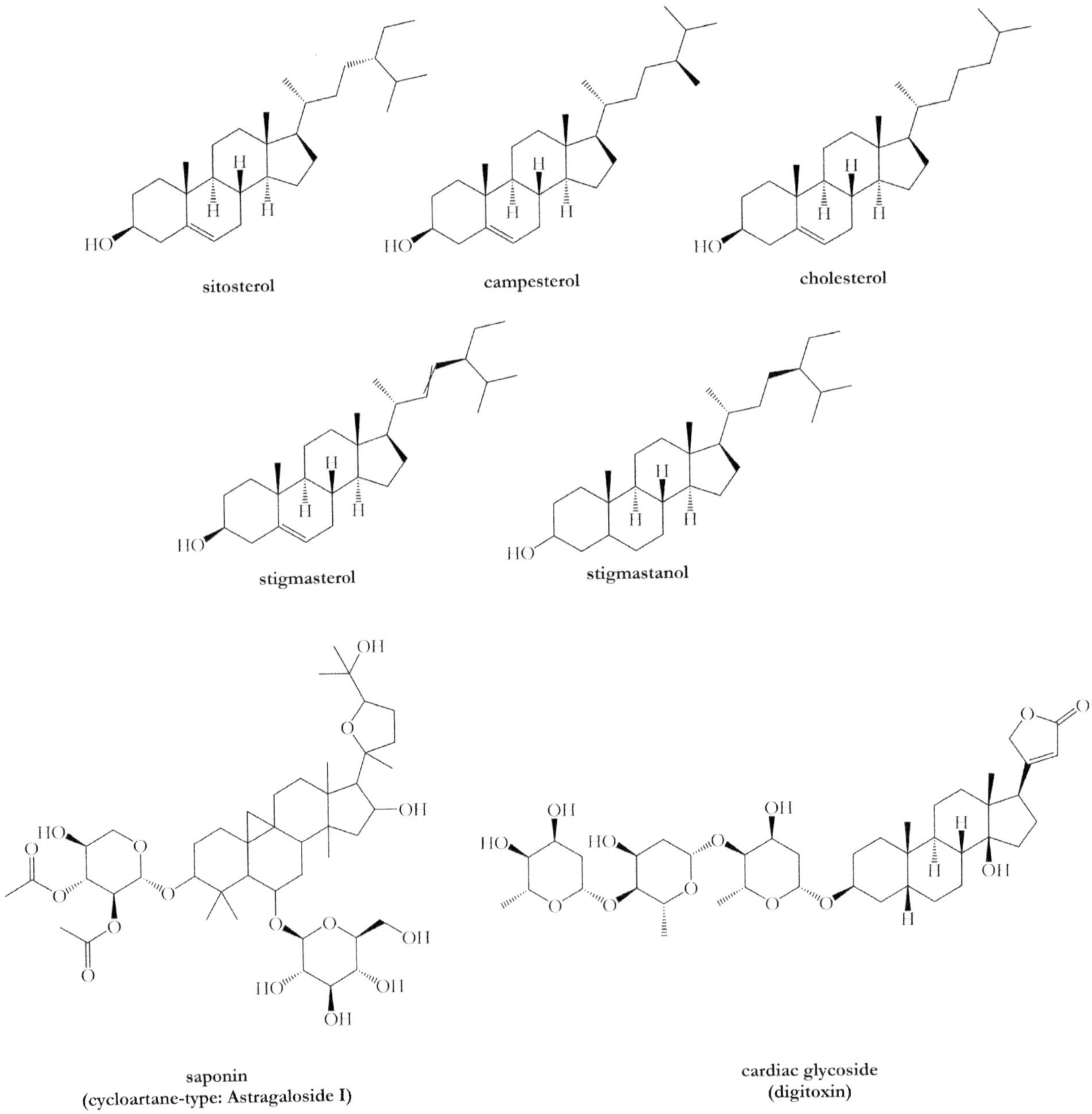

Fig. 3.13. Sterol-type and saponin-type infraclasses of the terpenoid-type triterpenoid subclass, with an example of a cycloartane-type saponin (Astragaloside I) and a cardiac glycoside-type saponin (digitoxin).

exhaustively detailed in the literature, particularly in introductory biology, biochemistry, biomedicine and botany textbooks, and lie outside the scope of this book. The remaining specialised triterpenoid contents of the plant metabolome are often found in a glycoconjugated form, which has been shown to confer pest/pathogen resistance to the plant.[5] The biosynthetic pathway of this large group of terpenoids is complex and involves a myriad of specialised enzyme systems. A significant branching point is the common precursor 2,3-oxidosqualene, which can initially assume one of two basic forms known as chair–boat–chair (CBC), which leads to the tetracyclic triterpenoids, or chair–chair–chair (CCC), which leads to the pentacyclic triterpenoids.[39] The triterpenoids have been implicated and thoroughly reviewed over the past 10 years with respect to their anti-inflammatory, antiviral, antineoplastic, anti-HIV, hypoglycaemic, antiparasitic, immunomodulatory, neuroprotective, anti-Alzheimer's, antidiabetic, neuropharmacological and insecticidal bioactivities.[41-46] The mechanisms implicated in these biological roles are too numerous to cover here. Further information on this topic is available in the many excellent published reviews that are referenced throughout this section and elsewhere.

Five specific infraclasses of triterpenoid are of particular interest with respect to their biological role either *in planta* or as potential pharmacophores: brassinosteroids, limonoids, quassinoids, cucurbitanes and saponins (cardiac glycosides). Brassinosteroids are considered a sixth infraclass-type of phytohormone (alongside auxins, gibberellins, cytokinins, abscisic acid and ethylene, the discussion of which can be found in any introductory plant-physiology textbook). They are growth promoters and stress protectors (drought, temperature, heavy metals, herbicidal injury and salinity) and are structurally similar to the ecdysteroids of the animal kingdom.[47] Brassinosteroids have been reported from the dicotyledonous Apetalae (Betulaceae, Cannabaceae, Caryophyllaceae, Fagaceae, Polygonaceae), Chloripetalae (Apiaceae, Brassicaceae, Fabaceae, Myrtaceae, Rosaceae, Rutaceae, Theaceae) and Sympetalae (Apocynaceae, Asteraceae, Boraginaceae, Convolvulaceae, Cucurbitaceae, Lamiaceae, Solanaceae), as well as the Gymnosperms (Cupressaceae, Ginkgoaceae, Pinaceae, Taxodiaceae) and the lower plants (Equisetaceae, Hydrodictyaceae, Marchantiaceae).[47] Limonoids were investigated initially as the bitter principle of *Citrus* fruit, but their incidental biological activity (insecticidal, antifeedant, antimicrobial, antiviral, antineoplastic, etc.) has resulted in further significant study.[19] The 300 reported limonoids are confined to the order Sapindales (syn. Rutales), particularly in the Meliaceae and Rutaceae, but also the Cneoraceae and Simaroubaceae.[19]

Quassinoids are the bitter principles from the Simaroubaceae family and are divided into five groups based on their carbon skeleton (C_{18}, C_{19}, C_{20}, C_{22}, C_{25}).[48] Interest in the quassinoids was predicated upon the discovery of their antileukaemic activity in the 1970s and since then they have shown significant antineoplastic, antimalarial, antiviral, anti-inflammatory, antifeedant, insecticidal, amoebicidal, antiulcer and herbicidal bioactivities.[48] Cucurbitanes (including the cucurbitacins) were originally isolated as the bitter principle of the Cucurbitaceae, but have since been isolated from the Brassicaceae, Scrophulariaceae, Begoniaceae, Elaeocarpaceae, Datiscaceae, Desfontainiaceae, Polemoniaceae, Primulaceae, Rubiaceae, Sterculiaceae, Rosaceae and Thymelaceae.[49] Cucurbitanes were initially of interest due to their marked cytotoxicity, but their non-specificity limited their medicinal application. Since that early research, the cucurbitanes have been well studied regarding their purgative, anti-inflammatory and antifertility activities.[49]

Finally, the saponins are in themselves a huge and structurally diverse group of triterpenoid/steroidal glycosides derived from the dicotyledonous Magneliopsida and the monocotyledonous Liliopsida plant classes.[5] The saponins are named due to the Latin word for soap (*sapo*), as these compounds have natural surfactant properties due to their mixed lipophilic/hydrophilic structure. The saponins are produced *in planta* as phytoalexins due to their antimicrobial and antifeedant properties, as well as being implicated in a huge range of biological activities.[50]

A significant infraclass grouped within the saponins are known as the cardiac glycosides, derived from the Apocynaceae (*Nerium* - oleander), Asparagaceae (*Convallaria* - lily of the valley), Moraceae (*Maquira* and Naucleopsis - South American poison dart genera) and Plantaginaceae (*Digitalis* - foxglove) plant families.[51] Plants from these families have been used in global traditional medicinal pharmacopoeia (Danish, German, Indian, Chinese, etc.) as heart tonics, diuretics and emetics, as well as for more sinister uses such as poison arrows/darts, rodent poisons and for murder/suicide.[51] Notable and well-studied examples are digitoxin (*Fig. 3.13*), digoxin, bufalin, oubain, oleandrin, thevetin, convallotoxin, peruvoside, proceraside, antiaroside, antiaritoxiosides, strophanthidin and antiarigenin, including crude extracts/mixtures of digitalis and digitalin. The cardiac glycosides manipulate the renin-angiotensin axis of the heart and can improve cardiac output through sodium-potassium ATPase affinity, but their safety/efficacy has come under serious scrutiny in modern medicine.[52]

REPRESENTATION

This section presents a synopsis of the specialised metabolic trends contained within the flora presented in *Tables 3.1 -3.4*, with a focus on any emergent trends in therapeutic use that correlate with reported biological activity in the sub/infraclass.

Monoterpenoids (acyclic)

A summary of the Australian customary medicinal flora reported to bear acyclic-type compounds of the monoterpenoid subclass is as follows: Apiales [Apiaceae

Table 3.1. Terpenoid-type (monoterpenoid) specialised metabolite distribution across the customary medicinal flora of Australia

Order	Family	Genus	Acyclic	Irregular	Pyrethrin	Iridoid	Secoiridoid	Cyclohexane	Megastigmane	Camphane	Bornane	Fenchane	Pinane	Carane	Thujane	Acyclic	Irregular
Apiales	Apiaceae	*Centella*[70,71]	+					+					+			+	
Apiales	Pittosporaceae	*Pittosporum*[72,73]	+			+							+			+	
Asparagales	Amaryllidaceae	*Crinum*[74–77]							+								
Asparagales	Orchidaceae	*Dendrobium*[78]	+					+								+	
Asterales	Asteraceae	*Acmella*[79]	+					+								+	
Asterales	Asteraceae	*Ageratum*[80]	+					+								+	
Asterales	Asteraceae	*Pterocaulon*[81,82]	+													+	
Asterales	Asteraceae	*Sphagneticolia*[83–86]						+									
Asterales	Goodeniaceae	*Goodenia*[87]				+											
Asterales	Goodeniaceae	*Scaevola*[88,89]				+											
Caryophyllales	Amaranthaceae	*Achyranthes*[90]	+			+		+								+	
Caryophyllales	Amaranthaceae	*Dysphania*[91,92]	+					+		+			+	+		+	
Caryophyllales	Portulaceae	*Portulaca*[93]	+													+	
Cucurbitales	Cucurbitaceae	*Trichosanthes*[94–98]						+	+								
Fabales	Fabaceae	*Erythrophleum*[99]							+								
Fabales	Fabaceae	*Tephrosia*[100–102]						+									
Fagales	Casuarinaceae	*Casuarina*[103]	+					+								+	
Gentianales	Apocynaceae	*Alstonia*[104]				+	+										
Gentianales	Apocynaceae	*Alyxia*[105]				+	+										
Gentianales	Apocynaceae	*Cerbera*[106]				+			+								
Gentianales	Apocynaceae	*Cynanchum*[107]	+													+	
Gentianales	Apocynaceae	*Tabernaemontana*[108–110]					+										
Gentianales	Gentianaceae	*Schenkia*[111–114]					+										
Gentianales	Loganiaceae	*Strychnos*[115–119]				+	+										
Gentianales	Rubiaceae	*Coelospermum*[120–123]				+											
Gentianales	Rubiaceae	*Morinda*[124–126]				+											
Lamiales	Lamiaceae	*Ajuga*[127]				+											
Lamiales	Lamiaceae	*Callicarpa*[128]	+			+							+			+	
Lamiales	Lamiaceae	*Clerodendrum*[129]				+			+								
Lamiales	Lamiaceae	*Mentha*[130]	+					+								+	
Lamiales	Lamiaceae	*Ocimum*[131]	+			+		+								+	
Lamiales	Lamiaceae	*Plectranthus*[132]	+					+		+			+	+	+	+	
Lamiales	Lamiaceae	*Prostanthera*[133–138]	+					+					+	+	+	+	
Lamiales	Lamiaceae	*Prunella*[139]								+			+				
Lamiales	Lamiaceae	*Teucrium*[140,141]	+			+		+					+			+	
Lamiales	Lamiaceae	*Vitex*[142–144]	+			+		+								+	
Lamiales	Scrophulariaceae	*Eremophila*[145]	+			+		+				+	+	+		+	
Lamiales	Scrophulariaceae	*Myoporum*[146–151]				+											

Order	Family	Genus	Acyclic	Irregular	Pyrethrin	Iridoid	Secoiridoid	Cyclohexane	Megastigmane	Camphane	Bornane	Fenchane	Pinane	Carane	Thujane	Acyclic	Irregular
Lamiales	Verbenaceae	*Verbena*[144,152,153]				+											
Laurales	Hernandiaceae	*Hernandia*[154,155]						+									
Laurales	Lauraceae	*Litsea*[156]	+	+				+		+	+	+	+		+	+	+
Magnoliales	Myristicaceae	*Myristica*[157–159]	+					+								+	
Malphigiales	Euphorbiaceae	*Acalypha*[160]						+		+			+				
Malphigiales	Euphorbiaceae	*Aleurites*[161]						+									
Malphigiales	Euphorbiaceae	*Croton*[162,163]						+	+				+				
Malphigiales	Euphorbiaceae	*Mallotus*[164,165]				+											
Malphigiales	Phyllanthaceae	*Breynia*[166,167]							+								
Malphigiales	Phyllanthaceae	*Phyllanthus*[168–172]	+					+								+	
Malphigiales	Violaceae	*Hybanthus*[173,174]						+							+		
Malvales	Malvaceae	*Hibiscus*[175–178]						+									
Malvales	Malvaceae	*Malva*[179,180]	+					+	+							+	
Malvales	Malvaceae	*Sida*[181]							+								
Myrtales	Myrtaceae	*Melaleuca*[58]	+					+					+		+	+	
Myrtales	Myrtaceae	*Syncarpia*[182–184]				+											
Myrtales	Myrtaceae	*Syzygium*[185–187]	+					+					+			+	
Pandanales	Pandanaceae	*Pandanus*[188,189]	+													+	
Piperales	Piperaceae	*Piper*[190,191]	+					+		+			+	+		+	
Poales	Cyperaceae	*Cyperus*[192]				+							+				
Poales	Poaceae	*Cymbopogon*[193]	+					+								+	
Polypodiales	Pteridaceae	*Adiantum*[194]											+				
Proteales	Nelumbonaceae	*Nelumbo*[195]	+					+	+	+	+		+	+	+	+	
Ranunculales	Menispermaceae	*Cissampelos*[196]						+									
Rosales	Moraceae	*Ficus*[197,198]	+					+		+			+			+	
Santalales	Santalaceae	*Santalum*[199,200]	+					+		+	+	+	+		+	+	
Sapindales	Burseraceae	*Canarium*[201]	+					+				+	+	+		+	
Sapindales	Meliaceae	*Melia*[202]							+								
Sapindales	Rutaceae	*Melicope*[203–209]							+								
Sapindales	Rutaceae	*Philotheca*[210–215]	+										+		+	+	
Sapindales	Rutaceae	*Zieria*[216–220]	+					+						+		+	
Sapindales	Sapindaceae	*Dodonaea*[221]	+					+					+			+	
Sapindales	Simaroubaceae	*Ailanthus*[222]				+											
Solanales	Convolvulaceae	*Convolvulus*[223]						+					+				
Solanales	Solanaceae	*Nicotiana*[224–226]	+					+		+	+		+		+	+	
Solanales	Solanaceae	*Solanum*[227]	+			+		+								+	
Vitales	Vitaceae	*Cissus*[228,229]				+											
Zingiberales	Zingiberaceae	*Curcuma*[230]	+					+					+			+	

Note: genera with no reported phytochemistry have been omitted see (www.cmfoa.info)

Table 3.2. Terpenoid-type (sesquiterpenoid) specialised metabolite distribution across the customary medicinal flora of Australia

Order	Family	Genus	Farnesane	Bisabolane	Germacrane	Humulane	Cedrane	Drimane	Himachalane	Guaiane	Pterosin	Cuparane	Trichothecane	Chamigrane	Vetivane	Miscellaneous
Apiales	Apiaceae	*Centella*[70,71]	+	+	+	+		+								
Apiales	Pittosporaceae	*Pittosporum*[72,73]	+	+	+	+		+		+						
Asparagales	Orchidaceae	*Dendrobium*[78]								+						+
Asterales	Asteraceae	*Acmella*[79]		+		+		+								+
Asterales	Asteraceae	*Ageratum*[80]		+		+										
Asterales	Asteraceae	*Centipeda*[231]								+						
Asterales	Asteraceae	*Chrysocephalum*[232,233]								+						
Asterales	Asteraceae	*Eclipta*[234,235]														+
Asterales	Asteraceae	*Elephantopus*[236]			+											+
Asterales	Asteraceae	*Gnaphalium*[237]								+						
Asterales	Asteraceae	*Helichrysum*[237]								+						
Asterales	Asteraceae	*Pterocaulon*[81,82]	+	+	+	+		+								
Asterales	Asteraceae	*Sphagneticolia*[83–86]			+			+		+						+
Asterales	Goodeniaceae	*Goodenia*[87]		+												
Boraginales	Boraginaceae	*Heliotropium*[238]	+													
Brassicales	Cleomaceae	*Cleome*[239]			+			+		+						
Canellales	Winteraceae	*Tasmannia*[240,241]						+		+						
Caryophyllales	Amaranthaceae	*Achyranthes*[90]								+						
Caryophyllales	Amaranthaceae	*Dysphania*[91,92]	+		+	+		+		+	+					
Caryophyllales	Polygonaceae	*Persicaria*[242–245]		+		+		+								
Fabales	Fabaceae	*Indigofera*[246]		+		+		+		+						
Fabales	Fabaceae	*Milettia*[247,248]				+		+								
Fabales	Fabaceae	*Tephrosia*[100–102]			+					+	+					
Fagales	Casuarinaceae	*Casuarina*[103]				+										
Gentianales	Apocynaceae	*Carissa*[249]			+			+								
Lamiales	Lamiaceae	*Ajuga*[127]						+					+			
Lamiales	Lamiaceae	*Callicarpa*[128]			+	+		+	+	+						
Lamiales	Lamiaceae	*Ocimum*[131]		+	+	+										+
Lamiales	Lamiaceae	*Plectranthus*[132]		+	+	+		+		+						
Lamiales	Lamiaceae	*Prostanthera*[133–138]			+	+		+	+	+						
Lamiales	Lamiaceae	*Prunella*[139]			+	+		+								
Lamiales	Lamiaceae	*Teucrium*[140,141]			+					+						
Lamiales	Lamiaceae	*Vitex*[142–144]						+		+						
Lamiales	Scrophulariaceae	*Eremophila*[145]	+	+			+	+		+						+
Lamiales	Scrophulariaceae	*Myoporum*[146–151]	+	+						+						+
Laurales	Lauraceae	*Litsea*[156]		+	+	+	+	+		+						
Malphigiales	Euphorbiaceae	*Chamaesyce*[250]	+			+										
Malphigiales	Euphorbiaceae	*Croton*[162,163]			+	+		+		+						+
Malphigiales	Euphorbiaceae	*Euphorbia*[250]	+			+										

Order	Family	Genus	Farnesane	Bisabolane	Germacrane	Humulane	Cedrane	Drimane	Himachalane	Guaiane	Pterosin	Cuparane	Trichothecane	Chamigrane	Vetivane	Miscellaneous
Malphigiales	Rhizophoraceae	*Rhizophora*[251,252]	+													
Myrtales	Lythraceae	*Ammannia*[253]	+													
Myrtales	Myrtaceae	*Melaleuca*[58]				+		+		+						
Myrtales	Myrtaceae	*Syzygium*[185–187]			+	+										
Pinales	Cupressaceae	*Callitris*[6,364]						+		+						
Piperales	Piperaceae	*Piper*[190,191]		+	+	+		+		+						+
Poales	Cyperaceae	*Scleria*[254–256]	+													
Polypodiales	Dennstaedtiaceae	*Pteridium*[257–260]				+		+								
Proteales	Nelumbonaceae	*Nelumbo*[195]		+	+			+		+						
Rosales	Moraceae	*Ficus*[197,198]		+	+	+		+		+						
Santalales	Santalaceae	*Santalum*[199,200]	+	+	+	+	+	+		+						+
Sapindales	Burseraceae	*Canarium*[201]	+		+	+		+		+						
Sapindales	Rutaceae	*Melicope*[203–209]	+													+
Sapindales	Rutaceae	*Philotheca*[210–215]						+		+						+
Sapindales	Rutaceae	*Zieria*[216–220]								+						
Sapindales	Sapindaceae	*Allophylus*[261]						+								
Sapindales	Sapindaceae	*Dodonaea*[221]				+										
Solanales	Convolvulaceae	*Convolvulus*[223]			+			+								
Solanales	Solanaceae	*Nicotiana*[224–226]		+	+	+	+			+					+	+
Solanales	Solanaceae	*Solanum*[227]	+			+		+		+					+	
Zingiberales	Zingiberaceae	*Curcuma*[230]	+	+	+	+		+		+						+

Note: grey rows have been taxonomically reclassified since respective publications, genera with no reported phytochemistry have been omitted; see (www.cmfoa.info).

Table 3.3. Terpenoid-type (diterpenoid) specialised metabolite distribution across the customary medicinal flora of Australia

Order	Family	Genus	Acyclic	Cembrane	Casbane	Dolabellane	Jatrophane	Daphnane	Taxane	Labdane	Halimane	Clerodane	Abietane	Pimarane	Rosane	Cassane	Kaurane
Apiales	Apiaceae	*Centella*[70,71]								+							+
Apiales	Pittosporaceae	*Pittosporum*[72,73]	+														+
Asterales	Asteraceae	*Ageratum*[80]	+														
Asterales	Asteraceae	*Chrysocephalum*[232,233]								+							+
Asterales	Asteraceae	*Gnaphalium*[237]								+							+
Asterales	Asteraceae	*Helichrysum*[237]								+							+
Asterales	Asteraceae	*Sphagneticolia*[83–86]															+
Asterales	Goodeniaceae	*Goodenia*[87]															+
Brassicales	Cleomaceae	*Cleome*[239]		+		+											
Fabales	Fabaceae	*Acacia*[262]								+						+	
Fabales	Fabaceae	*Caesalpinia*[263]													+	+	
Fabales	Fabaceae	*Cassia*[264]														+	

Table 3.3. continued

Order	Family	Genus	Acyclic	Cembrane	Casbane	Dolabellane	Jatrophane	Daphnane	Taxane	Labdane	Hallmane	Clerodane	Abietane	Pimarane	Rosane	Cassane	Kaurane
Fabales	Fabaceae	*Chamaecrista*[264]														+	
Fabales	Fabaceae	*Entada*[265,266]										+					
Fabales	Fabaceae	*Erythrophleum*[99]														+	
Fabales	Fabaceae	*Indigofera*[246]	+										+				
Fabales	Fabaceae	*Senna*[264]														+	
Lamiales	Lamiaceae	*Ajuga*[127]										+					
Lamiales	Lamiaceae	*Basilicum*[267,268]								+				+			
Lamiales	Lamiaceae	*Callicarpa*[128]								+		+	+	+			+
Lamiales	Lamiaceae	*Clerodendrum*[129]								+							
Lamiales	Lamiaceae	*Ocimum*[131]											+				
Lamiales	Lamiaceae	*Plectranthus*[132]											+				+
Lamiales	Lamiaceae	*Teucrium*[140,141]										+	+				
Lamiales	Lamiaceae	*Vitex*[142–144]								+	+	+	+	+			
Lamiales	Plantaginaceae	*Stemodia*[269–274]											+				
Lamiales	Scrophulariaceae	*Eremophila*[145]	+	+						+							
Malphigiales	Euphorbiaceae	*Aleurites*[161]									+	+					
Malphigiales	Euphorbiaceae	*Beyeria*[275]								+				+			+
Malphigiales	Euphorbiaceae	*Chamaesyce*[250]			+		+	+					+	+			+
Malphigiales	Euphorbiaceae	*Croton*[162,163]		+						+	+	+		+			+
Malphigiales	Euphorbiaceae	*Euphorbia*[250]			+		+	+					+	+			+
Malphigiales	Euphorbiaceae	*Excoecaria*[276]						+		+			+				+
Malphigiales	Euphorbiaceae	*Mallotus*[164,165]		+	+					+							
Malphigiales	Phyllanthaceae	*Phyllanthus*[168–172]															+
Malphigiales	Picrodendraceae	*Petalostigma*[277,278]												+			
Malphigiales	Rhizophoraceae	*Rhizophora*[251,252]								+			+	+			+
Malvales	Malvaceae	*Hibiscus*[175–178]	+														
Malvales	Thymelaceae	*Pimelea*[279,280]						+									
Malvales	Thymelaceae	*Wikstroemia*[281–285]		+													
Pinales	Araucariaceae	*Araucaria*[286]								+			+				+
Pinales	Cupressaceae	*Callitris*[63,64]								+			+				
Polypodiales	Pteridaceae	*Adiantum*[194]	+							+							
Ranunculales	Menispermaceae	*Tinospora*[287–289]										+					
Sapindales	Meliaceae	*Cedrela*[290–293]												+			
Sapindales	Meliaceae	*Melia*[202]											+				
Sapindales	Meliaceae	*Toona*[290–293]												+			
Sapindales	Sapindaceae	*Dodonaea*[221]								+		+					
Sapindales	Simaroubaceae	*Ailanthus*[222]		+													
Solanales	Convolvulaceae	*Ipomoea*[294]	+														
Solanales	Solanaceae	*Nicotiana*[224–226]	+	+						+							
Solanales	Solanaceae	*Solanum*[227]	+										+				+
Zingiberales	Zingiberaceae	*Curcuma*[230]								+							

Note: genera with no reported phytochemistry have been omitted; see (www.cmfoa.info).

Table 3.4. Terpenoid-type (triterpenoid) specialised metabolite distribution across the customary medicinal flora of Australia

Order	Family	Genus	Acyclic	Lanostane	Fusidane	Dammarane	Cycloartane	Brassinosteroid	Limonoid	Quassinoid	Lupane	Hopane	Ursane	Oleane	Steroidal	Saponin	Cardiac Glycoside	Acyclic	Lanostane
Alismatales	Araceae	*Amorphophallus*[295,296]				+					+				+	+			
Apiales	Pittosporaceae	*Pittosporum*[72,73]												+	+	+			
Arecales	Arecaceae	*Calamus*[297]											+			+			
Asparagales	Amaryllidaceae	*Crinum*[74–77]					+						+		+				
Asparagales	Asparagaceae	*Asparagus*[298]													+	+			
Asparagales	Orchidaceae	*Dendrobium*[78]													+				
Asterales	Asteraceae	*Acmella*[79]												+	+	+			
Asterales	Asteraceae	*Ageratum*[80]											+		+				
Asterales	Asteraceae	*Centipeda*[231]											+		+	+			
Asterales	Asteraceae	*Chrysocephalum*[232,233]	+															+	
Asterales	Asteraceae	*Eclipta*[234,235]									+		+	+	+	+	+		
Asterales	Asteraceae	*Elephantopus*[236]									+		+		+				
Asterales	Asteraceae	*Gnaphalium*[237]	+															+	
Asterales	Asteraceae	*Helichrysum*[237]	+															+	
Asterales	Asteraceae	*Pterocaulon*[81,82]	+												+			+	
Asterales	Asteraceae	*Sphagneticolia*[83–86]	+								+			+	+	+		+	
Asterales	Campanulaceae	*Lobelia*[299]											+	+	+				
Asterales	Goodeniaceae	*Goodenia*[87]											+	+					
Boraginales	Boraginaceae	*Ehretia*[300,301]									+				+				
Boraginales	Boraginaceae	*Heliotropium*[238]					+						+	+	+				
Boraginales	Boraginaceae	*Trichodesma*[302–304]									+		+	+	+				
Brassicales	Capparaceae	*Capparis*[305]					+						+	+	+				
Brassicales	Cleomaceae	*Cleome*[239]				+							+		+				
Caryophyllales	Amaranthaceae	*Achyranthes*[90]												+	+	+			
Caryophyllales	Amaranthaceae	*Atriplex*[306,307]													+	+			
Caryophyllales	Amaranthaceae	*Deeringia*[308]												+					
Caryophyllales	Amaranthaceae	*Dysphania*[91,92]													+	+			
Caryophyllales	Nyctaginaceae	*Boerhavia*[309]												+	+				
Caryophyllales	Plumbaginaceae	*Plumbago*[310]													+				
Caryophyllales	Polygonaceae	*Persicaria*[242–245]				+					+				+	+			
Caryophyllales	Portulaceae	*Portulaca*[93]									+			+	+	+			
Commelinales	Commelinaceae	*Cyanotis*[311,312]													+				
Cucurbitales	Cucurbitaceae	*Trichosanthes*[94–98]		+			+							+	+				+
Dioscoreales	Dioscoreaceae	*Dioscorea*[313,314]														+			
Ericales	Lecythidaceae	*Barringtonia*[315]									+			+	+	+			
Ericales	Lecythidaceae	*Planchonia*[316–320]											+	+	+				
Ericales	Primulaceae	*Aegiceras*[321]												+		+			

Table 3.4. continued

Order	Family	Genus	Acyclic	Lanostane	Fusidane	Dammarane	Cycloartane	Brassinosteroid	Limonoid	Quassinoid	Lupane	Hopane	Ursane	Oleane	Steroidal	Saponin	Cardiac Glycoside	Acyclic	Lanostane
Ericales	Sapotaceae	*Planchonella*[322–328]											+	+	+	+			
Fabales	Fabaceae	*Abrus*[329]					+					+		+		+			
Fabales	Fabaceae	*Acacia*[262]												+	+	+			
Fabales	Fabaceae	*Caesalpinia*[263]									+		+		+	+			
Fabales	Fabaceae	*Cassia*[264]													+				
Fabales	Fabaceae	*Chamaecrista*[264]													+				
Fabales	Fabaceae	*Entada*[265,266]									+					+			
Fabales	Fabaceae	*Erythrophleum*[99]		+										+	+	+			+
Fabales	Fabaceae	*Indigofera*[246]									+		+	+	+				
Fabales	Fabaceae	*Lysiphyllum*[330]													+				
Fabales	Fabaceae	*Milettia*[247,248]					+				+			+					
Fabales	Fabaceae	*Rhynchosia*[331]									+				+				
Fabales	Fabaceae	*Senna*[264]													+				
Fabales	Fabaceae	*Sesbania*[332,333]									+			+		+			
Fabales	Fabaceae	*Sophora*[334–336]									+		+	+	+	+			
Fabales	Fabaceae	*Tephrosia*[100–102]									+		+	+	+				
Fabales	Fabaceae	*Vigna*[337–339]													+				
Fagales	Casuarinaceae	*Casuarina*[103]									+				+				
Gentianales	Apocynaceae	*Alstonia*[104]									+		+	+	+	+	+		
Gentianales	Apocynaceae	*Carissa*[249]									+	+	+	+	+				
Gentianales	Apocynaceae	*Cerbera*[106]													+	+	+		
Gentianales	Apocynaceae	*Cynanchum*[107]									+		+	+	+				
Gentianales	Apocynaceae	*Marsdenia*[340]									+		+	+	+	+			
Gentianales	Apocynaceae	*Pentatropis*[341]					+												
Gentianales	Apocynaceae	*Tabernaemontana*[108–110]									+		+	+	+				
Gentianales	Loganiaceae	*Strychnos*[115–119]									+	+	+		+	+			
Gentianales	Rubiaceae	*Morinda*[124–126]													+				
Gentianales	Rubiaceae	*Nauclea*[342,343]	+										+		+			+	
Gentianales	Rubiaceae	*Timonius*[344,345]												+					
Lamiales	Lamiaceae	*Ajuga*[127]													+				
Lamiales	Lamiaceae	*Basilicum*[267,268]									+								
Lamiales	Lamiaceae	*Callicarpa*[128]									+		+	+					
Lamiales	Lamiaceae	*Clerodendrum*[129]									+		+	+	+	+			
Lamiales	Lamiaceae	*Ocimum*[131]											+	+	+				
Lamiales	Lamiaceae	*Plectranthus*[132]											+	+	+				
Lamiales	Lamiaceae	*Prostanthera*[133–138]											+		+				
Lamiales	Lamiaceae	*Prunella*[139]									+		+	+	+	+			

Order	Family	Genus	Acyclic	Lanostane	Fusidane	Dammarane	Cycloartane	Brassinosteroid	Limonoid	Quassinoid	Lupane	Hopane	Ursane	Oleane	Steroidal	Saponin	Cardiac Glycoside	Acyclic	Lanostane
Lamiales	Lamiaceae	*Teucrium*[140,141]						+			+		+	+	+				
Lamiales	Lamiaceae	*Vitex*[142–144]									+		+	+	+				
Lamiales	Plantaginaceae	*Bacopa*[346]				+										+			
Lamiales	Plantaginaceae	*Scoparia*[347–350]									+		+	+	+				
Lamiales	Plantaginaceae	*Stemodia*[269–274]									+				+	+			
Lamiales	Scrophulariaceae	*Eremophila*[145]											+	+	+				
Lamiales	Verbenaceae	*Verbena*[144,152,153,]											+	+	+				
Laurales	Lauraceae	*Cassytha*[351,352]													+				
Laurales	Lauraceae	*Litsea*[156]	+								+		+	+	+			+	
Liliales	Smilacaceae	*Smilax*[353–358]													+	+			
Malphigiales	Calophyllaceae	*Calophyllum*[359,360]	+								+		+	+	+			+	
Malphigiales	Euphorbiaceae	*Acalypha*[160]					+												
Malphigiales	Euphorbiaceae	*Aleurites*[161]										+	+	+	+				
Malphigiales	Euphorbiaceae	*Beyeria*[275]									+								
Malphigiales	Euphorbiaceae	*Chamaesyce*[250]	+	+							+		+	+	+	+		+	+
Malphigiales	Euphorbiaceae	*Croton*[162,163]													+				
Malphigiales	Euphorbiaceae	*Euphorbia*[250]	+	+							+		+	+	+	+		+	+
Malphigiales	Euphorbiaceae	*Excoecaria*[276]					+				+			+	+				
Malphigiales	Euphorbiaceae	*Mallotus*[164,165]	+								+	+	+	+	+		+	+	
Malphigiales	Phyllanthaceae	*Phyllanthus*[168–172]		+							+		+	+	+				+
Malphigiales	Rhizophoraceae	*Rhizophora*[251,252]									+	+	+	+	+				
Malphigiales	Violaceae	*Hybanthus*[173,174]										+			+				
Malvales	Malvaceae	*Grewia*[361–363]									+		+	+	+				
Malvales	Malvaceae	*Hibiscus*[175–178]													+				
Malvales	Malvaceae	*Sida*[181]											+		+				
Malvales	Malvaceae	*Sterculia*[364,365]					+				+			+	+				
Malvales	Malvaceae	*Thespesia*[366–369]													+				
Malvales	Thymelaceae	*Wikstroemia*[281–285]													+				
Myrtales	Lythraceae	*Ammannia*[253]									+			+	+				
Myrtales	Lythraceae	*Lythrum*[370]											+	+	+				
Myrtales	Myrtaceae	*Syncarpia*[182–184]											+	+	+				
Myrtales	Myrtaceae	*Syzygium*[185–187]												+	+				
Pandanales	Pandanaceae	*Pandanus*[188,189]		+			+								+				+
Poales	Cyperaceae	*Cyperus*[192]	+															+	
Poales	Poaceae	*Cymbopogon*[193]													+				
Polypodiales	Pteridaceae	*Adiantum*[194]		+							+	+	+	+		+			+
Proteales	Nelumbonaceae	*Nelumbo*[195]		+							+		+	+	+				+
Ranunculales	Menispermaceae	*Tinospora*[287–289]					+								+				

Table 3.4. continued

Order	Family	Genus	Acyclic	Lanostane	Fusidane	Dammarane	Cycloartane	Brassinosteroid	Limonoid	Quassinoid	Lupane	Hopane	Ursane	Oleane	Steroidal	Saponin	Cardiac Glycoside	Acyclic	Lanostane
Ranunculales	Ranunculaceae	*Clematis*[371]												+	+	+			
Rosales	Moraceae	*Ficus*[197,198]									+		+		+				
Rosales	Rhamnaceae	*Alphitonia*[372,373]				+					+				+				
Rosales	Rhamnaceae	*Ventilago*[374–380]									+				+				
Rosales	Rhamnaceae	*Ziziphus*[381–384]			+								+	+	+				
Rosales	Rosaceae	*Rubus*[385,386]											+	+	+	+			
Santalales	Loranthaceae	*Amyema*															+		
Santalales	Olacaceae	*Ximenia*[387,388]					+				+		+	+	+				
Santalales	Santalaceae	*Santalum*[199,200]				+									+				
Sapindales	Burseraceae	*Canarium*[201]											+	+	+	+			
Sapindales	Meliaceae	*Cedrela*[65,290–293]		+		+			+						+				+
Sapindales	Meliaceae	*Melia*[202]				+									+				
Sapindales	Meliaceae	*Owenia*[389,390]				+													
Sapindales	Meliaceae	*Toona*[65,290–293]		+		+			+						+				+
Sapindales	Rutaceae	*Melicope*[203–209]				+								+	+				
Sapindales	Sapindaceae	*Allophylus*[261]													+				
Sapindales	Sapindaceae	*Dodonaea*[221]									+			+	+	+			
Sapindales	Simaroubaceae	*Ailanthus*[67,222]				+	+			+	+	+	+		+				
Sapindales	Simaroubaceae	*Brucea*[66,391]				+				+					+				
Solanales	Convolvulaceae	*Convolvulus*[223]									+		+	+	+	+			
Solanales	Convolvulaceae	*Evolvulus*[392]													+	+			
Solanales	Convolvulaceae	*Ipomoea*[294]											+	+	+				
Solanales	Convolvulaceae	*Operculina*[393–396]		+		+	+								+	+			+
Solanales	Solanaceae	*Nicotiana*[224–226]													+	+			
Solanales	Solanaceae	*Solanum*[227]					+				+		+	+	+	+			
Vitales	Vitaceae	*Ampelocissus*													+				
Vitales	Vitaceae	*Cayratia*[397]											+						
Vitales	Vitaceae	*Cissus*[228,229]													+				
Zingiberales	Musaceae	*Musa*[398,399]													+				
Zygophyllales	Zygophyllaceae	*Tribulus*[400–405]													+	+			

Note: genera with no reported phytochemistry have been omitted; see (www.cmfoa.info).

(*Centella*)], Asparagales [Asparagaceae (*Asparagus*)], Asterales [Asteraceae (*Elephantopus*, *Gnaphalium*, *Helichrysum*)], Brassicales [Cleomaceae (*Cleome*)], Fabales [Fabaceae (*Canavalia*, *Entada*, *Senna*)], Liliales [Smilacaceae (*Smilax*)], Malphigiales [Euphorbiaceae (*Beyeria*)], Myrtales [Myrtaceae (*Angophora*, *Eucalyptus*, *Syncarpia*, *Syzygium*)], Oxalidales [Cunoniaceae (*Eucryphia*)], Pinales [Cupressaceae (*Callitris*)], Piperales [Piperaceae (*Piper*)], Polypodiales [Pteridaceae (*Adiantum*)], Rosales [Rhamnaceae (*Alphitonia*)], Santalales [Olacaceae (*Ximenia*)], Solanales [Convolvulaceae (*Convolvulus*)] and Zingiberales [Musaceae (*Musa*)]. There were no obvious trends in the customary uses over this disparate grouping of flora, with seven genera being used as tonics, six for analgesia, six for coughs/colds and six for stomach ailments.[53]

Monoterpenoids (monocyclic)

A summary of the Australian customary medicinal flora reported to bear iridoid-type compounds of the monoterpenoid subclass is as follows: Asparagales [Asparagaceae (*Asparagus*)], Caryophyllales [Polygonaceae (*Persicaria*), Portulaceae (*Portulaca*)], Fabales [Fabaceae (*Canavalia*)], Magnoliales [Myristicaceae (*Myristica*)], Malphigiales [Calophyllaceae (*Calophyllum*), Euphorbiaceae (*Aleurites*)], Malvales [Malvaceae (*Grewia, Hibiscus, Sida*)], Myrtales [Lythraceae (*Ammannia*), Myrtaceae (*Angophora, Corymbia, Syncarpia*)], Pinales [Cupressaceae (*Callitris*)], Piperales [Piperaceae (*Piper*)], Polypodiales [Pteridaceae (*Adiantum*)], Proteales [Nelumbonaceae (*Nelumbo*), Proteaceae (*Grevillea*)] and Sapindales [Rutaceae (*Philotheca*)]. A pattern emerges from the customary use of the flora reported to bear compounds of this infraclass, in that seven species were used for inflammation, seven for bactericidal applications and six for stomach ailments.[53] Considering the reported anti-inflammatory and antimicrobial bioactivities attributed to these compounds, this forms a meaningful correlation.[27,28]

A summary of the Australian customary medicinal flora reported to bear cyclohexane-type compounds of the monoterpenoid subclass is as follows: Apiales [Apiaceae (*Centella*)], Asterales [Asteraceae (*Elephantopus, Gnaphalium, Helichrysum*)], Canellales [Winteraceae (*Tasmannia*)], Fabales [Fabaceae (*Canavalia, Entada, Vigna*)], Laurales [Lauraceae (*Cassytha*)], Liliales [Smilacaceae (*Smilax*)], Myrtales [Myrtaceae (*Eucalyptus, Melalcuca, Syncarpia, Syzygium*)], Oxalidales [Cunoniaceae (*Eucryphia*)], Pinales [Cupressaceae (*Callitris*)], Piperales [Piperaceae (*Piper*)], Polypodiales [Pteridaceae (*Adiantum*)], Ranunculales [Menispermaceae (*Tinospora*)], Rosales [Rhamnaceae (*Alphitonia*)], Santalales [Olacaceae (*Ximenia*)], Sapindales [Anacardiaceae (*Semecarpus*), Burseraceae (*Canarium*), Meliaceae (*Toona*)], Solanales [Convolvulaceae (*Convolvulus*), Solanaceae (*Nicotiana*)], Vitales [Vitaceae (*Cayratia*)] and Zingiberales [Musaceae (*Musa*)]. The customary use trends that emerge from this group of monoterpenoids lends support to their conventional use in the general populous as useful first-aid compounds, being used as tonics, analgesics, bactericides and for inflammation, cough and stomach ailments. In general, the cyclohexane monoterpenoids have been suggested to have cardiovascular, antiulcer, analgesic, insecticidal and antimicrobial effects.[20,54-57] These biological activities lend support to the customary use of these native flora by Aboriginal and Torres Strait Islander peoples, particularly the aromatic members of the Myrtaceae.[58,59]

Monoterpenoids (bicyclic)

A summary of the Australian customary medicinal flora reported to bear bicyclic-type compounds of the monoterpenoid subclass is as follows: Apiales [Apiaceae (*Centella*)], Asparagales [Asparagaceae (*Asparagus*)], Fabales [Fabaceae (*Entada*)], Myrtales [Myrtaceae (*Angophora, Syzygium*)], Oxalidales [Cunoniaceae (*Eucryphia*)], Pandanales [Pandanaceae (*Pandanus*)], Pinales [Cupressaceae (*Callitris*)], Polypodiales [Pteridaceae (*Adiantum*)], Rosales [Rhamnaceae (*Alphitonia*)], Sapindales [Anacardiaceae (*Semecarpus*), Meliaceae (*Toona*)] and Solanales [Solanaceae (*Nicotiana*)]. The customary use of this infraclass of monoterpenoids resembles the others in this subclass, in that they were largely used as analgesics, bactericides and for stomach ailments.[53] The camphanes have demonstrated insecticidal, antimicrobial, antiviral, cardiovascular, anticoccidial, antinociceptive, antitussive and antineoplastic bioactivity, but toxicity is a concern, as both the camphanes and the bornanes have been implicated in the enhancement of blood–brain barrier permeability.[60,61]

Sesquiterpenoids

Summarising the representation of these groupings proved repetitive, as many of the plants reported to bear one infraclass often bear other sesquiterpenoid infraclasses, so an overall set of customary flora reported to bear sesquiterpenoids is as follows: Apiales [Apiaceae (*Centella*)], Asparagales [Asparagaceae (*Asparagus*), Asphodelaceae (*Dianella*)], Asterales [Asteraceae (*Elephantopus, Gnaphalium, Helichrysum, Pterocaulon, Sphagneticolia*), Goodeniaceae (*Goodenia, Scaevola*)], Boraginales [Boraginaceae (*Ehretia, Heliotropium*)], Brassicales [Cleomaceae (*Cleome*)], Canellales [Winteraceae (*Tasmannia*)], Caryophyllales [Polygonaceae (*Persicaria*)], Cucurbitales [Cucurbitaceae (*Trichosanthes*)], Fabales [Fabaceae (*Abrus, Caesalpinia, Canavalia, Entada, Milettia*)], Lamiales [Lamiaceae (*Teucrium*), Phrymaceae (*Mimulus*)], Laurales [Lauraceae (*Cassytha*)], Liliales [Smilacaceae (*Smilax*)], Malphigiales [Euphorbiaceae (*Acalypha*)], Myrtales [Lythraceae (*Ammannia*), Myrtaceae (*Angophora, Syncarpia, Syzygium*)], Oxalidales [Cunoniaceae (*Eucryphia*)], Pandanales [Pandanaceae (*Pandanus*)], Pinales [Cupressaceae (*Callitris*)], Piperales

[Piperaceae (*Piper*)], Polypodiales [Pteridaceae (*Adiantum*)], Proteales [Nelumbonaceae (*Nelumbo*)], Rosales [Rhamnaceae (*Alphitonia*)], Sapindales [Meliaceae (*Melia*, *Toona*), Rutaceae (*Bosistoa*)] and Solanales [Convolvulaceae (*Operculina*)]. The majority of the customary uses these plants were as tonics (30%), bactericidal agents (30%) analgesics (25%), anti-inflammatories (25%), cough remedies (25%) and for stomach ailments (25%).[53] A more detailed breakdown of the distribution the infraclasses of terpenoid-type specialised metabolites of the sesquiterpenoid subclass reported from the customary medicinal flora of Australia is presented in *Table 3.2*.

Diterpenoids

The structural diversity of the diterpenoids leads to a similar situation as for the sesquiterpenoids, in that there was substantial overlap between plants bearing one infraclass to another and summarising the representation of these groupings proved repetitive. An overall set of customary flora reported to bear diterpenoids is as follows: Alismatales [Araceae (*Amorphophallus*)], Apiales [Apiaceae (*Centella*)], Asparagales [Asparagaceae (*Asparagus*)], Asterales [Asteraceae (*Helichrysum*, *Sphagneticolia*)], Boraginales [Boraginaceae (*Ehretia*, *Heliotropium*)], Canellales [Winteraceae (*Tasmannia*)], Caryophyllales [Polygonaceae (*Persicaria*)], Fabales [Fabaceae (*Abrus*, *Senna*)], Gentianales [Gentianaceae (*Schenkia*)], Gentianales [Rubiaceae (*Morinda*)], Lamiales [Lamiaceae (*Ajuga*, *Callicarpa*, *Plectranthus*, *Prunella*, *Teucrium*), Plantaginaceae (*Stemodia*)], Myrtales [Lythraceae (*Ammannia*, *Lythrum*), Myrtaceae (*Angophora*, *Corymbia*, *Syncarpia*, *Syzygium*)], Pandanales [Pandanaceae (*Pandanus*)], Pinales [Cupressaceae (*Callitris*)], Piperales [Piperaceae (*Piper*)], Polypodiales [Dennstaedtiaceae (*Pteridium*), Pteridaceae (*Adiantum*)], Rosales [Moraceae (*Ficus*)], Sapindales [Burseraceae (*Canarium*), Meliaceae (*Cedrela*, *Melia*, *Toona*), Rutaceae (*Bosistoa*, *Geijera*, *Philotheca*)], Solanales [Convolvulaceae (*Convolvulus*, *Ipomoea*, *Operculina*)] and Vitales [Vitaceae (*Cayratia*)]. Approximately 30% of these plants were used customarily to treat inflammation, or as bactericidal agents, and 20% of these plants were used customarily for stomach ailments or diarrhoea.[53] Many of the more specific structural motifs of the diterpenoids are restricted to a single plant family or genus, so for more nuanced insights into the biological activities of these infraclasses readers will have to consult the literature. The most commonly reported biological activity across the subclass is antimicrobial, which would support their use as bactericidal agents and for gastrointestinal-related issues.[62] A more detailed breakdown of the distribution of terpenoid-type specialised metabolites of the diterpenoid subclass reported from the customary medicinal flora of Australia is presented in *Table 3.3*.

Triterpenoids

Brassinosteroids have only been reported from one customary medicinal plant genus in Australia, *Callitris* (Cupressaceae), and was used as an antiparasitic.[53,63,64] There are several members of the pertinent families in the Sapindales reported to bear limonoids that are used customarily in Australia: Meliaceae (*Cedrela*, *Melia*, *Toona*), Rutaceae (*Bosistoa*, *Geijera*, *Melicope*, *Philotheca*, *Zieria*) and Simaroubaceae (*Ailanthus*), but limonoids have only been isolated from the *Cedrela* and *Toona* genera. Liminoids have promising antimalarial and anti-inflammatory bioactivities and these genera were used customarily for fever, diarrhoea and menstrual issues.[53,65] Members of the Simaroubaceae (*Ailanthus*, *Brucea*) have been reported to be used customarily for coughs and fever, as a bactericide, analgesic, tonic and for stomach ailments and diarrhoea.[53] Both *Ailanthus* and *Brucea* have been reported to bear quassinoids.[66,67] Cucurbitanes have been reported from the Australian native *Momordica* (Cucurbitaceae), but in the literature this genus is not reported to be used customarily. Genera from families that have been reported to bear cucurbitanes do exist in the customary medicinal flora of Australia literature: Brassicaceae (*Rorippa*), Cucurbitaceae (*Diplocyclos*, *Trichosanthes*), Primulaceae (*Aegiceras*), Rosaceae (*Rubus*), Rubiaceae (*Coelospermum*, *Exallage*, *Morinda*, *Nauclea*, *Oldenlandia*, *Timonius*) and Thymelaeaceae (*Pimelea*, *Wikstroemia*), but it appears that *Trichosanthes* (Cucurbitaceae) is the only genus to be found bearing these compounds.[68,69]

The saponins are well represented among Australian customary medicinal flora: Alismatales [Araceae (*Amorphophallus*)], Asparagales [Asparagaceae (*Asparagus*), Asphodelaceae (*Dianella*)], Asterales [Asteraceae (*Acmella*, *Gnaphalium*, *Pterocaulon*), Goodeniaceae (*Goodenia*)], Canellales [Winteraceae (*Tasmannia*)], Fabales [Fabaceae (*Canavalia*, *Cassia*, *Entada*, *Milettia*, *Senna*)], Gentianales [Apocynaceae (*Carissa*, *Cerbera*, *Marsdenia*, *Pentatropis*, *Tylophora*), Gentianaceae (*Schenkia*), Rubiaceae (*Morinda*)], Lamiales [Lamiaceae (*Plectranthus*, *Prunella*), Scrophulariaceae (*Eremophila*), Verbenaceae (*Verbena*)], Magnoliales [Myristicaceae (*Myristica*)], Malphigiales [Euphorbiaceae (*Aleurites*,

Euphorbia)], Malvales [Malvaceae (*Grewia*)], Myrtales [Myrtaceae (*Corymbia*)], Pandanales [Pandanaceae (*Pandanus*)], Poales [Cyperaceae (*Schoenoplectus*)], Polypodiales [Dennstaedtiaceae (*Pteridium*)], Rosales [Rhamnaceae (*Ziziphus*)] and Sapindales [Meliaceae (*Melia*), Rutaceae (*Bosistoa*)]. There are no emergent trends in customary use across this wide range of flora. Cardiac glycoside-type triterpenoids have been reported from four Australian customary medicinal plant genera: Asterales [Goodeniaceae (*Goodenia*)], Magnoliales [Myristicaceae (*Myristica*)], Malphigiales [Euphorbiaceae (*Aleurites*)] and Sapindales [Rutaceae (*Philotheca*)]. However, no trends are apparent in the customary use of these four customary medicinal flora.

REFERENCES

1. Bhatla SC (2018) Secondary metabolites. In *Plant Physiology, Development and Metabolism*. (Eds SC Bhatla and MA Lal) pp. 1099–1166. Springer Nature Singapore, Singapore.
2. Bergman ME, Davis B, Phillips MA (2019) Medically useful plant terpenoids: Biosynthesis, occurrence, and mechanism of action. *Molecules (Basel, Switzerland)* **24**, 3961. doi:10.3390/molecules24213961
3. Demmig-Adams B (1990) Carotenoids and photoprotection in plants: a role for the xanthophyll zeaxanthin. *Biochimica et Biophysica Acta* **1020**, 1–24. doi:10.1016/0005-2728(90)90088-L
4. Krinsky NI, Johnson EJ (2005) Carotenoid actions and their relation to health and disease. *Molecular Aspects of Medicine* **26**, 459–516. doi:10.1016/j.mam.2005.10.001
5. Gunatilaka LA (2012) Plant natural products. In *Natural Products in Chemical Biology*. (Ed. N Civjan) pp. 3–31. John Wiley & Sons, NJ.
6. Bouvier F, Rahier A, Camara B (2005) Biogenesis, molecular regulation and function of plant isoprenoids. *Progress in Lipid Research* **44**, 357–429. doi:10.1016/j.plipres.2005.09.003
7. Douglas SE, Turner S (1991) Molecular evidence for the origin of plastids from a cyanobacterium-like ancestor. *Journal of Molecular Evolution* **33**, 267–273. doi:10.1007/BF02100678
8. Bohlmann J, Meyer-Gauen G, Croteau R (1998) Plant terpenoids synthases: molecular biology and phylogenetic analysis. *Proceedings of the National Academy of Sciences of the United States of America* **95**, 4126–4133. doi:10.1073/pnas.95.8.4126
9. Rasmann S, Kollner TG, Degenhardt J, Hiltpold I, Toepfer S *et al.* (2005) Recruitment of entomopathogenic nematodes by insect-damaged maize roots. *Nature* **434**, 732–737. doi:10.1038/nature03451
10. McCormick AC, Unsicker SB, Gershenzon J (2012) The specificity of herbivore-induced plant volatiles in attracting herbivore enemies. *Trends in Plant Science* **17**, 303–310. doi:10.1016/j.tplants.2012.03.012
11. Gershenzon J, Dudareva N (2007) The function of terpene natural products in the natural world. *Nature Chemical Biology* **3**, 408–414. doi:10.1038/nchembio.2007.5
12. Salminen A, Lehtonen M, Suuronen T, Kaarniranta K, Huuskonen J (2008) Terpenoids: natural inhibitors of NF-κB signalling with anti-inflammatory and anticancer potential. *Cellular and Molecular Life Sciences* **65**, 2979–2999. doi:10.1007/s00018-008-8103-5
13. Zhao L, Chang W-C, Xiao Y, Liu H-W, Liu P (2013) Methylerythritol phosphate pathway of isoprenoid biosynthesis. *Annual Review of Biochemistry* **82**, 497–530. doi:10.1146/annurev-biochem-052010-100934
14. Lange BM (2016) Online resources for gene discovery and biochemical research with aromatic and medicinal plants. *Phytochemistry Reviews* **15**, 489–510. doi:10.1007/s11101-015-9450-0
15. Bakkali F, Averbeck S, Averbeck D, Idaomar M (2008) Biological effects of essential oils: a review. *Food and Chemical Toxicology* **46**, 446–475. doi:10.1016/j.fct.2007.09.106
16. Meshnick SR (2002) Artemisinin: mechanisms of action, resistance and toxicity. *International Journal for Parasitology* **32**, 1655–1660. doi:10.1016/S0020-7519(02)00194-7
17. Sagara Y, Fernandez-Belda F, De Meis L, Inesi G (1992) Characterization of the inhibition of intracellular Ca2+ transport ATPases by thapsigargin. *The Journal of Biological Chemistry* **267**, 12606–12613. doi:10.1016/S0021-9258(18)42320-4
18. Schiff PB, Fant J, Horwitz SB (1979) Promotion of microtubule assembly *in vitro* by taxol. *Nature* **277**, 665–667. doi:10.1038/277665a0
19. Roy A, Saraf S (2006) Limonoids: overview of significant bioactive triterpenes distributed in plants kingdom. *Biological & Pharmaceutical Bulletin* **29**, 191–201. doi:10.1248/bpb.29.191
20. Maia MF, Moore SJ (2011) Plant-based insect repellents: a review of their efficacy, development and testing. *Malaria Journal* **10**, S11. doi:10.1186/1475-2875-10-S1-S11
21. Pertwee RG (2008) The diverse CB1 and CB2 receptor pharmacology of three plant cannabinoids: Δ9-tetrahydrocannabinol, cannabidiol and Δ9-tetrahydrocannabivarin. *British Journal of Pharmacology* **153**, 199–215. doi:10.1038/sj.bjp.0707442
22. Yeats RB (1979) Monoterpenoids. In *Terpenoids and Steroids*. (Ed. JR Hanson) pp. 3–80. J. W. Arrowsmith Ltd., Bristol, UK.
23. Mahmoud SS, Croteau RB (2002) Strategies for transgenic manipulation of monoterpene biosynthesis in plants. *Trends in Plant Science* **7**, 366–373. doi:10.1016/S1360-1385(02)02303-8
24. Lange BM (2015) Biosynthesis and biotechnology of high-value *p*-menthane monoterpenes, including menthol, carvone, and limonene. *Advances in Biochemical Engineering/Biotechnology* **148**, 319–353. doi:10.1007/10_2014_289
25. Grayson DH (2000) Monoterpenoids. *Natural Product Reports* **17**, 385–419. doi:10.1039/a804437f

26. Ghisalberti EL (1998) Biological and pharmacological activity of naturally occurring iridoids and secoiridoids. *Phytomedicine* **5**, 147–163. doi:10.1016/S0944-7113(98)80012-3
27. Dinda B, Debnath S, Harigaya Y (2007) Naturally occurring secoiridoids and bioactivity of naturally occurring iridoids and secoiridoids: a review, part 2. *Chemical & Pharmaceutical Bulletin* **55**, 689–728. doi:10.1248/cpb.55.689
28. Dinda B, Debnath S, Harigaya Y (2007) Naturally occurring iridoids: a review, part 1. *Chemical & Pharmaceutical Bulletin* **55**, 159–222. doi:10.1248/cpb.55.159
29. Leitereg TJ, Guadagni DG, Harris J, Mon TR, Teranishi R (1971) Chemical and sensory data supporting the difference between the odors of the enantiomeric carvones. *Journal of Agricultural and Food Chemistry* **19**, 785–787. doi:10.1021/jf60176a035
30. Rao AS (2017) Isolation, absolute configuration and bioactivities of megastigmanes or C13 isonorterpenoids. *Angewandte Chemie International Edition* **3**, 69–91.
31. Money T (1979) Sesquiterpenoids. In *Terpenoids and Steroids*. (Ed. JR Hanson) pp. 3–80. J. W. Arrowsmith Ltd., Bristol, UK.
32. Fraga BM (2013) Natural sesquiterpenoids. *Natural Product Reports* **30**, 1226–1264. doi:10.1039/c3np70047j
33. Zhan ZJ, Ying YM, Ma LF, Shan WG (2011) Natural disesquiterpenoids. *Natural Product Reports* **28**, 594–629. doi:10.1039/c0np00050g
34. Gao JM, Wu WJ, Zhang JW, Konishi Y (2007) The dihydro-β-agarofuran sesquiterpenoids. *Natural Product Reports* **24**, 1153–1189. doi:10.1039/b601473a
35. Zi J, Mafu S, Peters RJ (2014) To gibberellins and beyond! Surveying the evolution of (di)terpenoid metabolism. *Annual Review of Plant Biology* **65**, 259–286. doi:10.1146/annurev-arplant-050213-035705
36. Hanson JR (1979) Diterpenoids. In *Terpenoids and Steroids*. (Ed. JR Hanson) pp. 3–80. J. W. Arrowsmith Ltd., Bristol, UK.
37. Hanson JR (2013) Diterpenoids of terrestrial origin. *Natural Product Reports* **30**, 1346–1356. doi:10.1039/c3np70046a
38. Chinou I (2005) Labdanes of natural origin: biological activities. *Current Medicinal Chemistry* **12**, 1295–1317.
39. Thimmappa R, Geisler K, Louveau T, O'maille P, Osbourn A (2014) Triterpene biosynthesis in plants. *Annual Review of Plant Biology* **65**, 225–257. doi:10.1146/annurev-arplant-050312-120229
40. Domingo V, Arteaga JF, Quilez Del Moral JF, Barrero AF (2009) Unusually cyclized triterpenes: occurrence, biosynthesis and chemical synthesis. *Natural Product Reports* **26**, 115–134. doi:10.1039/B801470C
41. Hill RA, Connolly JD (2012) Triterpenoids. *Natural Product Reports* **29**, 780–818. doi:10.1039/c2np20027a
42. Hill RA, Connolly JD (2013) Triterpenoids. *Natural Product Reports* **30**, 1028–1065. doi:10.1039/C3NP70032A
43. Hill RA, Connolly JD (2015) Triterpenoids. *Natural Product Reports* **32**, 273–327. doi:10.1039/C4NP00101J
44. Hill RA, Connolly JD (2017) Triterpenoids. *Natural Product Reports* **34**, 90–122. doi:10.1039/C6NP00094K
45. Hill RA, Connolly JD (2018) Triterpenoids. *Natural Product Reports* **35**, 1294–1329. doi:10.1039/C8NP00029H
46. Hill RA, Connolly JD (2020) Triterpenoids. *Natural Product Reports* **37**, 962–998. doi:10.1039/C9NP00067D
47. Bajguz A, Tretyn A (2003) The chemical characteristic and distribution of brassinosteroids in plants. *Phytochemistry* **62**, 1027–1046. doi:10.1016/S0031-9422(02)00656-8
48. Guo Z, Vangapandu S, Sindelar RW, Walker LA, Sindelar RD (2005) Biologically active quassinoids and their chemistry: potential leads for drug design. *Current Medicinal Chemistry* **12**, 173–190. doi:10.2174/0929867053363351
49. Chen JC, Chiu MH, Nie RL, Cordell GA, Qiu SX (2005) Cucurbitacins and cucurbitane glycosides: structure and biological activities. *Natural Product Reports* **22**, 386–399. doi:10.1039/b418841c
50. Dinda B, Debnath S, Mohanta BC, Harigaya Y (2010) Naturally occurring triterpenoid saponins. *Chemistry & Biodiversity* **7**, 2327–2580. doi:10.1002/cbdv.200800070
51. Patel S (2016) Plant-derived cardiac glycosides: role in heart ailments and cancer management. *Biomedicine and Pharmacotherapy* **84**, 1036–1041. doi:10.1016/j.biopha.2016.10.030
52. Cherniavsky Lev M, Karlish SJD, Garty H (2015) Cardiac glycosides induced toxicity in human cells expressing a1-, a2-, or a3-isoforms of Na-K-ATPase. *American Journal of Physiology. Cell Physiology* **309**, C126–C135. doi:10.1152/ajpcell.00089.2015
53. Lassak EV, Mccarthy T (1983) *Australian Medicinal Plants*. Methuen, Sydney.
54. De Andrade TU, Brasil GA, Endringer DC, Da Nobrega FR, De Sousa DP (2017) Cardiovascular activity of the chemical constituents of essential oils. *Molecules (Basel, Switzerland)* **22**, 1539–1557. doi:10.3390/molecules22091539
55. De Assis Oliveira F, Andrade LN, De Sousa EBV, De Sousa DP (2014) Anti-ulcer activity of essential oil constituents. *Molecules (Basel, Switzerland)* **19**, 5717–5747. doi:10.3390/molecules19055717
56. De Sousa DP (2011) Analgesic-like activity of essential oils constituents. *Molecules (Basel, Switzerland)* **16**, 2233–2252. doi:10.3390/molecules16032233
57. Nerio LS, Olivero-Verbel J, Stashenko E (2010) Repellent activity of essential oils: a review. *Bioresource Technology* **101**, 372–378. doi:10.1016/j.biortech.2009.07.048
58. Sharifi-Rad J, Salehi B, Varoni EM, Sharopov F, Yousaf Z *et al.* (2017) Plants of the *Melaleuca* genus as antimicrobial agents: from farm to pharmacy. *Phytotherapy Research* **31**, 1475–1494. doi:10.1002/ptr.5880
59. Mulyaningsih S, Sporer F, Reichling J, Wink M (2011) Antibacterial activity of essential oils from *Eucalyptus* and of selected components against multidrug-resistant bacterial pathogens. *Pharmaceutical Biology* **49**, 893–899. doi:10.3109/13880209.2011.553625

60. Chen W, Vermaak I, Viljoen A (2013) Camphor: a fumigant during the Black Death and a coveted fragrant wood in ancient Egypt and Babylon: a review. *Molecules (Basel, Switzerland)* **18**, 5434–5454. doi:10.3390/molecules18055434
61. Zhang QL, Fu BM, Zhang ZJ (2017) Borneol, a novel agent that improves central nervous system drug delivery by enhancing blood-brain barrier permeability. *Drug Delivery* **24**, 1037–1044. doi:10.1080/10717544.2017.1346002
62. Mafu S, Zerbe P (2018) Plant diterpenoid metabolism for manufacturing the biopharmaceuticals of tomorrow: prospects and challenges. *Phytochemistry Reviews* **17**, 113–130. doi:10.1007/s11101-017-9513-5
63. Simoneit BRT, Cox RE, Oros DR, Otto A (2018) Terpenoid compositions of resins from *Callitris* species (Cupressaceae). *Molecules (Basel, Switzerland)* **23**, 3384. doi:10.3390/molecules23123384
64. Sadgrove NJ, Jones GL (2014) Medicinal compounds, chemically and biologically characterised from extracts of Australian *Callitris endlicheri* and *C. glaucophylla* (Cupressaceae): used traditionally in Aboriginal and colonial pharmacopoeia. *Journal of Ethnopharmacology* **153**, 872–883. doi:10.1016/j.jep.2014.03.054
65. De Leo M, Milella L, Braca A, De Tommasi N (2018) Cedrela and Toona genera: a rich source of bioactive limonoids and triterpenoids. *Phytochemistry Reviews* **17**, 751–783. doi:10.1007/s11101-018-9557-1
66. Elkhateeb A, Yamasaki M, Maede Y, Katakura K, Nabeta K *et al.* (2008) Anti-babesial quassinoids from the fruits of *Brucea javanica*. *Natural Product Communications* **3**, 1934578X0800300207.
67. kundu P, Laskar S (2010) A brief resume on the genus Ailanthus: chemical and pharmacological aspects. *Phytochemistry Reviews* **9**, 379–412. doi:10.1007/s11101-009-9157-1
68. Kanchanapoom T, Kasai R, Yamasaki K (2002) Cucurbitane, hexanorcucurbitane and octanorcucurbitane glycosides from fruits of Trichosanthes tricuspidata. *Phytochemistry* **59**, 215–228. doi:10.1016/S0031-9422(01)00430-7
69. Yang S-Y, Guo L-F, Liu Y, Zou J-B, Zhu H-Y *et al.* (2023) Trichosanates A–G and cucurbitacins W–Y, anticomplement monoterpenoids and cucurbitane-type triterpenoids from the pericarps of Trichosanthes kirilowii. *Bioorganic Chemistry* **139**, 106710. doi:10.1016/j.bioorg.2023.106710
70. Brinkhaus B, Lindner M, Schuppan D, Hahn EG (2000) Chemical, pharmacological and clinical profile of the East Asian medical plant *Centella aslatica*. *Phytomedicine* **7**, 427–448. doi:10.1016/S0944-7113(00)80065-3
71. James JT, Dubery IA (2009) Pentacyclic triterpenoids from the medicinal herb, *Centella asiatica* (L.) Urban. *Molecules (Basel, Switzerland)* **14**, 3922–3941. doi:10.3390/molecules14103922
72. Manase MJ, Mitaine-Offer AC, Miyamoto T, Tanaka C, Delemasure S *et al.* (2013) New triterpenoid estersaponins from the root barks of *Pittosporum verticillatum* subsp. *verticillatum* and evaluation of cytotoxicities. *Fitoterapia* **91**, 231–235. doi:10.1016/j.fitote.2013.09.002
73. Madikizela B, Mcgaw LJ (2017) *Pittosporum viridiflorum* Sims (Pittosporaceae): a review on a useful medicinal plant native to South Africa and tropical Africa. *Journal of Ethnopharmacology* **205**, 217–230. doi:10.1016/j.jep.2017.05.005
74. Refaat J, Kamel MS, Ramadan MA, Ali AA (2012) *Crinum*: an endless source of bioactive principles: a review. Part I: *Crinum* alkaloids: lycorine-type alkaloids. *International Journal of Pharmaceutical Sciences and Research* **3**, 1883–1890.
75. Refaat J, Kamel MS, Ramadan MA, Ali AA (2012) *Crinum*: an endless source of bioactive principles: a review. Part II: *Crinum* alkaloids: crinine-type alkaloids. *International Journal of Pharmaceutical Sciences and Research* **3**, 3091–3100.
76. Refaat J, Kamel MS, Ramadan MA, Ali AA (2012) *Crinum*: an endless source of bioactive principles: a review. Part III: *Crinum* alkaloids: belladine-, galanthamine-, lycorenine-, tazettine-type alkaloids and other minor types. *International Journal of Pharmaceutical Sciences and Research* **3**, 3630–3638.
77. Refaat J, Kamel MS, Ramadan MA, Ali AA (2013) *Crinum*: an endless source of bioactive principles: a review. Part IV: non alkaloidal constituents. *International Journal of Pharmaceutical Sciences and Research* **4**, 941–948.
78. Xu J, Han Q-B, Li S-L, Chen X-J, Wang X-N *et al.* (2013) Chemistry, bioactivity and quality control of *Dendrobium*, a commonly used tonic herb in traditional Chinese medicine. *Phytochemistry Reviews* **12**, 341–367. doi:10.1007/s11101-013-9310-8
79. Paulraj J, Govindarajan R, Palpu P (2013) The genus spilanthes ethnopharmacology, phytochemistry, and pharmacological properties: a review. *Advances in Pharmacological and Pharmaceutical Sciences* **2013**, 510298.
80. Yadav N, Ganie SA, Singh B, Chhillar AK, Yadav SS (2019) Phytochemical constituents and ethnopharmacological properties of *Ageratum conyzoides* L. *Phytotherapy Research* **33**, 2163–2178. doi:10.1002/ptr.6405
81. Macleod JK, Rasmussen HB (1999) A hydroxy-B-caryophyllene from *Pterocaulon serrulatum*. *Phytochemistry* **50**, 105–108. doi:10.1016/S0031-9422(98)00460-9
82. Medeiros-Neves B, Teixeira HF, Von Poser GL (2018) The genus *Pterocaulon* (Asteraceae): a review on traditional medicinal uses, chemical constituents and biological properties. *Journal of Ethnopharmacology* **224**, 451–464. doi:10.1016/j.jep.2018.06.012
83. Li X, Dong M, Liu Y, Shi Q-W, Kiyota H (2007) Structures and biological properties of the chemical constituents from the genus *Wedelia*. *Chemistry & Biodiversity* **4**, 823–836. doi:10.1002/cbdv.200790070
84. That QT, Jossang J, Jossang A, Kim PPN, Jaureguiberry G (2007) Wedelolides A and B: novel sesquiterpene sigma-lactones, (9R)-eudesman-9,12-olides, from *Wedelia trilobata*. *The Journal of Organic Chemistry* **72**, 7102–7105. doi:10.1021/jo070771m

85. Mizokami SS, Arakawa NS, Ambrosio SR, Zarpelon AC, Casagrande R *et al.* (2012) Kaurenoic acid from *Sphagneticola trilobata* inhibits inflammatory pain: effect on cytokine production and activation of the NO-cyclic GMP-protein kinase G-ATP-sensitive potassium channel signaling pathway. *Journal of Natural Products* **75**, 896–904. doi:10.1021/np200989t
86. Sun L, Wang Z, Wang Y, Xu J, He X (2020) Anti-proliferative and anti-neuroinflammatory eudesmanolides from *Wedelia* (*Sphagneticola trilobata* (L.) Pruski). *Fitoterapia* **142**, 104452. doi:10.1016/j.fitote.2019.104452
87. Ghisalberti EL (2004) The Goodeniaceae. *Fitoterapia* **75**, 429–446. doi:10.1016/j.fitote.2004.01.018
88. Suthiwong J, Thongsri Y, Yenjai C (2017) A new furanocoumarin from the fruits of *Scaevola taccada* and antifungal activity against *Pythium insidiosum*. *Natural Product Research* **31**, 453–459. doi:10.1080/14786419.2016.1188100
89. Liao G, Wu P, Liu Z, Xue J, Li H *et al.* (2020) 2H-pyranone and isocoumarin derivatives from the endophytic fungus *Pestalotiopsis microspora* SC3082 derived from *Scaevola taccada* (Gaertn.) Roxb. *Natural Product Research* **35**, 3644–3651.
90. He X, Wang X, Fang J, Chang Y, Ning N *et al.* (2017) The genus Achyranthes: a review on traditional uses, phytochemistry, and pharmacological activities. *Journal of Ethnopharmacology* **203**, 260–278. doi:10.1016/j.jep.2017.03.035
91. Kokanova-Nedialkova Z, Neialkov PT, Nikolov SD (2009) The genus *Chenopodium*: phytochemistry, ethnopharmacology and pharmacology. *Pharmacognosy Reviews* **3**, 280–306.
92. Koo KA, Lorent K, Gong W, Windsor P, Whittaker SJ *et al.* (2015) Biliatresone, a reactive natural toxin from *Dysphania glomulifera* and *D. littoralis*: discovery of the toxic moiety 1,2-diaryl-2-propenone. *Chemical Research in Toxicology* **28**, 1519–1521. doi:10.1021/acs.chemrestox.5b00227
93. Iranshahy M, Javadi B, Iranshahi M, Jahanbakhsh SP, Mahyari S *et al.* (2017) A review of traditional uses, phytochemistry and pharmacology of *Portulaca oleracea* L. *Journal of Ethnopharmacology* **205**, 158–172. doi:10.1016/j.jep.2017.05.004
94. Kasai R, Sasaki A, Hashimoto T, Kaneko T, Ohtani K *et al.* (1999) Cycloartane glycosides from *Trichosanthes tricuspidata*. *Phytochemistry* **51**, 803–808. doi:10.1016/S0031-9422(99)00078-3
95. Kanchanapoom T, Kasai R, Yamasaki K (2002) Cucurbitane, hexanorcucurbitane and octanorcucurbitane glycosides from fruits of *Trichosanthes tricuspidata*. *Phytochemistry* **59**, 215–228. doi:10.1016/S0031-9422(01)00430-7
96. Mai Le P, Guenard D, Franck M, Van TM, Gaspard C *et al.* (2002) New cytotoxic cucurbitacins from the pericarps of *Trichosanthes tricuspidata* fruits. *Natural Product Letters* **16**, 15–19. doi:10.1080/1057563029001/4755
97. Shaw PC, Lee KM, Wong KB (2005) Recent advances in trichosanthin, a ribosome-inactivating protein with multiple pharmacological properties. *Toxicon* **45**, 683–689. doi:10.1016/j.toxicon.2004.12.024
98. Yu X, Tang L, Wu H, Zhang X, Luo H *et al.* (2018) Trichosanthis Fructus: botany, traditional uses, phytochemistry and pharmacology. *Journal of Ethnopharmacology* **224**, 177–194. doi:10.1016/j.jep.2018.05.034
99. Son NT (2019) Genus *Erythrophleum*: botanical description, traditional use, phytochemistry and pharmacology. *Phytochemistry Reviews* **18**, 571–599. doi:10.1007/s11101-019-09640-0
100. Crombie L, Whiting DA (1998) Biosynthesis in the rotenoid group of natural products: applications of isotope methodology. *Phytochemistry* **49**, 1479–1507. doi:10.1016/S0031-9422(98)00178-2
101. Touqeer S, Saeed MA, Ajaib M (2013) A review on the phytochemistry and pharmacology of genus *Tephrosia*. *Phytopharmacology* **4**, 598–637.
102. Chen Y, Yan T, Gao C, Cao W, Huang R (2014) Natural products from the genus *Tephrosia*. *Molecules (Basel, Switzerland)* **19**, 1432–1458. doi:10.3390/molecules19021432
103. Mutharaj S, Muthusamy P, Ilango K (2019) Pharmacognostical, phytochemical and pharmacological review of *Casuarina equisetifolia* Linn. *World Journal of Pharmaceutical Research* **8**, 328–339.
104. Khyade MS, Kasote DM, Vaikos NP (2014) *Alstonia scholaris* (L.) R. Br. and *Alstonia macrophylla* Wall. ex G. Don: a comparative review on traditional uses, phytochemistry and pharmacology. *Journal of Ethnopharmacology* **153**, 1–18. doi:10.1016/j.jep.2014.01.025
105. Sriphana U, Thongsri Y, Ardwichai P, Poopasit K, Prariyachatigul C *et al.* (2013) New lignan esters from *Alyxia schlechteri* and antifungal activity against *Pythium insidiosum*. *Fitoterapia* **91**, 39–43. doi:10.1016/j.fitote.2013.08.005
106. Shen L-R, Jin S-M, Yin B-W, Du X-F, Wang Y-L *et al.* (2007) Chemical constituents of plants from the genus *Cerbera*. *Chemistry & Biodiversity* **4**, 1438–1449. doi:10.1002/cbdv.200790122
107. Han L, Zhou X, Yang M, Zhou L, Deng X *et al.* (2018) Ethnobotany, phytochemistry and pharmacological effects of plants in genus *Cynanchum* Linn. (Asclepiadaceae). *Molecules (Basel, Switzerland)* **23**, 1194. doi:10.3390/molecules23051194
108. Van Beek TA, Verpoorte R, Svendsen AB, Leeuwenberg AJM, Bisset NG (1984) *Tabernaemontana* L. (Apocynaceae): a review of its taxonomy, phytochemistry, ethnobotany and pharmacology. *Journal of Ethnopharmacology* **10**, 1–156. doi:10.1016/0378-8741(84)90046-1
109. Patchayasakul W, Pongchaidecha A, Chattipakorn N, Chattipakorn S (2008) Ethnobotany and ethnopharmacology of *Tabernaemontana divaricata*. *The Indian Journal of Medical Research* **127**, 317–335.
110. Abubakar IB, Loh HS (2016) A review on ethnobotany, pharmacology and phytochemistry of *Tabernaemontana corymbosa*. *The Journal of Pharmacy and Pharmacology* **68**, 423–432. doi:10.1111/jphp.12523

111. Shahat AA, Apers S, Van Miert S, Claeys M, Pieters L *et al.* (2001) Structure elucidation of three new acetylated flavonoid glycosides from *Centaurium spicatum*. *Magnetic Resonance in Chemistry* **39**, 625–629. doi:10.1002/mrc.900
112. Bibi H, Ali I, Sadozai SK, Atta Ur R (2006) Phytochemical studies and antibacterial activity of *Centaurium pulchellum* Druce. *Natural Product Research* **20**, 896–901. doi:10.1080/14786410500162047
113. Aberham A, Pieri V, Croom EM, Jr, Ellmerer E, Stuppner H (2011) Analysis of iridoids, secoiridoids and xanthones in *Centaurium erythraea*, *Frasera caroliniensis* and *Gentiana lutea* using LC-MS and RP-HPLC. *Journal of Pharmaceutical and Biomedical Analysis* **54**, 517–525. doi:10.1016/j.jpba.2010.09.030
114. El-Shanawany MA, Mohamed GA, Nafady AM, Ibrahim SRM, Radwan MM *et al.* (2011) A new xanthone from the roots of *Centaurium spicatum*. *Phytochemistry Letters* **4**, 126–128. doi:10.1016/j.phytol.2010.12.008
115. Itoh A, Tanaka Y, Nagakura N, Nishi T, Tanahashi T (2006) A quinic acid ester from *Strychnos lucida*. *Journal of Natural Medicines* **60**, 146–148. doi:10.1007/s11418-005-0021-3
116. Galliford CV, Scheidt KA (2007) Pyrrolidinyl-spirooxindole natural products as inspirations for the development of potential therapeutic agents. *Angewandte Chemie International Edition* **46**, 8748–8758. doi:10.1002/anie.200701342
117. Tchinda AT, Tamze V, Ngono ARN, Ayimele GA, Cao M *et al.* (2012) Alkaloids from the stem bark of *Strychnos icaja*. *Phytochemistry Letters* **5**, 108–113. doi:10.1016/j.phytol.2011.11.004
118. Guo R, Wang T, Zhou G, Xu M, Yu X *et al.* (2018) Botany, phytochemistry, pharmacology and toxicity of *Strychnos nux-vomica* L.: a review. *The American Journal of Chinese Medicine* **46**, 1–23. doi:10.1142/S0192415X18500015
119. Li W, Tang GH, Chen L, Tang YQ, Xu YK *et al.* (2018) New pyridocarbazole alkaloids from *Strychnos nitida*. *Natural Product Research* **32**, 1532–1536. doi:10.1080/14786419.2017.1385016
120. Bowie JH, Cooke RG, Wilkin PE (1962) Colouring matters of Australian plants. X: anthraquinones from *Coelospermum* species. *Australian Journal of Chemistry* **15**, 336–341. doi:10.1071/CH9620336
121. Lopez JL, Pusset J (1988) Plantes de Nouvelle-Caledonie, 115. Alcaloides monoterpeniques de *Coelospermum billardieri*. *Journal of Natural Products* **51**, 829–835. doi:10.1021/np50059a002
122. Benkrief R, Skaltsounis A-L, Tillequin F, Koch M (1991) Plantes de Nouvelle-Caledonie, 137. Iridoides de *Coelospermum billardieri*. *Journal of Natural Products* **54**, 532–541. doi:10.1021/np50074a028
123. Magiatis P, Skaltsounis A-L, Tillequin F, Seguin E, Cosson J-P (2002) Coelobillardin, an iridoid glucoside dimer from *Coelospermum billardieri*. *Phytochemistry* **60**, 415–418. doi:10.1016/S0031-9422(02)00118-8
124. Pawlus AD, Kinghorn AD (2007) Review of the ethnobotany, chemistry, biological activity and safety of the botanical dietary supplement *Morinda citrifolia* (noni)*. *The Journal of Pharmacy and Pharmacology* **59**, 1587–1609. doi:10.1211/jpp.59.12.0001
125. Potterat O, Hamburger M (2007) *Morinda citrifolia* (noni) fruit: phytochemistry, pharmacology, safety. *Planta Medica* **73**, 191–199. doi:10.1055/s-2007-967115
126. Zhang JH, Xin HL, Xu YM, Shen Y, He YQ *et al.* (2018) *Morinda officinalis* How.: a comprehensive review of traditional uses, phytochemistry and pharmacology. *Journal of Ethnopharmacology* **213**, 230–255. doi:10.1016/j.jep.2017.10.028
127. Luan F, Han K, Li M, Zhang T, Liu D *et al.* (2019) Ethnomedicinal uses, phytochemistry, pharmacology, and toxicology of species from the genus *Ajuga* L.: a systematic review. *The American Journal of Chinese Medicine* **47**, 959–1003. doi:10.1142/S0192415X19500502
128. Tu Y, Sun L, Guo M, Chen W (2013) The medicinal uses of *Callicarpa* L. in traditional Chinese medicine: an ethnopharmacological, phytochemical and pharmacological review. *Journal of Ethnopharmacology* **146**, 465–481. doi:10.1016/j.jep.2012.12.051
129. Wang JH, Luan F, He XD, Wang Y, Li MX (2018) Traditional uses and pharmacological properties of *Clerodendrum* phytochemicals. *Journal of Traditional and Complementary Medicine* **8**, 24–38. doi:10.1016/j.jtcme.2017.04.001
130. Mimica-Dukic N, Bozin B (2008) *Mentha* L. species (Lamiaceae) as promising sources of bioactive secondary metabolites. *Current Pharmaceutical Design* **14**, 3141–3150. doi:10.2174/138161208786404245
131. Singh D, Chaudhuri PK (2018) A review on phytochemical and pharmacological properties of Holy basil (*Ocimum sanctum* L.). *Industrial Crops and Products* **118**, 367–382. doi:10.1016/j.indcrop.2018.03.048
132. Abdel-Mogib M, Albar HA, Batterjee SM (2002) Chemistry of the genus *Plectranthus*. *Molecules (Basel, Switzerland)* **7**, 271–301. doi:10.3390/70200271
133. Southwell IA, Tucker DJ (1993) Cis-dihydroagarofuran from *Prostanthera* sp. aff. *ovalifolia*. *Phytochemistry* **33**, 857–862. doi:10.1016/0031-9422(93)85290-8
134. Dellar JE, Cole MD, Gray AI, Gibbons S, Waterman PG (1994) Antimicrobial sesquiterpenes from *Prostanthera* aff. *Melissifolia* and *P. rotundifolia*. *Phytochemistry* **36**, 957–960. doi:10.1016/S0031-9422(00)90471-0
135. Kisiel W, Piozzi F (1999) Phenylethanoid glycosides from *Prostanthera melissifolia*. *Phytochemistry* **51**, 1083–1085. doi:10.1016/S0031-9422(99)00167-3
136. Russell M, Southwell IA, Waterman PG (2001) α-Phellandren-8-ol, a rare monoterpenene alcohol from the oil of *Prostanthera staurophylla* F. Muell. *The Journal of Essential Oil Research* **13**, 446–447. doi:10.1080/10412905.2001.9699723
137. Collins TL, Jones GL, Sadgrove NJ (2014) Volatiles from the rare Australian desert plant *Prostanthera centralis* B.J.Conn (Lamiaceae): chemical composition and antimicrobial activity. *Agriculture* **4**, 308–316. doi:10.3390/agriculture4040308
138. Tang KSC, Konczak I, Zhao J (2017) Phenolic compounds of the Australian native herb *Prostanthera rotundifolia*

and their biological activities. *Food Chemistry* **233**, 530–539. doi:10.1016/j.foodchem.2017.04.137

139. Bai Y, Xia B, Xie W, Zhou Y, Xie J *et al.* (2016) Phytochemistry and pharmacological activities of the genus *Prunella*. *Food Chemistry* **204**, 483–496. doi:10.1016/j.foodchem.2016.02.047
140. Ulubelen A, Topcu G, Sonmez U (2000) Chemical and biological evaluation of genus *Teucrium*. *Studies in Natural Products Chemistry* **23**, 591–648. doi:10.1016/S1572-5995(00)80139-8
141. Bahramikia S, Yazdanparast R (2012) Phytochemistry and medicinal properties of *Teucrium polium* L. (Lamiaceae). *Phytotherapy Research* **26**, 1581–1593. doi:10.1002/ptr.4617
142. Filho JGS, Duringer J, Maia GLA, Tavares JF, Xavier HS *et al.* (2008) Ecdysteroids from *Vitex* species: distribution and compilation of their 13C-NMR spectral data. *Chemistry & Biodiversity* **5**, 707–713. doi:10.1002/cbdv.200890067
143. Rani A, Sharma A (2013) The genus *Vitex*: a review. *Pharmacognosy Reviews* **7**, 188–198. doi:10.4103/0973-7847.120522
144. Yao JL, Fang SM, Liu R, Oppong MB, Liu EW *et al.* (2016) A review on the terpenes from genus *Vitex*. *Molecules (Basel, Switzerland)* **21**, 1179. doi:10.3390/molecules21091179
145. Singab AN, Youssef FS, Ashour ML, Wink M (2013) The genus *Eremophila* (Scrophulariaceae): an ethnobotanical, biological and phytochemical review. *The Journal of Pharmacy and Pharmacology* **65**, 1239–1279. doi:10.1111/jphp.12092
146. Blackburne ID, Park RJ, Sutherland MD (1971) Terpenoid chemistry XVIII: myodesmone and isomyodesmone, toxic furanoid ketones from *Myoporum deserti* and *M. acuminatum*. *Australian Journal of Chemistry* **24**, 995–1007. doi:10.1071/CH9710995
147. Blackburne ID, Park RJ, Sutherland MD (1972) Terpenoid chemistry XX: myoporone and dehydromyoporone, toxic furanoid ketones from *Myoporum* and *Eremophila* species. *Australian Journal of Chemistry* **25**, 1787–1796. doi:10.1071/CH9721787
148. Metra PL, Sutherland MD (1983) Further skeletal variety in the toxic furanosesquiterpene ketones in the *Myoporum* genus. *Tetrahedron Letters* **24**, 1749–1752. doi:10.1016/S0040-4039(00)81761-X
149. Sutherland MD, Rodwell JL (1989) Terpenoid chemistry XXVIII: furanosesquiterpene B-ketols from *Myoporum betcheanum*, *M. deserti*, *M. montanum* and other Myoporaceae. *Australian Journal of Chemistry* **42**, 1995–2019. doi:10.1071/CH9891995
150. Menut C, Cabalion P, Hnawia E, Agnaniet H, Waikedre J *et al.* (2005) Two new furanosesquiterpenes from *Myoporum crassifolium* from New Caledonia. *Flavour and Fragrance Journal* **20**, 621–625. doi:10.1002/ffj.1509
151. Kanemoto M, Matsunami K, Otsuka H, Shinzato T, Ishigaki C *et al.* (2008) Chlorine-containing iridoid and iridoid glucoside, and other glucosides from leaves of *Myoporum bontioides*. *Phytochemistry* **69**, 2517–2522. doi:10.1016/j.phytochem.2008.07.002
152. Deepak M, Handa SS (1998) 3a,24-dihydroxy-urs-12-en-28-oic acid from *Verbena officinalis*. *Phytochemistry* **49**, 269–271. doi:10.1016/S0031-9422(97)01004-2
153. Deepak M, Handa SS (2000) Antiinflammatory activity and chemical composition of extracts of *Verbena officinalis*. *Phytotherapy Research* **14**, 463–465. doi:10.1002/1099-1573(200009)14:6<463::AID-PTR611>3.0.CO;2-G
154. Lakshmi V, Pandey K, Mishra SK, Srivastava S, Mishra M *et al.* (2009) An overview of family Hernandiaceae. *Records of Natural Products* **3**, 1–22.
155. Lai YW, Wang SW, Hu YY, Hwang TL, Cheng MJ *et al.* (2020) Anti-inflammatory alkaloids from the root bark of *Hernandia nymphaeifolia*. *Phytochemistry* **173**, 112326. doi:10.1016/j.phytochem.2020.112326
156. Wang YS, Wen ZQ, Li BT, Zhang HB, Yang JH (2016) Ethnobotany, phytochemistry, and pharmacology of the genus *Litsea*: An update. *Journal of Ethnopharmacology* **181**, 66–107. doi:10.1016/j.jep.2016.01.032
157. Kalbhen DA (1971) Nutmeg as a narcotic: a contribution to the chemistry and pharmacology of nutmeg (*Myristica fragrans*). *Angewandte Chemie International Edition* **10**, 370–374. doi:10.1002/anie.197103701
158. Valderrama JCM (2000) Distribution of flavonoids in the Myristicaceae. *Phytochemistry* **55**, 505–511. doi:10.1016/S0031-9422(00)00114-X
159. Nagja T, Vimal K, Sanjeev A (2016) *Myristica fragrans*: a comprehensive review. *International Journal of Pharmacy and Pharmaceutical Sciences* **8**, 27–30.
160. Seebaluck R, Gurib-Fakim A, Mahomoodally F (2015) Medicinal plants from the genus Acalypha (Euphorbiaceae): a review of their ethnopharmacology and phytochemistry. *Journal of Ethnopharmacology* **159**, 137–157. doi:10.1016/j.jep.2014.10.040
161. Dilpreet K, Amandeep K, Jaswinder K (2018) A review of *Aleurites moluccana*. *World Journal of Pharmacy and Pharmaceutical Sciences* **7**, 341–358.
162. Salatino A, Faria Salatino ML, Negri G (2007) Traditional uses, chemistry and pharmacology of *Croton* species (Euphorbiaceae). *Journal of the Brazilian Chemical Society* **18**, 11–33. doi:10.1590/S0103-50532007000100002
163. Gupta D, Bleakley B, Gupta RK (2008) Dragon's blood: botany, chemistry and therapeutic uses. *Journal of Ethnopharmacology* **115**, 361–380. doi:10.1016/j.jep.2007.10.018
164. Rivière C, Nguyen Thi Hong V, Tran Hong Q, Chataigné G, Nguyen Hoai N *et al.* (2010) *Mallotus* species from Vietnamese mountainous areas: phytochemistry and pharmacological activities. *Phytochemistry Reviews* **9**, 217–253. doi:10.1007/s11101-009-9152-6
165. Gangwar M, Goel RK, Nath G (2014) *Mallotus philippinensis* Muell. Arg (Euphorbiaceae): ethnopharmacology and phytochemistry review. *BioMed Research International* **2014**, 213973.
166. Morikawa H, Kasai R, Otsuka H, Hirata E, Shinzato T *et al.* (2004) Terpenic and phenolic glycosides from the leaves of *Breynia officinalis* Hemsl. *Chemical & Pharmaceutical Bulletin* **52**, 1086–1090. doi:10.1248/cpb.52.1086

167. Meng D, Chen W, Zhao W (2007) Sulfur-containing spiroketal glycosides from *Breynia fructicosa*. *Journal of Natural Products* **70**, 824–829. doi:10.1021/np0606300
168. Huang Y-L, Chen C-C, Hsu F-L, Chen C-F (1996) A new lignan from *Phyllanthus virgatus*. *Journal of Natural Products* **59**, 520–521. doi:10.1021/np9603380
169. Calixto JB, Santos ARS, Filho VC, Yunes RA (1998) A review of the plants of the genus *Phyllanthus:* their chemistry pharmacology and therapeutic potential. *Medicinal Research Reviews* **18**, 225–258. doi:10.1002/(SICI)1098-1128(199807)18:4<225::AID-MED2>3.0.CO;2-X
170. Huang Y-L, Chen C-C, Hsu F-L, Chen C-F (1998) Tannins, flavonol sulfonates, and a norlignan from *Phyllanthus virgatus*. *Journal of Natural Products* **61**, 1194–1197. doi:10.1021/np970336v
171. Bagalkotkar G, Sagineedu SR, Saad MS, Stanslas J (2006) Phytochemicals from *Phyllanthus niruri* Linn. and their pharmacological properties: a review. *The Journal of Pharmacy and Pharmacology* **58**, 1559–1570. doi:10.1211/jpp.58.12.0001
172. Patel JR, Tripathi P, Sharma V, Chauhan NS, Dixit VK (2011) *Phyllanthus amarus*: ethnomedicinal uses, phytochemistry and pharmacology: a review. *Journal of Ethnopharmacology* **138**, 286–313. doi:10.1016/j.jep.2011.09.040
173. Patel DK, Kumar R, Sairam K, Hemalatha S (2013) Hybanthus enneaspermus (L.) F. Muell: a concise report on its phytopharmacological aspects. *Chinese Journal of Natural Medicines* **11**, 199–206. doi:10.1016/S1875-5364(13)60017-5
174. Rajsekhar PB, Arvind Bharani RS, Angel J, Ramachandran M, Rajsekhar SPV (2016) *Hybanthus enneaspermus* (L) F. Muell: a phytopharmacological review on herbal medicine. *Journal of Chemical and Pharmaceutical Research* **8**, 351–355.
175. Wu P-L, Chuang T-H, He C-X, Wu T-S (2004) Cytotoxicity of phenylpropanoid esters from the stems of *Hibiscus taiwanensis*. *Bioorganic & Medicinal Chemistry* **12**, 2193–2197. doi:10.1016/j.bmc.2004.02.020
176. Wu P L, Wu T-S, He C-X, Su C-H, Lee K-H (2005) Constituents from the stems of *Hibiscus taiwanensis*. *Chemical & Pharmaceutical Bulletin* **53**, 56–59. doi:10.1248/cpb.53.56
177. Vasudeva N, Sharma SK (2008) Biologically active compounds from the genus *Hibiscus*. *Pharmaceutical Biology* **46**, 145–153. doi:10.1080/13880200701575320
178. Maganha EG, Halmenschlager RD, Rosa RM, Henriques Ja P, Ramos ALLD *et al.* (2010) Pharmacological evidences for the extracts and secondary metabolites from plants of the genus *Hibiscus*. *Food Chemistry* **118**, 1–10. doi:10.1016/j.foodchem.2009.04.005
179. Gasparetto JC, Martins CA, Hayashi SS, Otuky MF, Pontarolo R (2012) Ethnobotanical and scientific aspects of *Malva sylvestris* L.: a millennial herbal medicine. *The Journal of Pharmacy and Pharmacology* **64**, 172–189. doi:10.1111/j.2042-7158.2011.01383.x
180. Sharifi-Rad J, Melgar-Lalanne G, Hernandez-Alvarez AJ, Taheri Y, Shaheen S *et al.* (2020) *Malva* species: insights on its chemical composition towards pharmacological applications. *Phytotherapy Research* **34**, 546–567. doi:10.1002/ptr.6550
181. Dinda B, Das N, Dinda S, Dinda M, Silsarma I (2015) The genus *Sida* L.: a traditional medicine: its ethnopharmacological, phytochemical and pharmacological data for commercial exploitation in herbal drugs industry. *Journal of Ethnopharmacology* **176**, 135–176. doi:10.1016/j.jep.2015.10.027
182. Hodgson D, Ritchie E, Taylor WC (1960) Chemical studies of the Myrtaceae. II: the constituents of *Syncarpia laurifolia* Tenn. *Australian Journal of Chemistry* **13**, 385–389. doi:10.1071/CH9600385
183. Brophy JJ, Craig DC, Goldsack RJ, Fookes CJR (1994) Hillyl acetate, a keto-acetate from the leaf steam volatiles of *Syncarpia hillii*. *Phytochemistry* **37**, 1645–1647. doi:10.1016/S0031-9422(00)89584-9
184. Su Q, Dalal S, Goetz M, Cassera MB, Kingston DG (2016) Antiplasmodial phloroglucinol derivatives from *Syncarpia glomulifera*. *Bioorganic & Medicinal Chemistry* **24**, 2544–2548. doi:10.1016/j.bmc.2016.04.020
185. Baliga MS, Bhat HP, Baliga BRV, Wilson R, Palatty PL (2011) Phytochemistry, traditional uses and pharmacology of *Eugenia jambolana* Lam. (black plum): a review. *Food Research International* **44**, 1776–1789. doi:10.1016/j.foodres.2011.02.007
186. Ayyanar M, Subash-Babu P (2012) *Syzygium cumini* (L.) Skeels: a review of its phytochemical constituents and traditional uses. *Asian Pacific Journal of Tropical Biomedicine* **2**, 240–246. doi:10.1016/S2221-1691(12)60050-1
187. Cortés-Rojas DF, De Souza CRF, Oliveira WP (2014) Clove (*Syzygium aromaticum*): a precious spice. *Asian Pacific Journal of Tropical Biomedicine* **4**, 90–96. doi:10.1016/S2221-1691(14)60215-X
188. Nonato MG, Takayama H, Garson MJ (2008) *Pandanus* alkaloids: chemistry and biology. *The Alkaloids* **66**, 215–249.
189. Adkar PP, Bhaskar VH (2014) *Pandanus odoratissimus* (Kewda): a review on ethnopharmacology, phytochemistry, and nutritional aspects. *Advances in Pharmacological and Pharmaceutical Sciences* **2014**, 1–19. doi:10.1155/2014/120895
190. Parmar VS, Jain SC, Bisht KS, Jain R, Taneja P *et al.* (1997) Phytochemistry of the genus *Piper*. *Phytochemistry* **46**, 597–673. doi:10.1016/S0031-9422(97)00328-2
191. Xiang C-P, Shi Y-N, Liu F-F, Li H-Z, Zhang Y-J *et al.* (2016) A survey of the chemical compounds of *Piper* spp. (Piperaceae) and their biological activities. *Natural Product Communications* **11**, 1403–1408. doi:10.1177/1934578X1601100948
192. Peerzada AM, Ali HH, Naeem M, Latif M, Bukhari AH *et al.* (2015) *Cyperus rotundus* L.: traditional uses, phytochemistry, and pharmacological activities. *Journal of Ethnopharmacology* **174**, 540–560. doi:10.1016/j.jep.2015.08.012
193. Avoseh O, Oyedeji O, Rungqu P, Nkeh-Chungag B, Oyedeji A (2015) *Cymbopogon* species: ethnopharmacology, phytochemistry and the pharmacological importance. *Molecules (Basel, Switzerland)* **20**, 7438–7453. doi:10.3390/molecules20057438

194. Pan C, Chen YG, Ma XY, Jiang JH, He F *et al.* (2011) Phytochemical constituents and pharmacological activities of plants from the genus *Adiantum*: a review. *Tropical Journal of Pharmaceutical Research* **10**, 681–692. doi:10.4314/tjpr.v10i5.18
195. Sharma BR, Gautam LN, Adhikari D, Karki R (2017) A comprehensive review on chemical profiling of *Nelumbo nucifera*: potential for drug development. *Phytotherapy Research* **31**, 3–26. doi:10.1002/ptr.5732
196. Semwal DK, Semwal RB, Vermaak I, Viljoen A (2014) From arrow poison to herbal medicine: the ethnobotanical, phytochemical and pharmacological significance of *Cissampelos* (Menispermaceae). *Journal of Ethnopharmacology* **155**, 1011–1028. doi:10.1016/j.jep.2014.06.054
197. Ahmed F, Urooj A (2010) Traditional uses, medicinal properties, and phytopharmacology of *Ficus racemosa*: a review. *Pharmaceutical Biology* **48**, 672–681. doi:10.3109/13880200903241861
198. Singh D, Singh B, Goel RK (2011) Traditional uses, phytochemistry and pharmacology of *Ficus religiosa*: a review. *Journal of Ethnopharmacology* **134**, 565–583. doi:10.1016/j.jep.2011.01.046
199. Baldovini N, Delasalle C, Joulain D (2011) Phytochemistry of the heartwood from fragrant *Santalum* species: a review. *Flavour and Fragrance Journal* **26**, 7–26. doi:10.1002/ffj.2025
200. Kumar R, Anjum N, Tripathi YC (2015) Phytochemistry and pharmacology of *Santalum album* L.: a review. *World Journal of Pharmaceutical Research* **4**, 1842–1876.
201. Mogana R, Wiart C (2011) *Canarium* L.: a phytochemical and pharmacological review. *Journal of Pharmacy Research* **4**, 2482–2489.
202. Zhao L, Huo C-H, Shen L-R, Yang Y, Zhang Q *et al.* (2010) Chemical constituents of plants from the genus *Melia*. *Chemistry & Biodiversity* **7**, 839–859. doi:10.1002/cbdv.200900043
203. Simonsen HT, Adsersen A, Bremner P, Heinrich M, Wagner Smitt U *et al.* (2004) Antifungal constituents of *Melicope borbonica*. *Phytotherapy Research* **18**, 542–545. doi:10.1002/ptr.1482
204. Yamamoto M, Akita T, Koyama Y, Sueyoshi E, Matsunami K *et al.* (2008) Euodionosides A-G: megastigmane glucosides from leaves of *Euodia meliaefolia*. *Phytochemistry* **69**, 1586–1596. doi:10.1016/j.phytochem.2008.01.020
205. Zhao N, Li ZL, Li DH, Sun YT, Shan DT *et al.* (2015) Quinolone and indole alkaloids from the fruits of *Euodia rutaecarpa* and their cytotoxicity against two human cancer cell lines. *Phytochemistry* **109**, 133–139. doi:10.1016/j.phytochem.2014.10.020
206. Eliaser EM, Ho JH, Hashim NM, Rukayadi Y, Ee GCL *et al.* (2018) Phytochemical constituents and biological activities of *Melicope lunu-ankenda*. *Molecules (Basel, Switzerland)* **23**, 2708.
207. Fan S, Zhang C, Luo T, Wang J, Tang Y *et al.* (2019) Limonin: a review of its pharmacology, toxicity, and pharmacokinetics. *Molecules (Basel, Switzerland)* **24**, 3679. doi:10.3390/molecules24203679
208. Vu VT, Chen XL, Kong LY, Luo JG (2020) Melipatulinones A-C, three lignan-phloroglucinol hybrids from *Melicope patulinervia*. *Organic Letters* **22**, 1380–1384. doi:10.1021/acs.orglett.9b04680
209. Yao Q, Gao Y, Lai C, Wu C, Zhao CL *et al.* (2020) The phytochemistry, pharmacology and applications of *Melicope pteleifolia*: a review. *Journal of Ethnopharmacology* **251**, 112546. doi:10.1016/j.jep.2020.112546
210. Rashid MA, Armstrong JA, Gray AI, Waterman PG (1991) Pyranocoumarins as chemotaxonomic markers in *Eriostemon coccineus* and *Philotheca citrina*. *Phytochemistry* **30**, 4033–4035. doi:10.1016/0031-9422(91)83459-X
211. Ghisalberti EL (1998) Phytochemistry of the Australian Rutaceae: *Boronia*, *Eriostemon* and *Phebalium* species. *Phytochemistry* **47**, 163–176. doi:10.1016/S0031-9422(97)00571-2
212. Sultana N, Sarker SD, Armstrong JA, Wilson PG, Waterman PG (2003) The coumarins of *Philotheca* sensu lato: distribution and systematic significance. *Biochemical Systematics and Ecology* **31**, 681–691. doi:10.1016/S0305-1978(02)00226-0
213. Chlouchi A, Muyard F, Girard C, Waterman PG, Bévalot F (2005) Coumarins from the twigs of *Diplolaena mollis* P. G. Wilson (Rutaceae). *Biochemical Systematics and Ecology* **33**, 967–969. doi:10.1016/j.bse.2005.03.002
214. Chlouchi A, Girard C, Tillequin F, Bévalot F, Waterman PG *et al.* (2006) Coumarins and furoquinoline alkaloids from *Philotheca deserti* var. *deserti* (Rutaceae). *Biochemical Systematics and Ecology* **34**, 71–74. doi:10.1016/j.bse.2005.07.002
215. Chlouchi A, Girard C, Bévalot F, Waterman PG, Muyard F (2007) Taxonomically significant coumarins from three *Philotheca* species (Rutaceae). *Biochemical Systematics and Ecology* **35**, 251–254. doi:10.1016/j.bse.2006.09.011
216. Southwell IA (1981) Methoxystyrenes from the genus *Zieria*. *Phytochemistry* **20**, 1448–1450. doi:10.1016/0031-9422(81)80068-4
217. Flynn TM, Southwell IA (1987) Cyanogenesis in the genus *Zieria*. *Phytochemistry* **26**, 1669–1673. doi:10.1016/S0031-9422(00)82267-0
218. Flynn TM, Southwell IA (1987) Essential oil constituents of the genus *Zieria*. *Phytochemistry* **26**, 1673–1686. doi:10.1016/S0031-9422(00)82268-2
219. Southwell IA, Armstrong JA (1987) Chemical variation within the genus *Zieria*. *Phytochemistry* **26**, 1687–1692. doi:10.1016/S0031-9422(00)82269-4
220. Griffin SG, Leach DN, Markham JL, Johnstone R (1998) Antimicrobial activity of essential oils from *Zieria*. *The Journal of Essential Oil Research* **10**, 165–174. doi:10.1080/10412905.1998.9700870
221. Ghisalberti EL (1998) Ethnopharmacology and phytochemistry of *Dodonaea* species. *Fitoterapia* **69**, 99–113.
222. kundu P, Laskar S (2010) A brief resume on the genus *Ailanthus*: chemical and pharmacological aspects.

Phytochemistry Reviews **9**, 379–412. doi:10.1007/s11101-009-9157-1
223. Salehi B, Krochmal-Marczak B, Skiba D, Patra JK, Das SK *et al.* (2020) *Convolvulus* plant: a comprehensive review from phytochemical composition to pharmacy. *Phytotherapy Research* **34**, 315–328. doi:10.1002/ptr.6540
224. Nugroho LH, Verpoorte R (2002) Secondary metabolism in tobacco. *Plant Cell, Tissue and Organ Culture* **68**, 105–125. doi:10.1023/A:1013853909494
225. Dewey RE, Xie J (2013) Molecular genetics of alkaloid biosynthesis in *Nicotiana* tabacum. *Phytochemistry* **94**, 10–27. doi:10.1016/j.phytochem.2013.06.002
226. Jassbi AR, Zare S, Asadollahi M, Schuman MC (2017) Ecological roles and biological activities of secialized metabolites from the genus *Nicotiana*. *Chemical Reviews* **117**, 12227–12280. doi:10.1021/acs.chemrev.7b00001
227. Kaunda JS, Zhang YJ (2019) The genus *Solanum*: an ethnopharmacological, phytochemical and biological properties review. *Natural Products and Bioprospecting* **9**, 77–137. doi:10.1007/s13659-019-0201-6
228. Stohs SJ, Ray SD (2013) A review and evaluation of the efficacy and safety of *Cissus quadrangularis* extracts. *Phytotherapy Research* **27**, 1107–1114. doi:10.1002/ptr.4846
229. Siddiqua A, Mittapally S (2017) A review of *Cissus quadrangularis*. *Journal of Pharmaceutical Innovation* **6**, 329–334.
230. Nahar L, Sarker SD (2007) Phytochemistry of the genus *Curcuma*. In *Turmeric: The Genus Curcuma*. (Eds PN Ravindran, K Nirmal Babu and K Sivaraman) pp. 71–106. CRC Press, Boca Raton, FL.
231. Wu P, Su MX, Wang Y, Wang GC, Ye WC *et al.* (2012) Supercritical fluid extraction assisted isolation of sesquiterpene lactones with antiproliferative effects from *Centipeda minima*. *Phytochemistry* **76**, 133–140. doi:10.1016/j.phytochem.2012.01.003
232. Zdero C, Bohlmann F, King RM (1992) *Ent*-labdanes, manoyloxide and helipterol derivatives from *Chrysocephalum ambiguum*. *Phytochemistry* **31**, 1631–1638. doi:10.1016/0031-9422(92)83119-J
233. Lourens AC, Viljoen AM, Van Heerden FR (2008) South African *Helichrysum* species: a review of the traditional uses, biological activity and phytochemistry. *Journal of Ethnopharmacology* **119**, 630–652. doi:10.1016/j.jep.2008.06.011
234. Chung IM, Rajakumar G, Lee JH, Kim SH, Thiruvengadam M (2017) Ethnopharmacological uses, phytochemistry, biological activities, and biotechnological applications of *Eclipta prostrata*. *Applied Microbiology and Biotechnology* **101**, 5247–5257. doi:10.1007/s00253-017-8363-9
235. Feng L, Zhai YY, Xu J, Yao WF, Cao YD *et al.* (2019) A review on traditional uses, phytochemistry and pharmacology of *Eclipta prostrata* (L.) L. *Journal of Ethnopharmacology* **245**, 112109. doi:10.1016/j.jep.2019.112109
236. Hiradeve SM, Rangari VD (2014) A review on pharmacology and toxicology of *Elephantopus scaber* Linn. *Natural Product Research* **28**, 819–830. doi:10.1080/14786419.2014.883394
237. Zheng X, Wang W, Piao H, Xu W, Shi H *et al.* (2013) The genus *Gnaphalium* L. (Compositae): phytochemical and pharmacological characteristics. *Molecules (Basel, Switzerland)* **18**, 8298–8318. doi:10.3390/molecules18078298
238. Goyal N, Sharma SK (2014) Bioactive phytoconstituents from genus Heliotropium. *International Journal of Green Pharmacy* **8**, 217–225.
239. Singh H, Mishra A, Mishra AK (2018) The chemistry and pharmacology of *Cleome* genus: a review. *Biomedicine and Pharmacotherapy* **101**, 37–48. doi:10.1016/j.biopha.2018.02.053
240. Asakawa Y, Ludwiczuk A, Harinantenaina L, Toyota M, Nishiki M *et al.* (2012) Distribution of drimane sesquiterpenoids and tocopherols in liverworts, ferns and higher plants: Polygonaceae, Canellaceae, and Winteraceae species. *Natural Product Communications* **7**, 685–692. doi:10.1177/1934578X1200700601
241. Cock IE (2013) The phytochemistry and chemotherapeutic potential of *Tasmannia lanceolata* (Tasmanian pepper): a review. *Pharmacognosy Communications* **3**, 1–13. doi:10.5530/pc.2013.2.1
242. Peng W, Qin R, Li X, Zhou H (2013) Botany, phytochemistry, pharmacology, and potential application of *Polygonum cuspidatum* Sieb.et Zucc.: a review. *Journal of Ethnopharmacology* **148**, 729–745. doi:10.1016/j.jep.2013.05.007
243. Lin L, Ni B, Lin H, Zhang M, Li X *et al.* (2015) Traditional usages, botany, phytochemistry, pharmacology and toxicology of *Polygonum multiflorum* Thunb.: a review. *Journal of Ethnopharmacology* **159**, 158–183. doi:10.1016/j.jep.2014.11.009
244. Ayaz M, Ahmad I, Sadiq A, Ullah F, Ovais M *et al.* (2020) *Persicaria hydropiper* (L.) Delarbre: a review on traditional uses, bioactive chemical constituents and pharmacological and toxicological activities. *Journal of Ethnopharmacology* **251**, 112516. doi:10.1016/j.jep.2019.112516
245. Gou KJ, Zeng R, Ma Y, Li AN, Yang K *et al.* (2020) Traditional uses, phytochemistry, and pharmacology of *Persicaria orientalis* (L.) Spach: a review. *Journal of Ethnopharmacology* **249**, 112407. doi:10.1016/j.jep.2019.112407
246. Gerometta E, Grondin I, Smadja J, Frederich M, Gauvin-Bialecki A (2020) A review of traditional uses, phytochemistry and pharmacology of the genus *Indigofera*. *Journal of Ethnopharmacology* **253**, 112608. doi:10.1016/j.jep.2020.112608
247. Harrison JJEK, Dankyi E, Kingsford-Adaboh R, Ishida H (2011) In search of new leads: a closer look at the therapeutic potential of the constituents of *Millettia thonningii*, *Millettia machycarpa* and their structural analogues. *International Journal of Pharmacy and Pharmaceutical Sciences* **3**, 71–81.
248. Al Muqarrabun LM, Ahmat N, Ruzaina SA, Ismail NH, Sahidin I (2013) Medicinal uses, phytochemistry and pharmacology of *Pongamia pinnata* (L.) Pierre: a review. *Journal of Ethnopharmacology* **150**, 395–420. doi:10.1016/j.jep.2013.08.041

249. Kaunda JS, Zhang YJ (2017) The genus *Carissa*: an ethnopharmacological, phytochemical and pharmacological review. *Natural Products and Bioprospecting* **7**, 181–199. doi:10.1007/s13659-017-0123-0
250. Shi Q-W, Su X-H, Kiyota H (2008) Chemical and pharmacological research of the plants in genus *Euphorbia*. *Chemical Reviews* **108**, 4295–4327. doi:10.1021/cr078350s
251. Nebula M, Harisankar HS, Chandramohanakumar N (2013) Metabolites and bioactivities of Rhizophoraceae mangroves. *Natural Products and Bioprospecting* **3**, 207–232. doi:10.1007/s13659-013-0012-0
252. Bibi SN, Fawzi MM, Gokhan Z, Rajesh J, Nadeem N *et al.* (2019) Ethnopharmacology, phytochemistry, and global distribution of mangroves: a comprehensive review. *Marine Drugs* **17**, 231.
253. Poornima V, Sharanya M, Jeyam M (2014) An ethnomedical, pharmacological and phytochemical review of *Ammannia baccifera* L. *World Journal of Pharmaceutical Research* **3**, 1771–1789.
254. Efange SMN, Brun R, Wittlin S, Connolly JD, Hoye TR *et al.* (2009) Okundoperoxide, a bicyclic cyclofarnesylsesquiterpene endoperoxide from *Scleria striatinux* with antiplasmodial activity. *Journal of Natural Products* **72**, 280–283. doi:10.1021/np800338p
255. Nyongbela KD, Ndjoko KI, Brun R, Wittlin S, Mbah JA *et al.* (2009) Occurrence of sesquiterpene derivatives in *Scleria striatonux* De Wild (Cyperaceae). *Natural Product Communications* **4**, 5–8. doi:10.1177/1934578X0900400103
256. Nyongbela KD, Makolo FL, Hoye TR, Efange SMN (2016) Isolation and characterisation of sclerienone C from *Scleria striatinux*. *Natural Product Communications* **11**, 5–6. doi:10.1177/1934578X1601100102
257. Agnew MP, Lauren DR (1991) Determination of ptaquiloside in bracken fern (*Pteridium esculentum*). *Journal of Chromatography. A* **538**, 462–468. doi:10.1016/S0021-9673(01)88870-1
258. Tanaka N, Yuhara H, Wada H, Murakami T, Cambie RC *et al.* (1993) Phenolic constituents of *Pteridium esculentum*. *Phytochemistry* **32**, 1037–1039. doi:10.1016/0031-9422(93)85251-L
259. Yamada K, Ojika M, Kigoshi H (2007) Ptaquiloside, the major toxin of bracken, and related terpene glycosides: chemistry, biology and ecology. *Natural Product Reports* **24**, 798–813. doi:10.1039/b614160a
260. Fletcher MT, Hayes PY, Somerville MJ, De Voss JJ (2010) Ptesculentoside, a novel norsesquiterpene glucoside from the Australian bracken fern *Pteridium esculentum*. *Tetrahedron Letters* **51**, 1997–1999. doi:10.1016/j.tetlet.2010.02.032
261. Chavan RB, Gaikwad DK (2016) The ethnobotany, phytochemistry and biological properties of *Allophylus* species used in traditional medicine: a review. *World Journal of Pharmacy and Pharmaceutical Sciences* **5**, 664–682.
262. Seigler DS (2003) Phytochemistry of Acacia—sensu lato. *Biochemical Systematics and Ecology* **31**, 845–873. doi:10.1016/S0305-1978(03)00082-6
263. Zanin JL, De Carvalho BA, Martineli PS, Dos Santos MH, Lago JH *et al.* (2012) The genus *Caesalpinia* L. (Caesalpiniaceae): phytochemical and pharmacological characteristics. *Molecules (Basel, Switzerland)* **17**, 7887–7902. doi:10.3390/molecules17077887
264. Yadav JP, Arya V, Yadav S, Panghal M, Kumar S *et al.* (2010) *Cassia occidentalis* L.: a review on its ethnobotany, phytochemical and pharmacological profile. *Fitoterapia* **81**, 223–230. doi:10.1016/j.fitote.2009.09.008
265. Singh O, Ali M, Akhtar N (2011) Phenolic acid glucosides from the seeds of *Entada phaseoloides* Merill. *Journal of Asian Natural Products Research* **13**, 682–687. doi:10.1080/10286020.2011.584444
266. Sugimoto S, Matsunami K, Otsuka H (2018) Biological activity of *Entada phaseoloides* and *Entada rheedei*. *Journal of Natural Medicines* **72**, 12–19. doi:10.1007/s11418-017-1146-x
267. Tan YP, Houston SD, Modhiran N, Savchenko AI, Boyle GM *et al.* (2019) Stachyonic acid: a dengue virus inhibitor from Basilicum polystachyon. *Chemistry (Basel, Switzerland)* **25**, 5664–5667.
268. Tan YP, Xue Y, Savchenko AI, Houston SD, Modhiran N *et al.* (2019) Basimarols A, B, and C, highly oxygenated pimarane diterpenoids from Basilicum polystachyon. *Journal of Natural Products* **82**, 2828–2834. doi:10.1021/acs.jnatprod.9b00522
269. Manchand PS, White JD, Wright H, Clardy J (1973) Structures of stemodin and stemodinone. *Journal of the American Chemical Society* **95**, 2705–2706. doi:10.1021/ja00789a060
270. Chamy MC, Piovano M, Barbarino JA, Gambaro V (1991) Stemodane diterpenoids from *Stemodia chilensis*. *Phytochemistry* **30**, 1719–1721. doi:10.1016/0031-9422(91)84245-N
271. Hufford CD, Oguntimein BO, Muhammad I (1992) New stemodane diterpenes from *Stemodia maritima*. *Journal of Natural Products* **55**, 48–52. doi:10.1021/np50079a007
272. Allen JG, Colegate SM, Mitchell AA, Mulder RJ, Raisbeck MF (2006) The bioactivity-guided isolation and structural identification of toxic cucurbitacin steroidal glucosides from *Stemodia kingii*. *Phytochemical Analysis* **17**, 226–235. doi:10.1002/pca.914
273. Rodrigues FE, Lima JQ, de Oliveira MDCFD, Vasconcelos JN, Santiago GM *et al.* (2010) Diterpene and other constituents from *Stemodia maritima* (Scrophulariaceae). *Journal of the Brazilian Chemical Society* **21**, 1581–1586. doi:10.1590/S0103-50532010000800025
274. Leonelli F, Mostarda A, De Angelis L, Lamba D, Demitri N *et al.* (2016) Proof of the structure of the *Stemodia chilensis* tetracyclic diterpenoid (+)-19-acetoxystemodan-12-ol by synthesis from (+)-podocarpic acid: X-ray structure determination of a key intermediate. *Journal of Natural Products* **79**, 1155–1159. doi:10.1021/acs.jnatprod.5b00834
275. Bakker HJ, Ghisalberti EL, Jefferies PR (1972) Biosynthesis of diterpenes in *Beyeria leschenaultii*. *Phytochemistry* **11**, 2221–2231. doi:10.1016/S0031-9422(00)88381-8

276. Yin B-W, Shen L-R, Zhang M-L, Zhao L, Wang Y-L *et al.* (2008) Chemical constituents of plants from the genus *Excoecaria*. *Chemistry & Biodiversity* **5**, 2356–2371. doi:10.1002/cbdv.200890201
277. Grace MH, Jin Y, Wilson GR, Coates RM (2006) Structures, biogenetic relationships, and cytotoxicity of pimarane-derived diterpenes from *Petalostigma pubescens*. *Phytochemistry* **67**, 1708–1715. doi:10.1016/j.phytochem.2005.09.026
278. Ilanko A, Cock IE (2019) The interactive antimicrobial activity of conventional antibiotics and *Petalostigma* spp. extracts against bacterial triggers of some autoimmune inflammatory diseases. *Pharmacognosy Journal* **11**, 292–309. doi:10.5530/pj.2019.11.45
279. Hayes PY, Chow S, Somerville MJ, De Voss JJ, Fletcher MT (2009) Pimelotides A and B, diterpenoid ketal-lactone orthoesters with an unprecedented skeleton from *Pimelea elongata*. *Journal of Natural Products* **72**, 2081–2083. doi:10.1021/np900573k
280. Hayes PY, Chow S, Somerville MJ, Fletcher MT, De Voss JJ (2010) Daphnane- and tigliane-type diterpenoid esters and orthoesters from *Pimelea elongata*. *Journal of Natural Products* **73**, 1907–1913. doi:10.1021/np1005746
281. Duh C-Y, Phoebe CHJ, Pezzuto JM, Kinghorn AD, Farnsworth NR (1986) Plant anticancer agents. XLII: cytotoxic constituents from *Wikstroemia elliptica*. *Journal of Natural Products* **49**, 706–709. doi:10.1021/np50046a031
282. Liao S-G, Wu Y, Yue J-M (2006) Lignans from *Wikstroemia hainanensis*. *Helvetica Chimica Acta* **89**, 73–80.
283. Wang G-C, Zhang X-L, Wang Y-F, Li G-Q, Ye W-C *et al.* (2012) Four new dilignans from the roots of *Wikstroemia indica*. *Chemical & Pharmaceutical Bulletin* **60**, 920–923. doi:10.1248/cpb.c12-00132
284. Fan Q, Jiang Y-P, Zhu D-Q, Xu W, Huang W-Q *et al.* (2018) Phenols from the rhizome of *Wikstroemia indica*. *Biochemical Systematics and Ecology* **78**, 59–62. doi:10.1016/j.bse.2018.04.004
285. Li SF, Jiao YY, Zhang ZQ, Chao JB, Jia J *et al.* (2018) Diterpenes from buds of *Wikstroemia chamaedaphne* showing anti-hepatitis B virus activities. *Phytochemistry* **151**, 17–25. doi:10.1016/j.phytochem.2018.01.021
286. Aslam MS, Choudhary BA, Uzair M, Ijaz AS (2013) Phytochemical and ethno-pharmacological review of the genus *Araucaria*: review. *Tropical Journal of Pharmaceutical Research* **12**, 651–659. doi:10.4314/tjpr.v12i4.31
287. Hungerford NL, Sands DPA, Kitching W (1998) Isolation and structure of some constituents of the Australian medicinal plant *Tinospora smilacina* ('snakevine'). *Australian Journal of Chemistry* **51**, 1103–1111. doi:10.1071/C98034
288. Pathak AK, Jain DC, Sharma RP (2008) Chemistry and biological activities of the genera *Tinospora*. *International Journal of Pharmacology* **33**, 277–287.
289. Singh D, Chaudhuri PK (2017) Chemistry and pharmacology of *Tinospora cordifolia*. *Natural Product Communications* **12**, 299–308. doi:10.1177/1934578X1701200240
290. Negi JS, Bisht VK, Bandari AK, Bharti MK, Sundriyal RC (2011) Chemical and pharmacological aspects of *Toona* (Meliaceae). *Research Journal of Phytochemistry* **5**, 14–21. doi:10.3923/rjphyto.2011.14.21
291. Divakar PR (2017) Phytopharmacology of *Toona ciliata*: a review. *International Journal of Pharmaceutics* **8**, 30–35.
292. Peng W, Liu Y, Hu M, Zhang M, Yang J *et al.* (2019) *Toona sinensis*: a comprehensive review on its traditional usages, phytochemisty, pharmacology and toxicology. *Revista Brasileira de Farmacognosia-Brazilian Journal of Pharmacognosy* **29**, 111–124. doi:10.1016/j.bjp.2018.07.009
293. Almubayedh H, Ahmad R (2020) Ethnopharmacology, phytochemistry, biological activities, and therapeutic applications of *Cedrela serrata* Royle: a mini review. *Journal of Ethnopharmacology* **246**, 112206. doi:10.1016/j.jep.2019.112206
294. Meira M, Da Silva EP, David JM, David JP (2012) Review of the genus *Ipomoea*: traditional uses, chemistry and biological activities. *Revista Brasileira de Farmacognosia-Brazilian Journal of Pharmacognosy* **22**, 682–713. doi:10.1590/S0102-695X2012005000025
295. Mallik J, Das J, Banik RK (2018) Pharmacognostic profile and pharmacological activity of different parts of *Amorphophallus campanulatus* (Roxb.) Blume: a complete overview. *Asian Journal of Pharmaceutical Research & Development* **6**, 4–8. doi:10.22270/ajprd.v6i1.354
296. Padeepika C, Selvakumar R, Krishnakumar T, Sajad UNN, Sajeev MS (2018) Pharmacology and phytochemistry of underexploited tuber crops: a review. *Journal of Pharmacognosy and Phytochemistry* **7**, 1007–1019.
297. Agostini-Costa TDS (2018) Bioactive compounds and health benefits of some palm species traditionally used in Africa and the Americas: a review. *Journal of Ethnopharmacology* **224**, 202–229. doi:10.1016/j.jep.2018.05.035
298. Negi JS, Singh P, Joshi GP, Rawat MS, Bisht VK (2010) Chemical constituents of *Asparagus*. *Pharmacognosy Reviews* **4**, 215–220. doi:10.4103/0973-7847.70921
299. Folquitto DG, Swiech JND, Pereira CB, Bobek VB, Halila Possagno GC *et al.* (2019) Biological activity, phytochemistry and traditional uses of genus *Lobelia* (Campanulaceae): a systematic review. *Fitoterapia* **134**, 23–38. doi:10.1016/j.fitote.2018.12.021
300. Li L, Peng Y, Yao X, Xu L-J, Wulan T-N *et al.* (2010) Chemical constituents and biological activities of plants from the genus *Ehretia* Linn. *Chinese Herbal Medicines* **2**, 106–111.
301. Shukla A, Kaur A (2018) A systematic review of traditional uses bioactive phytoconstituents of genus *Ehretia*. *Asian Journal of Pharmaceutical and Clinical Research* **11**, 88–100. doi:10.22159/ajpcr.2018.v11i6.25178
302. Omar M, Defeo J, Youngken HWJ (1983) Chemical and toxicity studies of *Trichodesma africanum* L. *Journal of Natural Products* **46**, 153–156. doi:10.1021/np50026a001
303. Hosamani KM (1994) Ricinoleic and cyclopropene acids in *Trichodesma zeylanicum* seed oil. *Phytochemistry* **37**, 1621–1624. doi:10.1016/S0031-9422(00)89578-3

304. Mangai AS (2018) Pharmacological efficacy of *Trichodesma indicum* (Linn.) R. Br., in folk medicine: an updated review. *Asian Journal of Pharmaceutical and Clinical Research* **11**, 24–27. doi:10.22159/ajpcr.2018.v11i3.24412
305. Tlili N, Elfalleh W, Saadaoui E, Khaldi A, Triki S *et al.* (2011) The caper (*Capparis* L.): ethnopharmacology, phytochemical and pharmacological properties. *Fitoterapia* **82**, 93–101. doi:10.1016/j.fitote.2010.09.006
306. Clauser M, Dall'acqua S, Loi MC, Innocenti G (2013) Phytochemical investigation on *Atriplex halimus* L. from Sardinia. *Natural Product Research* **27**, 1940–1944. doi:10.1080/14786419.2013.793684
307. Mroczek A (2015) Phytochemistry and bioactivity of triterpene saponins from Amaranthaceae family. *Phytochemistry Reviews* **14**, 577–605. doi:10.1007/s11101-015-9394-4
308. Sati OP, Bahuguna S, Uniyal S, Sakakibara J, Kaiya T *et al.* (1990) A new saponin from *Deeringia amaranthoides*. *Journal of Natural Products* **53**, 466–469. doi:10.1021/np50068a029
309. Patil KS, Bhalsing SR (2016) Ethnomedicinal uses, phytochemistry and pharmacological properties of the genus *Boerhavia*. *Journal of Ethnopharmacology* **182**, 200–220. doi:10.1016/j.jep.2016.01.042
310. Garg V, Jyoti, Dutt R (2018) The genus *Plumbago*: an update. *International Journal of Green and Herbal Chemistry* **7**, 313–326.
311. Tan CY, Wang JH, Zhang H, Li X (2002) A new phytosterone from *Cyanotis arachnoidea*. *Journal of Asian Natural Products Research* **4**, 7–11. doi:10.1080/10286020290019631
312. Tan CY, Wang JH, Li X (2003) Phytoecdysteroid constituents from *Cyanotis arachnoidea*. *Journal of Asian Natural Products Research* **5**, 237–240. doi:10.1080/1028602031000105830
313. Kovacs A, Vasas A, Hohmann J (2008) Natural phenanthrenes and their biological activity. *Phytochemistry* **69**, 1084–1110. doi:10.1016/j.phytochem.2007.12.005
314. Lv H, She G (2010) Naturally occurring diarylheptanoids. *Natural Product Communications* **5**, 1687–1708. doi:10.1177/1934578X1000501035
315. Mohanty A, Das D, Ghosh G, Sahu PK (2016) *Barringtonia acutangula*: ethnobotanical and phytopharmacological review. *International Journal of Pharmaceutical Sciences Review and Research* **38**, 22–29.
316. Khong PW, Lewis KG (1977) New chemical constituents of *Planchonia careya*. *Australian Journal of Chemistry* **30**, 1311–1322. doi:10.1071/CH9771311
317. Khong PW, Lewis KG (1979) New chemical constituents of *Planchonia careya*. II: bark constituents soluble in light petroleum. *Australian Journal of Chemistry* **32**, 1621–1626. doi:10.1071/CH9791621
318. Crublet M-L, Long C, Sévenet T, Hadi HA, Lavaud C (2003) Acylated flavonol glycosides from leaves of *Planchonia grandis*. *Phytochemistry* **64**, 589–594. doi:10.1016/S0031-9422(03)00215-2
319. McRae JM, Yang Q, Crawford RJ, Palombo EA (2008) Antibacterial compounds from *Planchonia careya* leaf extracts. *Journal of Ethnopharmacology* **116**, 554–560. doi:10.1016/j.jep.2008.01.007
320. McRae JM, Yang Q, Crawford RJ, Palombo EA (2008) Acylated flavonoid tetraglycoside from *Planchonia careya* leaves. *Phytochemistry Letters* **1**, 99–102. doi:10.1016/j.phytol.2008.04.003
321. Rajeswari K, Bhaskara Rao T (2015) *Aegicera corniculatum* Linn (Myrsinaceae). *Journal of Chemical and Pharmaceutical Research* **7**, 305–316.
322. Hart NK, Lamberton JA (1966) Pyrrolizidine alkaloids from *Planchonella* species (family Sapotaceae). *Australian Journal of Chemistry* **19**, 1259–1264. doi:10.1071/CH9661259
323. Hart NK, Johns SR, Lamberton JA (1968) Alkaloids of the Sapotaceae: trans-B-methylthioacrylate and tiglate esters of (-)-isoretronecanol from a *Planchonella* species. *Australian Journal of Chemistry* **21**, 1393–1395. doi:10.1071/CH9681393
324. Eade RA, Hunt K, Simes JJH, Stern W (1969) Extractives of Australian timbers IX: the saponin of *Planchonella pohlmanianum*. *Australian Journal of Chemistry* **22**, 2703–2707. doi:10.1071/CH9692703
325. Cambie RC, Ser NA, Kokubun T (1997) Heartwood constituents of *Planchonella vitiensis*. *Biochemical Systematics and Ecology* **25**, 677–678. doi:10.1016/S0305-1978(97)00063-X
326. Lee T-H, Juang S-H, Hsu F-L, Wu C-Y (2005) Triterpene acids from the leaves of *Planchonella duclitan* (Blanco) Bakhuizan. *Journal of the Chinese Chemical Society (Taipei)* **52**, 1275–1280. doi:10.1002/jccs.200500184
327. Lee S-S, Tseng C-C, Chen C-K (2010) Three new benzophenone glucosides from the leaves of *Planchonella obovata*. *Helvetica Chimica Acta* **93**, 522–529. doi:10.1002/hlca.200900264
328. Chen H-Y, Guh J-H, Chan S-H, Lee S-S (2015) Cytotoxic protobassic acid glycosides from *Planchonella obovata* leaf. *Phytochemistry Letters* **11**, 229–235. doi:10.1016/j.phytol.2015.01.005
329. Garaniya N, Bapodra A (2014) Ethno botanical and Phytopharmacological potential of Abrus precatorius L.: a review. *Asian Pacific Journal of Tropical Biomedicine* **4**, S27–S34. doi:10.12980/APJTB.4.2014C1069
330. Farag MA, Sakna ST, El-Fiky NM, Shabana MM, Wessjohann LA (2015) Phytochemical, antioxidant and antidiabetic evaluation of eight *Bauhinia* L. species from Egypt using UHPLC-PDA-qTOF-MS and chemometrics. *Phytochemistry* **119**, 41–50. doi:10.1016/j.phytochem.2015.09.004
331. Rammohan A, Reddy GM, Bhaskar BV, Gunasekar D, Zyryanov GV (2020) Phytochemistry and pharmacological activities of the genus *Rhynchosia*: a comprehensive review. *Planta* **251**, 9. doi:10.1007/s00425-019-03311-2
332. Hasan N, Osman H, Mohamad S, Chong WK, Awang K *et al.* (2012) The chemical components of *Sesbania grandiflora* root and their antituberculosis activity. *Pharmaceuticals* **5**, 882–889. doi:10.3390/ph5080882

333. Noviany N, Samadi A, Yuliyan N, Hadi S, Aziz M *et al.* (2020) Structure characterization and biological activity of 2-arylbenzofurans from an Indonesian plant, *Sesbania grandiflora* (L.) Pers. *Phytochemistry Letters* **35**, 211–215. doi:10.1016/j.phytol.2019.12.008
334. Krishna PM, Rao KNV, Sandhya S, Banji D (2012) A review on phytochemical, ethnomedical and pharmacological studies on genus *Sophora*, Fabaceae. *Revista Brasileira de Farmacognosia-Brazilian Journal of Pharmacognosy* **22**, 1145–1154. doi:10.1590/S0102-695X2012005000043
335. He X, Fang J, Huang L, Wang J, Huang X (2015) *Sophora flavescens* Ait.: traditional usage, phytochemistry and pharmacology of an important traditional Chinese medicine. *Journal of Ethnopharmacology* **172**, 10–29. doi:10.1016/j.jep.2015.06.010
336. He X, Bai Y, Zhao Z, Wang X, Fang J *et al.* (2016) Local and traditional uses, phytochemistry, and pharmacology of *Sophora japonica* L.: a review. *Journal of Ethnopharmacology* **187**, 160–182. doi:10.1016/j.jep.2016.04.014
337. Onyilagha JC, Islam S, Ntamatungiro S (2009) Comparative phytochemistry of eleven species of *Vigna* (Fabaceae). *Biochemical Systematics and Ecology* **37**, 16–19. doi:10.1016/j.bse.2008.11.013
338. Leu YL, Hwang TL, Kuo PC, Liou KP, Huang BS *et al.* (2012) Constituents from *Vigna vexillata* and their anti-inflammatory activity. *International Journal of Molecular Sciences* **13**, 9754–9768. doi:10.3390/ijms13089754
339. Tang D, Dong Y, Ren H, Li L, He C (2014) A review of phytochemistry, metabolite changes, and medicinal uses of the common food mung bean and its sprouts (*Vigna radiata*). *Chemistry Central Journal* **8**, 4. doi:10.1186/1752-153X-8-4
340. Wang P, Yang J, Zhu Z, Zhang X (2018) *Marsdenia tenacissima*: a review of traditional uses, phytochemistry and pharmacology. *The American Journal of Chinese Medicine* **46**, 1449–1480. doi:10.1142/S0192415X18500751
341. Rasool N, Ahmad VU, Malik A (1991) Terpenoids from *Pentatropis spiralis*. *Phytochemistry* **30**, 1331–1332. doi:10.1016/S0031-9422(00)95230-0
342. Boucherle B, Haudecoeur R, Queiroz EF, De Waard M, Wolfender JL *et al.* (2016) *Nauclea latifolia*: biological activity and alkaloid phytochemistry of a West African tree. *Natural Product Reports* **33**, 1034–1043. doi:10.1039/C6NP00039H
343. Haudecoeur R, Peuchmaur M, Peres B, Rome M, Taiwe GS *et al.* (2018) Traditional uses, phytochemistry and pharmacological properties of African *Nauclea* species: A review. *Journal of Ethnopharmacology* **212**, 106–136. doi:10.1016/j.jep.2017.10.011
344. Johns SR, Lamberton JA (1970) The identification of a new alkaloid from *Timonius kaniensis* (Rubiaceae) as dihydrocupreine. *Australian Journal of Chemistry* **23**, 211–212. doi:10.1071/CH9700211
345. Khan IA, Sticher O, Rali T (1993) New triterpenes from the leaves of *Timonius timon*. *Journal of Natural Products* **56**, 2163–2165. doi:10.1021/np50102a019
346. Russo A, Borrelli F (2005) Bacopa monniera, a reputed nootropic plant: an overview. *Phytomedicine* **12**, 305–317. doi:10.1016/j.phymed.2003.12.008
347. Hayashi T, Kishi MP, Kawasaki M, Arisawa M, Shimizu M *et al.* (1987) Scopadulcic acid -A and -B, new diterpenoids with a novel skeleton, from a Paraguayan crude drug "Typycha Kuratu" (*Scoparia dulcis* L.). *Tetrahedron Letters* **28**, 3693–3696. doi:10.1016/S0040-4039(00)96357-3
348. Kawasaki M, Hayashi T, Arisawa M, Morita N, Berganza LH (1988) 8-hydroxytricetin 7-glucuronide, a B-glucuronidase inhibitor from *Scoparia dulcis*. *Phytochemistry* **27**, 3709–3711. doi:10.1016/0031-9422(88)80811-2
349. Hayashi T, Kawasaki M, Miwa Y, Taga T, Morita N (1990) Antiviral agents of plant origin. III. Scopadulin, a novel tetracyclic diterpene from *Scoparia dulcis* L. *Chemical & Pharmaceutical Bulletin* **38**, 945–947. doi:10.1248/cpb.38.945
350. Mishra MR, Behera RK, Jha S, Panda AK, Mishra A *et al.* (2011) A brief review on phytoconstituents and ethnopharmacology of *Scoparia dulcis* Linn. (Scrophulariaceae). *International Journal of Phytotherapy and Phytopharmacology* **3**, 422–438.
351. Chang F-R, Chao Y-C, Teng C-M, Wu Y-C (1998) Chemical constituents from *Cassytha filiformis* II. *Journal of Natural Products* **61**, 863–866.
352. Tsai T-H, Wang G-J, Lin L-C (2008) Vasorelaxing alkaloids and flavonoids from *Cassytha filiformis*. *Journal of Natural Products* **71**, 289–291. doi:10.1021/np070564h
353. Cox SD, Jayasinghe KC, Markham JL (2005) Antioxidant activity in Australian native sarsaparilla (*Smilax glyciphylla*). *Journal of Ethnopharmacology* **101**, 162–168. doi:10.1016/j.jep.2005.04.005
354. Wang WX, Li TX, Ma H, Zhang JF, Jia AQ (2013) Tumoral cytotoxic and antioxidative phenylpropanoid glycosides in *Smilax riparia* A. DC. *Journal of Ethnopharmacology* **149**, 527–532. doi:10.1016/j.jep.2013.07.011
355. Chen W, Shou X-A, Chen Y, Qin N, Qiao W *et al.* (2014) A new aurone from *Smilax riparia*. *Chemistry of Natural Compounds* **50**, 989–993. doi:10.1007/s10600-014-1143-1
356. Shu J, Liang F, Zhu G, Liu X, Yu J *et al.* (2017) Lignan glycosides from the rhizomes of *Smilax trinervula* and their biological activities. *Phytochemistry Letters* **20**, 1–8. doi:10.1016/j.phytol.2017.03.002
357. Tian LW, Zhang Z, Long HL, Zhang YJ (2017) Steroidal saponins from the genus *Smilax* and their biological activities. *Natural Products and Bioprospecting* **7**, 283–298. doi:10.1007/s13659-017-0139-5
358. Hua S, Zhang Y, Liu J, Dong L, Huang J *et al.* (2018) Ethnomedicine, phytochemistry and pharmacology of *Smilax glabra*: an important traditional Chinese medicine. *The American Journal of Chinese Medicine* **46**, 261–297. doi:10.1142/S0192415X18500143
359. Su X-H, Zhang M-L, Li L-G, Huo C-H, Gu Y-C *et al.* (2008) Chemical constituents of the plants of the genus *Calophyllum*. *Chemistry & Biodiversity* **5**, 2579–2608. doi:10.1002/cbdv.200890215

360. Cechinel Filho V, Meyre-Silva C, Niero R (2009) Chemical and pharmacological aspects of the genus *Calophyllum*. *Chemistry & Biodiversity* **6**, 313–327. doi:10.1002/cbdv.200800082
361. Goyal PK (2012) Phytochemical and pharmacological properties of the genus *Grewia*: a review. *International Journal of Pharmacy and Pharmaceutical Sciences* **4**, 72–78.
362. Ullah W, Uddin G, Siddiqui BS (2012) Ethnic uses, pharmacological and phytochemical profile of genus *Grewia*. *Journal of Asian Natural Products Research* **14**, 186–195. doi:10.1080/10286020.2011.639764
363. Zia-Ul-Haq M, Stankovic MS, Rizwan K, Feo VD (2013) *Grewia asiatica* L., a food plant with multiple uses. *Molecules (Basel, Switzerland)* **18**, 2663–2682. doi:10.3390/molecules18032663
364. Al Muqarrabun LM, Ahmat N (2015) Medicinal uses, phytochemistry and pharmacology of family Sterculiaceae: a review. *European Journal of Medicinal Chemistry* **92**, 514–530. doi:10.1016/j.ejmech.2015.01.026
365. Thabet AA, Youssef FS, El-Shazly M, Singab ANB (2018) *Sterculia* and *Brachychiton*: a comprehensive overview on their ethnopharmacology, biological activities, phytochemistry and the role of their gummy exudates in drug delivery. *The Journal of Pharmacy and Pharmacology* **70**, 450–474. doi:10.1111/jphp.12876
366. Vadivel V, Sriram S, Brindha P (2016) Distribution of flavonoids among Malvaceae family members - a review. *International Journal of Green Pharmacy* **10**, s33–s45.
367. Hiteksha SP, Mamta BS (2017) *Thespesia populnea* Linn.: a review. *International Journal of Pharmaceutics* **4**, 1–5.
368. Reddy MVB (2017) Review on chemistry, biological activity and biosynthesis of naturally occurring compounds of cadalene type sesquiterpenoids from Malvaceae. *Der Pharma Chemica* **9**, 40–49.
369. Suvarna CHM, Sriya P, Pavan KA (2018) A review on phytochemical and pharmacological properties of *Thespesia populnea*. *Journal of Drug Delivery and Therapeutics* **8**, 1–4. doi:10.22270/jddt.v8i4.1718
370. Piwowarski JP, Granica S, Kiss AK (2015) *Lythrum salicaria* L.: underestimated medicinal plant from European traditional medicine: a review. *Journal of Ethnopharmacology* **170**, 226–250. doi:10.1016/j.jep.2015.05.017
371. Chawla R, Kumar S, Sharma A (2012) The genus *Clematis* (Ranunculaceae): chemical and pharmacological perspectives. *Journal of Ethnopharmacology* **143**, 116–150. doi:10.1016/j.jep.2012.06.014
372. Muhammad D, Lalun N, Bobichon H, Le Magrex Debar E, Gangloff SC *et al.* (2016) Triterpenoids from the leaves of Alphitonia xerocarpus Baill and their biological activity. *Phytochemistry* **129**, 45–57. doi:10.1016/j.phytochem.2016.07.005
373. Raju R, Gunawardena D, Ahktar MA, Low M, Reddell P *et al.* (2016) Anti-inflammatory chemical profiling of the Australian rainforest tree *Alphitonia petriei* (Rhamnaceae). *Molecules (Basel, Switzerland)* **21**, 1521. doi:10.3390/molecules21111521
374. Cooke RG, Johnson BL (1963) Colouring matters of Australian plants. XI: quinones from *Ventilago viminalis*. *Australian Journal of Chemistry* **16**, 695–702. doi:10.1071/CH9630695
375. Rao BK, Hanumaiah T, Rao CP, Rao GSR, Rao KVJ *et al.* (1983) Anthraquinones in *Ventilago* species. *Phytochemistry* **22**, 2583–2585. doi:10.1016/0031-9422(83)80170-8
376. Hanumaiah T, Marshall DS, Rao BK, Rao CP, Rao GSR *et al.* (1985) Benzoisochromanquinones in *Ventilago* species. *Phytochemistry* **24**, 2373–2378. doi:10.1016/S0031-9422(00)83044-7
377. Hanumaiah T, Rao BK, Rao CP, Rao GSR, Rao JUM *et al.* (1985) Naphthalenes and naphthoquinones from *Ventilago* species. *Phytochemistry* **24**, 1811–1815. doi:10.1016/S0031-9422(00)82557-1
378. Jammula SR, Pepalla SB, Hanumaiah T, Rao KVJ, Thomson RH (1991) Benzisochromanquinones from *Ventilago goughii*. *Phytochemistry* **30**, 3741–3744. doi:10.1016/0031-9422(91)80100-F
379. Lin L-C, Chou C-J, Kuo Y-C (2001) Cytotoxic principles from *Ventilago leiocarpa*. *Journal of Natural Products* **64**, 674–676. doi:10.1021/np000569d
380. Periyasamy K, Kaliyaperumal S (2016) Ethnobotanical, phytochemical and pharmaceutical studies of medicinal plant, *Ventilago maderaspatana* Gaertn. (red creeper): a review. *International Journal of Current Pharmaceutical Research* **8**, 16–18.
381. Cassels BK, Eckhardt G, Kaussmann EU, Tschesche R (1974) Cyclopeptide alkaloids of *Zizyphus oenoplia*. *Tetrahedron* **30**, 2461–2466. doi:10.1016/S0040-4020(01)97117-0
382. Mahajan RT, Chopda MZ (2009) Phytopharmacology of *Ziziphus jujuba* Mill.: a plant review. *Pharmacognosy Reviews* **3**, 320–329.
383. Gao QH, Wu CS, Wang M (2013) The jujube (*Ziziphus jujuba* Mill.) fruit: a review of current knowledge of fruit composition and health benefits. *Journal of Agricultural and Food Chemistry* **61**, 3351–3363. doi:10.1021/jf4007032
384. Prabhavathi S, Vajayalakshmi S (2015) New hydroxycarboxylic acid from the root bark of *Ziziphus oenoplia*. *Journal of Pharmacognosy and Phytochemistry* **3**, 150–152.
385. Rocabado GO, Bedoya LM, Abad MJ, Bermejo P (2008) *Rubus*: a review of its phytochemical and pharmacological profile. *Natural Product Communications* **3**, 423–436. doi:10.1177/1934578X0800300319
386. Lee J, Dossett M, Finn CE (2012) *Rubus* fruit phenolic research: the good, the bad, and the confusing. *Food Chemistry* **130**, 785–796. doi:10.1016/j.foodchem.2011.08.022
387. Le NH, Malterud KE, Diallo D, Paulsen BS, Nergard CS *et al.* (2012) Bioactive polyphenols in *Ximenia americana* and the traditional use among Malian healers. *Journal of Ethnopharmacology* **139**, 858–862. doi:10.1016/j.jep.2011.12.031

388. Maroyi A (2016) *Ximenia caffra* Sond. (Ximeniaceae) in sub-Saharan Africa: a synthesis and review of its medicinal potential. *Journal of Ethnopharmacology* **184**, 81–100. doi:10.1016/j.jep.2016.02.052
389. Mulholland DA, Taylor DAH (1992) Limonoids from Australian members of the Meliaceae. *Phytochemistry* **31**, 4163–4166. doi:10.1016/0031-9422(92)80434-G
390. Mulholland DA, Kotsos M, Mahomed HA, Taylor DAH (1998) Triterpenoids from *Owenia cepiodora*. *Phytochemistry* **49**, 2457–2460. doi:10.1016/S0031-9422(98)00307-0
391. Liu J-H, Jin H-Z, Zhang W-D, Yan S-K, Shen Y-H (2009) Chemical constituents of plants from the genus *Brucea*. *Chemistry & Biodiversity* **6**, 57–70. doi:10.1002/cbdv.200700409
392. Austin DF (2008) *Evolvulus alsinoides* (Convolvulaceae): an American herb in the Old World. *Journal of Ethnopharmacology* **117**, 185–198. doi:10.1016/j.jep.2008.01.038
393. Ding W, Zeng F, Xu L, Chen Y, Wang Y *et al.* (2011) Bioactive dammarane-type saponins from *Operculina turpethum*. *Journal of Natural Products* **74**, 1868–1874. doi:10.1021/np200274m
394. Ding W, Jiang ZH, Wu P, Xu L, Wei X (2012) Resin glycosides from the aerial parts of *Operculina turpethum*. *Phytochemistry* **81**, 165–174. doi:10.1016/j.phytochem.2012.05.010
395. Gupta S, Ved A (2017) *Operculina turpethum* (Linn.) Silva Manso as a medicinal plant species: a review on bioactive components and pharmacological properties. *Pharmacognosy Reviews* **11**, 158–166.
396. Lira-Ricardez J, Pereda-Miranda R, Castaneda-Gomez J, Fragoso-Serrano M, Simas RC *et al.* (2019) Resin glycosides from the roots of *Operculina macrocarpa* (Brazilian Jalap) with purgative activity. *Journal of Natural Products* **82**, 1664–1677. doi:10.1021/acs.jnatprod.9b00222
397. Kumar D, Kumar S, Gupta J, Arya R, Gupta A (2011) A review on chemical and biological properties of *Cayratia trifolia* Linn. (Vitaceae). *Pharmacognosy Reviews* **5**, 184–188. doi:10.4103/0973-7847.91117
398. Pereira A, Maraschin M (2015) Banana (*Musa* spp) from peel to pulp: ethnopharmacology, source of bioactive compounds and its relevance for human health. *Journal of Ethnopharmacology* **160**, 149–163. doi:10.1016/j.jep.2014.11.008
399. Mathew NS, Negi PS (2017) Traditional uses, phytochemistry and pharmacology of wild banana (*Musa acuminata* Colla): a review. *Journal of Ethnopharmacology* **196**, 124–140. doi:10.1016/j.jep.2016.12.009
400. Achenbach H, Hubner H, Brandt W, Reiter M (1994) Cardioactive steroid saponins and other constituents from the aerial parts of *Tribulus cistoides*. *Phytochemistry* **35**, 1527–1543. doi:10.1016/S0031-9422(00)86890-9
401. Achenbach H, Hubner H, Reiter M (1996) Cholestane- and pregnane-type glycosides from the roots of *Tribulus cistoides*. *Phytochemistry* **41**, 907–917. doi:10.1016/0031-9422(95)00622-2
402. Kostova I, Dinchev D (2005) Saponins in *Tribulus terrestris*: chemistry and bioactivity. *Phytochemistry Reviews* **4**, 111–137. doi:10.1007/s11101-005-2833-x
403. Chhatre S, Nesari T, Somani G, Kanchan D, Sathaye S (2014) Phytopharmacological overview of *Tribulus terrestris*. *Pharmacognosy Reviews* **8**, 45–51. doi:10.4103/0973-7847.125530
404. Neychev V, Mitev V (2016) Pro-sexual and androgen enhancing effects of *Tribulus terrestris* L.: fact or fiction. *Journal of Ethnopharmacology* **179**, 345–355. doi:10.1016/j.jep.2015.12.055
405. Santos HO, Howell S, Teixeira FJ (2019) Beyond tribulus (*Tribulus terrestris* L.): the effects of phytotherapics on testosterone, sperm and prostate parameters. *Journal of Ethnopharmacology* **235**, 392–405. doi:10.1016/j.jep.2019.02.033

4
Alkaloid-type specialised metabolites

INTRODUCTION

The term 'alkaloid' was coined in the early 19th century by German apothecary C. F. W. Meißner to denote the discovery of basic ('alkali-like') plant substances being isolated at the time, with the term 'alkali' itself coming from the Arabic ('al qaly' - lit. ashes of the saltwort).[1,2] The definition of an alkaloid has changed with time and scope; a biologist regards them as bioactive nitrogen heterocycles, a pharmacologist treats them as plant-derived complex nitrogenous substances, slightly soluble in water with intense physiological action, and a chemist understands them to be biogenically derived materials, generally based on *N*-heterocyclic compounds with alkaline chemical properties and significant bioactivity.[1,2] These definitions are similar but vague; it's the exceptions that make the rule, as, in general, amino acids, peptides, nucleosides and many antibiotics often fit within the definition but are not considered to be alkaloids. A convention adopted throughout the 20th century (which has since fallen out of favour, but will be adopted here for its structural clarity) is the division of alkaloids into true alkaloids (biogenically synthesised *N*-heterocycles derived from amino acids), protoalkaloids (derived from amino acid biosynthesis, but the amino nitrogen is not incorporated in the heterocycle) and pseudoalkaloids (alkaloid-like compounds with non-amino origins).[1,2] The number of reported natural alkaloids encompassing all three of these categories is increasing, with 200 being reported in 1939, 8,000–12,000 being variously reported in the 2000s and > 40,000 currently recorded in the Dictionary of Alkaloids (a subset of the Dictionary of Natural Products online database).[1-7] This chapter will proceed through an introduction to the class itself, followed by more detailed breakdowns of each of the terpenoid-type metabolic subclasses it contains. The end of the chapter will present a summary of the representation of these metabolites reported to be present across the customary medicinal flora of Australia, detailed in *Table 4.1*. This information can also be explored using the associated website (www.cmfoa.info).

The true alkaloids are split into subclasses based on the structure of their heterocyclic ring and these are further delineated based on their amino-acid origins: L-ornithine (pyrrolidine, pyrrolizidine, tropane), L-lysine (piperidine, quinolizidine, indolizidine), L-tryptophan (simple indole, non-terpenoid indole derivatives [e.g. β-carboline]), L-tyrosine (isoquinoline) and L-histidine (imidazole) (*Fig. 4.1*).[1,2] The protoalkaloids (phenylethylamine, benzylamine, etc.) and the pseudoalkaloids derived from anthranilic acid (quinazoline), nicotinic acid (pyridine) and purine (methylxanthine) are also included in this chapter (*Fig. 4.1*). The large and highly significant category of alkaloterpenes [monoterpenoid indole alkaloids (MIA), also referred to as terpenoid indole alkaloids (TIAs): hemiterpene (ergot-type), monoterpene (five 'recognised' groups of significant diversity), sesquiterpene alkaloids, diterpene alkaloids, steroidal alkaloids, etc. placed within the meroterpenes)] and other categories formed through the combination of alkaloids with other sub-/infraclasses will be addressed in Chapter 5. L-arginine is responsible for the biosynthesis of many of the true alkaloids that are derived from marine invertebrates; these compounds are highly diverse and biomedically significant, but lie outside the scope of this book.[1,2]

The reported percentage of the terrestrial plant kingdom that produces alkaloids has changed over the decades from 20% to 40%, likely due to increasingly sophisticated analytical techniques becoming available (chemical detection techniques such as Dragendorff's reagent or ninhydrin versus instrumental techniques such as HPLC-PDA, LC-HRESIMS or NMR spectroscopy).[1,2] Some notable terrestrial sources are the following higher plant families: Apocynaceae [indoles (ajmalicine, ibogaine, *Vinca* alkaloids)], Asteraceae [pyrrolizidines (*Senecio* alkaloids)], Boraginaceae [pyrrolizidines (indicine)], Erythoxylaceae [tropanes (cocaine)], Fabaceae [pyrrolizidines (lupinine), quinolizidines (sparteine), indoles (tryptamines)], Menispermaceae [curare-type isoquinoline and indole alkaloids (poison arrow toxins)], Nitrariaceae [β-carboline-type indoles (harmala alkaloids)], Papaveraceae [isoquinolines (morphine)], Rubiaceae [methylxanthines (caffeine), indoles (quinine-type MIAs)] and Solanaceae [tropanes (hyoscyamine), pyridine (nicotine), benzylamine (capsaicin)].[1,2 4-7]

The role of alkaloids in the plant kingdom is thought to be more obvious than that of the phenolic- and terpenoid-type classes discussed earlier in this chapter. The alkaloids are constitutive phytoanticipin defence compounds that are bitter (often toxic) antifeedant compounds, primarily used against the herbivorous Euarthropoda and Chordata animal phyla.[8-11] This relationship between plants and the organisms that feed on them is likely as old as terrestrial plant life itself and as such this co-evolutionary process has resulted in significant specificity for the disruption of biological systems, a factor that is paramount in their role in medicine both historically and today.[9-11] The plants store alkaloids in tissues that are of physiological importance (fruits, seeds, young leaves, roots, bark, etc.) and protect themselves from their own chemical arsenal by confining the intracellular alkaloid content in vacuoles, away from the organic machinery they are designed to interfere with.[8-11] The biochemical basis for the plant–animal ecological relationship surrounding alkaloids has also proven to be successful in allelopathic (plant–plant) and antimicrobial settings.[9-11] In addition to this secondary metabolic role, the important alkaloid-type plant growth hormones such as the auxins [e.g. indole acetic acid (IAA)] and cytokinins [(CKs) e.g. zeatin] are considered primary metabolites and will not be considered here for that reason.

When it comes to medicinal applications, the alkaloids are by far the most significant natural product (NP) class. A selection of notable examples, with the active nitrogenous component listed in brackets, are: atropine (tropane), berberine (isoquinoline - benzylisoquinoline), boldine (isoquinoline - aporphine), caffeine (methylxanthine pseudoalkaloid), cathine (phenylethylamine protoalkaloid), codeine (isoquinoline - morphine), colchicine (isoquinoline protoalkaloid), emetine (isoquinoline), ephedrine (phenylethylamine protoalkaloid), eserine/physostigmine (non-terpenoid indole derivative), galanthamine (isoquinoline-derived), hydrastine (isoquinoline), hyoscine/scopolamine (tropane), hyoscyamine (tropane), lobeline (pyridine pseudoalkaloid), morphine (isoquinoline - morphine), narceine (isoquinoline protoalkaloid - morphine-derived), nicotine (pyridine), noscapine/narcotine (isoquinoline - benzylisoquinoline), papaverine (isoquinoline - benzylisoquinoline), pilocarpine (imidazole), quinidine (quinoline), quinine (quinoline), rescinnamine (indole - *Vinca* alkaloid), sanguinarine (isoquinoline), sparteine (quinolizidine), theobromine (methylxanthine), theophylline (methylxanthine), tubocurarine (isoquinoline dimeric macrocycle), vinblastine (indole dimer - *Vinca* alkaloids), vincamine (indole dimer - *Vinca* alkaloid) and vincristine (indole dimer - *Vinca* alkaloids) (*Fig. 4.2* and *Fig. 4.3*).[1,2,12]

It must be emphasised from the outset that the alkaloids are an incredibly diverse, profoundly bioactive class of compounds and for every subclass there are infraclasses, and for many infraclasses there are sub-infraclasses and so on, all with varying physiological effects. This chapter is only able to give a cursory description of this illustrious group of NPs and at some point a limit to the detail must be made. Further details for specific subclasses and infraclasses are given in any number of textbooks (referenced here and elsewhere), if the reader is interested to pursue this topic.[1-3,7,13-17] A representative distribution of alkaloid-type specialised metabolites in the customary medicinal plant genera of Australia is given in *Table 4.1*.

Table 4.1. Phenolic-type specialised metabolite distribution across the customary medicinal flora of Australia

Order	Family	Genus	Pyrrolidine	Pyrrolizidine	Tropane	Piperidine	Quinolizidine	Indolizidine	Indole	β-carboline	Physostigmine	Canthin-6-one	Carbazole	Indolenine	Simple Amine	Betalain	Isoquinoline	Benzylisoquinoline	Imidazole	Quinazoline	Quinoline	Pyridine	Methylxanthine
Alismatales	Araceae	*Alocasia*[103,104]				+			+														
Alismatales	Araceae	*Amorphophallus*[105,106]																				+	
Alismatales	Araceae	*Rhaphidophora*[107,108]							+														
Asparagales	Amaryllidaceae	*Crinum*[109-112]							+	+							+	+					
Asparagales	Asparagaceae	*Asparagus*[113]		+																			
Asparagales	Orchidaceae	*Dendrobium*[114]	+					+											+				
Asterales	Asteraceae	*Ageratum*[115]		+																			
Asterales	Campanulaceae	*Lobelia*[116]				+																	
Asterales	Goodeniaceae	*Goodenia*[117]																				+	
Boraginales	Boraginaceae	*Ehretia*[118,119]		+																			
Boraginales	Boraginaceae	*Heliotropium*[120]		+																			
Boraginales	Boraginaceae	*Trichodesma*[121-123]		+																			
Brassicales	Capparaceae	*Capparis*[124]	+						+														
Brassicales	Cleomaceae	*Cleome*[125]																				+	
Caryophyllales	Amaranthaceae	*Achyranthes*[126]	+												+			+					
Caryophyllales	Amaranthaceae	*Atriplex*[127,128]							+						+								
Caryophyllales	Amaranthaceae	*Dysphania*[129,130]			+	+			+													+	
Caryophyllales	Nyctaginaceae	*Boerhavia*[131]					+									+							
Caryophyllales	Portulaceae	*Portulaca*[132]				+			+								+					+	
Cucurbitales	Cucurbitaceae	*Trichosanthes*[133-137]								+													
Ericales	Primulaceae	*Aegiceras*[138]	+						+	+					+								
Ericales	Sapotaceae	*Planchonella*[139-145]		+																			
Fabales	Fabaceae	*Abrus*[146]							+													+	
Fabales	Fabaceae	*Acacia*[147]							+	+					+								
Fabales	Fabaceae	*Cassia*[148]				+																+	
Fabales	Fabaceae	*Chamaecrista*[148]				+																+	

Order	Family	Genus	Pyrrolidine	Pyrrolizidine	Tropane	Piperidine	Quinolizidine	Indolizidine	Indole	β-carboline	Physostigmine	Canthin-6-one	Carbazole	Indolenine	Simple Amine	Betalain	Isoquinoline	Benzylisoquinoline	Imidazole	Quinazoline	Quinoline	Pyridine	Methylxanthine
Fabales	Fabaceae	*Crotalaria*[149,150]		+																			
Fabales	Fabaceae	*Cynometra*[151]																	+				
Fabales	Fabaceae	*Entada*[152,153]							+														
Fabales	Fabaceae	*Erythrina*[154,155]															+	+					
Fabales	Fabaceae	*Erythrophleum*[156]							+														
Fabales	Fabaceae	*Indigofera*[157]							+														
Fabales	Fabaceae	*Mucuna*[158–160]	+						+	+							+					+	
Fabales	Fabaceae	*Senna*[148]				+																+	
Fabales	Fabaceae	*Sophora*[161–163]					+																
Fabales	Fabaceae	*Swainsona*[164–166]						+															
Fabales	Fabaceae	*Templetonia*[167–169]					+															+	
Fabales	Fabaceae	*Vigna*[170–172]							+														
Fagales	Casuarinaceae	*Casuarina*[173]		+																			
Gentianales	Apocynaceae	*Alstonia*[174]							+														
Gentianales	Apocynaceae	*Cynanchum*[175]															+						
Gentianales	Apocynaceae	*Tabernaemontana*[176–178]							+														
Gentianales	Apocynaceae	*Tylophora*[179–182]						+															
Gentianales	Loganiaceae	*Strychnos*[183–187]							+														
Gentianales	Rubiaceae	*Coelospermum*[188–191]																				+	
Gentianales	Rubiaceae	*Nauclea*[192,193]							+	+												+	
Gentianales	Rubiaceae	*Timonius*[194,195]							+												+		
Lamiales	Plantaginaceae	*Bacopa*[196]																				+	
Lamiales	Plantaginaceae	*Scoparia*[197–200]							+														
Laurales	Atherospermataceae	*Atherosperma*[201]															+	+					
Laurales	Atherospermataceae	*Daphnandra*[201–203]																+					
Laurales	Atherospermataceae	*Doryphora*[201,204]															+	+					

Table 4.1. continued

Order	Family	Genus	Pyrrolidine	Pyrrolizidine	Tropane	Piperidine	Quinolizidine	Indolizidine	Indole	β-carboline	Physostigmine	Canthin-6-one	Carbazole	Indolenine	Simple Amine	Betalain	Isoquinoline	Benzylisoquinoline	Imidazole	Quinazoline	Quinoline	Pyridine	Methylxanthine
Laurales	Hernandiaceae	*Gyrocarpus*[205]																+					
Laurales	Hernandiaceae	*Hernandia*[206,207]															+	+					
Laurales	Lauraceae	*Cassytha*[208,209]																+					
Laurales	Lauraceae	*Litsea*[210]						+									+	+					
Lycopodiales	Lycopodiaceae	*Lycopodium*[211]				+	+																
Malphigiales	Euphorbiaceae	*Acalypha*[212]																					+
Malphigiales	Euphorbiaceae	*Croton*[213,214]								+								+					
Malphigiales	Euphorbiaceae	*Euphorbia*[215]																					
Malphigiales	Euphorbiaceae	*Excoecaria*[216]				+																	
Malphigiales	Phyllanthaceae	*Flueggea*[217-219]						+	+														
Malphigiales	Phyllanthaceae	*Phyllanthus*[220-224]		+				+															
Malphigiales	Phyllanthaceae	*Securinega*[217-219]						+	+														
Malphigiales	Violaceae	*Hybanthus*[225,226]																					+
Malvales	Malvaceae	*Grewia*[227-229]							+														
Malvales	Malvaceae	*Sida*[230]						+	+						+					+	+		
Malvales	Malvaceae	*Sterculia*[231,232]																			+		
Pandanales	Pandanaceae	*Pandanus*[233,234]	+			+																	
Piperales	Piperaceae	*Piper*[235,236]	+			+											+					+	
Poales	Cyperaceae	*Cyperus*[237]															+						
Polypodiales	Pteridaceae	*Adiantum*[238]													+								
Proteales	Nelumbonaceae	*Nelumbo*[239]																+					
Ranunculales	Menispermaceae	*Cissampelos*[240]															+	+			+		
Ranunculales	Menispermaceae	*Tinospora*[241-243]	+														+	+			+		
Ranunculales	Ranunculaceae	*Clematis*[244]							+												+		
Rosales	Rhamnaceae	*Ziziphus*[245-248]	+																				
Sapindales	Anacardiaceae	*Buchanania*[249]											+										

Order	Family	Genus	Pyrrolidine	Pyrrolizidine	Tropane	Piperidine	Quinolizidine	Indolizidine	Indole	β-carboline	Physostigmine	Canthin-6-one	Carbazole	Indolenine	Simple Amine	Betalain	Isoquinoline	Benzylisoquinoline	Imidazole	Quinazoline	Quinoline	Pyridine	Methylxanthine
Sapindales	Meliaceae	*Melia*[250]								+													
Sapindales	Rutaceae	*Bosistoa*[251]																			+		+
Sapindales	Rutaceae	*Flindersia*[252]							+														
Sapindales	Rutaceae	*Melicope*[253–259]							+												+		
Sapindales	Rutaceae	*Philotheca*[260–265]																			+		
Sapindales	Rutaceae	*Sarcomelicope*[266–271]																			+		
Sapindales	Simaroubaceae	*Ailanthus*[272]								+		+											
Sapindales	Simaroubaceae	*Brucea*[273]							+	+		+									+		
Solanales	Convolvulaceae	*Convolvulus*[274]			+																		
Solanales	Convolvulaceae	*Evolvulus*[275]	+		+																		
Solanales	Convolvulaceae	*Ipomoea*[276]			+			+															
Solanales	Convolvulaceae	*Xenostegia*[277–279]	+	+	+																		
Solanales	Solanaceae	*Duboisia*[22,280,281]			+																	+	
Solanales	Solanaceae	*Nicotiana*[282–284]	+																			+	
Solanales	Solanaceae	*Solanum*[285]	+																				
Zygophyllales	Zygophyllaceae	*Tribulus*[286–291]								+													

Note: genera with no reported phytochemistry have been omitted see (www.cmfca.info)

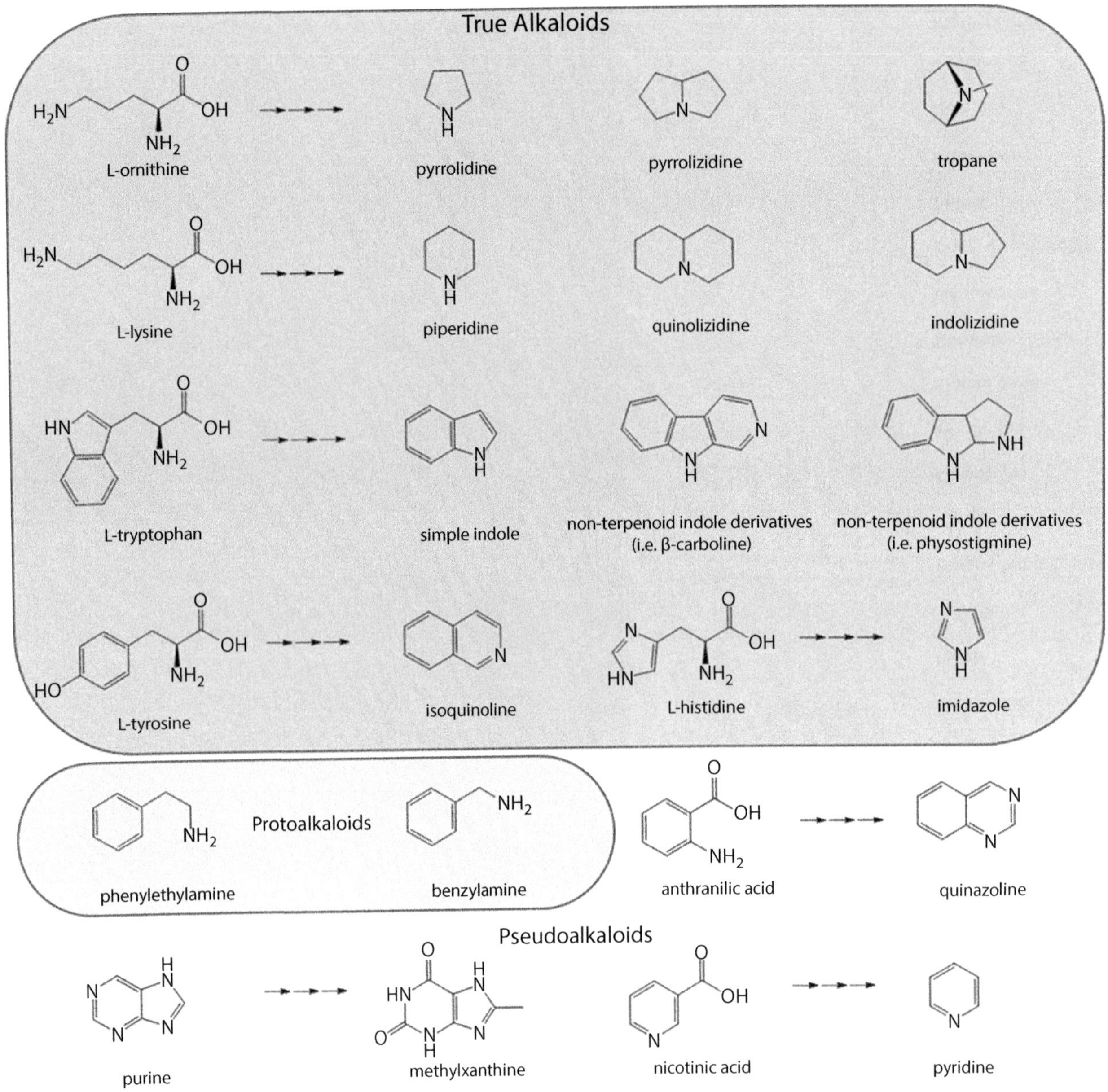

Fig. 4.1. Alkaloid-type specialised metabolite subclass scaffolds, showing true alkaloids (dark green), protoalkaloids (light green) and pseudoalkaloids (uncoloured).

L-ORNITHINE-DERIVED ALKALOIDS

L-Ornithine (a non-proteinogenic amino acid) is considered a plant primary metabolite and is biosynthesised from L-glutamic acid (L-glutamate) derived in turn from the Krebs cycle. L-Ornithine is biosynthetically converted during plant secondary metabolism to give the pyrrolidine (C_4N), pyrrolizidine (C_7N) and tropane (C_4NC_5) alkaloidal scaffolds (*Fig. 4.4*). L-Ornithine is decarboxylated by ornithine decarboxylase facilitated by coenzyme pyridoxal phosphate (PLP) to putrescine, at which point it can branch off to form the linear polyamines (spermine, spermidine, etc.) or take the first committed step of pyrrolidine biosynthesis to *N*-methylputrescine via putrescine *N*-methyl transferase (PMT). From here, the newly methylated amino moiety is oxidised to an aldehyde by the action of diamino oxidase (DAO) to give 4-aminobutanal, which spontaneously cyclises to the *N*-methyl-Δ^1-pyrrolinium cation from which the pyrrolidine alkaloids arise (*Fig. 4.4*).[1,17–19] Feeding and inhibition experiments in *Nicotiana* spp. show that production of pyrrolidine (viz. all L-ornithine derived) alkaloids is regulated at the level of putrescine – with enzyme activity influenced by nutrients, water supply and herbivorous stress.[17,20] It should be noted that this is not the only available route to putrescine, as L-arginine can also proceed through agmatine via arginine decarboxylase (ADC) and putrescine synthase, but L-arginine itself is largely a postcursor of L-ornithine, although the validity and extent of this hypothesis is still under investigation.[21] The biosynthesis of the pyrrolizidine core (a necine base) is also still under

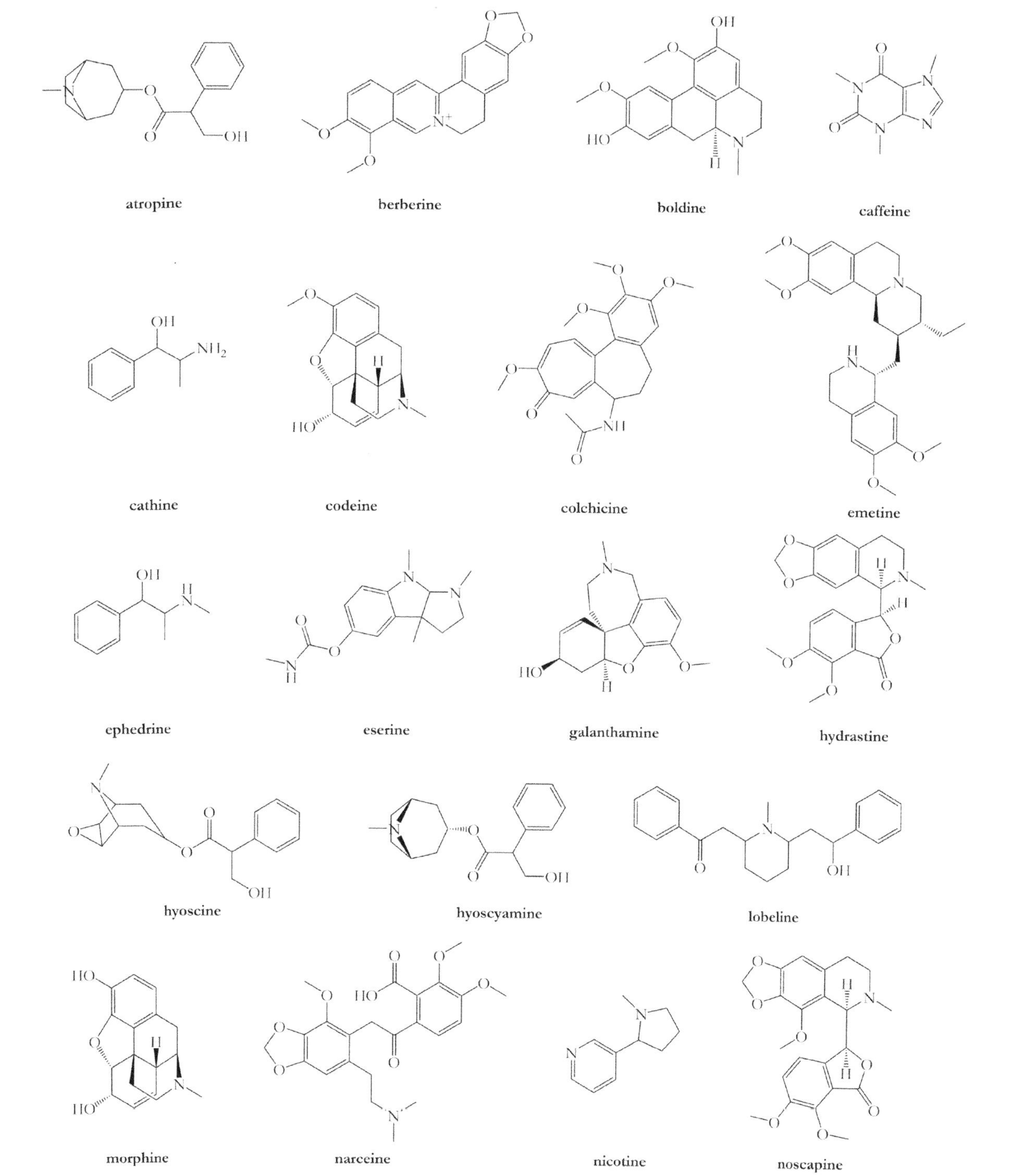

Fig. 4.2. Key representative specialised metabolites of the alkaloid-type class with medicinal applications.

investigation, but so far it has been demonstrated to be the result of condensation of a putrescine unit with a spermidine unit to form homospermidine via the action of homospermidine synthase and subsequent cyclisation to retronecine via pyrrolizidine-1-carbaldehyde, without ever forming pyrrolidine (*Fig. 4.4*).[20] Conversely, tropane alkaloids proceed from the *N*-methylpyrrolinium cation reacting with a C_4 acetoacetyl-CoA unit to form 2-*exo*-carboxytropinone (a common branching point for the tropane alkaloids) via a Mannich condensation. 2-*exo*-Carboxytropinone can be reduced and esterified to form ecgonine (specifically in *Erythroxylum coca* (Lam., 1786)

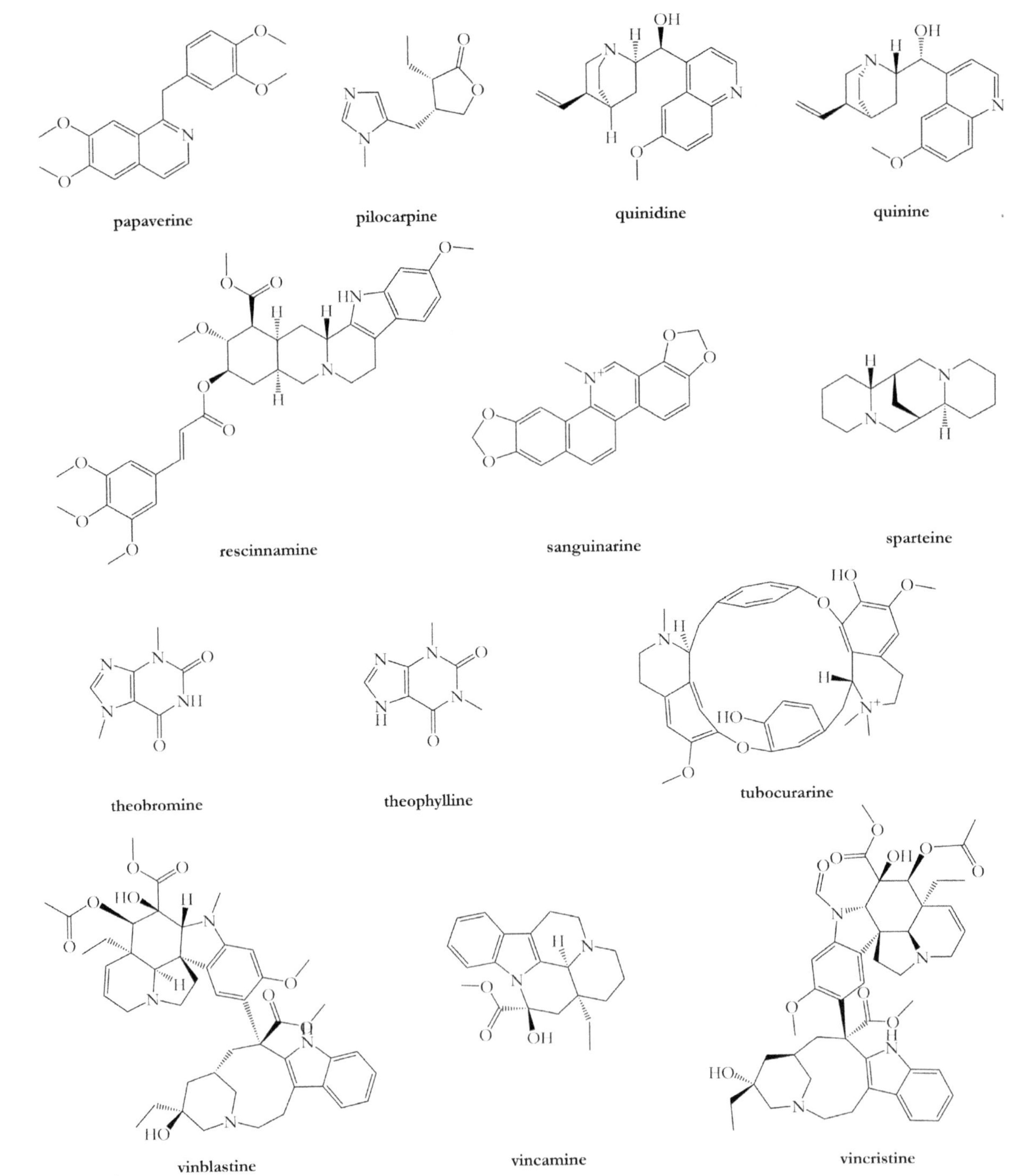

Fig. 4.3. Key representative specialised metabolites of the alkaloid-type class with medicinal applications.

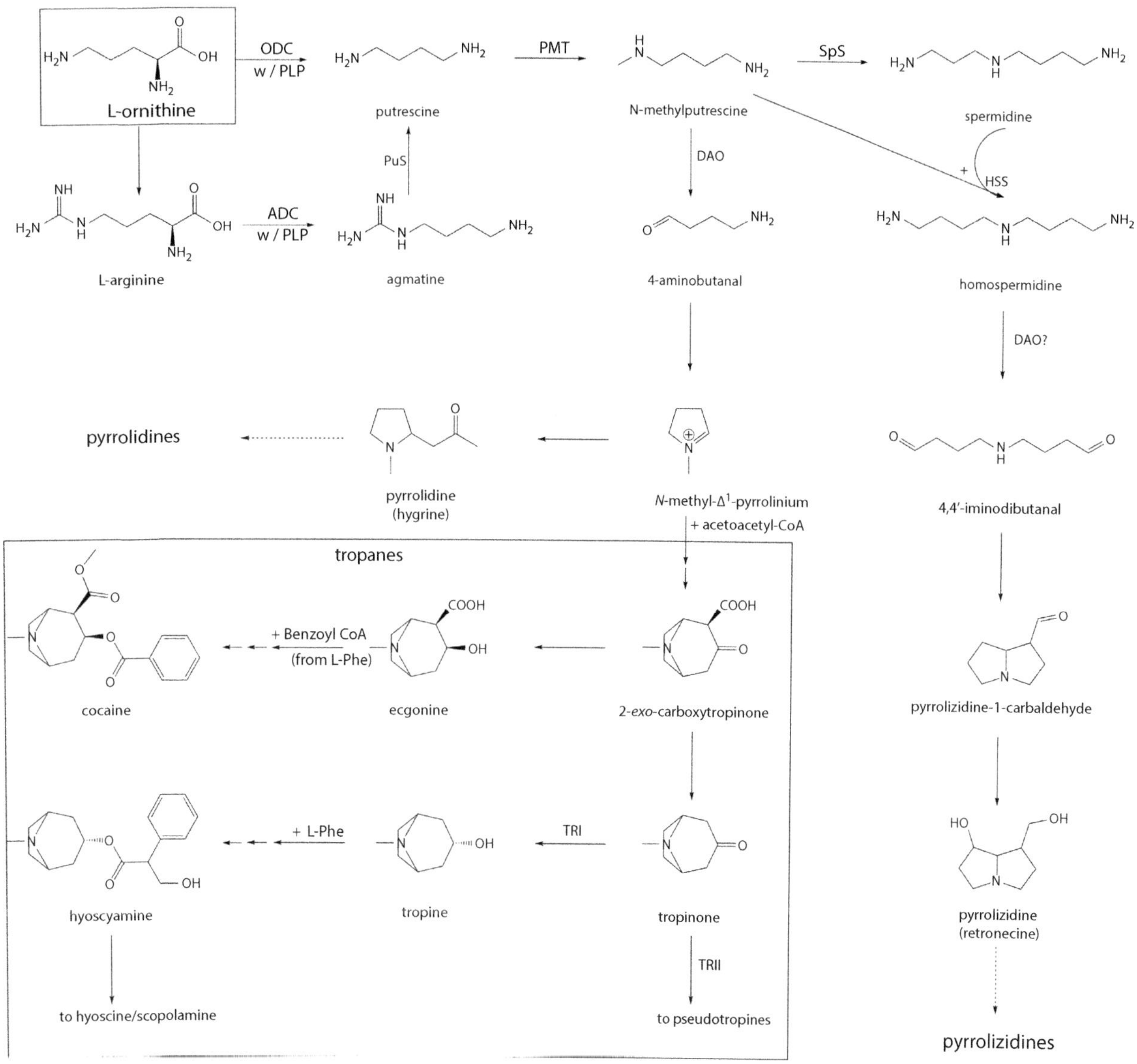

Fig. 4.4. Biosynthetic pathway of the L-ornithine-derived alkaloid subclasses. ADC = arginine decarboxylase, DAO = diamino oxidase, HSS = homospermidine synthase, ODC = ornithine decarboxylase, PLP = pyridoxal phosphate, PMT = putrescine N-methyl transferase, PuS = putrescine synthase, TRI = tropinone reductase 1, TRII = tropinone reductase 2.

and *E. granatense* (D. Morris (Hieron, 1895)), forming cocaine and its derivatives or decarboxylated to tropinone and then reduced via tropinone reductase I to form tropine (and hyoscyamine, hyoscine, etc. through condensation with phenylalanine-derived phenyllactic acid), then tropinone reductase II to form pseudotropine (and calystegine, etc.) (*Fig. 4.4*).[1,17,22]

Pyrrolidine

The ~80 known pyrrolidine alkaloids follow a relatively simple biosynthetic pathway and so are frequently encountered in diverse plant families with no particular trends, often featuring as minor products in some of the more notable alkaloid-producing plant families (e.g. Solanaceae).[23,24] Some key examples of pyrrolidine-type compounds are nicotine – condensed with a pyridine pseudoalkaloid derived from nicotinic acid (*Nicotiana*), hygrine/cuscohygrine (*Erythroxylum*), bgugaine (*Arisarum*) and 1,4-dideoxy-1,4-imino-*D*-ribitol (*Morus*) (*Fig. 4.5*).[1] At least 37 pharmaceutical drugs in clinical use contain a pyrrolidine moiety; some examples are procyclidine, bepridil and the racetams (piracetam, aniracetam) (*Fig. 4.5*).[25] Plant families known to contain pyrrolidine NPs are the Araceae [*Arisarum* ((bgugaine)],

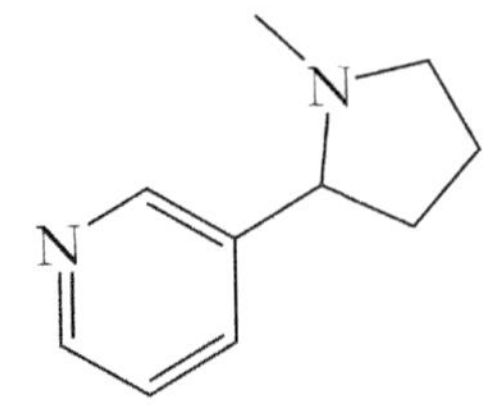

nicotine

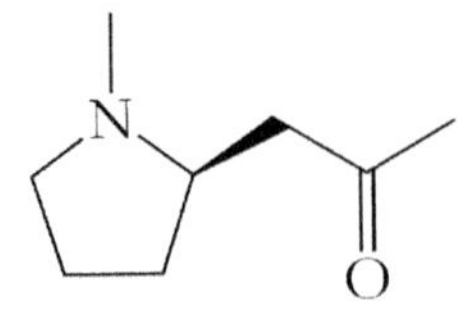

hygrine

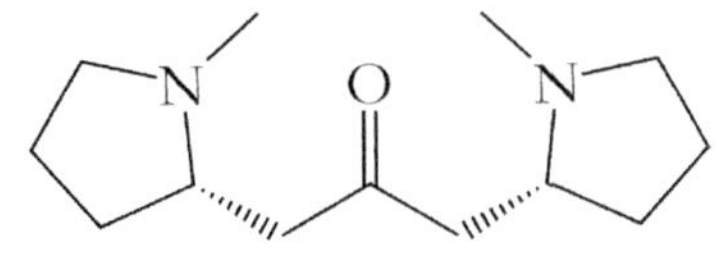

cuscohygrine

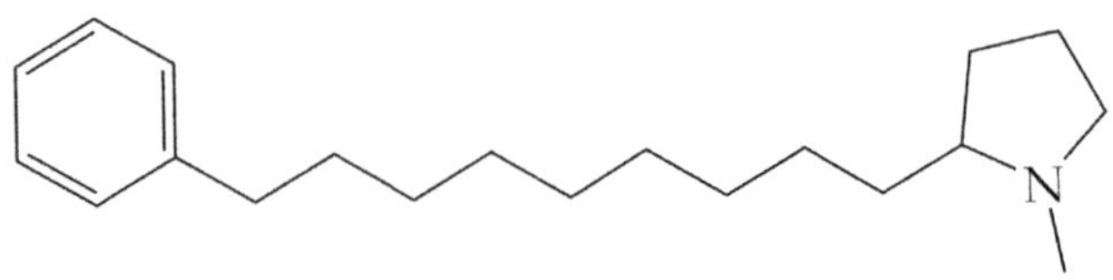

irniine

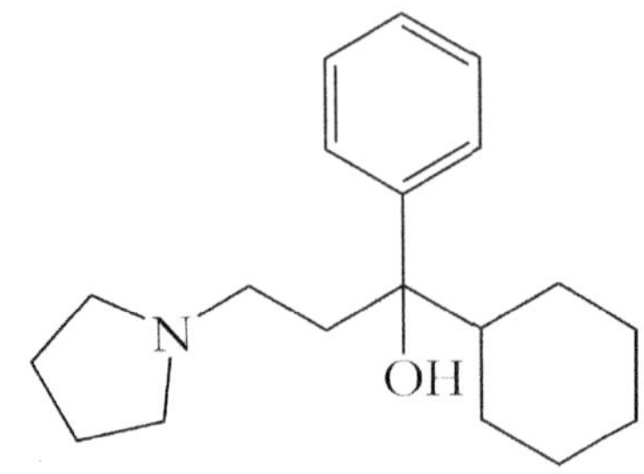

procyclidine

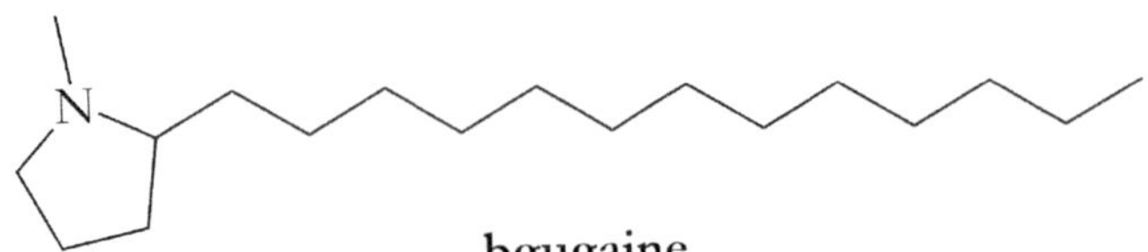

bgugaine

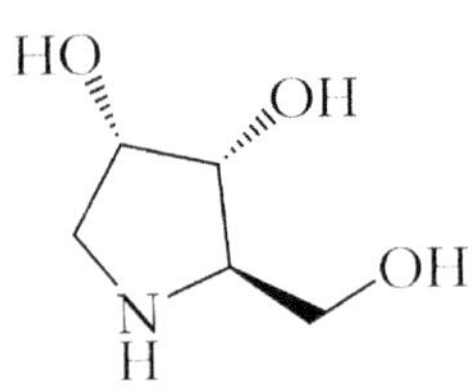

1,4-dideoxy-1,4-imino-D-ribitol

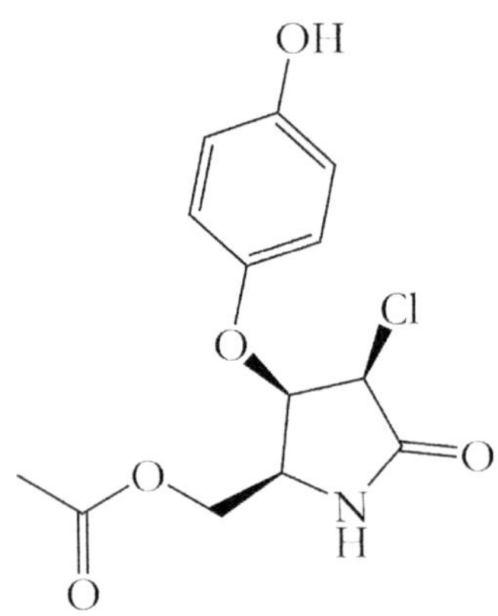

aegyptolidine A

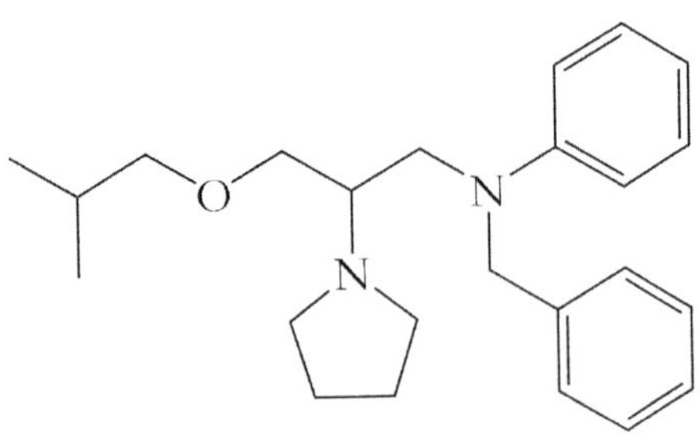

bepridil

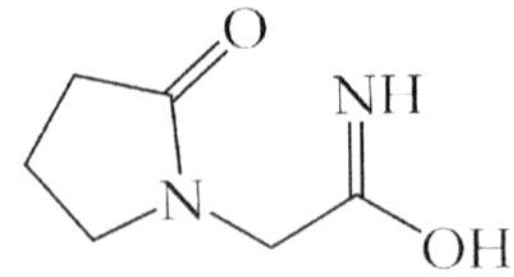

piracetam

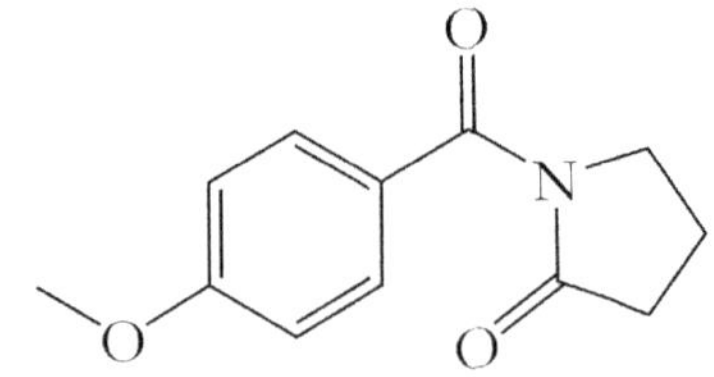

aniracetam

Fig. 4.5. Key representatives of the pyrrolidine-type infraclass of the L-ornithine derived subclass of alkaloids (both natural products and pharmaceutical drugs).

Capparidaceae [Boscia (stachydrine)], Erythroxylaceae [*Erythroxylum* (hygrine/cuscohygrine in the coca plant)], Moraceae (Morus) and the Solanaceae [*Atropa*, *Nicotiana* (nicotine in tobacco)].[1,17]

NP bearing a pyrrolidine moiety have been implicated in antioxidant, anti-inflammatory, antimicrobial (e.g. bgugaine), antiparasitic, anthelmintic, antineoplastic (e.g. aegyptolidine A) and neuropharmacological bioactivities.[26] The compounds bgugaine and irniine have been shown to be strongly hepatotoxic in rat and human cell models.[27] The pyrrolidine-pyridine alkaloid nicotine is well known for its neuroactivity and consequent global use as a recreational drug.[26] Recent interest has been shown regarding the globular 3D conformation available to the sp^3-hybridised pyrrolidines relative to the 2D sp^2-aromatic pyrroles.[25] This enables both heightened 'pseudorotation' around each atomic axis, but also permits four stereogenic centres (and 16 different stereoisomers), which make the pyrrolidines promising as scaffolds for the development of new therapeutic drugs.[25] The distribution of alkaloid-type specialised metabolites of the pyrrolidine infraclass reported from the customary medicinal flora of Australia is presented in *Table 4.1.*

Pyrrolizidine

Pyrrolizidine alkaloids, typically isolated as *N*-oxides or esters, are much more common than the pyrrolidines [> 660 compounds of four major types (platynecine, heliotridine, retronecine, otonecine) reported from 6,000 spp., and estimated to be present in 3% of Angiosperms] (*Fig. 4.6*).[28,29] The pyrrolizidines are known to be hepatotoxic (causing pyrollizidine alkaloidosis, which presents as hepatic veno-occlusive disease, liver cirrhosis, megalocystosis and cancer), tumorigenic, neurotoxic, cytotoxic and genotoxic.[20,29] Hence, the human use of pyrrolizidine-containing medicinal herbs has raised concerns, particularly those from the Boraginaceae (borage, comfrey) and Asteraceae (coltsfoot) families. Toxic pyrrolizidines are known to be used as a defence mechanism among arthropods that have developed resistance to the ingestion of this alkaloid infraclass (e.g. the caterpillar/moth *Utetheisa*). Pyrrolizidines are widespread; plant families known to have high concentrations of pyrrolizidines are the Apocynaceae (*Parsonia*), Asteraceae (*Adenostyles, Ageratum, Crassocephalum, Eupatorium, Farfugium, Gynura, Ligularia, Neurolaena, Petasites, Senecio, Tussilago*), Boraginaceae (*Anchusa, Borago, Cordia, Cynoglossum, Echium, Heliotropium, Lithospermum, Symphytum*), Fabaceae (*Crotalaria, Lotononis, Buchenroedera*), Orchidaceae (*Liparis*), Poaceae (*Festuca, Lolium*) and Solanaceae (*Chromolaena, Datura, Essium, Hyoscyamus, Withania*).[1,2,17,20,28–30] Other families that have been reported to contain pyrrolizidines are the Celastraceae, Euphorbiaceae, Santalanaceae, Sapotaceae, Scrophulariaceae, Ranunculaceae and Rhizophoraceae.[31] The necine-basicity of the pyrrolizidine core typically results in esterification with a necic acid, as seen in the macrolide-type pyrrolizidine senecionines derived from *Senecio* (*Fig. 4.6*).[32]

The variety of combinations available through mixing the necine bases and necic acids has resulted in a diverse range of pyrrolizidine alkaloids found across the plant kingdom.[20,32] Pyrrolizidines are thought to be synthesised in the roots of a plant and then transported via the phloem, aided by the hydrophilicity of their *N*-oxide state.[20,32] The 'toxic bioactivity of the pyrrolizidines derived from across a broad range of phylogenetically unrelated genera suggests an evolutionary advantage to their formation, particularly in defence from insect herbivores and to a lesser degree pathogenic microbes.[20] Their wide distribution and environmental persistence

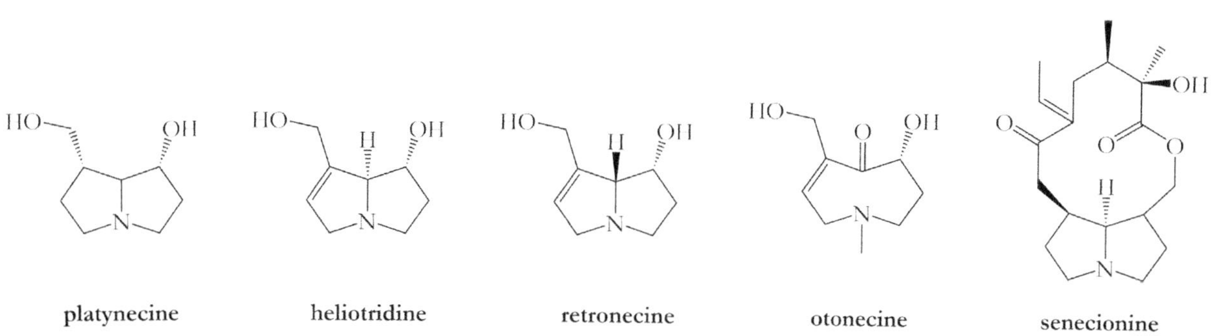

Fig. 4.6. Key representatives of the pyrrolizidine-type infraclass of the L-ornithine derived subclass of alkaloids (both natural products and pharmaceutical drugs).

have led to a high risk of poisoning through consumption of contaminated products such as grains, honey, milk, tea and medicine. Toxicity primarily arises from the 1,2-unsaturated necine base moiety of pyrrolizidines of the retronecine-, otonecine- and heliotridine-types, which are dehydrogenated by cytochrome p450 monooxygenases in the liver to yield reactive dehydropyrrolizidine intermediates (pyrroles, pyrrolic esters). These reactive metabolites are electrophilic and can bind -OH, -SH, and -NH nucleophilic groups of important macromolecules such as glutathione, DNA and proteins.[33–37] Some pyrrolizidines possess beneficial properties, such as glycosidase inhibition, which is implicated in antidiabetic, anticancer, antimicrobial and antiviral effects.[38–40] The distribution of alkaloid-type specialised metabolites of the pyrrolizidine infraclass reported from the customary medicinal flora of Australia is presented in *Table 4.1.*

Tropane

Approximately 300 different tropane alkaloids have been isolated as NPs, with notable examples being of the tropine-type [e.g. L-(−)-hyoscyamine (and its stable L(−)/D(+) racemic derivative atropine] and hyoscine/scopolamine, named differently in UK and USA English, respectively) or ecgonine-type (e.g. cocaine) (*Fig. 4.4* and *Fig. 4.7*).[1,22,41,42] Iboga-alkaloids bear a tropane-like moiety, but they also bear an indole moiety and have been condensed with a monoterpene to form a MIA, with distribution and biological activity more in conformance with that miscellaneous-type NP subclass, as such they will be addressed in Chapter 5. Tropanes are present in some of the oldest herbal medicines (and famous poisons or narcotic drugs) known to humans: deadly nightshade [*Atropa bella-donna* (L., 1753)], mandrake [*Atropa mandragora* (L., 1759)], henbane [*Hyoscyamus niger* (L., 1753), *H. albus* (L., 1753)], jimsonweed/thornapple [*Datura stramonium* (L., 1753)], scopola [*Scopolia carniolica* (Jacq., 1764)] and cocaine/coca leaf [*Erythroxylum coca* (Lam., 1786) and *E. granatense* (D. Morris (Hieron, 1895))], largely due to their anticholinergic, stimulating and hallucinogenic/dissociative effects on the human nervous system.[1,22] Plant families known for their tropane alkaloid content are the Asteraceae (*Eupatorium*, *Senecio*), Brassicaceae (*Cochlearia*), Dioscoreaceae, Elaeocarpaceae, Erythroxylaceae (*Erythroxylon*), Euphorbiaceae (*Phyllanthus*), Orchidaceae, Proteaceae, Rhizophoraceae and Solanaceae (*Atropa*, *Brugmansia*, *Datura*, *Duboisia*, *Hyoscyamus*, *Scopolia*).[1,2]

Humans have a longstanding relationship with the biological activity of the various naturally sourced tropane alkaloids. The tropine-type tropanes, such as hyoscyamine/atropine and hyoscine/scopolamine, have found widespread use throughout history as medicines, poisons and narcotics.[1,2] Hyoscyamine/atropine has useful medical applications due to its effect on respiratory tract secretory activity and bronchodilation, heart rate modification (bradycardia at low doses, tachycardia at high doses), reduced gastric secretions, inhibition of sweating (and raised body temperature) and as an antidote to organophosphorous pesticide poisoning (and

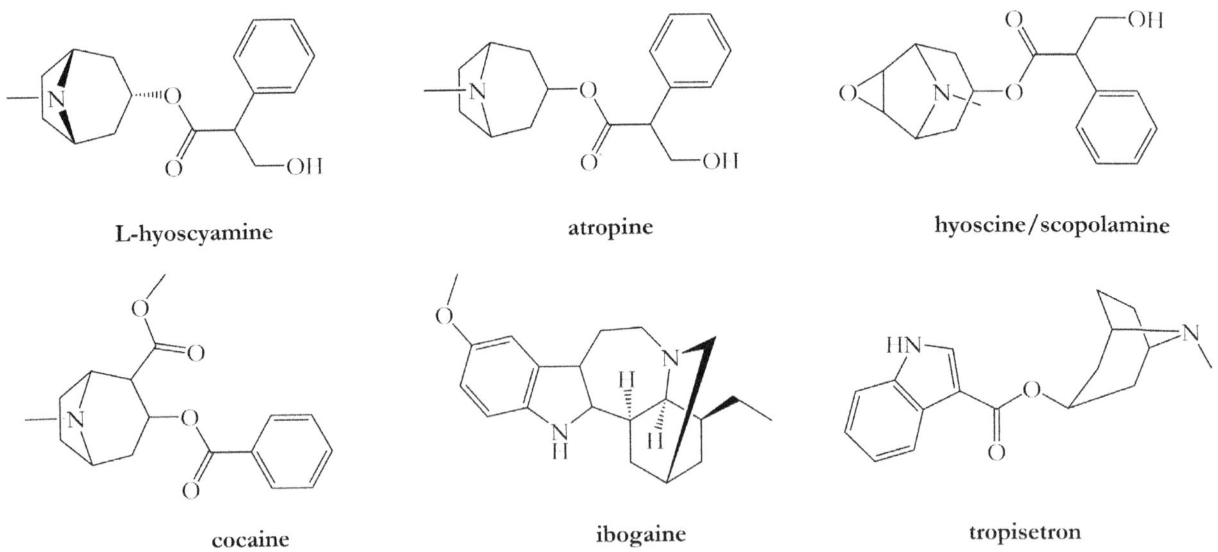

Fig. 4.7. Key representatives of the tropane-type infraclass of the L-ornithine derived subclass of alkaloids (both natural products and pharmaceutical drugs).

some nerve gases).[22] The tropine-type tropane alkaloids have also found a use in optometry/ophthalmology due to their significant mydriatic (pupil dilating) properties, which was considered historically to be a mark of beauty or fashion.[43] Hence, it was used in the parlours and politics of early modern Europe as 'Belladonna' (from *Atropa bella-donna* L., 1753, so named for the Greek sister of fate *Atropos* who cuts the thread of life). Topical application of even a minute amount of these compounds results in mydriasis and its associated cyclopegia (lens impairment), which facilitates detailed ophthalmic examinations. The stable atropine (as a dilute eye drop solution) is the favoured choice for this application because it can be standardised and lacks central nervous system (CNS) activity at these minute doses.[44] The use of hyoscyamine/atropine, hysocine/scopolamine is hardly eclipsed in its applications in modern medicine, which range from antiemesis, antispasmodic and resuscitation, to its nefarious use as an interrogative aid (by the Nazis in World War 2 or security agencies throughout the 20th century as a 'truth serum', unsurprisingly poorly documented) or as a dissociative hallucinogen (typically obtained from the flowers of plants from the genus *Datura*, similarly poorly documented).[22] Ecgonine derivatives (mainly cocaine) have a similarly chequered history in terms of their human-use value. The Inca Empire of the Peruvian highlands considered the coca plant (*Erythroxylon coca* Lam., 1786) as a 'gift from the Gods', due to the stimulating, energising effect of chewing its leaves.[22] This same effect has taken on a global significance since coca leaf's appearance in the raw form (in 1860, used as a topical anaesthetic) or as an elucidated isolate: cocaine (in 1898, with its complete synthesis achieved in 1901).[22] From the turn of the 20th century onwards, cocaine spread across the Western world, sold as a medicine (famously used and abused by lauded neurologist/psychoanalyst Sigmund Freud, author Robert Louis Stevenson, author/journalist Charles Baudelaire, industrialist/inventor Thomas Edison, playwright Tennessee Williams and philosopher Jean-Paul Sartre, among many others) or as a beverage additive (e.g. Coca Cola, only fleetingly - but this globally famous brand still bears its tropane-origins in its registered trademark). However, it only took 15 years before the addictive properties of cocaine as a recreational drug-of-abuse became glaringly apparent, an issue we still face a century later, with 1.5 million registered cocaine addicts in the USA alone.[45,46]

Beyond these more general applications of the tropane-type alkaloids across society, they have genuinely substantial uses within modern pharmacology. Their initial uses as mydriatics, spasmolytics and local anaesthetics have been improved upon over the past century by detailed structure-activity-relationship (SAR) research and pharmacophore, ligand-docking and stereoselectivity experiments.[22] The anticholinergic activity of the tropanes was implicated in the conceptual development of a 'receptor' in the historical development of pharmacology. Atropine and many other tropane alkaloids are competitive antagonists of muscarinic receptor acetylcholine (ACh) action, with stereoselectivity between the *S*-(-)- and *R*-(+)- forms of hyoscyamine varying from 30- to 300-fold, respectively.[47,48] ACh receptors are widespread throughout the human system and are involved in the CNS (cognition, sensory motor systems, cardiovascular and respiratory activity and stress responses) and peripheral nervous system (PNS: heart rate, smooth muscle contraction, glandular secretion and vasodilation).[47,48]

ACh disfunction has been implicated in depression, epilepsy, Parkinson's disease and Alzheimer's disease; thus tropane-type alkaloids are still under substantial investigation.[22,47,48] Many of the actions that have been attributed to tropane-type alkaloids in this section are associated with their interaction with ACh receptors (AChRs) in the CNS and PNS, but this is not their only mechanism of action. For example, the synthetic derivative of tropine, tropisetron, is a selective serotonin type-3 receptor antagonist, an α1 glycine receptor potentiator and an α7 nicotinic AChR selective partial agonist, as well as targeting the calcineurin pathway implicated in the inflammatory response.[22,49] Ecgonine-type tropane alkaloids bind and inhibit the dopamine uptake carrier, with the consequent rise in synaptic dopamine being largely responsible for their use as a recreational drug of abuse.[50]

The distribution of alkaloid-type specialised metabolites of the tropane infraclass reported from the customary medicinal flora of Australia is presented in *Table 4.1*. One genus of the customary medicinal flora of Australia, *Duboisia* (Solanaceae), is of particular significance regarding the tropane-type alkaloidal constituents. Members of *Duboisia*, such as *D. hopwoodii* (F.Muell.) F.Muell., 1874, *D. leichhardtii* (F.Muell.) F.Muell., 1882 and *D. myoporoides* R.Br., 1810 are reportedly used as narcotics due to the giddy, soporific effect that aqueous extracts of the bark of these trees have on human physiology.[51] This became pharmacologically significant during World War 2 when soldiers were given hyoscine tablets (isolated from *Duboisia* and exported to the

Western Front) to treat the nausea and vomiting associated with seasickness.[51,52] This anti-emetic effect is due to antagonism of different human muscarinic-receptor subtypes (notably M_3 and M_5) that mediate the cholinergic activity pathways through the brain stem that integrate/initiate the vomiting reflex [e.g. the nucleus tractus solaritus].[52]

L-LYSINE-DERIVED ALKALOIDS

L-Lysine is a proteinogenic amino acid involved in both primary and secondary metabolism in plants. As with the L-ornithine-derived alkaloids discussed above, the L-lysine-derived alkaloids have their origins in the Krebs cycle. As L-glutamic acid led to L-ornithine and the 5-membered heterocyclic pyrrolidine alkaloids, L-aspartic acid leads to L-lysine, which in turn gives rise to the 6-membered heterocyclic piperidine (C_5N), quinolizidine (C_5NC_4) and indolizidine (C_5NC_3) alkaloidal scaffolds (*Fig. 4.8*).[1,2] The biogenic pathway arising from L-lysine deviates immediately, with the piperidine and quinolizidine infraclasses forming via a pyridoxal phosphate (PLP)-bound intermediate that is decarboxylated via lysine decarboxylase to a PLP-bound cadaverine branching point (currently under investigation), while the indolizidine infraclass arises from L-pipecolic acid (*Fig. 4.8*).[1,2]

Piperidine-type alkaloids form in much the same way as the pyrrolidine alkaloids from L-ornithine, being transformed via an oxidase enzyme [thought until recently to be DAO, but since revised to copper amino oxidase (CuAO)] to 5-aminopentanal, that spontaneously cyclises to the Schiff base Δ^1-piperideine, which is protonated to a Δ^1-piperidinium cation and can then bifurcate to yield (*R*)- and (*S*)-piperidine-type alkaloids (*Fig. 4.8*).[1,15,18,19,31]

Quinolizidine-type alkaloids also arise from PLP-bound cadaverine, but in a much more complex process go through the same enzymatic oxidation (DAO/CuAO) to 5-aminopentanal and spontaneous formation of the

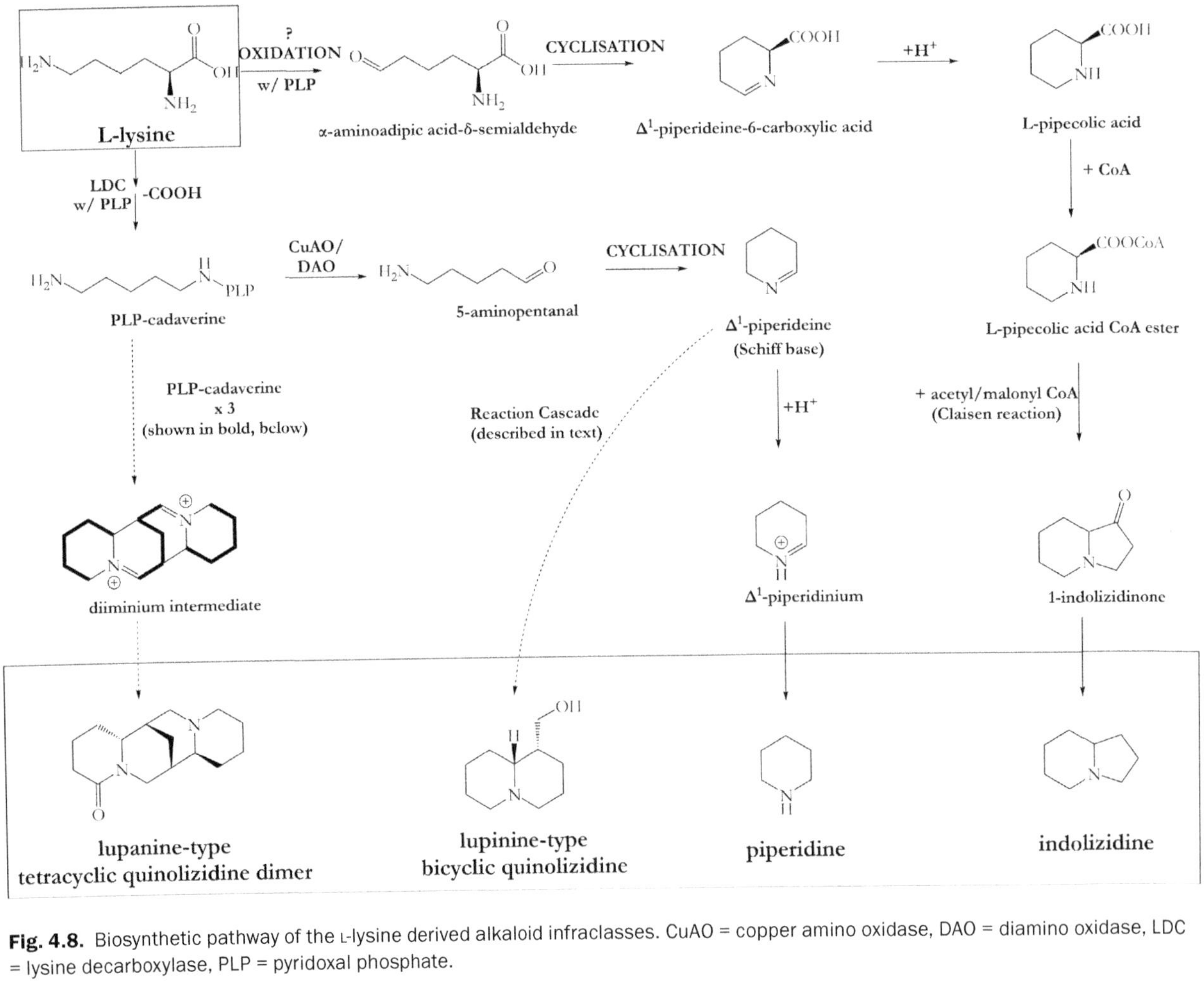

Fig. 4.8. Biosynthetic pathway of the L-lysine derived alkaloid infraclasses. CuAO = copper amino oxidase, DAO = diamino oxidase, LDC = lysine decarboxylase, PLP = pyridoxal phosphate.

Schiff base Δ^1-piperideine, followed putatively by four further minor reactions (aldol formation, hydrolysis of imine, oxidation and another Schiff base formation, that are yet to be fully substantiated) to give the bicyclic lupinine-type quinolizidines, while the tetracyclic sparteine- and lupanine-type quinolizidines are thought to arise from the condensation of three PLP-bound cadaverines directly, without releasing free intermediates in a membrane-associated enzyme channeling process (*Fig. 4.8*).[1,2,17,53,54]

Indolizidine biosynthesis arises from the PLP-mediated oxidation of L-lysine to α-aminoadipic acid-δ-semialdehyde, which cyclises to Δ^1-piperideine-6-carboxylic acid and then protonated to form L-pipecolic acid, forming a CoA-ester and undergoing a Claisen reaction with acetyl/malonyl CoA before undergoing an amide/imine ring closure to yield 1-indolizidinone, which is reduced to give the various indolizidines (*Fig. 4.8*).[1,2,17,55] The L-lysine derived alkaloids are generally thought to be synthesised in the chloroplasts of mesophyll cells, with movement to epidermal (and subterranean) accumulation points mediated via carrier proteins.[1] The fact that L-lysine is an exogenous amino acid in animals often results in the detrimental biological effects observed across the L-lysine-derived alkaloids, leading to a historical relationship between these compounds and humans in the form of common and severe livestock poisoning.[1,8] The distribution of alkaloid-type specialised metabolites of the L-lysine derived infraclasses reported from the customary medicinal flora of Australia is presented in *Table 4.1*.

Piperidine

The 700 known piperidine alkaloids have a limited distribution in the plant kingdom, with their most notable source (and namesake) being the Piperaceae [e.g. piperine from black pepper (*Piper nigrum* L., 1753)].[31,56] Outside the simpler piperidine alkaloids of the Piperaceae, a wide variety of different structural motifs have been reported, such as the polyhydroxy piperidine alkaloids of the Aspidaceae, Fabaceae (*Derris*, *Lonchocarpus*), Moraceae (*Morus*) and Polygonaceae (*Fagopyrum*).[1,56] Piperidine alkaloids have been found to condense with several other metabolic sub-/infraclasses to give rise to diverse hybrid alkaloids such as the tobacco alkaloids [Solanaceae (*Nicotiana* spp.)] based on the anabasine/anatabine scaffold (condensed with a nicotinic acid derivative). The 6-membered tropane-type piperidine analogues, such as anaferine and anahygrine (condensed with acetyl/malonyl-CoA) from the Solanaceae (*Withania*), Punicaceae (*Punica*) or *Sedum*-type (Crassulaceae) piperidine alkaloids (e.g. sedamine), involve condensation with phenylpropanoid moieties (derived from phenylalanine), as do *Lobelia*-type (Campanulaceae) piperidine alkaloids (e.g. lobeline) (*Fig. 4.9*).[1] A grouping of complex pentacyclic piperidine-infraclass alkaloids (e.g. lycopodine) is derived from primitive seedless vascular plants of the Lycopodaceae (*Lycopodium*), a notable observation as alkaloids are uncommon in the lower plants.[1,2,17] In addition, certain polyketide-derived piperidine-type (pseudo)alkaloids are known, such as the toxic coniine-type pseudoalkaloids of the Apiaceae (*Conium*), Sarraceniaceae (*Sarracenia*) and Liliaceae (*Aloe*), pinidine-type piperidine pseudoalkaloids of the Pinaceae (*Pinus*), *Galbulimina*-type piperidine pseudoalkaloids of the Himatandraceae and the *Elaeocarpus*-type piperidine pseudoalkaloids of the Eleocarpaceae.[1,2,17] The specialised metabolites derived from the polyketide pathway will be expanded upon in Chapter 5.

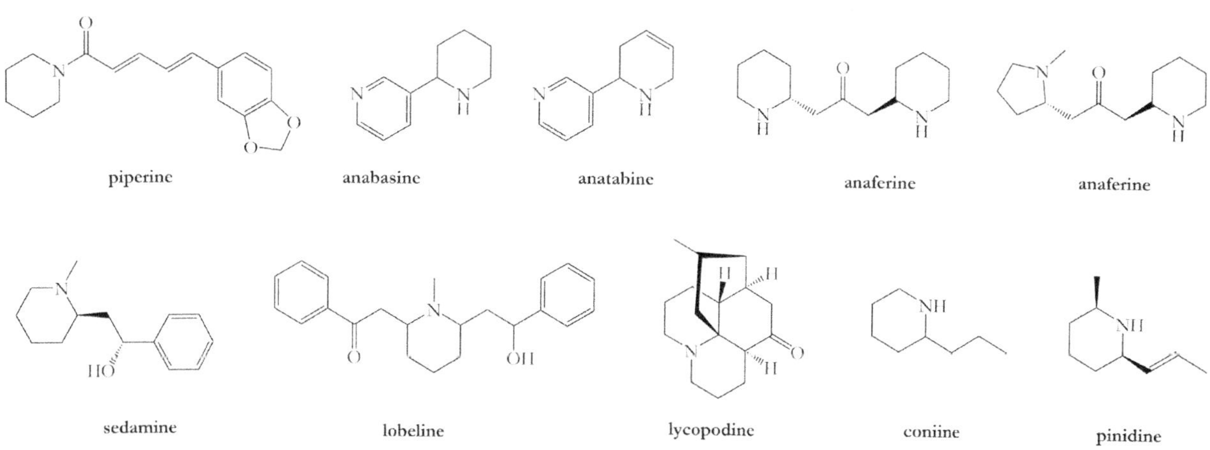

Fig. 4.9. Key representatives of the piperidine-type infraclass of the L-lysine-derived subclass of alkaloid-type specialised metabolites.

Some of the most studied piperidine-type alkaloids are the coniine-type piperidine pseudoalkaloids, isolated from hemlock [*Conium maculatum* L., 1753 - (Apiaceae)] and famously used in the state-sanctioned demise of the Greek philosopher Socrates in 399 BCE [150–300 mg of coniine (6–8 leaves) is fatal to adult humans].[57] Toxic coniine-type piperidine alkaloids, and by inference the piperidine moiety, are considered neurotoxic and teratogenic, having a primary mechanism of action through stimulation and subsequent desensitisation of nicotinic AChRs in the CNS and PNS, leading to respiratory depression and death, which is similar to the action of the curare alkaloids discussed in the benzylisoquinoline-type alkaloid section below.[1,8,57] Hemlock has a long history of traditional use in the Western pharmacopoeia, as a topical treatment for herpes, erysipelas, tumours, joint pain, rabies, gout, ulcers, as a tincture for its sedative, antispasmodic and analgesic effects (rather dangerous) or as an antidote to strychnine poisoning (presumably as a last resort).[1,8,57] Problems with standardisation and frequent poisoning led to its clinical discontinuation in 1938. Piperidine moieties abound in synthetic pharmaceutical drugs such as antipsychotics (e.g. droperidol, melperone, thioridazine, etc.), opioids (e.g. fentanyl, loperamide, prodine, etc.), selective serotonin reuptake inhibitors (e.g. paroxetine], selective estrogenic receptor modulators [e.g. raloxifene], arylcyclohexamines (e.g. phenylcyclohexyl piperidine), stimulants (e.g. ethylphenidate), vasodilators (e.g. minoxidil) and even certain weapons of chemical warfare (e.g. Ditran, *N*-methyl-3-piperidyl benzilate) (*Fig. 4.10*).[58] The distribution of alkaloid-type specialised metabolites of the piperidine infraclass reported from the customary medicinal flora of Australia is presented in *Table 4.1*.

Quinolizidine

Unlike the L-ornithine-derived pyrrolizidines, the 570 quinolizidines that have been reported in the literature usually occur in a free state (not esterified or as an *N*-oxide).[31] The quinolizidines are abundant in the Fabaceae [10 primitive Papilonoideae tribes, particularly the Genisteae (27 genera), the Podalyrieae (9 genera) and

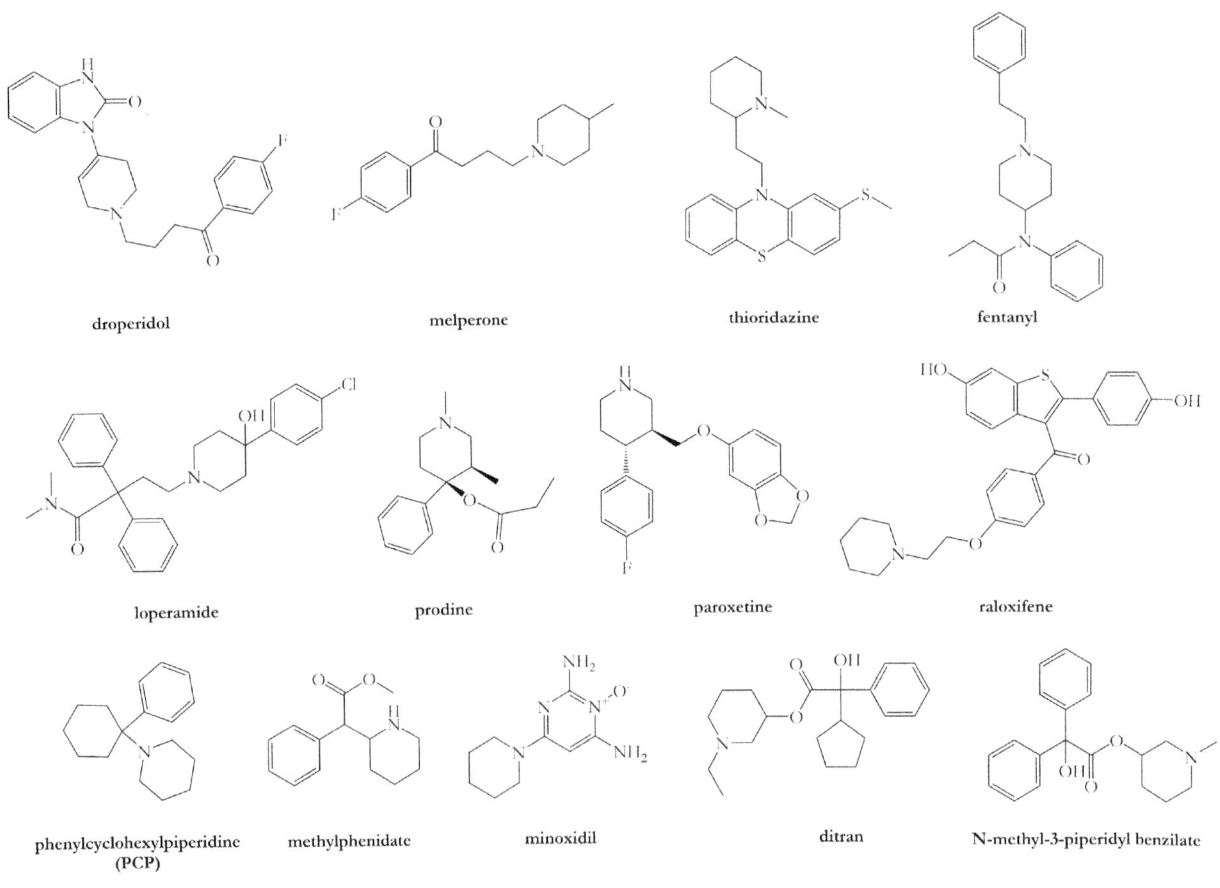

Fig. 4.10. Key representatives of the piperidine-type infraclass of the L-lysine-derived subclass of alkaloid-type that have been developed into pharmaceutical drugs.

the Sophorae (16 genera), where they are the typically the most abundant alkaloids present]. Some other miscellaneous groups of alkaloids with coincidental quinolizidine ring systems (derived from metabolic pathways other than L-lysine) have been isolated from various plant families: Lythraceae, Nymphaceae (*Nuphar*), Ericaceae (*Vaccinium*), Euphorbiaceae (*Poranthera*).[1,2] Outside the Fabaceae, 'true' quinolizidines (from L-lysine) have been reported from the Berberidaceae (*Leontice*), Chenopodiaceae (*Anabasis*), Orobanchaceae (*Orobanche*), Papaveraceae (*Chelidonium*), Ranunculaceae, Santalaceae (*Osyris*), Scrophulariaceae (*Castilleja*) and Solanaceae (*Solanum*).[59] The quinolizidine-type alkaloid infraclass can be divided simply into bicyclic, tricyclic and tetracyclic divisions or more specifically into sub-infraclasses based on structure (simple (lupinine-type), leontidine-type, sparteine/lupanine-type, sparteine/lupanine-type esters, tricyclic sparteine/lupanin-type degradation products, pyridine base-type, matrine-type, ormosia-type and others) (*Fig. 4.11*).[60] Lupinine-type (*Lupinus*) are the simplest and most common quinolizidine-type alkaloids and have shown light-dependent variation *in planta* throughout a 24-h period.[1]

The quinolizidine-type alkaloids demonstrate pronounced toxicity through their inhibition of phenylalanine tRNA binding to ribosomes or during aspects of translation.[59,61] Cytisine-type compounds are notably teratogenic, while the quinolizidine infraclass in general has shown antimicrobial, allelopathic, antifeedant and antiparasitic biological activities, again substantiating their proposed ecological role as phytoalexin/phytoanticipin defence compounds.[59,61] Livestock poisoning is commonly attributed to the quinolizidine content of grazing plants (LD_{50} of 10–600 mg/kg). Medical applications of drugs bearing a quinolizidine moiety (e.g. matrine, cytisine) are amoebicidal and antineoplastic (Ehrlich ascites).[59,61] The presence of cytisine in the seeds of *Sophora secundiflora* (Ortega) Lag. ex DC., 1813 – (Fabaceae) has been implicated in the hallucinogenic effects of the mescal bean in the traditional cultures of Mexico and the south-west USA.[1] The distribution of alkaloid-type specialised metabolites of the quinolizidine infraclass reported from the customary medicinal flora of Australia is presented in *Table 4.1*.

Indolizidine

Over 170 compounds of the indolizidine infraclass have been reported.[31] Their intermediary nature between pyrrolizidines and quinolizidines, and similarity to indoles, has led them to be somewhat neglected as a clearly delineated group.[62] Additionally, they often do not react with Dragondorff/ninhydrin and thus may be missed in screening programs.[1,2] Coincidental indolizidine-like ring structures abound among other alkaloid sub-/infraclasses (*Aspidosperma* indole alkaloids, *Erythrina* benzylisoquinoline alkaloids, Amaryllidaceae alkaloids (phenylalanine/tyrosine derived), Solanaceae steroidal alkaloids), further complicating the disambiguation

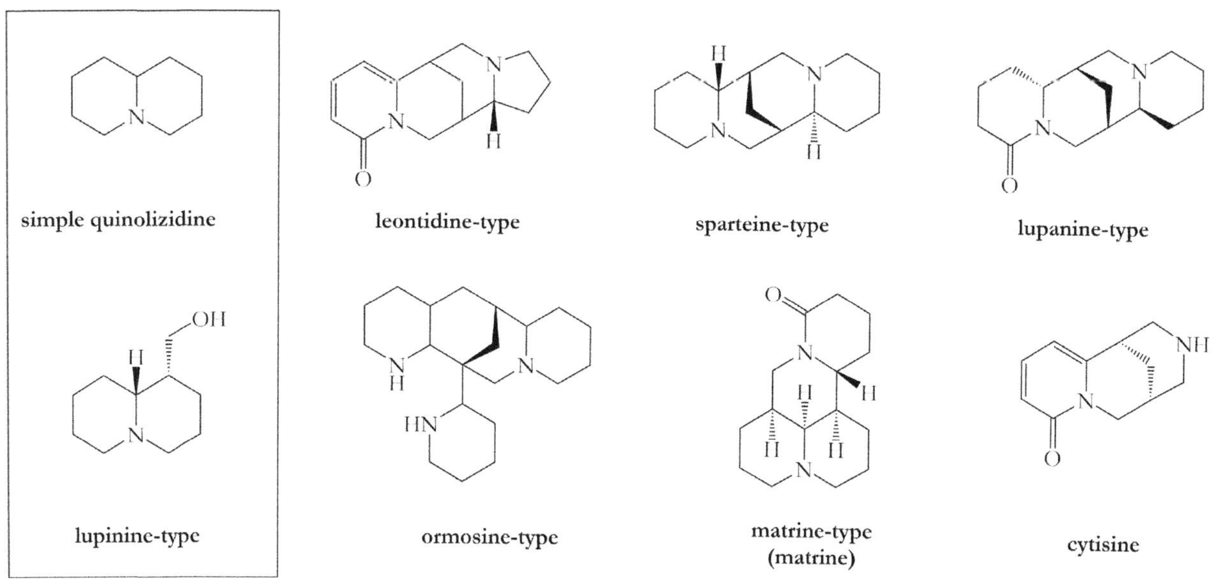

Fig. 4.11. Key representatives of the quinolizidine-type infraclass of the L-lysine-derived subclass of alkaloids (both natural products and pharmaceutical drugs).

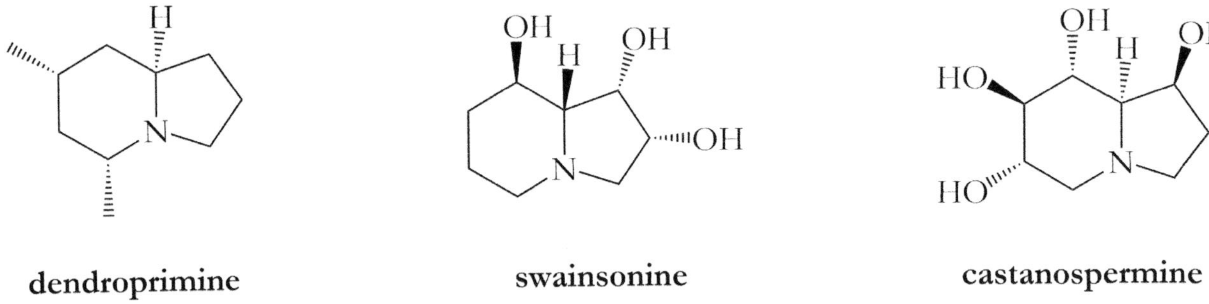

Fig. 4.12. Key representatives of the 'true' indolizidine-type infraclass of the L-lysine-derived subclass of alkaloid-type specialised metabolites.

process.[1,2] 'True' indolizidines (L-lysine derived) have been reported from the Orchidaceae [*Dendrobium* (e.g. dendroprimine)], Fabaceae [*Astragalus*, *Castanospermum* (e.g. castanospermine), *Oxytropis*, *Swainsona* (e.g. swainsonine)] (*Fig. 4.12*).[1,2,31]

In animal systems, glycosidase enzymes are integral in the digestion of carbohydrates (catabolism) and the production of membranes and glycoproteins (anabolism).[39,40] Indolizidines (e.g. castanospermine) interfere with glycoprotein processing, root elongation and the proliferation of cytomegaloviruses and retroviruses and are potent selective inhibitors of emulsin β-glucosidase (complete inhibition at 50 μg/mL) while leaving many other glycosidases uninhibited at any concentration. Indolizidines are demonstrated allelopathic phytotoxins (e.g. canastospermine shows complete inhibition at 300 ppb for dicots and 200 ppb for monocots).[1] Swainsonine has been shown to selectively inhibit α-mannosidase, one of the three types of animal mannosidases, again showing increased specificity.[55] This selectivity provides opportunities for glycoprotein SAR research and the role of this and other indolizidines in immune response, cancer metastasis, intracellular transport and viral infectivity.[1] Changes in glycoprotein processing leads to defective lysosomal function, which causes defective cytoplasmic vacuoles in the brain and in turn causes breakdown of the dopaminergic pathways of the CNS.

Beyond these activities related to glycosidase inhibition, indolizidines have been implicated in antineoplastic activity (melanoma, Ehrlich ascites carcinoma, colourectal cancer, human liver cancer, lymphoma and leukaemia) through promotion of apoptosis, immune system enhancement, increased intracellular (Ca^{2+}), inactivation of the phosphoinositide 3-kinase (PI3K)/ protein kinase B (AKT)/mammalian target of rapamycin (mTOR) (PI3K/AKT/mTOR) signalling pathway and repression of miR-92a (pancreatic cancer cell line).[1] As with the quinolizidines, the indolizidines are potent ecological phytoalexin/phytoanticipin defence compounds, with significant antimalarial, antimicrobial, antiviral, antiparasitic and antifeedant (particularly insecticidal) properties.[1,62] Indolizidines have shown anti-inflammatory activity, studied with regard to the treatment of neurological disorders (Alzheimer's) and cardiovascular disease.[1] The distribution of alkaloid-type specialised metabolites of the indolizidine infraclass reported from the customary medicinal flora of Australia is presented in *Table 4.1*.

L-TRYPTOPHAN-DERIVED ALKALOIDS

The alkaloids that are derived from L-tryptophan are difficult to reconcile, either due to placement within plant primary metabolism (e.g. the phytohormone IAA) or through overlaps with other alkaloidal sub-/infraclasses of differing origins; for example, the quinoline 'pseudoalkaloids' that arise from anthranilic acid, although this is ambiguous, as L-tryptophan also has its origins in anthranilic acid (see below). Additionally, converging biosynthetic pathways (e.g. indoles arising from the ring closure of simple aromatic amines) or significant derivatisation occurring beyond the indole core (e.g. combining indoles with terpenoids) can further complicate the matter.[1,2] Thus, here discussion will be limited to the secondary metabolites that conform to the stipulation of indoles derived from L-tryptophan *sensu stricto*, but some digression must occur out of necessity. Of the more than 4100 reported structures that fit the loose category of L-tryptophan-derived alkaloids, the relatively limited set of infraclasses that will be discussed here are the simple indoles and the non-terpenoid indole derivatives β-carboline, physostigmine and others (canthin-6-one, carbazole, indolenine, etc.) (*Fig. 4.13*).[1,2]

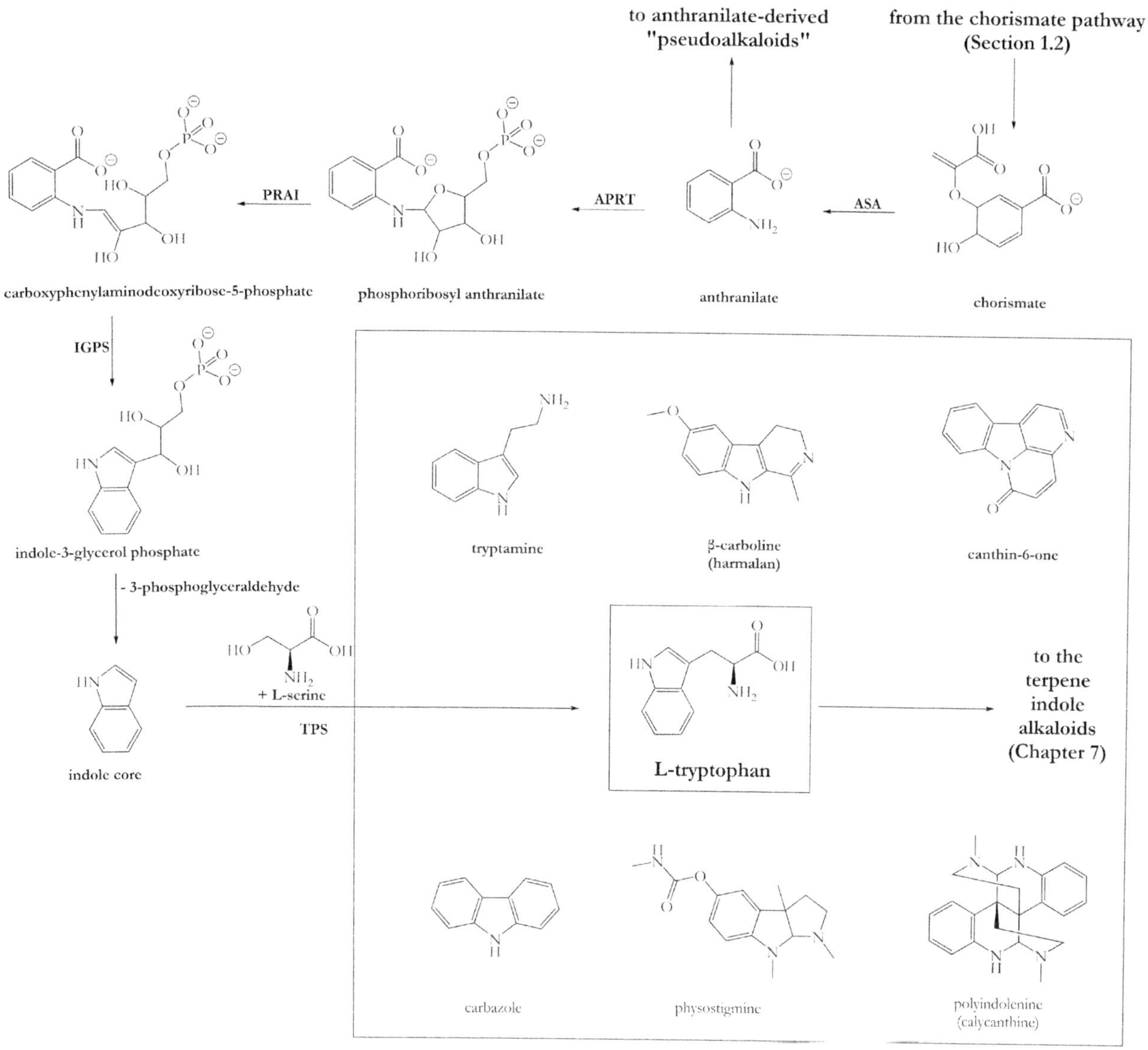

Fig. 4.13. Biosynthetic pathway of the infraclasses of the L-tryptophan-derived alkaloid subclass. ASA = anthranilate synthase, APRT = anthranilate phosphoribosyl transferase, IGPS = indole glycerol phosphate synthase, PRAI = phosphoribosyl anthranilate isomerase, TPS = tryptophan synthase.

The alkaloids derived from L-tryptophan discussed here follow relatively simple biosynthetic pathways because the indole core of these compounds is already in place in the amino acid itself and any further biosynthesis of these compounds is straightforward enough to be addressed in the specific sections. It is worth outlining how L-tryptophan is produced *in planta*, as it will resolve some of the ambiguity caused by trying to delineate clear sub-/infraclasses within this complex NP class. Chorismate is transformed to anthranilate through the action of anthranilate synthase (ASA), before having a phosphoribosyl group attached to the new amino moiety by anthranilate phosphoribosyl transferase (APRT) *Fig. 4.13*).[18,19,63] The ribose moiety of phosphoribosyl anthranilate is then opened by phosphoribosyl anthranilate isomerase (PRAI) to form carboxyphenylaminodeoxyribose-5-phosphate, which is reductively decarboxylated by indole glycerol phosphate synthase (IGPS) to form indole-3-glycerol phosphate.[63] The 3-phosphoglyceraldehyde moiety is then cleaved from indole-3-glycerol phosphate to form the indole core, then converted to L-tryptophan by tryptophan synthase (TPS) through condensation with L-serine (*Fig. 4.13*).[18,19,63] The distribution of alkaloid-type specialised metabolites of the L-tryptophan derived infraclasses reported from the customary medicinal flora of Australia is presented in *Table 4.1*.

Simple indoles

The simple indoles are an infraclass of alkaloids with notable bioactivity, based on the core structure of a decarboxylated L-tryptophan, tryptamine. Tryptamine is the precursor to the ubiquitous plant primary metabolite IAA, an auxin that is implicated in the induction of cell elongation and division, as well as coordinating the growth and development of plant organs (*Fig. 4.14*). The monoamine neurotransmitter serotonin [5-hydroxytryptamine (5-HT)] is widespread in the plant kingdom and has a complex role in humans, modulating mood, cognition, reward, learning, memory and many physiological processes.[1,64–67] The compound *N,N*-dimethyltryptamine (and many of the tryptamine derivatives such as the fungal-derived psilocin and psilocybin) are notable and potent hallucinogens, with associated religious and spiritual use spread across global traditional and modern cultures.[65,66]

The efficacy of these compounds is commensurate with serotonin's relationship with the 5-HT G protein-coupled receptors (and ligand-gated ion channels) of the CNS and PNS.[64] The 5-HT receptors are modulators of many neurotransmitters (glutamate, GABA, dopamine, epinephrine, ACh, etc.) and hormones (oxytocin, prolactin, vasopressin, cortisol, corticotropin, etc.) in animal nervous systems.[64–66] Unsurprisingly, interference with the regular functioning of these integral modulation channels has widespread biological and neurological implications; hence compounds that bear a tryptamine scaffold are of significant interest in pharmacology with regard to the development of new drugs with antidepressant, antipsychotic, anorectic, antiemetic, gastroprokinetic, antimigraine or even hallucinogenic/entactogenic properties (for use alongside therapy).[64–66]

Because they are produced from a simple decarboxylation of a common amino acid (L-tryptophan) and are

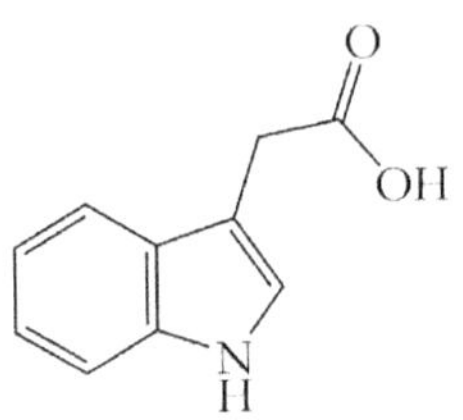

indole-3-acetic acid (IAA)

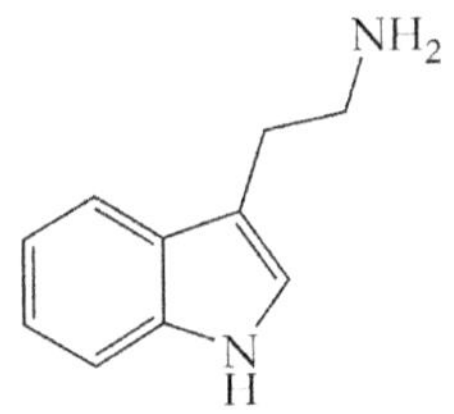

tryptamine

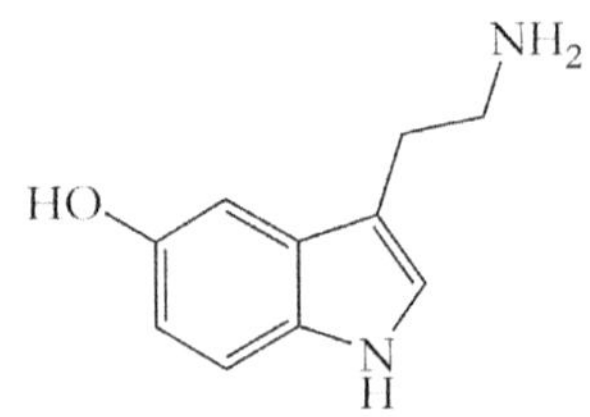

serotonin

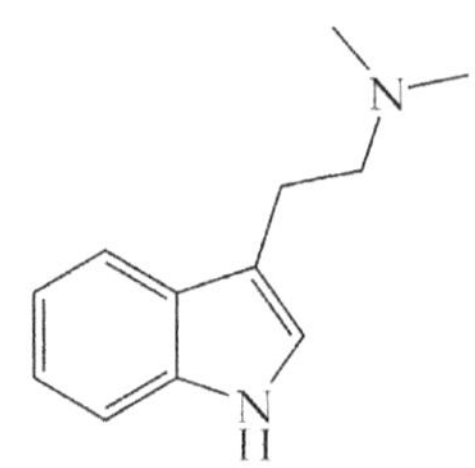

***N,N*-dimethyltryptamine**

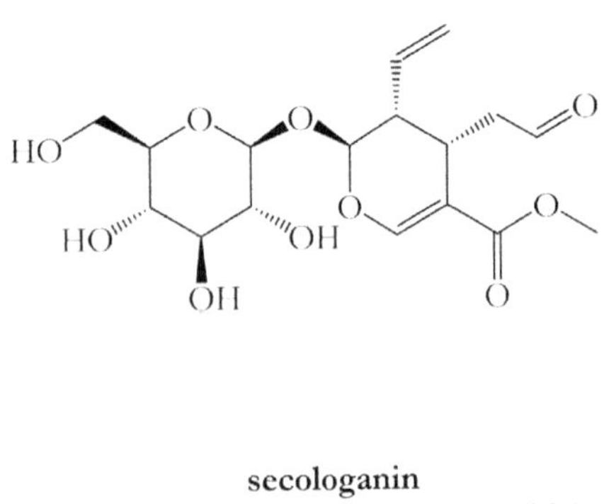

secologanin (a secoiridoid terpenoid glycoside)

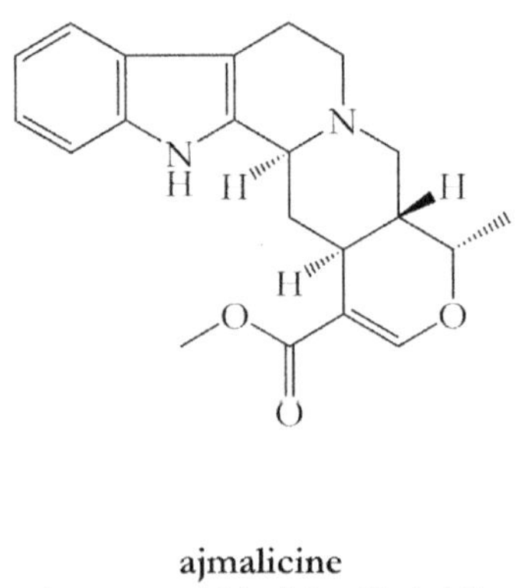

ajmalicine (a terpenoid indole alkaloid)

Fig. 4.14. Key representatives of the simple indole-type infraclass of the L-tryptophan-derived subclass of alkaloid-type specialised metabolites (and two representative terpenoid indole alkaloid derivatives).

implicated in plant primary metabolism (forming IAA), simple indole tryptamines are widespread in the plant kingdom and certain plant families (Fabaceae, Poaceae) possess these compounds in relative abundance.[68,69] The distribution of alkaloid-type specialised metabolites of the simple indole infraclass reported from the customary medicinal flora of Australia is presented in *Table 4.1*.

Non-terpenoid indole derivatives

There are several subinfraclasses of L-tryptophan-derived alkaloids that all originate from the simple indole (tryptamine) scaffold and are grouped under the infraclass, non-terpenoid indole derivatives: β-carboline (e.g. harmala), physostigmine (e.g. galantamine, rivastigmine) and others [e.g. canthin-6-one, carbazole (e.g. calycanthine) and indolenine (polyindolenine)] (*Fig. 4.15*).[1]

β-Carboline

The β-carbolines are an uncommon but significantly bioactive group of indole alkaloids, originally isolated from *Peganum harmala* L., 1753 - (Nitrariaceae), from which their trivial name 'harmala/harmane alkaloids' arises.[70] The β-carbolines are biosynthesised through the condensation/cyclisation of a tryptamine unit with a three-carbon pyruvate unit, before decarboxylation to give the harmalan core from which the other β-carbolines arise (*Fig. 4.15*). The β-carbolines are most abundant in the Rubiaceae, but are also present in the Acanthaceae (*Adhatoda*), Apocynaceae (*Aspidosperma*, *Alstonia*, *Picrasma*), Eleagnaceae (*Eleagnus*), Fabaceae (*Gastrolobium*, *Petalostylis*), Loganiaceae (*Strychnos*), Malphigiaceae (*Banisteriopsis*), Passifloraceae (*Passiflora*), Ranunculaceae (*Thalictrum*), Rutaceae (Clausena, *Murraya*), Simaroubaceae (*Ailanthus*), Solanaceae (*Vestia*) and Zygophyllaceae (*Peganum*).[1,64,67–69] The distribution of alkaloid-type specialised metabolites of the β-carboline subinfraclass reported from the customary medicinal flora of Australia is presented in *Table 4.1*.

The β-carbolines are known to be monoamine oxidase inhibitors (MAOIs), which prevent the enzyme-mediated metabolic breakdown of endogenous and exogenous neurotransmitters and drugs (such as tryptamine-based simple indole alkaloids), a phenomenon that has been discovered numerous times spatially and temporally in human history, perhaps most famously as the psychedelic tryptamine/β-carboline mixture known as *ayahuasca* used ceremonially by various South American shamanic cultures.[65,66,70] Beyond this MAOI

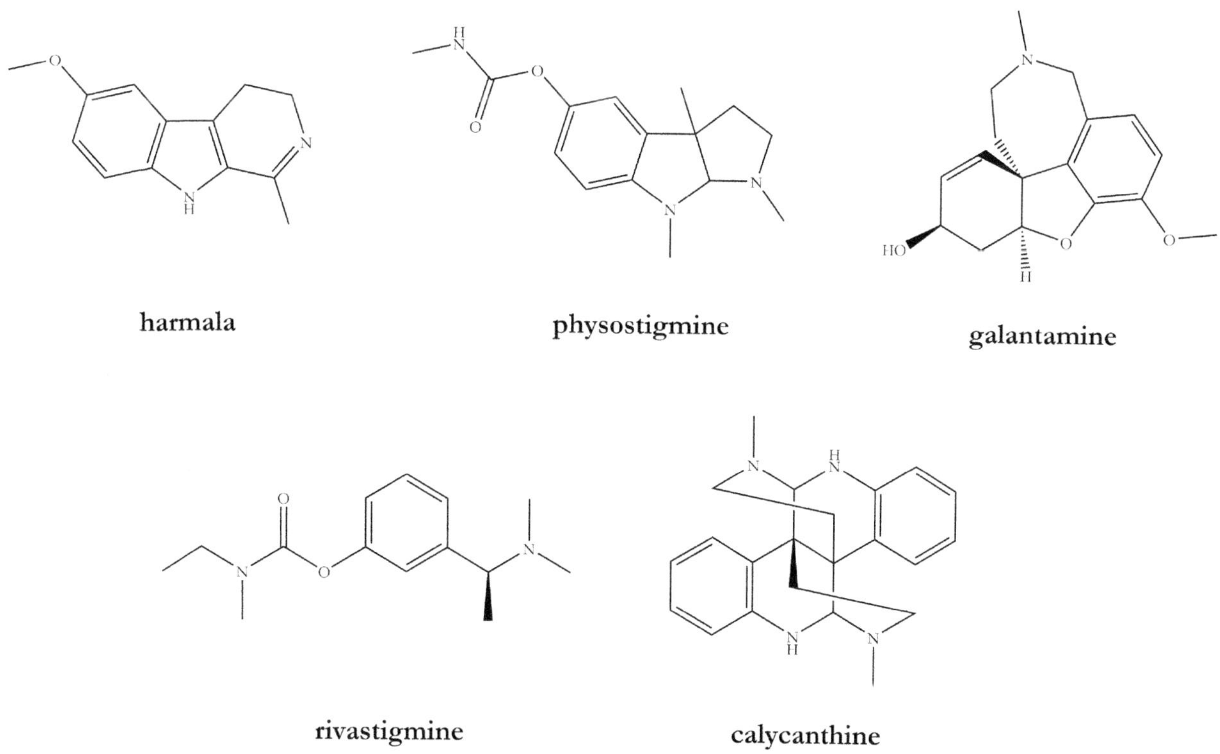

Fig. 4.15. Key representatives of the simple non-terpenoid indole-type infraclass of the L-tryptophan-derived subclass of alkaloids (both natural products and pharmaceutical drugs).

activity, the β-carbolines inhibit membrane transport, inhibit (Na^+, K^+)-ATPase activity, bind to DNA and demonstrate potent antiprotozoal activity [against *Trypanosoma cruzi* Chagas, 1909 – (2–5 µg/mL)], anthelmintic activity [against *Leishmania amazonensis* Lainson & Shaw, 1972 – (24 µg/mL)], antimalarial activity (against MR *Plasmodium falciparum* Welch, 1897), antifeedant activity and phototoxicity.[1,64,70]

Physostigmine

Physostigmine (eserine) alkaloids are structural relatives of the indoles, bearing a second pentacyclic nitrogen heterocycle adjacent to the first indole heterocycle, forming a pyrrolo(2,3b)indole system (*Fig. 4.15*).[71] The biosynthetic origin of physostigmine has only been determined relatively recently, with *C*-methylation of an indole precursor followed by an unusual acetylation-dependent reaction cascade to complete the second heterocycle.[72] These compounds are toxic to mammals (LD_{50} of 4.5 mg/kg in mice) and were originally isolated from the calabar bean [*Physostigma venenosum* Balf., 1861 – (Fabaceae)] of Eastern Nigeria, but have also been isolated from the Euphorbiaceae (*Hippomane*).[1,71,73] In a manner similar to the imidazoles, physostigmines (marketed as antilirium or isotopo eserine) are cholinesterase inhibitors, causing acetylcholine accumulation and continuous cholinergic activation, stimulating secretions and myotic (pupil contraction) effects.[71] These potent biological activities have been of interest to modern medicine since their applications were first suggested in 1862 by Thomas Richard Fraser of the University of Edinburgh.[73] Its ability to cross the blood–brain barrier has resulted in physostigmine drugs being historical antidotes for anticholinergic drug overdoses [from *Datura stramonium* L., 1753, *Atropa bella-donna* L., 1753, or γ-hydroxybutyric acid (GHB)], but its severe side effects have led to its replacement by more tolerable synthetic drugs such as donepezil, galantamine and rivastigmine (*Fig. 4.15*).[74]

Other non-terpenoid indole derivatives

The canthin-6-ones are a small group of indole alkaloids that are structurally related to the β-carbolines, with an additional heterocyclic ring bearing a ketone moiety attached to the indole heterocycle.[1] The canthin-6-ones are predominantly sourced from the Rutaceae (*Pentaceras*, *Zanthoxylum*) and Simaroubaceae, but are also present in the Amaranthaceae, Caryophyllaceae and Malvaceae.[31] Plants from genera belonging to these families have been used traditionally for a variety of ailments, fever, intestinal disease, dysentery, glandular swelling and malaria.[1] Certain canthin-6-ones have demonstrated cytotoxicity against the KB (nasopharynx carcinoma) and P-388 cell lines.[1] Carbazoles are similar in structure to the β-carbolines, but lack the aromatic nitrogen heterocycle. These compounds have been isolated from the Meliaceae (*Ekebergia*) and the Rutaceae (*Clausena*, *Glycosmis*, *Micromelum*, *Murraya*). Certain carbazoles have demonstrated cytotoxicity against KB cells.[75] Multinuclear indole compounds [with monomers often structurally similar to the pyrrolo(2,3b)indole physostigmines], known as indolenines (or polyindolenines (e.g. calycanthine; *Fig. 4.15*) have been isolated from the Calycanthaceae (*Calycanthus*, *Chimonanthus*), Celastraceae (*Bhesa*) and Rubiaceae (*Borreria*, *Hodgkinsonia*, *Palicourea*, *Psychotria*).[1]

L-TYROSINE- OR L-PHENYLALANINE-DERIVED ALKALOIDS

Several of the alkaloid-type NP infraclasses belonging to the subclass derived from aromatic amino acids such as L-tyrosine or L-phenylalanine do not possess a nitrogen heterocycle and are thus classed as simple amines. Besides these, over 4,000 alkaloid-type NPs bearing an isoquinoline moiety have been reported, rivaling the indoles as the largest alkaloid sub-/infraclass.[1,31] Most of the isoquinoline-type alkaloids are derived from 3,4-dihydroxytyramine (dopamine), which is combined with a carbonyl-bearing moiety (of many origins) and condensed to give the generic isoquinoline scaffold.[1,2,17] Although it is known that both L-phenylalanine and L-tyrosine can function as the source for this infraclass, the interconversion between these two structurally similar amino acids makes this disambiguation difficult, although L-tyrosine is largely considered to be the predominant precursor. The isoquinoline infraclass has two significant subinfraclasses: simple isoquinoline and benzylisoquinoline (BIQ) [with further division: tetrahydrobenzylisoquinoline (THBIQ), morphine-type BIQ, berberine-type BIQ, bisbenzylisoquinoline (BBIQ), etc.] (*Fig. 4.16*).[1,2,17] The isoquinolines that have condensed with the iridoid monoterpene secologanin will be addressed together with the MIAs in Chapter 5. Many of the more minor infraclasses (hasubanan promorphinanes, protopines, benzophenanthridines, phthalideisoquinolines, rhoedanes, dibenzopyrrocolines, pavines/isopavines, cularines, erythrina-type, ipecacuanha-type,

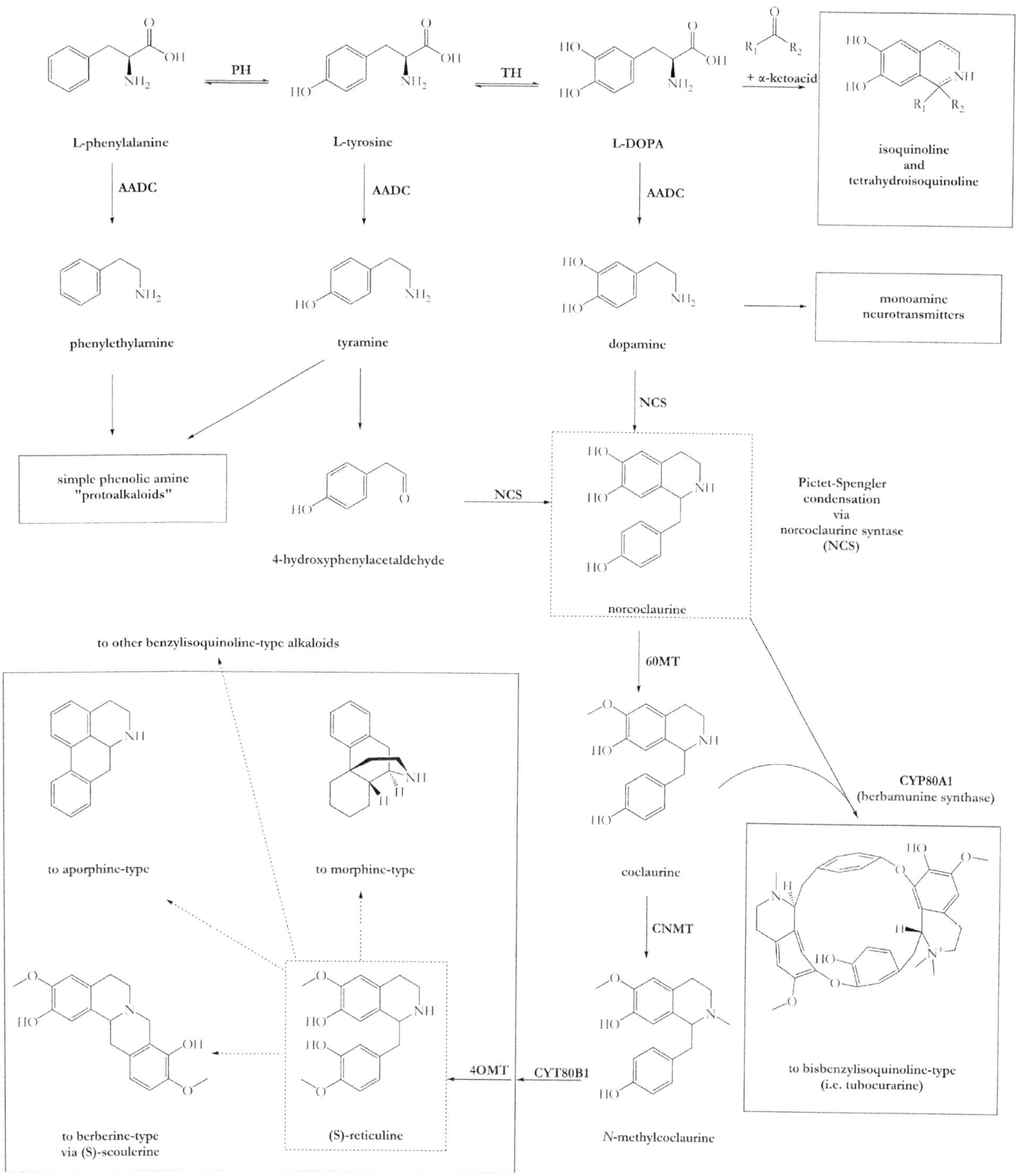

Fig. 4.16. Biosynthetic pathway of the infraclasses of the L-tyrosine-derived alkaloid subclass. 4OMT = 4-O-methyltransferase, 6OMT = 6-O-methyltransferase, AADC = aromatic amino-acid decarboxylase, CNMT = coclaurine N-methyltransferase, CYP80A1 = P450 enzyme (berbamunine synthase), CYT80B1 = P450 hydroxylase, NCS = norcoclaurine synthase, PH = phenylalanine hydroxylase, TH = tyrosine hydroxylase.

etc.) are too numerous and specific to go into detail in this book. Finally, reference must be made to tetrahydroisoquinoline (THIQ), which is an important alkaloid subinfraclass in its own right but will be addressed as it arises within the broader isoquinoline infraclass.

A more detailed breakdown will be provided when addressing each sub-infraclass, but the general distribution of isoquinoline-type alkaloid natural products across global plant families can be given here: Alangiaceae, Amaryllidaceae (*Crinum* – these are of a type peculiar to this family), Annonaceae, Apiaceae (*Heracleum*), Araceae (*Lysichiton*), Araliaceae (*Hedera*), Aristolochiaceae, Berberidaceae, Cactaceae (Lophophora), Caprifoliaceae (*Symphoricarpos*), Chenopodiaceae, Combretaceae, Euphorbiaceae (*Croton*), Eupomatiaceae, Fabaceae (*Andrachne, Erythrina*), Fumariaceae, Hernandiaceae, Lauraceae, Magnoliaceae, Menispermaceae, Monimiaceae, Musaceae, Nelumbonaceae, Nymphaceae, Papaveraceae, Phellinaceae (*Phelline*), Ranunculaceae, Rhamnaceae (*Colletia, Colubrina, Discaria, Phylica, Rhamnus, Retanilla, Talguena, Zizyphus*), Rubiaceae (*Cephaelis*), Sterculiaceae, Symplocaceae (*Symplocos*) and Winteraceae.[1,2,17,31] The most prolific isoquinoline producers belong to the Magnoliales order (Annonaceae) and the Ranunculales order (Berberidaceae, Menispermaceae, Papaveraceae, Ranunculaceae) and some significant work has gone into the chemotaxonomy of these orders and families based on this fact.[76] The distribution of alkaloid-type specialised metabolites of the L-tyr/L-phe derived infraclasses reported from the customary medicinal flora of Australia is presented in *Table 4.1.*

Simple phenolic amines

The simple phenolic amines are a large and disparate group of which ~1100 have been isolated, although disambiguation of this group is difficult, due to overlap with other alkaloid infraclasses such as indole, pyridine, isoquinoline, etc.[1,77] Most of these compounds are involved in plant primary metabolism, forming the metabolic pools for further alkaloid production. Their biosynthetic route is uncommonly simple, with one enzyme aromatic amino-acid decarboxylase (AADC), possessing the ability to decarboxylate a large proportion of the bioavailable amino acids to form the aliphatic and phenolic amine protoalkaloids. Due to their simplicity, the aliphatic amines (e.g. methylamine, ethylamine, etc.) are widespread in the plant kingdom, but are particularly abundant in the Caryophylaceae, Rosaceae and Staphylaceae, with many of these compounds demonstrating insecticidal activity (*Fig. 4.17*).[1] The phenolic amines primarily based upon phenylethylamine and tyramine (derived from the enzymatic decarboxylation of phenylalanine and tyrosine, respectively), will be the focus of this section. The phenolic amines are widespread, but are particularly abundant in the Cactaceae, Fabaceae, Rosaceae and Rutaceae where they play a role in pathogen protection, as they generally possess antifungal bioactivity, although they have also been implicated in pollination.[1]

Phenylethylamine has been shown to be poisonous to livestock and many phenylethylamine/tyramine derivatives (e.g.; dopamine, epinephrine, norepinephrine, etc.) are significant monoamine neurotransmitters in the animal kingdom.[77] Despite the significant interest that the monoamine neurotransmitters attract, even a cursory discussion of this type of metabolite lies outside the scope of this book, although some examples of aromatic amine natural products can be mentioned here. Hordenine, from barley [Poaceae (*Hordeum vulgare* L., 1753)] is a methylated tyramine derivative that is allelopathic and a demonstrated antifeedant towards grasshoppers. Ephedrine, from the traditional Chinese medicine *ma huang* [Ephedraceae (*Ephedra sinica* Stapf, 1972)], is a phenylethylamine derivative that stimulates the CNS and elevates blood pressure through its adrenergic interactions. The compound forms the core of many synthetic amphetamines, which are both significant psychiatric medicines [e.g. dextroamphetamine (Benzedrine, Adderall, etc.)] and common drugs of abuse (e.g. methamphetamine, etc.) (*Fig. 4.17*).[1,2,8,17] Cathinones, from the North African/Middle Eastern Khat tree [Celastraceae (*Catha edulis* (Vahl) Endl., 1841)], are derivatives of the ephedrines, with similar CNS and cardiostimulative effects and similar use (and abuse) occurs in societies where these plants are native.[66] Mescaline, widespread across the Cactaceae and particularly from Peyote (*Lophophora williamsii* Lem. ex J.F.Cels, 1894), is a well-known historical hallucinogen that plays a significant role in the religious life of the traditional and customary cultures of the Americas (as well as modern enthusiasts), along with the indole-type alkaloid psilocybin (derived from the fungal kingdom) mentioned in the indole section above.[1,66] The distribution of alkaloid-type specialised metabolites of the simple amine infraclass reported from the customary medicinal flora of Australia is presented in *Table 4.1.*

methylamine

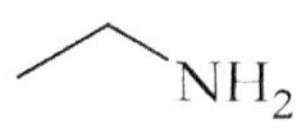

ethylamine

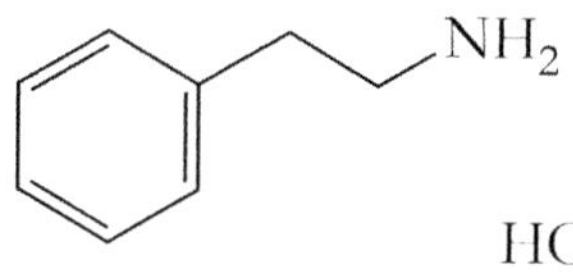

phenylethylamine

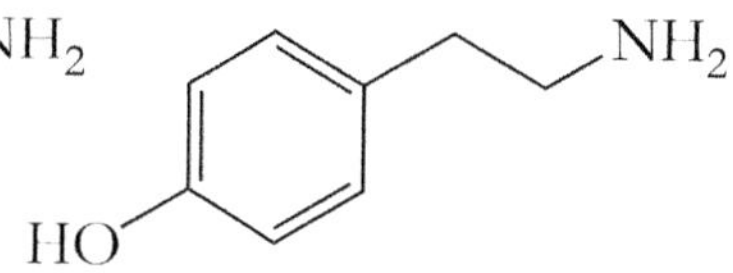

tyramine

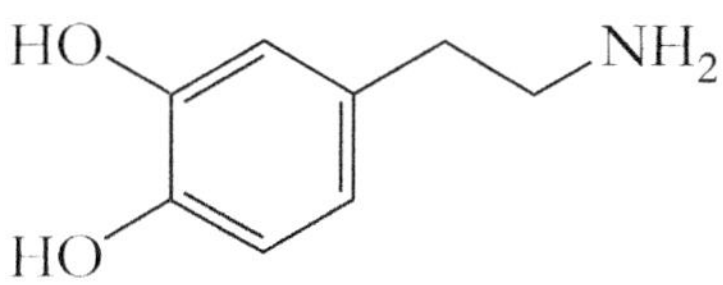

dopamine

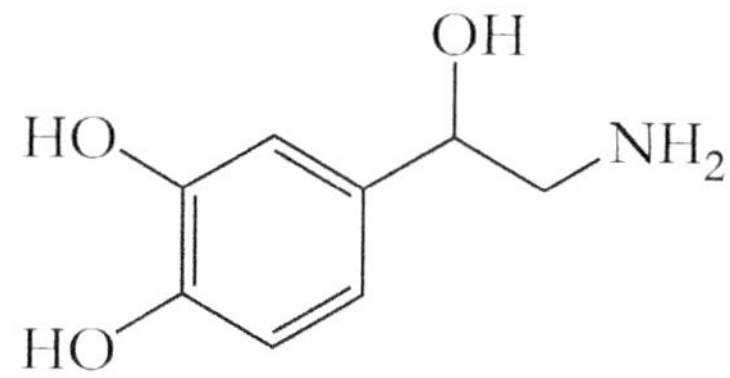

norepinephrine

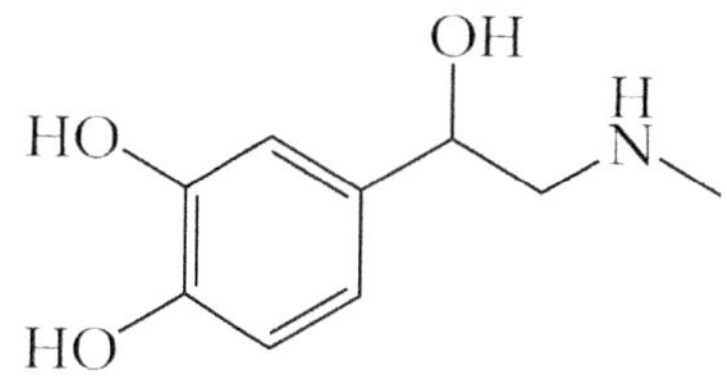

epinephrine

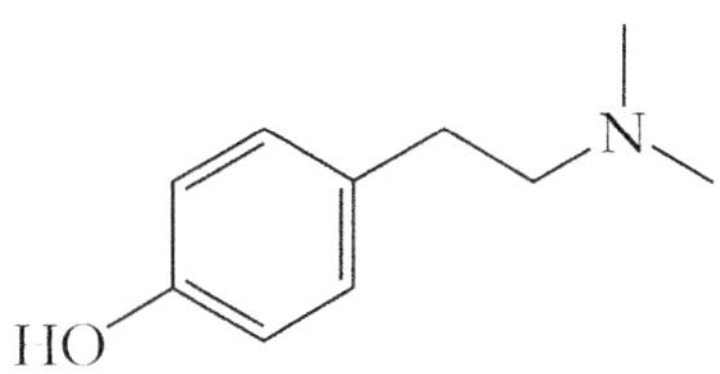

hordenine

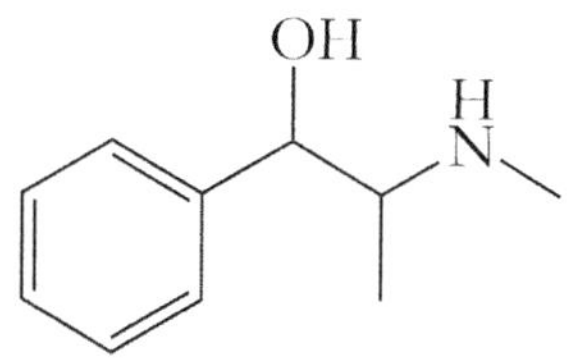

ephedrine

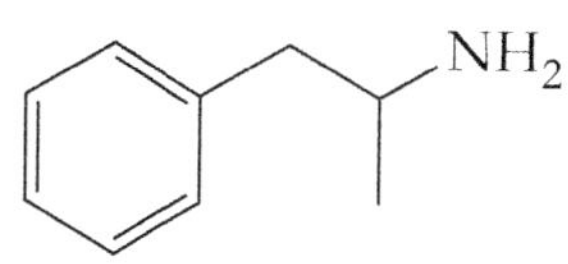

benzedrine

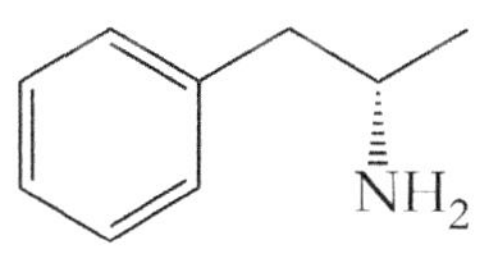

dextroamphetamine

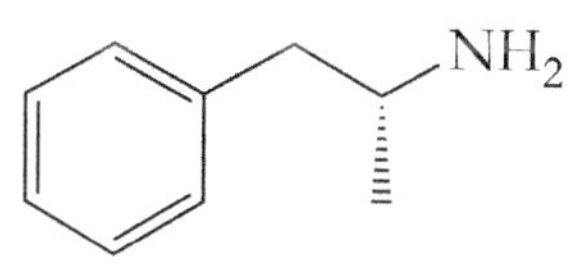

levoamphetamine

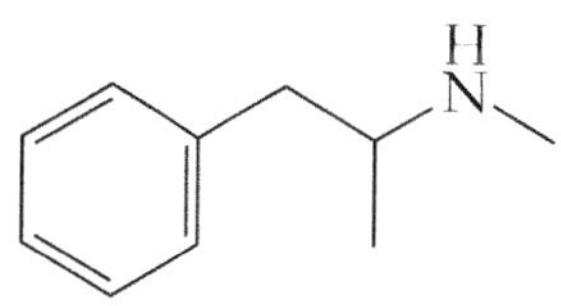

methamphetamine

(mixture in a 3:1 ratio gives Adderall)

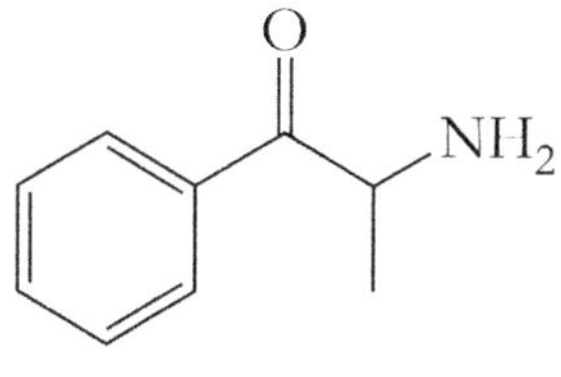

cathinone

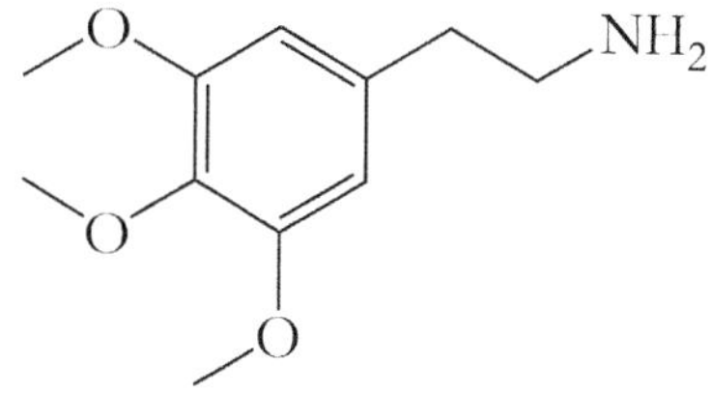

mescaline

Fig. 4.17. Key representatives of the simple- and aromatic amine-type infraclasses of the L-tyrosine/L-phenylalanine-derived subclass of alkaloids (both natural products and pharmaceutical drugs).

Betalains

A curious form of water-soluble, pigmented phenolic amines were characterised in the mid-20th century: they are known as the betalains and ~80 have been isolated from families in the Caryophyllales order (Aizoaceae, Amaranthaceae, Basellaceae, Cactaceae, Caryophyllaceae Chenopodiaceae, Didieriaceae, Molluginaceae, Nyctaginaceae, Phytolaccaceae, Portulacaceae, Stegnospermaceae), as well as three fungal genera (*Amanita, Hygrocybe, Hygrophorus*).[1,78] Due to the deep red colours of plants from these families, the betalains were initially confused with the anthocyanins until it was recognised that these compounds are nitrogenous and it was noted that they were responsible for the red-violet colour of these plants [e.g. beetroot, amaranth, many cacti (*Opuntia, Eulychnia, Hylocereus*)].[1] The betalains have been isolated as two major structural types, the reddish betacyanins (e.g. betanin) and the more yellow betaxanthins (e.g. vulgaxanthin) that have a common piperidine/dihydropyridine moiety (betalamic acid) (*Fig. 4.18*).[78] Despite bearing indole- (typically from L-tryptophan) and piperidine/dihydropyridine moieties (typically from L-lysine), the betacyanins have been categorised here due to both nitrogen heterocycles arising from L-tyrosine.[1] Labelling studies have shown a reasonably simple intramolecular cyclisation of L-DOPA (generated from L-tyrosine by cytochrome P450 enzyme CYP76AD1) to give the cyclo-DOPA indole core, while a ring cleavage (again of L-DOPA) mediated by 3,4-DOPA dioxygenase gives rise to the pendant betalamic acid moiety seen in the most well studied betacyanin, betanidin, which exists *in planta* as its glucoside betanin (*Fig. 4.18*).[78] Betaxanthins arise from the condensation of betalamic acid with simple amino acids (e.g. proline for indicaxanthin, glutamine for vulgaxanthin) (*Fig. 4.18*).[78]

The presence of these red-yellow pigments rather than the near-ubiquitous anthocyanins in the Caryophyllales, which also all bear proteinaceous sieve-tube cells, suggests an early divergence from the rest of the Angiospermae; notably the Caryophyllaceae and Molluginaceae appear to have recovered the ability to produce anthocyanins after losing it in favour of the betalains.[1] The betalains are good electron donors and thus have commensurate antioxidant properties, in a manner similar to other key physiological antioxidants such as the anthocyanins and the quinones.[78] In addition to their antioxidant activities, betalains have been shown to have antimicrobial, antiparasitic, antineoplastic, antilipidemic, hepatoprotective, neuroprotective, cardioprotective, antidiabetic and anti-inflammatory effects.[78]

Isoquinoline

The ~130 isolated simple isoquinoline NPs arise from condensation of L-DOPA (from L-tyrosine) with an α-ketoacid (rather than the aldehydes typical of the tetrahydroisoquinolines and benzylisoquinolines), although some variance is observed (such as methionine, leucine, mevalonic acid, pyruvic acid, peyoruvic acid, or 4-methoxy-2-oxopentanoic acid).[1,31] Isoquinoline monomers have been known to dimerise or oligomerise via an as-yet

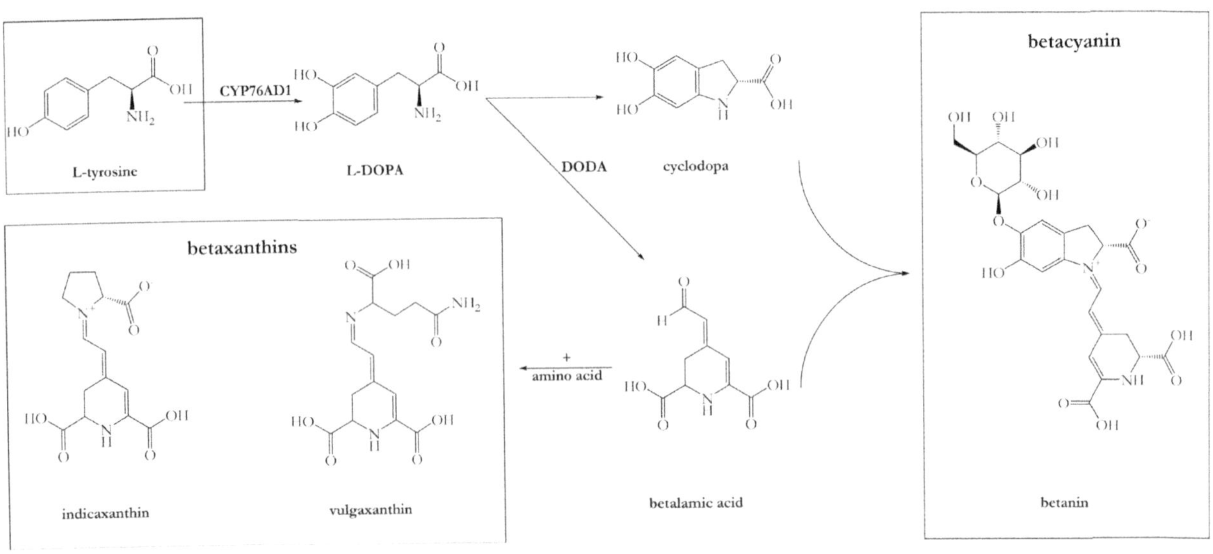

Fig. 4.18. Biosynthetic pathway of the betalain infraclass of the L-tyrosine-derived alkaloid subclass, showing structural variance. CYP76AD1 = cytochrome P450 enzyme, DODA = 4,5-DOPA dioxygenase, L-DOPA = L-3,4-dihydroxyphenylalanine.

uncharacterised free-radical mechanism. The isoquinolines predominate in the Cactaceae (*Lophophora*), Chenopodiaceae and Fabaceae plant families, but have also been reported from the Alangiaceae, Musaceae, Nymphaceae and Sterculiaceae.[31] Many of the genera that have isoquinolines of any kind are also abundant in their often psychoactive β-phenylethylamine protoalkaloid precursors (e.g. mescaline in the *Lophophora*).[65,66,77] Conversely, isoquinolines are not known to be psychoactive, although have been implicated in cardiovascular and antineoplastic activity.[1] The distribution of alkaloid-type specialised metabolites of the isoquinoline infraclass reported from the customary medicinal flora of Australia is presented in *Table 4.1*.

Benzylisoquinoline

The more than 4,000 isolated BIQs arise from the condensation of dopamine and 4-hydroxyphenylacetaldehyde (which contributes this sub-infraclasses pendant benzyl moiety) to give the important precursor norcoclaurine via a Pictet-Spengler condensation facilitated by norcoclaurine synthase.[1,19] Norcoclaurine is then sequentially *O*-methylated, *N*-methylated, hydroxylated and *O*-methylated again by the enzymes 4-*O*-methyltransferase (4OMT), coclaurine methyltransferase (COMT), a P450 hydroxylase (CYT80B1) and 4-*O*-methyltransferase (4OMT), respectively, to yield the significant biogenic branching point (*S*)-reticuline (*Fig. 4.16*). This large sub-infraclass of often toxic compounds is further divided into smaller groupings, the simplest of which are the THBIQs, of which 100 have been reported and often form the branching point for more complex representatives of the isoquinoline infraclass.

The THBIQs are isolated primarily from the Annonaceae (*Annona*), Berberidaceae (*Berberis*), Lauraceae (*Cinnamomum*), Papaveraceae (*Argemone*, *Papaver*) and Rhamnaceae (*Phylica*).[1] The most important (and widespread) THBIQ is (*S*)-reticuline, which is the precursor to the aporphine-type, morphine-type, berberine-type THBIQs and several other isoquinoline sub-infraclasses (*Fig. 4.16*).[1,19] The simplest THBIQ derivatives are the aporphines, of which ~650 have been isolated and are present in generally the same plant families as the other BIQs, particularly the Annonaceae, Aristocholaceae, Euphorbiaceae, Eupomatiaceae, Lauraceae, Magnoliaceae, Menispermaceae, Monimiaceae, Nelumbonaceae, Papaveraceae and the Ranunculaceae.[1] The aporphines are generally toxic (like most of the BIQs), largely as a result of their antagonistic effects on the dopaminergic system receptors, causing narcosis and convulsions The distribution of alkaloid-type specialised metabolites of the BIQ infraclass reported from the customary medicinal flora of Australia is presented in *Table 4.1*.

Morphine-type benzylisoquinolines

The morphine-type BIQs derived from a select few members of the genus *Papaver* of the Papaveraceae family (primarily *Papaver somniferum* L., 1753 – the opium poppy) are some of the most significant NPs in humanity's historical relationship with nature. Classical authors such as Theophrastus (371–287 BCE) and Dioscorides (40–90 CE) make mention of the intoxicating effects of this peculiar plant species. The 16th century CE Swiss-German alchemist Paracelsus is credited with the preparation of the early analgesic pill known as 'laudanum', that was notable for its dubious ingredients (pearls, musk, amber, nutmeg and 'possibly opium') and probable lack of morphine-type BIQs, an oversight that was rectified over a century later by English physician Thomas Sydenham, who standardised the opiate content of the tincture (but retained the name).[14] In 1805 CE, Friedrich Sertürner was the first scientist to isolate a pharmacologically active pure NP from nature, morphine.[79] Crude opium (12–14% morphine) is obtained by lacerating the immature capsules of the opium poppy, which allows the latex to exude and then dry. Over 1,000 tons of opium is legally obtained in this way each year, of which ~65% is used to produce codeine and ~20% is used to produce medical-grade morphine. Numerous attempts have been made to transition the natural laboratory that is the opium poppy into the realm of biotechnological factories, but none of these have been successful to date. The oxidative coupling seen to arise in the aporphine-type BIQs was the clue needed to understand the biosynthetic pathway of the morphine alkaloids thebaine, codeine and morphine, from the (-)-(*R*)-reticuline precursor mentioned above (*Fig. 4.16* and *Fig. 4.19*).

The myriad biological effects of the morphine-type BIQ alkaloids on the CNS are challenging to summarise, stimulating some pathways and depressing others. The classic symptoms of morphine intoxication (besides its notable antitussive properties) are analgesia (without loss of consciousness), drowsiness, mood change and mental haze, with death due to respiratory arrest if the lethal dose is exceeded. Most of these symptoms are due to interactions with opiate receptors in the brain that mimic the endogenous opioid-type ligands of the vertebrate nervous system. Three classes of opioid-type G

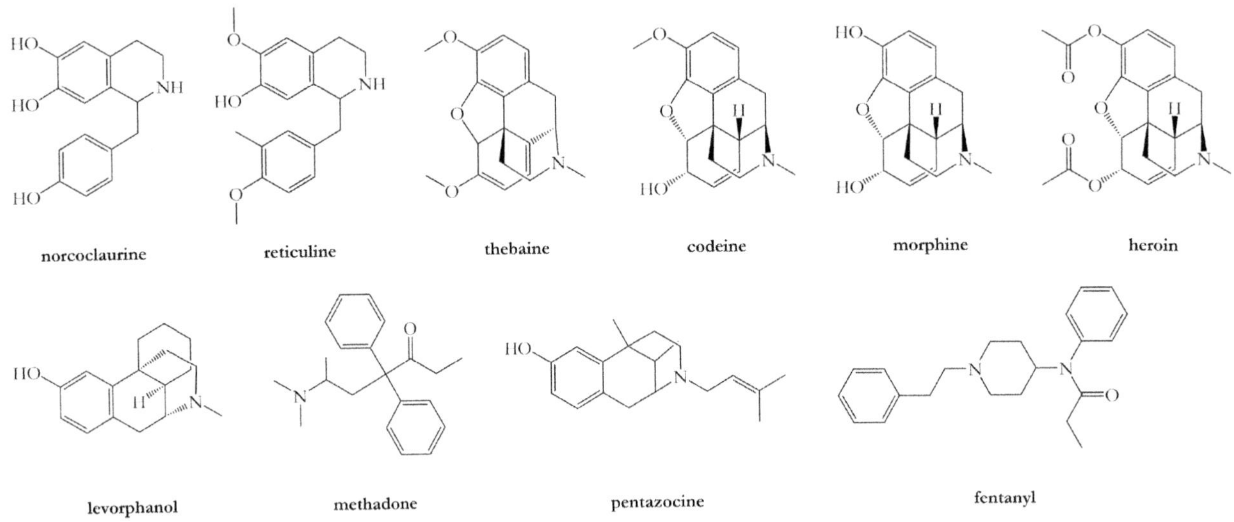

Fig. 4.19. Key representatives of the benzylisoquinoline infraclasses of the L-tyrosine/L-phenylalanine-derived subclass of alkaloids (both natural products and pharmaceutical drugs).

protein-coupled receptors are distinguished, originally named delta, kappa and mu, renamed in 1996 to OP1, OP2 and OP3, respectively, and again in 2000 to DOP/δ, KOP/κ and MOP/μ, respectively.[80] The opioid receptors are widely distributed in the CNS and more specifically in the PNS [vas deferens, knee joint, gastrointestinal (GI) tract, heart and immune systems, etc.]. The discovery of the endogenous ligands of the opiate receptors and their three pro-hormone precursors (proenkephalin, which breaks down to form the DOP-ligands met- and leu-enkephalin; prodynorphin, which breaks down to form the KOP-agonists Dynorphin A and B; pro-opiomelanocortin, which breaks down to form the MOP-agonist β-endorphin, which can also bind DOP and KOP receptors) has greatly advanced our mechanistic understanding of how these potent drugs work in vivo.[80] The notable drug of abuse, heroin, a synthetic diacetate of morphine, crosses the blood-brain barrier more readily than the natural morphine-type isoquinolines, which leads to intense euphoria, analgesia and consequent addiction.[80] Synthetic derivatives of the opioids proliferated during the 20th century and are typically divided into four categories with wide-ranging and often potent biological activities: morphinan (e.g. levorphanol), diphenylheptane (e.g. methadone), benzomorphan (e.g. pentazocine) and phenylpiperidine (e.g. fentanyl) (*Fig. 4.19*).[80]

Berberine-type benzylisoquinolines

The ~100 berberine-type (and protoberberine-type) BIQs are a widely distributed form of the BIQ sub-infraclass of alkaloid-type NPs. These specific BIQs have been reported from the Alangiaceae, Annonaceae, Berberidaceae, Fabaceae (*Erythrina*), Fumariaceae, Juglandaceae, Lauraceae, Menispermaceae, Papaveraceae, Ranunculaceae, Rubiaceae and Rutaceae.[1,2] These BIQs differ from the sub-infraclasses previously described because they arise biosynthetically from (+)-(*S*)-reticuline and are notable in that the most primitive berberine-type BIQ (*S*)-scoulerine and its derivatives feature a so-called 'berberine bridge' derived from methionine (*Fig. 4.19*).[19] Berberine-type BIQs are notable antiviral/antiparasitic compounds (IC_{50} against *Plasmodium falciparum* Welch 1897 of 0.14–0.36 μg/mL) and have also demonstrated a wide range of biological activities: antimicrobial, trypanocidal, antiamoebic, antineoplastic, antidiarrheal, antifeedant and tuberculostatic.[1,2] Mechanistically, the berberine-type BIQs have been shown to inhibit reverse transcriptase, intercalate DNA, inhibit aldol reductase, acetylcholinesterase, alcohol dehydrogenase, aldehyde reductase, diamine oxidase, tyrosine decarboxylase, HIV reverse transcriptase and RNA synthesis.[1,2]

Bisbenzylisoquinolines

Approximately 400 BIQ dimers (BBIQs) have been isolated from the plant kingdom, with intermolecular coupling (typically one or two diphenylether bridges) between different monomeric isoquinoline units being observed (an example of which, proceeding through the P450 berbamine synthase enzyme (CYP80A1), is shown in *Fig. 4.19*).[1,2,31] BBIQs are isolated from many of the same families as the BIQs described above. BBIQs have been submitted to a range of biological activity assays, with moderate to weak antimicrobial activity across the sub-infraclass.[1] Aporphine-BIQ BBIQs have shown antineoplastic activity, but with significant toxicity.[1] These side effects are very likely the same sort of toxicity that is being exploited in the poison known as curare used to tip arrows in South and Central America. Curare has now become a stand-in term for any poison used to tip hunting implements worldwide, but the original term comes from the Carib language of Guyana.[1] Three types of curare are now delineated: tube/bamboo [primarily BBIQs (tubocurarine, etc.) derived from the genus *Chondrodendron* (Menispermaceae) and the most toxic type based on LD_{50}], pot curare [mixed compounds from the Menispermaceae (BBIQs) and the Loganiaceae (active ingredients are strychnine-type MIAs addressed in Chapter 5] and ground curare (strychnine-type MIAs derived from Loganiaceae alone, also addressed in Chapter 5). Tubocurarine is inactive orally, but injection of it, or many of the curare-type BBIQs, rapidly interferes with the CNS, causing respiratory failure and death (LD_{50} of 33.2 mg/kg in mice).[1]

L-HISTIDINE-DERIVED ALKALOIDS

Imidazole

The imidazole-type alkaloids arise directly from L-histidine and are a relatively small group of alkaloids in the plant kingdom, although they are abundant in microorganisms and marine invertebrates.[81] The most notable imidazoles are the ubiquitous hormone/neurotransmitter histamine and the peripheral stimulant of the parasympathetic nervous system pilocarpine, which is derived from *Pilocarpus jaborandi* Holmes, 1892 - (Rutaceae) and is used clinically as a myotic (pupil constrictor, typically to counter the mydriatic effect of atropine) and to stimulate saliva/sweat to treat the dry mouth associated with Sjögren syndrome or radiation therapy (*Fig. 4.20*).[1] Imidazoles have been reported in the Acanthaceae (*Macrorungia*), Cactaceae (*Dolicothele*), Euphorbiaceae (*Alchornea*, *Glochidion*), Fabaceae (*Cassia*, *Cynometra*), Orchidaceae (*Dendrobium*), Rutaceae (*Casimiroa*, *Pilocarpus*) and Urticaceae (*Cypholophus*) plant families.[1,82] The biosynthetic origin of the imidazoles is through a simple decarboxylation of the proteinogenic amino acid L-histidine, which already bears an imidazole scaffold, to the important biomolecule histamine.[1] Synthetic drugs that incorporate the imidazole moiety are the azole antifungals (e.g. ketoconazole), which inhibit the conversion of lanosterol to ergosterol in fungi by inhibition of lanosterol 14α-demethylase.[83] The distribution of alkaloid-type specialised metabolites of the L-histidine derived subclass reported from the customary medicinal flora of Australia is presented in *Table 4.1*.

'PSEUDOALKALOIDS'

The term pseudoalkaloid is no longer in use, because the stipulation that these alkaloids 'did not originate from an amino acid' is merely a matter of perspective, as many of these metabolic pools are interrelated and what was once an amino acid (such as L-tryptophan or L-aspartate) can be broken down and recycled into precursors that would not be considered an amino acid (such as anthranilic acid or nicotinic acid). Despite this fact, it is convenient to discuss these compounds here (rather than with their amino acid group), as their biosynthesis, distribution in the plant kingdom and biological activities are diverse enough to warrant separate categorisation. The three main metabolic sources of 'pseudoalkaloids' are: anthranilic acid, nicotinic acid and purine.[1,2]

Anthranilic acid-derived pseudoalkaloids

Excepting the fact that L-tryptophan is biosynthetically derived from anthranilic acid, several other important pseudoalkaloid-type alkaloids arise directly from anthranilic acid, particularly in the Rutaceae.[1,2,84] The most clearly defined infraclass is the quinazolines, but also certain quinolines arise directly from anthranilic acid (rather than the quinine-type terpenoid indole alkaloid with coincidental quinoline scaffolds, derived from L-tryptophan and discussed in Chapter 5), as well as several other anthranilic acid-derived alkaloids (quinolone, furoquinoline, pyranoquinoline, acridone, etc.). Approximately 80 quinazoline [including pyrroloquinazoline and quinazolinocarboline (Rutaceae only)] alkaloids have been reported from the following plant families: Acanthaceae (*Anisotes*, *Justicia*), Araliaceae (*Mackinlaya*), Arecaceae (*Daemonorops*), Fabaceae (*Galega*), Hydrangeaceae (*Dichroa*, *Hydrangea*), Malvaceae (*Sida*), Rutaceae (*Euxylophora*, *Glycosmis*, *Hortia*, *Tetradium*, *Zanthophylum*), Scrophulariaceae (*Linaria*) and Zygophyllaceae (*Peganum*; co-occurring with the harmaline-type β-carbolines typical of this genus).[13,85] The remaining subtypes of anthranilic acid-derived alkaloids number in the hundreds and are largely derived from the Rutaceae, with the major groups being quinolone, furoquinoline, pyranoquinoline and acridone (*Fig. 4.21*).

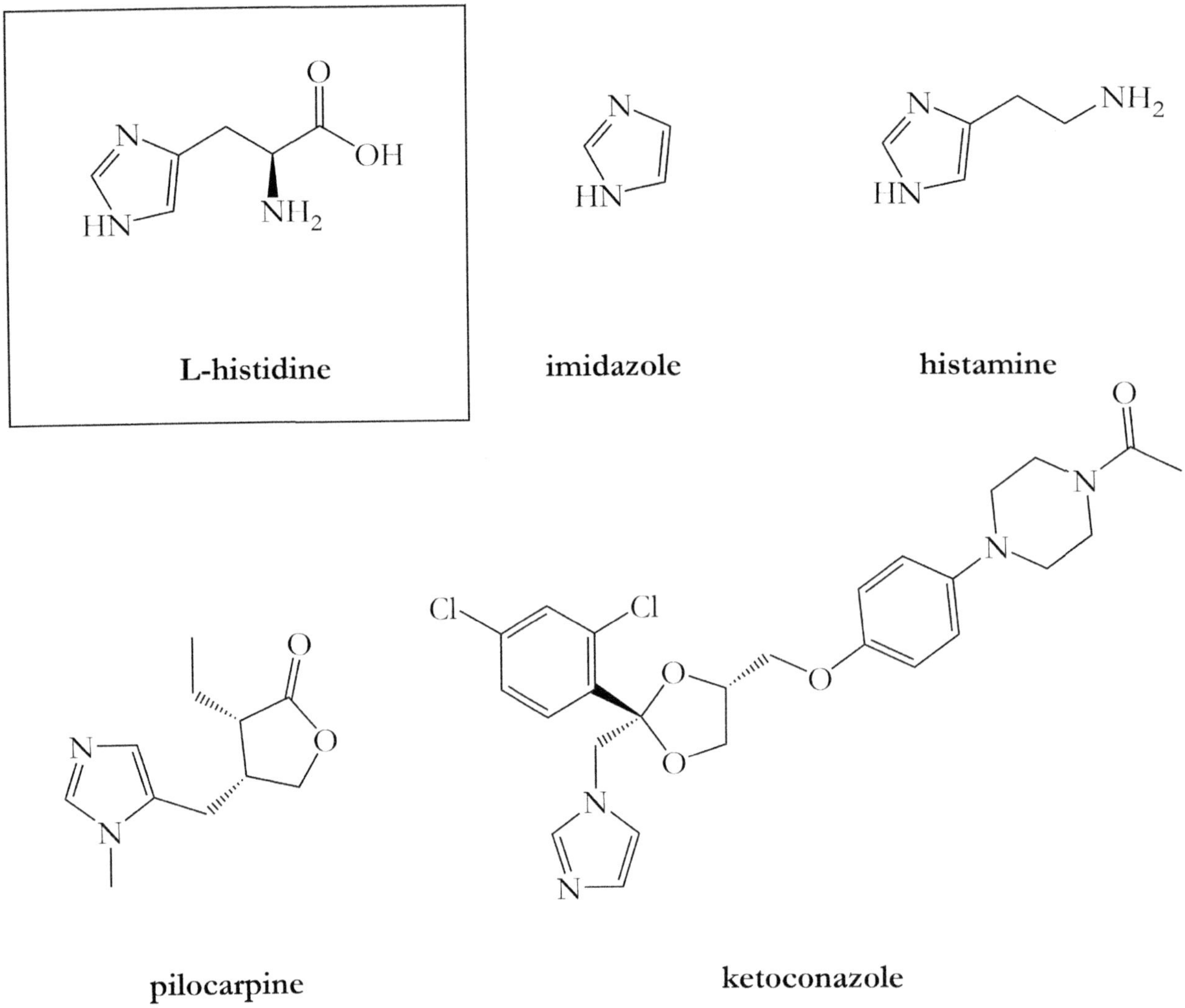

Fig. 4.20. Key representatives L-histidine derived subclass of alkaloids (both natural products and pharmaceutical drugs).

The distribution of these anthranilic acid-derived alkaloids (as well as BIQs, imidazoles and various indole-based meroterpenoids) among the Rutaceae has led to significant chemosystematic phylogeny, which lies outside the scope of this book.[16] The biosynthesis of these compounds is simple, with anthranilic acid providing the core and an extension group or groups (condensing with the carboxylic acid moiety, before cyclisation) providing the rest of the structure. The quinazolines have been shown to obtain their second nitrogen heterocycle from aspartic acid, although other nitrogen-bearing candidates have been proposed, while the quinolines/quinolones (and their prenylated derivatives), as well as the acridones arise from the introduction (and subsequent cyclisation) of one or multiple units of acetate/malonate (*Fig. 4.21*).[1,13]

In terms of biological activity, quinazolines are the most significant type of anthranilic acid-derived 'pseudoalkaloid', with over 50 quinazoline derivatives in clinical use. The vasicine/vasicinone quinazoline skeleton, isolated primarily from *Justicia adhatoda* L., 1753 – (Acanthaceae), has a significant effect on the respiratory and circulatory systems, as well as antimicrobial, anthelmintic and antiperiodic bioactivities (*Fig. 4.22*).[1,13] The quinazoline, febrifugine, has ~100-fold the antimalarial activity of quinine against *Plasmodium* spp., but its toxicity precludes its use clinically. Although it is derived from the animal kingdom (e.g. from *fugu* fish of the *Diodon*, *Lagocephalus*, *Takifugu*, *Sphoeroides* genera), tetrodotoxin features a quinazoline moiety and is highly significant as one of the most potent low molecular weight neurotoxins known to science. The mechanism

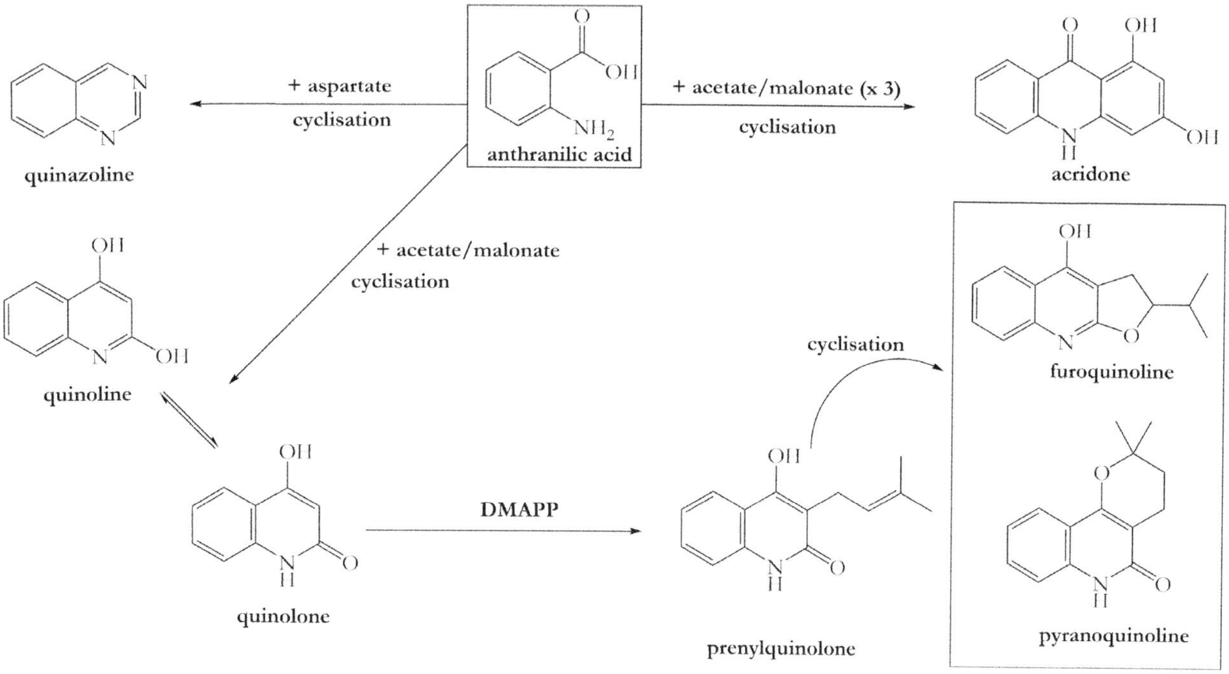

Fig. 4.21. Biosynthetic pathway of the anthranilic acid derived infraclasses of the 'pseudoalkaloid' subclass, showing structural variance. DMAPP = dimethylallylpyrophosphate.

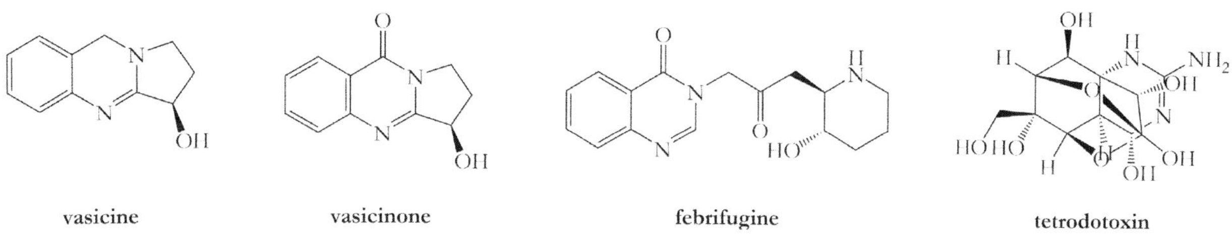

Fig. 4.22. Key representatives of the quinazoline-type infraclass of the anthranilic acid-derived 'pseudoalkaloid' subclass of alkaloids (both natural products and pharmaceutical drugs).

of action of febrifugine involves binding to voltage-gated sodium channels, preventing the passage of sodium ions and effectively stopping neural information and muscle contraction (*Fig. 4.22*).[1,13]

Anthranilic acid-derived quinolones show similar bioactivities to the quinolines discussed at length in Chapter 5. Pyranoquinolines and furoquinolines have demonstrated antifeedant activities, DNA intercalation and photoaddition and phototoxicity to both bacteria and fungi, while acridones have shown antineoplastic activity across several cell lines, interfering with nucleoside transport across plasma membranes, together with antimalarial and antimicrobial activities.[1] The quinazolines are poorly represented among the customary medicinal plant genera of Australia. The distribution of alkaloid-type specialised metabolites of the quinoline infraclass reported from the customary medicinal flora of Australia is presented in *Table 4.1*.

Nicotinic acid-derived pseudoalkaloids

The pyridine ring found in the ~250 pyridine alkaloids isolated from the plant kingdom can result from the direct decarboxylation of nicotinic acid (also known as the essential human nutrient, niacin), aromatisation of the piperidine alkaloids and some other more obscure pathways that lie outside the scope of this book (*Fig. 4.23*).[1,31] The main pathway involved in pyridine biosynthesis proceeds from glyceraldehyde-3-phosphate and aspartic acid, which condense to form quinolinic acid and then on to the important primary metabolic coenzyme nicotinamide adenine dinucleotide, which is the origin of the pyridine alkaloids.[1,17] The most significant member of the greater

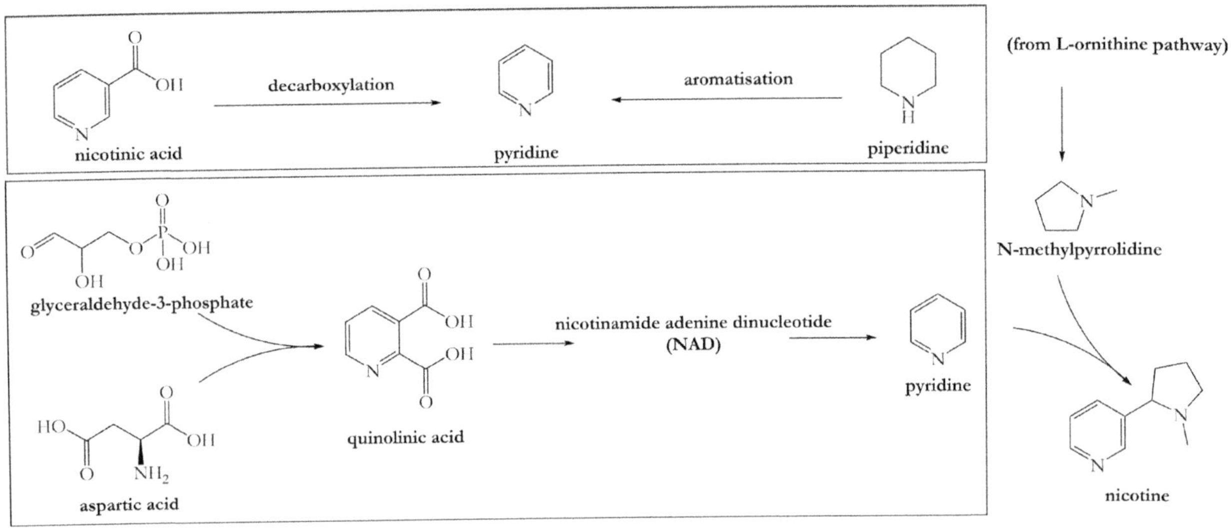

Fig. 4.23. Biosynthetic pathway of the nicotinic acid derived infraclasses of the 'pseudoalkaloid' subclass, showing structural variance.

pyridine alkaloids is nicotine, which is widely distributed but primarily associated with the Solanaceae (*Nicotiana* spp. where it can represent up to 10% of the plant's total metabolism).[86] Nicotine is formed from the condensation of a nicotinic acid unit with an *N*-methylpyrrolidine unit and functions as an antifeedant in the form of both a constitutive phytoanticipin and a readily induced phytoalexin (4-fold increase after leaf-damage).[1]

Several other significant pyridine alkaloids are anabasine [Solanaceae (*Duboisia, Nicotiana*)], ricinine [Euphorbiaceae (*Ricinus*)], arecoline [Arecaceae (*Areca*)] and dioscorine [Dioscoreaceae (*Dioscorea*)], which are all toxic to animals due to their inhibition of the nicotinic AChR in the CNS.[87] This CNS activity is behind the ~8,000 years of historical use of tobacco (*Nicotiana* spp.) among Native American groups and mostly undocumented but similarly ancient traditional use of pituri (*Duboisia* spp.) among Indigenous Peoples.[87,88] Importantly, this CNS activity attributed to pyridine alkaloids may have significance regarding the development of new drugs to treat CNS disorders such as neurodegenerative disease, psychiatric conditions, epilepsy, multiple sclerosis, etc.[87] The distribution of alkaloid-type specialised metabolites of the pyridine infraclass reported from the customary medicinal flora of Australia is presented in *Table 4.1.*

Purine-derived pseudoalkaloids

Both purine and pyrimidine bases have been isolated in their free form from the plant kingdom, as well as glycosylated NPs. Purine 'pseudoalkaloid' NPs are the more important of the two, typically deriving from a xanthine possessing an *N*-alkyl substituent. The cytokinins are a group of plant primary metabolites that are structurally and metabolically represented by zeatin (*Fig. 4.24*). Three common purine-derived methylxanthine NPs that are of great importance in human society are caffeine (1,3,7-trimethylxanthine), theophylline (1,3-dimethylxanthine) and theobromine (3,7-dimethylxanthine) (*Fig. 4.24*).[1] Caffeine is the world's most widely consumed psychoactive drug, primarily sourced from either coffee [Rubiaceae (*Coffea*)], tea [Theaceae (*Camellia*)], cocoa [Malvaceae (*Theobroma*)], cola [Malvaceae (*Cola*)] and yerba mate [Aquafoliaceae (*Ilex*)].[89,90]

Theophylline and theobromine are demethylated breakdown products of the trimethylxanthine, caffeine, a phenomenon that occurs both *in planta* and in the human liver [alongside 1,7-dimethylxanthine (paraxanthine)]. These three methylxanthines arise biosynthetically from 7-methylxanthine through the action of various methyltransferases utilising either methionine or S-adenosylmethionine as the methyl-source.[90] Xanthine derivatives show five major pharmacological activities (CNS/respiratory stimulation, skeletal muscle stimulation, diuresis, cardiac stimulation, smooth muscle relaxation), with these effects being mechanistically rationalised via both competitive non-selective phosphodiesterase inhibitors (inflammatory cascade inhibition) and non-selective adenosine receptor antagonists (neurostimulatory and cardiac effects).[1,90]

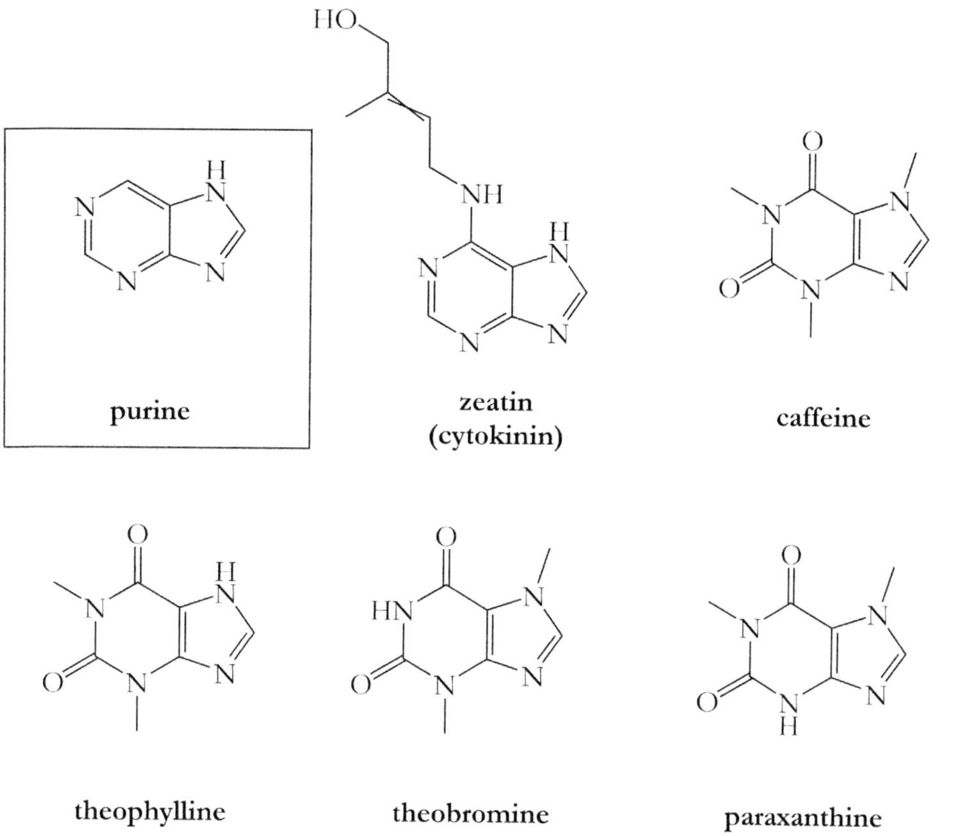

Fig. 4.24. Key representative specialised metabolites of the purine-derived 'pseudoalkaloid' subclass of alkaloids.

REPRESENTATION

This section presents a synopsis of the specialised metabolic trends contained within the flora presented in *Table 4.1*, with a focus on any emergent trends in therapeutic use that correlate with reported biological activity in the sub-/infraclass.

L-Ornithine derived

Pyrrolidine

A summary of the Australian customary medicinal flora reported to bear pyrrolidine-type compounds of the L-ornithine-derived subclass of alkaloids is as follows: Asparagales [Orchidaceae (*Dendrobium*)], Brassicales [Capparaceae (*Capparis*)], Caryophyllales [Amaranthaceae (*Achyranthes*)], Ericales [Primulaceae (*Aegiceras*)], Fabales [Fabaceae (*Mucuna*)], Pandanales [Pandanaceae (*Pandanus*)], Piperales [Piperaceae (*Piper*)], Ranunculales [Menispermaceae (*Tinospora*)], Rosales [Rhamnaceae (*Ziziphus*)] and Solanales [Convolvulaceae (*Evolvulus*, *Xenostegia*), Solanaceae (*Nicotiana*, *Solanum*)]. The trend in customary use across these plant genera was moderate, with ~40% of these plants used for inflammation or other analgesic applications.[51] Various therapeutic agents bearing a pyrrolidine moiety have been reported to bear anti-inflammatory properties and appear in 37 FDA-approved drugs in clinical use today. Steric factors have a strong bearing on the bioactivity of these compounds and the enantioselectivity of proteins causes substantial SAR differentiation between the derivatives of this alkaloid subclass.[91]

Pyrrolizidine

A summary of the Australian customary medicinal flora reported to bear pyrrolizidine-type compounds of the L-ornithine-derived subclass of alkaloids is as follows: Asparagales [Asparagaceae (*Asparagus*)], Asterales [Asteraceae (*Ageratum*)], Boraginales [Boraginaceae (*Ehretia*, *Heliotropium*, *Trichodesma*)], Ericales [Sapotaceae (*Planchonella*)], Fabales [Fabaceae (*Crotalaria*)], Fagales [Casuarinaceae (*Casuarina*)], Malphigiales [Phyllanthaceae (*Phyllanthus*)] and Solanales [Convolvulaceae (*Xenostegia*)].

The only trend in customary use evident across plants from these genera is a 50% use-rate in bactericidal applications.[51] Pyrrolizidines are noted for their acute toxicity (hepato-, geno-, neuro-) and are widely studied with regard to their antimicrobial activity, particularly from the genera *Heliotropium*, *Crotalaria* and *Senecio*. A particular synthetic compound of this subclass (PA-1) demonstrated broadspectrum antimicrobial activity, with MBC as low as 0.0078 mg/mL against *Escherichia coli* and *Staphylococcus aureus*. It is proposed that this compound's mechanism of action was through interaction with membrane phospholipids, compromising the integrity of the cells protective barrier and leading to lysis.[92]

Tropane

A summary of the Australian customary medicinal flora reported to bear tropane-type compounds of the L-ornithine-derived subclass of alkaloids is as follows: Asparagales [Asparagaceae (*Asparagus*)], Asterales [Asteraceae (*Ageratum*)], Boraginales [Boraginaceae (*Ehretia*, *Heliotropium*, *Trichodesma*)], Ericales [Sapotaceae (*Planchonella*)], Fabales [Fabaceae (*Crotalaria*)], Fagales [Casuarinaceae (*Casuarina*)], Malphigiales [Phyllanthaceae (*Phyllanthus*)] and Solanales [Convolvulaceae (*Xenostegia*)]. As with the genera bearing pyrrolizidine-type L-ornithine derived alkaloids, these plants have a 50% customary use-trend as bactericidal agents.[51] Tropane-bearing therapeutic agents have been notably used as antispasmodics (hyoscine butylbromide, trospium chloride), mydriatics (atropine), antivirals (maraviroc) and bronchodilators (ipratropium). Importantly, concerning their customary use, the antibacterial activity of compounds bearing tropane moieties such as peuromotlin and retapamulin has been well substantiated. Peuromotlin has been shown to inhibit bacterial protein production and its safer derivative, retapamulin, is used topically to treat impetigo.[93]

L-Lysine derived

Piperidine

A summary of the Australian customary medicinal flora reported to bear piperidine-type compounds of the L-lysine-derived subclass of alkaloids is as follows: Alismatales [Araceae (*Alocasia*)], Asterales [Campanulaceae (*Lobelia*)], Caryophyllales [Amaranthaceae (*Dysphania*), Portulaceae (*Portulaca*)], Fabales [Fabaceae (*Cassia*, *Chamaecrista*, *Senna*)], Lycopodiales [Lycopodiaceae (*Lycopodium*)], Malphigiales [Euphorbiaceae (*Excoecaria*)], Pandanales [Pandanaceae (*Pandanus*)] and Piperales [Piperaceae (*Piper*)]. There appears to be no clear trends in the customary use of plants from these genera. Clinical applications and research into drugs bearing piperidine moieties varies, as they are one of the most common fragments appearing in FDA-approved drugs, with evidence of antineoplastic, antimicrobial, analgesic, antiinflamatory and antipsychotic activities.[94]

Quinolizidine

A summary of the Australian customary medicinal flora reported to bear quinolizidine-type compounds of the L-lysine-derived subclass of alkaloids is as follows: Caryophyllales [Nyctaginaceae (*Boerhavia*)], Fabales [Fabaceae (*Sophora*, *Templetonia*)] and Lycopodiales [Lycopodiaceae (*Lycopodium*)]. Two of these genera (*Boerhavia*, *Sophora*) have been reported to be in use customarily for stomach ailments or digestive issues. Quinolizidine-bearing therapeutics have been implicated in a wide range of pharmacological effects, but their implicit toxicity is a concern, particularly regarding their teratogenicity.[95]

Indolizidine

A summary of the Australian customary medicinal flora reported to bear indolizidine-type compounds of the L-lysine-derived subclass of alkaloids is as follows: Asparagales [Orchidaceae (*Dendrobium*)], Fabales [Fabaceae (*Swainsona*)], Gentianales [Apocynaceae (*Tylophora*)], Laurales [Lauraceae (*Litsea*)], Malphigiales [Phyllanthaceae (*Flueggea*, *Phyllanthus*, *Securinega*)], Malvales [Malvaceae (*Sida*)] and Solanales [Convolvulaceae (*Ipomoea*)]. A moderate trend is observed across the customary use applications of plants from these genera, with 56% being used for their analgesic properties.[51,96]

L-Tryptophan derived

Indole

A summary of the Australian customary medicinal flora reported to bear indole-type compounds of the L-tryptophan-derived subclass of alkaloids is as follows: Alismatales [Araceae (*Alocasia*, *Rhaphidophora*)], Asparagales [Amaryllidaceae (*Crinum*)], Brassicales [Capparaceae (*Capparis*)], Caryophyllales [Amaranthaceae (*Atriplex*, *Dysphania*), Portulaceae (*Portulaca*)], Ericales [Primulaceae (*Aegiceras*)], Fabales [Fabaceae (*Abrus*, *Acacia*, *Entada*, *Erythrophleum*, *Indigofera*, *Mucuna*, *Vigna*)], Gentianales [Apocynaceae (*Alstonia*, *Tabernaemontana*), Loganiaceae (*Strychnos*), Rubiaceae (*Nauclea*, *Timonius*)], Lamiales [Plantaginaceae (*Scoparia*)], Malphigiales [Phyllanthaceae (*Flueggea*, *Securinega*)],

Malvales [Malvaceae (*Grewia*, *Sida*)], Ranunculales [Ranunculaceae (*Clematis*)] and Sapindales [Rutaceae (*Flindersia*, *Melicope*), Simaroubaceae (*Brucea*)]. This broad set of genera does not have any outstanding trends in customary use, but there is a moderate level of application as bactericidal (40%), anti-inflammatory (37%) and analgesic (30%) agents.[51] The ability of the indole scaffold to bind with high affinity to multiple receptors has led to their widespread use in the development of novel therapeutics with broad pharmacological application. Antimicrobial, anti-inflammatory, analgesic, antiviral and antineoplastic biological activities are investigated substantially and reviewed in the literature.[97]

β-Carboline

A summary of the Australian customary medicinal flora reported to bear β-carboline-type compounds of the L-tryptophan-derived subclass of alkaloids is as follows: Asparagales [Amaryllidaceae (*Crinum*)], Cucurbitales [Cucurbitaceae (*Trichosanthes*)], Ericales [Primulaceae (*Aegiceras*)], Fabales [Fabaceae (*Acacia*, *Mucuna*)], Gentianales [Rubiaceae (*Nauclea*)], Malphigiales [Euphorbiaceae (*Croton*)], Sapindales [Meliaceae (*Melia*), Simaroubaceae (*Ailanthus*, *Brucea*)] and Zygophyllales [Zygophyllaceae (*Tribulus*)]. There were no obvious trends across the recorded customary uses of these genera in the literature. Compounds bearing this moiety have been reported to bear a range of biological activities (antineoplastic, antibacterial, anti-inflammatory, etc.), with SAR studies showing specific activity enhancement with the modification of various structural centres (C1, C3, N2, N9).[98]

Physostigmine

None of the plant genera reported to have customary use in the literature have been found to bear compounds of the physostigmine-type of the L-tryptophan-derived subclass of alkaloids.

Other non-terpenoid indole derivatives

Canthin-6-one

A summary of the Australian customary medicinal flora reported to bear canthin-6-one-type compounds of the L-tryptophan-derived subclass of alkaloids is as follows: Sapindales [Simaroubaceae (*Ailanthus*, *Brucea*). These two plant genera are both reported to be in use customarily for fever and diarrhoea.[51] The documented antileishmanial, antimalarial and antimicrobial activities of these compounds could be implicated in these applications. Mechanistically, their antifungal properties are attributed to interference with fungal fatty acid metabolism, their antibacterial properties are attributed to targeting bacterial type-II topoisomerase DNA gyrase (GyrA) and their antiviral properties are attributed to the inhibition of viral replication/transcription through binding Mpro/3CLpro and PLpro.[99]

Carbazole

A summary of the Australian customary medicinal flora reported to bear carbazole-type compounds of the L-tryptophan-derived subclass of alkaloids is as follows: Sapindales [Anacardiaceae (*Buchannia*)]. This plant genus is reported to be used customarily as an emollient (moisturiser), for toothache and for eye ailments, uses that are difficult to substantiate based on biological activity.[51] There are a variety of carbazole-based drugs on the market (carvedilol, carprofen, carazostatin, ellipticine, ollivacine, datelliptium, alectinib, celiptium, etc.). Three carbazole-derivatives (ellipticine, alectinib, midostaurin) have been applied as antineoplastic therapeutics due to their downregulation of DNA topoisomerase-II. The carbazomycins are notable antibiotic agents through their inhibition of 5-lipoxygenase.[100]

Indolenine

None of the plant genera reported to have customary use in the literature have been found to bear compounds of the indolenine-type of the L-tryptophan-derived subclass of alkaloids.

L-Tyrosine- or L-phenylalanine derived

Simple amine

A summary of the Australian customary medicinal flora reported to bear simple amine-type compounds of the L-tyr/phe-derived subclass of alkaloids is as follows: Caryophyllales [Amaranthaceae (*Achyranthes*, *Atriplex*)], Ericales [Primulaceae (*Aegiceras*)], Fabales [Fabaceae (*Acacia*)], Malvales [Malvaceae (*Sida*)] and Polypodiales [Pteridaceae (*Adiantum*)]. Two-thirds of these plants have been reported to be in use customarily to treat coughs and colds and half of them have been reported to be used to treat stomach ailments and inflammation.[51] The wide variety of these compounds precludes any overarching summary of their biological activities, with a discussion of specific notable members of this infraclass outlined earlier.

Betalain

A summary of the Australian customary medicinal flora reported to bear betalain-type compounds of the L-tyr/phe-derived subclass of alkaloids is as follows: Caryophyllales [Nyctaginaceae (*Boerhavia*)]. The genus *Boerhavia* has been reported to be in use customarily as a treatment for stomach ailments and as a diuretic. Betalains are notable antioxidants and an extract of betalain-rich *Opuntia ficus indica* fruit attenuated the ulcerogenic effect of ethanol on the stomach in a rat study and had diuretic effects comparable to that of the clinically applied hydrochlorothiazide.[101]

Isoquinoline

A summary of the Australian customary medicinal flora reported to bear isoquinoline-type compounds of the L-tyr/phe-derived subclass of alkaloids is as follows: Asparagales [Amaryllidaceae (*Crinum*)], Caryophyllales [Portulaceae (*Portulaca*)], Fabales [Fabaceae (*Erythrina*, *Mucuna*)], Gentianales [Apocynaceae (*Cynanchum*)], Laurales [Atherospermataceae (*Atherosperma*, *Doryphora*), Hernandiaceae (*Hernandia*), Lauraceae (*Litsea*)], Piperales [Piperaceae (*Piper*)], Poales [Cyperaceae (*Cyperus*)] and Ranunculales [Menispermaceae (*Cissampelos*, *Tinospora*)]. There were no notable trends across the customary uses of these plant genera.

Benzylisoquinoline

A summary of the Australian customary medicinal flora reported to bear BIQ-type compounds of the L-tyr/phe-derived subclass of alkaloids is as follows: Asparagales [Amaryllidaceae (*Crinum*)], Caryophyllales [Amaranthaceae (*Achyranthes*)], Fabales [Fabaceae (*Erythrina*)], Laurales [Atherospermataceae (*Atherosperma*, *Daphnandra*, *Doryphora*), Hernandiaceae (*Gyrocarpus*, *Hernandia*), Lauraceae (*Cassytha*, *Litsea*)], Malphigiales [Euphorbiaceae (*Croton*)], Proteales [Nelumbonaceae (*Nelumbo*)] and Ranunculales [Menispermaceae (*Cissampelos*, *Tinospora*)]. There were no notable trends across the customary uses of these plant genera.

L-Histidine derived

Imidazole

A summary of the Australian customary medicinal flora reported to bear imidazole-type compounds of the L-histidine-derived subclass of alkaloids is as follows: Asparagales [Orchidaceae (*Dendrobium*)] and Fabales [Fabaceae (*Cynometra*)]. There was no overlap between the customary uses of these two plant genera.

'Pseudoalkaloids'

Quinazoline

A summary of the Australian customary medicinal flora reported to bear quinazoline-type compounds of the 'pseudoalkaloid' subclass of alkaloids is as follows: Malvales [Malvaceae (*Sida*)]. The genus *Sida* has been reported to be used customarily for inflammation, cough, stomach ailments, diarrhoea and bites/stings.[51] Therapeutic compounds bearing quinazoline moieties have been implicated in a wide range of biological activities, yet correlation between these activities and the customary use of these plants is not substantiated at this point.[102]

Quinoline

A summary of the Australian customary medicinal flora reported to bear quinoline-type compounds of the 'pseudoalkaloid' subclass of alkaloids is as follows: Gentianales [Rubiaceae (*Timonius*)], Malvales [Malvaceae (*Sida*, *Sterculia*)], Ranunculales [Menispermaceae (*Cissampelos*, *Tinospora*), Ranunculaceae (*Clematis*)] and Sapindales [Rutaceae (*Bosistoa*, *Melicope*, *Philotheca*, *Sarcomelicope*), Simaroubaceae (*Brucea*)]. There were no notable trends across the customary uses of these plant genera.

Pyridine

A summary of the Australian customary medicinal flora reported to bear pyridine-type compounds of the 'pseudoalkaloid' subclass of alkaloids is as follows: Alismatales [Araceae (*Amorphophallus*)], Asterales [Goodeniaceae (*Goodenia*)], Brassicales [Cleomaceae (*Cleome*)], Caryophyllales [Amaranthaceae (*Dysphania*), Portulaceae (*Portulaca*)], Fabales [Fabaceae (*Abrus*, *Cassia*, *Chamaecrista*, *Mucuna*, *Senna*, *Templetonia*)], Gentianales [Rubiaceae (*Coelospermum*, *Nauclea*)], Lamiales [Plantaginaceae (*Bacopa*)], Piperales [Piperaceae (*Piper*)] and Solanales [Solanaceae (*Duboisia*, *Nicotiana*)]. The reported customary use of these plant genera show two weak-moderate trends in therapeutic application, with 35% of these plants being used for inflammation (*Abrus*, *Bacopa*, *Cleome*, *Dysphania*, *Mucuna*, *Nauclea*) and a separate 35% being used as a narcotic (*Amorphophallus*, *Cassia*, *Chamaecrista*, *Duboisia*, *Nicotiana*, *Senna*), with no overlap between these two groups.[51]

Methylxanthine

A summary of the Australian customary medicinal flora reported to bear methylxanthine-type compounds of the 'pseudoalkaloid' subclass of alkaloids is as follows: Malphigiales [Euphorbiaceae (*Acalypha*), Violaceae (*Hybanthus*)] and Sapindales [Rutaceae (*Bosistoa*)]. There were no notable trends across the customary uses of these plant genera.

REFERENCES

1. Seigler DS (1998) *Plant Secondary Metabolism.* Springer, New York, NY.
2. Aniszewski T (2007) Definition, typology and occurrence of alkaloids. In *Alkaloids: Secrets of Life: Alkaloid Chemistry, Biological Significance, Applications and Ecological Role.* pp. 1–60. Elsevier, Oxford, UK.
3. Buckingham J (2010) *Dictionary of Alkaloids.* CRC Press, Boca Raton FL..
4. Wink M (2010) Introduction: biochemistry physiology and ecological functions of secondary metabolites. In *Biochemistry of Plant Secondary Metabolism.* (Ed. M Wink) pp. 1–19. Wiley-Blackwell, Sussex, UK.
5. Brielmann HL, Setzer WN, Kaufman PB, Kirakosyan A, Cseke LJ (2006) Phytochemicals: the chemical components of plants. In *Natural Products from Plants.* (Eds LJ Cseke A Kirakosyan, PB Kaufman, SL Warber, JA Duke *et al.*) pp. 1–50. CRC Press, Boca Raton, FL.
6. Roberts MF, Wink M (1998) Introduction. In *Alkaloids: Biochemistry, Ecology, and Medical Applications.* (Eds MF Roberts and M Wink) pp. 1–10. Springer Science+Business, New York, NY.
7. Wink M (1998) A short history of alkaloids. In *Alkaloids: Biochemistry, Ecology, and Medical Applications.* (Eds MF Roberts and M Wink) pp. 11–44. Springer Science+Business, New York, NY.
8. Aniszewski T (2007) Biological significance of alkaloids. In *Alkaloids: Secrets of Life: Alkaloid Chemistry, Biological Significance, Applications and Ecological Role.* pp. 141–180. Elsevier, Oxford, UK.
9. Wink M (1998) Chemical ecology of alkaloids. In *Alkaloids: Biochemistry, Ecology, and Medical Applications.* (Eds MF Roberts and M Wink) pp. 265–300. Springer Science+Business, New York, NY.
10. Wink M (1998) Modes of action of alkaloids. In *Alkaloids: Biochemistry, Ecology, and Medical Applications.* (Eds MF Roberts and M Wink) pp. 301–326. Springer Science+Business, New York, NY.
11. Wink M (1998) Antimicrobially active alkaloids. In *Alkaloids: Biochemistry, Ecology, and Medical Applications.* (Eds MF Roberts and M Wink) pp. 397–434. Springer Science+Business, New York, NY.
12. Schmeller T, Wink M (1998) Utilisation of alkaloids in modern medicine. In *Alkaloids: Biochemistry, Ecology, and Medical Applications.* (Eds MF Roberts and M Wink) pp. 435–460. Springer Science+Business, New York, NY.
13. Johne S (1984) The quinazoline alkaloids. In *Progress in the Chemistry of Organic Natural Products.* (Eds W Herz, H Grisebach, GW Kirby and C Tamm) pp. 159–219. Springer-Verlag, New York, NY.
14. Hodgson B (2001) *In the Arms of Morpheus: The Tragic History of Laudanum, Morphine, and Patent Medicines.* Firefly Books, Buffalo, NY.
15. Torrsell KBG (1983) *Natural Product Chemistry. A Mechanistic and Biosynthetic Approach to Secondary Metabolism.* Wiley & Sons, Chichester, UK.
16. Waterman PG, Grundon MF (1983) *Chemistry and Chemical Taxonomy of the Rutales.* Academic Press, London, UK.
17. Aniszewski T (2007) Alkaloid chemistry. In *Alkaloids: Secrets of Life: Alkaloid Chemistry, Biological Significance, Applications and Ecological Role.* pp. 61–140. Elsevier, Oxford, UK.
18. Roberts MF (1998) Enzymology of alkaloid biosynthesis. In *Alkaloids: Biochemistry, Ecology, and Medical Applications.* (Eds MF Roberts and M Wink) pp. 109–146. Springer Science+Business, New York, NY.
19. Roberts MF, Strack D, Wink M (2010) Biosynthesis of alkaloids and betalains. In *Biochemistry of Plant Secondary Metabolism.* (Ed. M Wink) pp. 20–91. Wiley-Blackwell, Sussex, UK.
20. Schramm S, Kohler N, Rozhon W (2019) Pyrrolizidine alkaloids: biosynthesis, biological activities and occurrence in crop plants. *Molecules (Basel, Switzerland)* **24**, 498. doi:10.3390/molecules24030498
21. Spenser ID (1985) Stereochemical aspects of the biosynthetic routes leading to the pyrrolizidine and quinolizidine alkaloids. *Pure and Applied Chemistry* **57**, 453–470. doi:10.1351/pac198557030453
22. Grynkiewicz G, Gadzikowska M (2008) Tropane alkaloids as medicinally useful natural products and their synthetic derivatives as new drugs. *Pharmacological Reports* **60**, 439–463.
23. Massiot G, Delaude C (1986) Pyrrolidine alkaloids. In *The Alkaloids.* (Ed. A Brossi) pp. 270–322. Academic Press, New York, USA.
24. Pinder AR (1993) Pyrrole, pyrrolidine, pyridine, piperidine, and related alkaloids. In *Alkaloids and Sulphur Compounds.* (Ed. PG Waterman) pp. 241–269. Academic Press, London, UK.
25. Li Petri G, Raimondi MV, Spano V, Holl R, Barraja P *et al.* (2021) Pyrrolidine in drug discovery: a versatile scaffold for novel biologically active compounds. *Topics in Current Chemistry* **379**, 34. doi:10.1007/s41061-021-00347-5
26. Islam MT, Mubarak MS (2020) Pyrrolidine alkaloids and their promises in pharmacotherapy. *Advances in Traditional Medicine* **20**, 13–22. doi:10.1007/s13596-019-00419-4
27. Rakba N, Melhaoui A, Loyer P, Delcros JG, Morel I *et al.* (1999) Bgugaine, a pyrrolidine alkaloid from *Arisarum vulgare*, is a strong hepatotoxin in rat and human liver cell cultures. *Toxicology Letters* **104**, 239–248. doi:10.1016/S0378-4274(98)00375-0

28. Smith LW, Culvenor CCJ (1981) Plant sources of hepatotoxic pyrrolizidine alkaloids. *Journal of Natural Products* **44**, 129–152. doi:10.1021/np50014a001
29. Wink M (2019) Quinolizidine and pyrrolizidine alkaloid chemical ecology: a mini review on their similarities and differences. *Journal of Chemical Ecology* **45**, 109–115. doi:10.1007/s10886-018-1005-6
30. Wesseling A-M, Demetrowitsch TJ, Schwarz K, Ober D (2017) Variability of pyrrolizidine alkaloid occurrence in species of the grass subfamily Pooideae (Poaceae). *Frontiers in Plant Science* **8**, 2046. doi:10.3389/fpls.2017.02046
31. Verpoorte R, Van Der Heuden R, Van Gulik WM, Ten Hoopen HJG (1991) Plant biotechnology for the production of alkaloids. In *The Alkaloids*. (Ed. A Brossi) pp. 1–188. Academic Press, New York, NY.
32. Hartmann T, Ehmke A, Eilert U, Von Borstel K, Theuring C (1989) Sites of synthesis, translocation and accumulation of pyrrolizidine N-oxides in *Senecio vulgaris* L. *Planta* **177**, 98–107. doi:10.1007/BF00392159
33. Fu PP, Xia Q, Lin G, Chou MW (2004) Pyrrolizidine alkaloids: genotoxicity, metabolism enzymes, metabolic activation, and mechanisms. *Drug Metabolism Reviews* **36**, 1–55. doi:10.1081/DMR-120028426
34. Edgar JA, Molyneux RJ, Colegate SM (2015) Pyrrolizidine alkaloids: potential role in the etiology of cancers, pulmonary hypertension, congenital anomalies, and liver disease. *Chemical Research in Toxicology* **28**, 4–20. doi:10.1021/tx500403t
35. Fu PP (2017) Pyrrolizidine alkaloids: metabolic activation pathways leading to liver tumor initiation. *Chemical Research in Toxicology* **30**, 81–93. doi:10.1021/acs.chemrestox.6b00297
36. Li N, Xia Q, Ruan J, Fu PP, Lin G (2011) Hepatotoxicity and tumorigenicity induced by metabolic activation of pyrrolizidine alkaloids in herbs. *Current Drug Metabolism* **12**, 823–834. doi:10.2174/138920011797470119
37. Pereira TN, Webb RI, Reilly PE, Seawright AA, Prakash AS (1998) Dehydromonocrotaline generates sequence-selective N-7 guanine alkylation and heat and alkali stable multiple fragment DNA crosslinks. *Nucleic Acids Research* **26**, 5441–5447. doi:10.1093/nar/26.23.5441
38. Lahiri R, Ansari AA, Vankar YD (2013) Recent developments in design and synthesis of bicyclic azasugars, carbasugars and related molecules as glycosidase inhibitors. *Chemical Society Reviews* **42**, 5102–5118. doi:10.1039/c3cs35525j
39. Winchester B, Fleet GWJ (1992) Amino-sugar glycosidase inhibitors: versatile tools for glycobiologists. *Glycobiology* **2**, 199–210. doi:10.1093/glycob/2.3.199
40. Asano N, Nash RJ, Molyneux RJ, Fleet GWJ (2000) Sugar-mimic glycosidase inhibitors: natural occurrence, biological activity and prospects for therapeutic application. *Tetrahedron* **11**, 1645–1680. doi:10.1016/S0957-4166(00)00113-0
41. Leete E (1990) Recent developments in the biosynthesis of the tropane alkaloids. *Planta Medica* **56**, 339–352. doi:10.1055/s-2006-960979
42. Wooley JG (1993) Tropane alkaloids. In *Alkaloids and Sulphur Compounds*. (Ed. PG Waterman) pp. 133–173. Academic Press, New York, NY.
43. Sneader W (2005) Plant products analogues and compounds derived from them. In *Drug Discovery: A History*. (Ed. W Sneader) pp. 115–150. Wiley & Sons, Chichester, UK.
44. Gadzikowska M, Grynkiewicz G (2002) Tropane alkaloids in pharmaceutical and phytochemical analysis Tropane alkaloids in pharmaceutical and phytochemical analysis. *Acta Poloniae Pharmaceutica* **59**, 149–160.
45. Kreek MJ, Bart G, Lilly C, Laforge KS, Nielsen DA (2005) Pharmacogenetics and human molecular genetics of opiate and cocaine addictions and their treatment. *Pharmacological Reviews* **57**, 1–26. doi:10.1124/pr.57.1.1
46. O'Brien MS, Anthony JC (2005) Risk of becoming cocaine dependent: epidemiological estimates for the United States, 2000–2001. *Neuropsychopharmacology* **30**, 1006–1018. doi:10.1038/sj.npp.1300681
47. Gualtieri F (2000) Cholinergic receptors and neurodegenerative diseases. *Pharmaceutica Acta Helvetiae* **74**, 85–89. doi:10.1016/S0031-6865(99)00021-7
48. Christen P, Bieri S, Veuthey JL (2007) Analysis of tropane alkaloids in biological matrices. In *Modern Alkaloids: Structure, Isolation, Synthesis and Biology*. (Eds E Fatorusso and O Taglialatela-Scafati) pp. 339–367. Wiley-VCH, New York, NY.
49. Simpson K, Spencer CM, Mcclellan KJ (2000) Tropisetron: an update of its use in the prevention of chemotherapy-induced nausea and vomiting. *Drugs* **59**, 1297–1315. doi:10.2165/00003495-200059060-00008
50. Sora I, Hall FS, Andrews AM, Itokawa M, Li XF *et al.* (2001) Molecular mechanism of cocaine reward: combined dopamine and serotonin transporter knockouts eliminate cocaine place preference. *Proceedings of the National Academy of Sciences of the United States of America* **98**, 5300–5305. doi:10.1073/pnas.091039298
51. Lassak EV, Mccarthy T (1983) *Australian Medicinal Plants*. Methuen, Sydney.
52. Sanger GJ, Andrews PLR (2018) A history of drug discovery for treatment of nausea and vomiting and the implications for future research. *Frontiers in Pharmacology* **9**, 913. doi:10.3389/fphar.2018.00913
53. Bunsupa S, Yamazaki M, Saito K (2012) Quinolizidine alkaloid biosynthesis: recent advances and future prospects. *Frontiers in Plant Science* **3**, 1–7. doi:10.3389/fpls.2012.00239
54. Frick KM, Kamphuis LG, Siddique KH, Singh KB, Foley RC (2017) Quinolizidine alkaloid biosynthesis in lupins and prospects for grain quality improvement. *Frontiers in Plant Science* **8**, 1–12. doi:10.3389/fpls.2017.00087
55. Harris CM, Schneider MJ, Ungemach FS, Hill JE, Harris TM (1988) Biosynthesis of the toxic indolizidine alkaloids slaframine and swainsonine in *Rhizoctonia leguminicola*: metabolism of 1-hydroxyindolizidines. *Journal of the American Chemical Society* **110**, 940–949. doi:10.1021/ja00211a039

56. Green BT, Lee ST, Panter KE, Brown DR (2012) Piperidine alkaloids: human and food animal teratogens. *Food and Chemical Toxicology* **50**, 2049–2055. doi:10.1016/j.fct.2012.03.049
57. Hotti H, Rischer H (2017) The killer of Socrates: coniine and related alkaloids in the plant kingdom. *Molecules (Basel, Switzerland)* **22**, 1962. doi:10.3390/molecules22111962
58. Vitaku E, Smith DT, Njardarson JT (2014) Analysis of the structural diversity, substitution patterns, and frequency of nitrogen heterocycles among U.S. FDA approved pharmaceuticals. *Journal of Medicinal Chemistry* **57**, 10257–10274. doi:10.1021/jm501100b
59. Wink M (1987) Quinolizidine alkaloids: biochemistry, metabolism, and function in plants and cell suspension cultures. *Planta Medica* **53**, 509–514. doi:10.1055/s-2006-962797
60. Kinghorn AD, Balandrin MF (1984) Quinolizidine alkaloids of the Leguminosae: structural types, analysis, chemotaxonomy and biological activities. In *Alkaloids: Chemical and Biological Perspectives.* (Ed. SW Pelletier) pp. 105–148. Wiley, New York, NY.
61. Wink M (1993) Quinolizidine alkaloids. In *Alkaloids and Sulphur Compounds.* (Ed. PG Waterman) Academic Press, London, UK.
62. Zhang J, Morris-Natschke SL, Ma D, Shang X-F, Yang C-J *et al.* (2021) Biologically active indolizidine alkaloids. *Medicinal Research Reviews* **41**, 928–960. doi:10.1002/med.21747
63. Radwanski ER, Last RL (1995) Tryptophan biosynthesis and metabolism: biochemical and molecular genetics. *The Plant Cell* **7**, 921–934.
64. Ruddick JP, Evans AK, Nutt DJ, Lightman SL, Rook GA *et al.* (2006) Tryptophan metabolism in the central nervous system: medical implications. *Expert Reviews in Molecular Medicine* **8**, 1–27. doi:10.1017/S1462399406000068
65. Fantegrossi WE, Murnane KS, Reissig CJ (2008) The behavioral pharmacology of hallucinogens. *Biochemical Pharmacology* **75**, 17–33. doi:10.1016/j.bcp.2007.07.018
66. Arunotayanun W, Gibbons S (2012) Natural product 'legal highs'. *Natural Product Reports* **29**, 1304–1316. doi:10.1039/c2np20068f
67. Ishikura M, Abe T, Choshi T, Hibino S (2015) Simple indole alkaloids and those with a nonrearranged monoterpenoid unit. *Natural Product Reports* **32**, 1389–1471. doi:10.1039/C5NP00032G
68. Smith TA (1977) Tryptamine and related compounds in plants. *Phytochemistry* **16**, 171–175. doi:10.1016/S0031-9422(00)86778-3
69. Wink M (2013) Evolution of secondary metabolites in legumes (Fabaceae). *South African Journal of Botany* **89**, 164–175. doi:10.1016/j.sajb.2013.06.006
70. Patel K, Gadewar M, Tripathi R, Prasad SK, Patel DK (2012) A review on medical importance, pharmacological activity and biological aspects of beta-carboline alkaloid "Harmine". *Asian Pacific Journal of Tropical Biomedicine* **2**, 660–664. doi:10.1016/S2221-1691(12)60116-6
71. Batiha GE-S, Alkazmi LM, Nadwa EH, Rashwan EK, Beshbishy AM *et al.* (2020) Physostigmine: a plant alkaloid isolated from *Physostigma venenosum*: a review on pharmacokinetics, pharmacological and toxicological activities. *Journal of Drug Delivery and Therapeutics* **10**, 187–190. doi:10.22270/jddt.v10i1-s.3866
72. Liu J, Ng T, Rui Z, Ad O, Zhang W (2014) Unusual acetylation-dependent reaction cascade in the biosynthesis of the pyrroloindole drug physostigmine. *Angewandte Chemie International Edition* **53**, 136–139. doi:10.1002/anie.201308069
73. Proudfoot A (2006) The early toxicology of physostigmine: a tale of beans, great men and proud egos. *Toxicological Reviews* **25**, 99–138. doi:10.2165/00139709-200625020-00004
74. Mehta M, Adem A, Sabbagh M (2012) New acetylcholinesterase inhibitors for Alzheimer's disease. *International Journal of Alzheimer's Disease* **2012**, 728983. doi:10.1155/2012/728983
75. Bhattacharyya P, Chakraborty DP (1987) Carbazole alkaloids. *Fortschritte der Chemie Organischer Naturstoffe* **52**, 159–209. doi:10.1007/978-3-7091-8906-1_4
76. Waterman PG (1998) *Alkaloids: Biochemistry, Ecology, and Medical Applications.* (Eds MF Roberts and M Wink) pp. 87–108. Springer Science+Business, New York, NY.
77. Smith TA (1977) Phenethylamine and related compounds in plants. *Phytochemistry* **16**, 9–18. doi:10.1016/0031-9422(77)83004-5
78. Sadowska-Bartosz I, Bartosz G (2021) Biological properties and applications of betalains. *Molecules (Basel, Switzerland)* **26**, 2520. doi:10.3390/molecules26092520
79. Yuan H, Ma Q, Piao G (2016) The traditional medicine and modern medicine from natural products. *Molecules (Basel, Switzerland)* **21**, 559. doi:10.3390/molecules21050559
80. Pathan H, Williams J (2012) Basic opioid pharmacology: an update. *British Journal of Pain* **6**, 11–16. doi:10.1177/2049463712438493
81. Jin Z (2006) Imidazole, oxazole and thiazole alkaloids. *Natural Product Reports* **23**, 464–496. doi:10.1039/b502166a
82. Maat L, Beyerman HC (1983) The imidazole alkaloids. In *The Alkaloids.* (Ed. A Brossi) pp. 282–334. Academic Press, New York, NY.
83. Sheehan DJ, Hitchcock CA, Sibley CM (1999) Current and emerging azole antifungal agents. *Clinical Microbiology Reviews* **12**, 40–79. doi:10.1128/CMR.12.1.40
84. Wink M, Botschen F, Gosmann C, Schafer H, Waterman PG (2010) Chemotaxonomy seen from a phylogenetic perspective and evolution of secondary metabolism. *Annual Review of Plant Biology* **40**, 364–433.
85. Groger D (1980) Alkaloids derived from tryptophan and anthranilic acid. In *Secondary Plant Products.* (Eds EA Bell and BV Charlwood) pp. 128–159. Springer-Verlag, Berlin.
86. Robinson T (1974) Metabolism and function of alkaloids in plants. *Science* **184**, 430–435. doi:10.1126/science.184.4135.430

87. Lin SX, Curtis MA, Sperry J (2020) Pyridine alkaloids with activity in the central nervous system. *Bioorganic & Medicinal Chemistry* **28**, 115820. doi:10.1016/j.bmc.2020.115820
88. Ratsch A, Steadman KJ, Bogossian F (2010) The pituri story: a review of the historical literature surrounding traditional Australian Aboriginal use of nicotine in Central Australia. *Journal of Ethnobiology and Ethnomedicine* **6**, 26. doi:10.1186/1746-4269-6-26
89. Nehlig A, Daval J-L, Debry G (1992) Caffeine and the central nervous system: mechanisms of action, biochemical, metabolic and psychostimulant effects. *Brain Research. Brain Research Reviews* **17**, 139–170. doi:10.1016/0165-0173(92)90012-B
90. Arnaud MJ (2010) Pharmacokinetics and metabolism of natural methylxanthines in animal and man. In *Methylxanthines*. (Ed. BB Fredholm) pp. 33–91. Springer-Verlag, Berlin.
91. Li Petri G, Raimondi MV, Spanò V, Holl R, Barraja P *et al.* (2021) Pyrrolidine in drug discovery: a versatile scaffold for novel biologically active compounds. *Topics in Current Chemistry* **379**, 34. doi:10.1007/s41061-021-00347-5
92. Wei X, Ruan W, Vrieling K (2021) Current knowledge and perspectives of pyrrolizidine alkaloids in pharmacological applications: a mini-review. *Molecules (Basel, Switzerland)* **26**, 1970. doi:10.3390/molecules26071970
93. Farghaly TA, Masaret GS, Abdulwahab HG (2023) The patent review of the biological activity of tropane containing compounds. *Expert Opinion on Therapeutic Patents* **33**, 875–899. doi:10.1080/13543776.2023.2299349
94. Abdelshaheed MM, Fawzy IM, El-Subbagh HI, Youssef KM (2021) Piperidine nucleus in the field of drug discovery. *Future Journal of Pharmaceutical Sciences* **7**, 188. doi:10.1186/s43094-021-00335-y
95. Cely-Veloza W, Kato MJ, Coy-Barrera E (2023) Quinolizidine-type alkaloids: chemodiversity, occurrence, and bioactivity. *ACS Omega* **8**, 27862–27893. doi:10.1021/acsomega.3c02179
96. Zhang J, Morris-Natschke SL, Ma D, Shang XF, Yang CJ *et al.* (2021) Biologically active indolizidine alkaloids. *Medicinal Research Reviews* **41**, 928–960. doi:10.1002/med.21747
97. Kumar S, Ritika (2020) A brief review of the biological potential of indole derivatives. *Future Journal of Pharmaceutical Sciences* **6**, 121. doi:10.1186/s43094-020-00141-y
98. Abinaya R, Srinath S, Soundarya S, Sridhar R, Balasubramanian KK *et al.* (2022) Recent developments on synthesis strategies, SAR studies and biological activities of β-Carboline derivatives: an update. *Journal of Molecular Structure* **1261**, 132750. doi:10.1016/j.molstruc.2022.132750
99. Farouil L, Sylvestre M, Fournet A, Cebrián-Torrejón G (2022) Review on canthin-6-one alkaloids: distribution, chemical aspects and biological activities. *European Journal of Medicinal Chemistry Reports* **5**, 100049. doi:10.1016/j.ejmcr.2022.100049
100. Kasim SM, Al-Dabbagh BM, Mustafa YF (2022) A review on the biological potentials of carbazole and its derived products. *Eurasian Chemical Communications* **4**, 495–512.
101. Sadowska-Bartosz I, Bartosz G (2021) Biological properties and applications of betalains. *Molecules (Basel, Switzerland)* **26**, 2520. doi:10.3390/molecules26092520
102. Karan R, Agarwal P, Sinha M, Mahato N (2021) Recent advances on quinazoline derivatives: a potential bioactive scaffold in medicinal chemistry. *ChemEngineering* **5**, 73. doi:10.3390/chemengineering5040073
103. Huang W, Li C, Wang Y, Yi X, He X (2017) Anti-inflammatory lignanamides and monoindoles from Alocasia macrorrhiza. *Fitoterapia* **117**, 126–132. doi:10.1016/j.fitote.2017.01.014
104. Huang W, Yi X, Feng J, Wang Y, He X (2017) Piperidine alkaloids from Alocasia macrorrhiza. *Phytochemistry* **143**, 81–86. doi:10.1016/j.phytochem.2017.07.012
105. Mallik J, Das J, Banik RK (2018) Pharmacognostic profile and pharmacological activity of different parts of *Amorphophallus campanulatus* (Roxb.) Blume: a complete overview. *Asian Journal of Pharmaceutical Research & Development* **6**, 4–8. doi:10.22270/ajprd.v6i1.354
106. Padeepika C, Selvakumar R, Krishnakumar T, Sajad UNN, Sajeev MS (2018) Pharmacology and phytochemistry of underexploited tuber crops: a review. *Journal of Pharmacognosy and Phytochemistry* **7**, 1007–1019.
107. Zhang H-J, Tamez PA, Hoang VD, Tan GT, Hung NV *et al.* (2001) Antimalarial compounds from *Rhaphidophora decursiva*. *Journal of Natural Products* **64**, 772–777. doi:10.1021/np010037c
108. Zhang H, Qiu S, Tamez P, Tan GT, Aydogmus Z *et al.* (2002) Antimalarial agents from plants. II: decursivine, a new antimalarial indole alkaloid from *Rhaphidophora decursiva*. *Pharmaceutical Biology* **40**, 221–224. doi:10.1076/phbi.40.3.221.5832
109. Refaat J, Kamel MS, Ramadan MA, Ali AA (2012) *Crinum*: an endless source of bioactive principles: a review. Part I: *Crinum* alkaloids: lycorine-type alkaloids. *International Journal of Pharmaceutical Sciences and Research* **3**, 1883–1890.
110. Refaat J, Kamel MS, Ramadan MA, Ali AA (2012) *Crinum*: an endless source of bioactive principles: a review. Part II: *Crinum* alkaloids: crinine-type alkaloids. *International Journal of Pharmaceutical Sciences and Research* **3**, 3091–3100.
111. Refaat J, Kamel MS, Ramadan MA, Ali AA (2012) *Crinum*: an endless source of bioactive principles: a review. part III; *Crinum* alkaloids: belladine-, galanthamine-, lycorenine-, tazettine-type alkaloids and other minor types. *International Journal of Pharmaceutical Sciences and Research* **3**, 3630–3638.
112. Refaat J, Kamel MS, Ramadan MA, Ali AA (2013) *Crinum*; an endless source of bioactive principles: a review. Part IV: non alkaloidal constituents. *International Journal of Pharmaceutical Sciences and Research* **4**, 941–948.
113. Negi JS, Singh P, Joshi GP, Rawat MS, Bisht VK (2010) Chemical constituents of *Asparagus*. *Pharmacognosy Reviews* **4**, 215–220. doi:10.4103/0973-7847.70921

114. Xu J, Han Q-B, Li S-L, Chen X-J, Wang X-N *et al.* (2013) Chemistry, bioactivity and quality control of *Dendrobium*, a commonly used tonic herb in traditional Chinese medicine. *Phytochemistry Reviews* **12**, 341-367. doi:10.1007/s11101-013-9310-8
115. Yadav N, Ganie SA, Singh B, Chhillar AK, Yadav SS (2019) Phytochemical constituents and ethnopharmacological properties of *Ageratum conyzoides* L. *Phytotherapy Research* **33**, 2163–2178. doi:10.1002/ptr.6405
116. Folquitto DG, Swiech JND, Pereira CB, Bobek VB, Halila Possagno GC *et al.* (2019) Biological activity, phytochemistry and traditional uses of genus *Lobelia* (Campanulaceae): a systematic review. *Fitoterapia* **134**, 23–38. doi:10.1016/j.fitote.2018.12.021
117. Ghisalberti EL (2004) The Goodeniaceae. *Fitoterapia* **75**, 429–446. doi:10.1016/j.fitote.2004.01.018
118. Li L, Peng Y, Yao X, Xu L-J, Wulan T-N *et al.* (2010) Chemical constituents and biological activities of plants from the genus *Ehretia* Linn. *Chinese Herbal Medicines* **2**, 106–111.
119. Shukla A, Kaur A (2018) A systematic review of traditional uses bioactive phytoconstituents of genus *Ehretia*. *Asian Journal of Pharmaceutical and Clinical Research* **11**, 88–100. doi:10.22159/ajpcr.2018.v11i6.25178
120. Goyal PK (2012) Phytochemical and pharmacological properties of the genus *Grewia*: a review. *International Journal of Pharmacy and Pharmaceutical Sciences* **4**, 72–78.
121. Omar M, Defeo J, Youngken HWJ (1983) Chemical and toxicity studies of *Trichodesma africanum* L. *Journal of Natural Products* **46**, 153–156. doi:10.1021/np50026a001
122. Hosamani KM (1994) Ricinoleic and cyclopropene acids in *Trichodesma zeylanicum* seed oil. *Phytochemistry* **37**, 1621-1624. doi:10.1016/S0031-9422(00)89578-3
123. Mangai AS (2018) Pharmacological efficacy of *Trichodesma indicum* (Linn.) R. Br., in folk medicine: an updated review. *Asian Journal of Pharmaceutical and Clinical Research* **11**, 24–27. doi:10.22159/ajpcr.2018.v11i3.24412
124. Tlili N, Elfalleh W, Saadaoui E, Khaldi A, Triki S, Nasri N (2011) The caper (*Capparis* L.): ethnopharmacology, phytochemical and pharmacological properties. *Fitoterapia* **82**, 93–101. doi:10.1016/j.fitote.2010.09.006
125. Singh H, Mishra A, Mishra AK (2018) The chemistry and pharmacology of *Cleome* genus: a review. *Biomedicine and Pharmacotherapy* **101**, 37–48. doi:10.1016/j.biopha.2018.02.053
126. He X, Wang X, Fang J, Chang Y, Ning N *et al.* (2017) The genus Achyranthes: a review on traditional uses, phytochemistry, and pharmacological activities. *Journal of Ethnopharmacology* **203**, 260–278. doi:10.1016/j.jep.2017.03.035
127. Clauser M, Dall'acqua S, Loi MC, Innocenti G (2013) Phytochemical investigation on *Atriplex halimus* L. from Sardinia. *Natural Product Research* **27**, 1940–1944. doi:10.1080/14786419.2013.793684
128. Mroczek A (2015) Phytochemistry and bioactivity of triterpene saponins from Amaranthaceae family. *Phytochemistry Reviews* **14**, 577–605. doi:10.1007/s11101-015-9394-4
129. Kokanova-Nedialkova Z, Neialkov PT, Nikolov SD (2009) The genus *Chenopodium*: phytochemistry, ethnopharmacology and pharmacology. *Pharmacognosy Reviews* **3**, 280–306.
130. Koo KA, Lorent K, Gong W, Windsor P, Whittaker SJ *et al.* (2015) Biliatresone, a reactive natural toxin from *Dysphania glomulifera* and *D. littoralis*: discovery of the toxic moiety 1,2-diaryl-2-propenone. *Chemical Research in Toxicology* **28**, 1519-1521. doi:10.1021/acs.chemrestox.5b00227
131. Patil KS, Bhalsing SR (2016) Ethnomedicinal uses, phytochemistry and pharmacological properties of the genus *Boerhavia*. *Journal of Ethnopharmacology* **182**, 200–220. doi:10.1016/j.jep.2016.01.042
132. Iranshahy M, Javadi B, Iranshahi M, Jahanbakhsh SP, Mahyari S *et al.* (2017) A review of traditional uses, phytochemistry and pharmacology of *Portulaca oleracea* L. *Journal of Ethnopharmacology* **205**, 158–172. doi:10.1016/j.jep.2017.05.004
133. Kasai R, Sasaki A, Hashimoto T, Kaneko T, Ohtani K *et al.* (1999) Cycloartane glycosides from *Trichosanthes tricuspidata*. *Phytochemistry* **51**, 803–808. doi:10.1016/S0031-9422(99)00078-3
134. Kanchanapoom T, Kasai R, Yamasaki K (2002) Cucurbitane, hexanorcucurbitane and octanorcucurbitane glycosides from fruits of *Trichosanthes tricuspidata*. *Phytochemistry* **59**, 215–228. doi:10.1016/S0031-9422(01)00430-7
135. Mai LP, Guenard D, Franck M, Van TM, Gaspard C *et al.* (2002) New cytotoxic cucurbitacins from the pericarps of *Trichosanthes tricuspidata* fruits. *Natural Product Letters* **16**, 15–19. doi:10.1080/1057563029001/4755
136. Shaw PC, Lee KM, Wong KB (2005) Recent advances in trichosanthin, a ribosome-inactivating protein with multiple pharmacological properties. *Toxicon* **45**, 683–689. doi:10.1016/j.toxicon.2004.12.024
137. Yu X, Tang L, Wu H, Zhang X, Luo H *et al.* (2018) Trichosanthis Fructus: botany, traditional uses, phytochemistry and pharmacology. *Journal of Ethnopharmacology* **224**, 177–194. doi:10.1016/j.jep.2018.05.034
138. Rajeswari K, Bhaskara Rao T (2015) *Aegicera corniculatum* Linn (Myrsinaceae). *Journal of Chemical and Pharmaceutical Research* **7**, 305–316.
139. Hart NK, Lamberton JA (1966) Pyrrolizidine alkaloids from *Planchonella* species (family Sapotaceae). *Australian Journal of Chemistry* **19**, 1259–1264. doi:10.1071/CH9661259
140. Hart NK, Johns SR, Lamberton JA (1968) Alkaloids of the Sapotaceae: trans-B-methylthioacrylate and tiglate esters of (-)-isoretronecanol from a *Planchonella* species. *Australian Journal of Chemistry* **21**, 1393–1395. doi:10.1071/CH9681393
141. Eade RA, Hunt K, Simes JJH, Stern W (1969) Extractives of Australian timbers IX: the saponin of *Planchonella pohlmanianum*. *Australian Journal of Chemistry* **22**, 2703–2707. doi:10.1071/CH9692703

142. Cambie RC, Ser NA, Kokubun T (1997) Heartwood constituents of *Planchonella vitiensis*. *Biochemical Systematics and Ecology* **25**, 677–678. doi:10.1016/S0305-1978(97)00063-X
143. Lee T-H, Juang S-H, Hsu F-L, Wu C-Y (2005) Triterpene acids from the leaves of *Planchonella duclitan* (Blanco) Bakhuizan. *Journal of the Chinese Chemical Society (Taipei)* **52**, 1275–1280. doi:10.1002/jccs.200500184
144. Lee S-S, Tseng C-C, Chen C-K (2010) Three new benzophenone glucosides from the leaves of *Planchonella obovata*. *Helvetica Chimica Acta* **93**, 522–529. doi:10.1002/hlca.200900264
145. Chen H-Y, Guh J-H, Chan S-H, Lee S-S (2015) Cytotoxic protobassic acid glycosides from *Planchonella obovata* leaf. *Phytochemistry Letters* **11**, 229–235. doi:10.1016/j.phytol.2015.01.005
146. Garaniya N, Bapodra A (2014) Ethno botanical and phytopharmacological potential of Abrus precatorius L.: a review. *Asian Pacific Journal of Tropical Biomedicine* **4**, S27–S34. doi:10.12980/APJTB.4.2014C1069
147. Seigler DS (2003) Phytochemistry of Acacia—sensu lato. *Biochemical Systematics and Ecology* **31**, 845–873. doi:10.1016/S0305-1978(03)00082-6
148. Yadav JP, Arya V, Yadav S, Panghal M, Kumar S *et al.* (2010) *Cassia occidentalis* L.: a review on its ethnobotany, phytochemical and pharmacological profile. *Fitoterapia* **81**, 223–230. doi:10.1016/j.fitote.2009.09.008
149. Pitanga BP, Nascimento RP, Silva VD, Costa SL (2012) The role of astrocytes in metabolism and neurotoxicity of the pyrrolizidine alkaloid monocrotaline, the main toxin of *Crotalaria retusa*. *Frontiers in Pharmacology* **3**, 144. doi:10.3389/fphar.2012.00144
150. Al-Snafi AE (2016) The contents and pharmacology of *Crotalaria juncea*: a review. *IOSR Journal of Pharmacy* **6**, 77–86.
151. Waterman PG, Faulkner DF (1981) Imidazole alkaloids from *Cynometra hankei*. *Phytochemistry* **20**, 2765–2767. doi:10.1016/0031-9422(81)85283-1
152. Singh O, Ali M, Akhtar N (2011) Phenolic acid glucosides from the seeds of *Entada phaseoloides* Merill. *Journal of Asian Natural Products Research* **13**, 682–687. doi:10.1080/10286020.2011.584444
153. Sugimoto S, Matsunami K, Otsuka H (2018) Biological activity of *Entada phaseoloides* and *Entada rheedei*. *Journal of Natural Medicines* **72**, 12–19. doi:10.1007/s11418-017-1146-x
154. Majinda RRT, Wanjala CCW, Juma BF (2005) Bioactive non-alkaloidal constituents from the genus *Erythrina*. *Studies in Natural Products Chemistry* **32**, 821–853. doi:10.1016/S1572-5995(05)80070-5
155. Parsons AF, Palframan MJ (2010) *Erythrina* and related alkaloids. *The Alkaloids. Chemistry and Biology* **68**, 39–81. doi:10.1016/S1099-4831(10)06802-1
156. Son NT (2019) Genus *Erythrophleum*: botanical description, traditional use, phytochemistry and pharmacology. *Phytochemistry Reviews* **18**, 571–599. doi:10.1007/s11101-019-09640-0
157. Gerometta E, Grondin I, Smadja J, Frederich M, Gauvin-Bialecki A (2020) A review of traditional uses, phytochemistry and pharmacology of the genus *Indigofera*. *Journal of Ethnopharmacology* **253**, 112608. doi:10.1016/j.jep.2020.112608
158. Sathiyanarayanan L, Arulmozhi S (2007) *Mucuna pruriens* Linn.: a comprehensive review. *Pharmacognosy Reviews* **1**, 157–162.
159. Tse GG, Kim BB, Mcmurtray AM, Nakamoto BK (2013) Case of levodopa toxicity from ingestion of *Mucuna gigantea*. *Hawai'i Journal of Medicine & Public Health : a Journal of Asia Pacific Medicine & Public Health* **72**, 157–160.
160. Vora R, Joshi AN, Joshi NC (2018) Comparison of L-Dopa content in three species of genus Mucuna by different extraction techniques. *Annals of Plant Sciences* **7**, 1973–1977. doi:10.21746/aps.2018.7.1.10161. Krishna PM, Rao KNV, Sandhya S, Banji D (2012) A review on phytochemical, ethnomedical and pharmacological studies on genus *Sophora*, Fabaceae. *Revista Brasileira de Farmacognosia-Brazilian Journal of Pharmacognosy* **22**, 1145–1154. doi:10.1590/S0102-695X2012005000043
162. He X, Fang J, Huang L, Wang J, Huang X (2015) *Sophora flavescens* Ait.: traditional usage, phytochemistry and pharmacology of an important traditional Chinese medicine. *Journal of Ethnopharmacology* **172**, 10–29. doi:10.1016/j.jep.2015.06.010
163. He X, Bai Y, Zhao Z, Wang X, Fang J *et al.* (2016) Local and traditional uses, phytochemistry, and pharmacology of *Sophora japonica* L.: a review. *Journal of Ethnopharmacology* **187**, 160–182. doi:10.1016/j.jep.2016.04.014
164. Colegate SM, Dorling PR, Huxtable CR (1979) A spectroscopic investigation of swainsonine: an alpha-mannosidase inhibitor isolated from *Swainsona canescens*. *Australian Journal of Chemistry* **32**, 2257–2264. doi:10.1071/CH9792257
165. Dorling PR, Huxtable CR, Colegate SM (1980) Inhibition of lysosomal alpha-mannosidase by swainsonine, an indolizidine alkaloid isolated from *Swainsona canescens*. *The Biochemical Journal* **191**, 649–651. doi:10.1042/bj1910649
166. Grum DS, Cook D, Baucom D, Mott IW, Gardner DR *et al.* (2013) Production of the alkaloid swainsonine by a fungal endophyte in the host *Swainsona canescens*. *Journal of Natural Products* **76**, 1984–1988. doi:10.1021/np400274n
167. Cannon JR, Williams JR, Arbain D, Brossi A, Blount JF *et al.* (1991) Alkaloids of *Templetonia retusa*: chemistry, structure and absolute configuration of (-)-templetine. *Australian Journal of Chemistry* **44**, 509–523. doi:10.1071/CH9910509
168. Greinwald R, Henrichs C, Veen G, Ross JH, Witte L *et al.* (1995) A survey of alkaloids in *Templetonia biloba* (Benth.) Polhill (Fabaceae: Brongniartieae). *Biochemical Systematics and Ecology* **23**, 649–654. doi:10.1016/0305-1978(95)00035-6
169. Greinwald R, Ross JH, Witte L, Czygan F-C (1996) Alkaloids of *Templetonia incana*. *Biochemical Systematics and Ecology* **24**, 423–426. doi:10.1016/0305-1978(96)00024-5

170. Onyilagha JC, Islam S, Ntamatungiro S (2009) Comparative phytochemistry of eleven species of *Vigna* (Fabaceae). *Biochemical Systematics and Ecology* **37**, 16–19. doi:10.1016/j.bse.2008.11.013
171. Leu YL, Hwang TL, Kuo PC, Liou KP, Huang BS *et al.* (2012) Constituents from *Vigna vexillata* and their anti-inflammatory activity. *International Journal of Molecular Sciences* **13**, 9754–9768. doi:10.3390/ijms13089754
172. Tang D, Dong Y, Ren H, Li L, He C (2014) A review of phytochemistry, metabolite changes, and medicinal uses of the common food mung bean and its sprouts (*Vigna radiata*). *Chemistry Central Journal* **8**, 4. doi:10.1186/1752-153X-8-4
173. Mutharaj S, Muthusamy P, Ilango K (2019) Pharmacognostical, phytochemical and pharmacological review of *Casuarina equisetifolia* Linn. *World Journal of Pharmaceutical Research* **8**, 328–339.
174. Khyade MS, Kasote DM, Vaikos NP (2014) *Alstonia scholaris* (L.) R. Br. and *Alstonia macrophylla* Wall. ex G. Don: a comparative review on traditional uses, phytochemistry and pharmacology. *Journal of Ethnopharmacology* **153**, 1–18. doi:10.1016/j.jep.2014.01.025
175. Han L, Zhou X, Yang M, Zhou L, Deng X *et al.* (2018) Ethnobotany, phytochemistry and pharmacological effects of plants in genus *Cynanchum* Linn. (Asclepiadaceae). *Molecules (Basel, Switzerland)* **23**, 1194. doi:10.3390/molecules23051194
176. Van Beek TA, Verpoorte R, Svendsen AB, Leeuwenberg AJM, Bisset NG (1984) *Tabernaemontana* L. (Apocynaceae): a review of its taxonomy, phytochemistry, ethnobotany and pharmacology. *Journal of Ethnopharmacology* **10**, 1–156. doi:10.1016/0378-8741(84)90046-1
177. Patchayasakul W, Pongchaidecha A, Chattipakorn N, Chattipakorn S (2008) Ethnobotany and ethnopharmacology of *Tabernaemontana divaricata*. *The Indian Journal of Medical Research* **127**, 317–335.
178. Abubakar IB, Loh HS (2016) A review on ethnobotany, pharmacology and phytochemistry of *Tabernaemontana corymbosa*. *The Journal of Pharmacy and Pharmacology* **68**, 423–432. doi:10.1111/jphp.12523
179. Ali M, Bhutani KK (1989) Alkaloids from *Tylophora indica*. *Phytochemistry* **28**, 3513–3517. doi:10.1016/0031-9422(89)80376-0
180. Abe F, Iwase Y, Yamauchi T, Honda K, Hayashi N (1995) Phenathroindolizidine alkaloids from *Tylophora tanakae*. *Phytochemistry* **39**, 695–699. doi:10.1016/0031-9422(95)00034-5
181. Gurav S, Senthilkumar DGP, Tembare R, Mani T (2011) *Tylophora indica*: a review on its ethnobotany, phytochemical and pharmacological profile. *Asian Journal of Biochemical and Pharmaceutical Research* **3**, 405–414.
182. Zhao JY, Wang XJ, Liu Z, Meng FX, Sun SF *et al.* (2019) Nonadride and spirocyclic anhydride derivatives from the plant endophytic fungus *Talaromyces purpurogenus*. *Journal of Natural Products* **82**, 2953–2962. doi:10.1021/acs.jnatprod.9b00210
183. Itoh A, Tanaka Y, Nagakura N, Nishi T, Tanahashi T (2006) A quinic acid ester from *Strychnos lucida*. *Journal of Natural Medicines* **60**, 146–148. doi:10.1007/s11418-005-0021-3
184. Galliford CV, Scheidt KA (2007) Pyrrolidinyl-spirooxindole natural products as inspirations for the development of potential therapeutic agents. *Angewandte Chemie International Edition* **46**, 8748–8758. doi:10.1002/anie.200701342
185. Tchinda AT, Tamze V, Ngono ARN, Ayimele GA, Cao M *et al.* (2012) Alkaloids from the stem bark of *Strychnos icaja*. *Phytochemistry Letters* **5**, 108–113. doi:10.1016/j.phytol.2011.11.004
186. Guo R, Wang T, Zhou G, Xu M, Yu X *et al.* (2018) Botany, phytochemistry, pharmacology and toxicity of *Strychnos nux-vomica* L.: a review. *The American Journal of Chinese Medicine* **46**, 1–23. doi:10.1142/S0192415X18500015
187. Li W, Tang GH, Chen L, Tang YQ, Xu YK *et al.* (2018) New pyridocarbazole alkaloids from *Strychnos nitida*. *Natural Product Research* **32**, 1532–1536. doi:10.1080/14786419.2017.1385016
188. Bowie JH, Cooke RG, Wilkin PE (1962) Colouring matters of Australian plants. X: anthraquinones from *Coelospermum* species. *Australian Journal of Chemistry* **15**, 336–341. doi:10.1071/CH9620336
189. Lopez JL, Pusset J (1988) Plantes de Nouvelle-Caledonie, 115. Alcaloides monoterpeniques de *Coelospermum billardieri*. *Journal of Natural Products* **51**, 829–835. doi:10.1021/np50059a002
190. Benkrief R, Skaltsounis A-L, Tillequin F, Koch M (1991) Plantes de Nouvelle-Caledonie, 137. Iridoides de *Coelospermum billardieri*. *Journal of Natural Products* **54**, 532–541. doi:10.1021/np50074a028
191. Magiatis P, Skaltsounis A-L, Tillequin F, Seguin E, Cosson J-P (2002) Coelobillardin, an iridoid glucoside dimer from *Coelospermum billardieri*. *Phytochemistry* **60**, 415–418. doi:10.1016/S0031-9422(02)00118-8
192. Boucherle B, Haudecoeur R, Queiroz EF, De Waard M, Wolfender JL *et al.* (2016) *Nauclea latifolia*: biological activity and alkaloid phytochemistry of a West African tree. *Natural Product Reports* **33**, 1034–1043. doi:10.1039/C6NP00039H
193. Haudecoeur R, Peuchmaur M, Peres B, Rome M, Taiwe GS *et al.* (2018) Traditional uses, phytochemistry and pharmacological properties of African *Nauclea* species: a review. *Journal of Ethnopharmacology* **212**, 106–136. doi:10.1016/j.jep.2017.10.011
194. Johns SR, Lamberton JA (1970) The identification of a new alkaloid from *Timonius kaniensis* (Rubiaceae) as dihydrocupreine. *Australian Journal of Chemistry* **23**, 211–212. doi:10.1071/CH9700211
195. Khan IA, Sticher O, Rali T (1993) New triterpenes from the leaves of *Timonius timon*. *Journal of Natural Products* **56**, 2163–2165. doi:10.1021/np50102a019
196. Russo A, Borrelli F (2005) Bacopa monniera, a reputed nootropic plant: an overview. *Phytomedicine* **12**, 305–317. doi:10.1016/j.phymed.2003.12.008

197. Hayashi T, Kishi MP, Kawasaki M, Arisawa M, Shimizu M *et al.* (1987) Scopadulcic acid -A and -B, new diterpenoids with a novel skeleton, from a Paraguayan crude drug "Typycha Kuratu" (*Scoparia dulcis* L.). *Tetrahedron Letters* **28**, 3693–3696. doi:10.1016/S0040-4039(00)96357-3
198. Kawasaki M, Hayashi T, Arisawa M, Morita N, Berganza LH (1988) 8-hydroxytricetin 7-glucuronide, a Β-glucuronidase inhibitor from *Scoparia dulcis. Phytochemistry* **27**, 3709–3711. doi:10.1016/0031-9422(88)80811-2
199. Hayashi T, Kawasaki M, Miwa Y, Taga T, Morita N (1990) Antiviral agents of plant origin. III: scopadulin, a novel tetracyclic diterpene from *Scoparia dulcis* L. *Chemical & Pharmaceutical Bulletin* **38**, 945–947. doi:10.1248/cpb.38.945
200. Mishra MR, Behera RK, Jha S, Panda AK, Mishra A *et al.* (2011) A brief review on phytoconstituents and ethnopharmacology of *Scoparia dulcis* Linn. (Scrophulariaceae). *International Journal of Phytotherapy and Phytopharmacology* **3**, 422–438.
201. Urzua AM, Cassels BK (1978) Alkaloid chemosystematics, chemotaxonomy and biogenesis in the Atherospermataceae. *Lloydia* **41**, 98–113.
202. Bick IRC, Leow H-M (1978) Alkaloids of *Daphnandra johnsonii. Australian Journal of Chemistry* **31**, 2539–2544. doi:10.1071/CH9782539
203. Bick IRC, Sotheeswaran S (1978) Alkaloids of *Daphnandra apatela. Australian Journal of Chemistry* **31**, 2077–2083. doi:10.1071/CH9782077
204. Buchanan MS, Davis RA, Duffy S, Avery VM, Quinn RJ (2009) Antimalarial benzylisoquinoline alkaloid from the rainforest tree *Doryphora sassafras. Journal of Natural Products* **72**, 1541–1543. doi:10.1021/np9002564
205. Conserva LM, Pereira Ca B, Barbosa-Filho JM (2005) Alkaloids of the Hernandiaceae: occurrence and a compilation of their biological activities. *The Alkaloids. Chemistry and Biology* **62**, 175–243. doi:10.1016/S1099-4831(05)62003-2
206. Lakshmi V, Pandey K, Mishra SK, Srivastava S, Mishra M *et al.* (2009) An overview of family Hernandiaceae. *Records of Natural Products* **3**, 1–22.
207. Lai YW, Wang SW, Hu YY, Hwang TL, Cheng MJ *et al.* (2020) Anti-inflammatory alkaloids from the root bark of *Hernandia nymphaeifolia. Phytochemistry* **173**, 112326. doi:10.1016/j.phytochem.2020.112326
208. Chang F-R, Chao Y-C, Teng C-M, Wu Y-C (1998) Chemical constituents from *Cassytha filiformis* II. *Journal of Natural Products* **61**, 863–866.
209. Tsai T-H, Wang G-J, Lin L-C (2008) Vasorelaxing alkaloids and flavonoids from *Cassytha filiformis. Journal of Natural Products* **71**, 289–291. doi:10.1021/np070564h
210. Wang YS, Wen ZQ, Li BT, Zhang HB, Yang JH (2016) Ethnobotany, phytochemistry, and pharmacology of the genus *Litsea*: an update. *Journal of Ethnopharmacology* **181**, 66–107. doi:10.1016/j.jep.2016.01.032
211. Ma X, Gang DR (2004) The Lycopodium alkaloids. *Natural Product Reports* **21**, 752–772. doi:10.1039/b409720n
212. Seebaluck R, Gurib-Fakim A, Mahomoodally F (2015) Medicinal plants from the genus Acalypha (Euphorbiaceae): a review of their ethnopharmacology and phytochemistry. *Journal of Ethnopharmacology* **159**, 137–157. doi:10.1016/j.jep.2014.10.040
213. Salatino A, Faria Salatino ML, Negri G (2007) Traditional uses, chemistry and pharmacology of *Croton* species (Euphorbiaceae). *Journal of the Brazilian Chemical Society* **18**, 11–33. doi:10.1590/S0103-50532007000100002
214. Gupta D, Bleakley B, Gupta RK (2008) Dragon's blood: botany, chemistry and therapeutic uses. *Journal of Ethnopharmacology* **115**, 361–380. doi:10.1016/j.jep.2007.10.018
215. Shi Q-W, Su X-H, Kiyota H (2008) Chemical and pharmacological research of the plants in genus *Euphorbia. Chemical Reviews* **108**, 4295–4327. doi:10.1021/cr078350s
216. Yin B-W, Shen L-R, Zhang M-L, Zhao L, Wang Y-L *et al.* (2008) Chemical constituents of plants from the genus *Excoecaria. Chemistry & Biodiversity* **5**, 2356–2371. doi:10.1002/cbdv.200890201
217. Raj D, Luczkiewicz M (2008) Securinega suffruticosa. *Fitoterapia* **79**, 419–427. doi:10.1016/j.fitote.2008.02.011
218. Wehlauch R, Gademann K (2017) Securinega alkaloids: complex structures, potent bioactivities, and efficient total syntheses. *Asian Journal of Organic Chemistry* **6**, 1146–1159. doi:10.1002/ajoc.201700142
219. Wang XF, Liu FF, Zhu Z, Fang QQ, Qu SJ *et al.* (2019) Flueggenoids A–E, new dinorditerpenoids from *Flueggea virosa. Fitoterapia* **133**, 96–101. doi:10.1016/j.fitote.2018.12.025
220. Huang Y-L, Chen C-C, Hsu F-L, Chen C-F (1996) A new lignan from *Phyllanthus virgatus. Journal of Natural Products* **59**, 520–521. doi:10.1021/np9603380
221. Calixto JB, Santos ARS, Filho VC, Yunes RA (1998) A review of the plants of the genus *Phyllanthus:* their chemistry pharmacology and therapeutic potential. *Medicinal Research Reviews* **18**, 225–258. doi:10.1002/(SICI)1098-1128(199807)18:4<225::AID-MED2>3.0.CO;2-X
222. Huang Y-L, Chen C-C, Hsu F-L, Chen C-F (1998) Tannins, flavonol sulfonates, and a norlignan from *Phyllanthus virgatus. Journal of Natural Products* **61**, 1194–1197. doi:10.1021/np970336v
223. Bagalkotkar G, Sagineedu SR, Saad MS, Stanslas J (2006) Phytochemicals from *Phyllanthus niruri* Linn. and their pharmacological properties: a review. *The Journal of Pharmacy and Pharmacology* **58**, 1559–1570. doi:10.1211/jpp.58.12.0001
224. Patel JR, Tripathi P, Sharma V, Chauhan NS, Dixit VK (2011) *Phyllanthus amarus*: ethnomedicinal uses, phytochemistry and pharmacology: a review. *Journal of Ethnopharmacology* **138**, 286–313. doi:10.1016/j.jep.2011.09.040
225. Patel DK, Kumar R, Sairam K, Hemalatha S (2013) Hybanthus enneaspermus (L.) F. Muell: a concise report

on its phytopharmacological aspects. *Chinese Journal of Natural Medicines* **11**, 199–206. doi:10.1016/S1875-5364(13)60017-5

226. Rajsekhar PB, Arvind Bharani RS, Angel J, Ramachandran M, Rajsekhar SPV (2016) *Hybanthus enneaspermus* (L) F. Muell: a phytopharmacological review on herbal medicine. *Journal of Chemical and Pharmaceutical Research* **8**, 351–355.
227. Goyal PK (2012) Phytochemical and pharmacological properties of the genus *Grewia*: a review. *International Journal of Pharmacy and Pharmaceutical Sciences* **4**, 72–78.
228. Ullah W, Uddin G, Siddiqui BS (2012) Ethnic uses, pharmacological and phytochemical profile of genus *Grewia*. *Journal of Asian Natural Products Research* **14**, 186–195. doi:10.1080/10286020.2011.639764
229. Zia-Ul-Haq M, Stankovic MS, Rizwan K, Feo VD (2013) *Grewia asiatica* L., a food plant with multiple uses. *Molecules (Basel, Switzerland)* **18**, 2663–2682. doi:10.3390/molecules18032663
230. Dinda B, Das N, Dinda S, Dinda M, Silsarma I (2015) The genus *Sida* L.: a traditional medicine: Its ethnopharmacological, phytochemical and pharmacological data for commercial exploitation in herbal drugs industry. *Journal of Ethnopharmacology* **176**, 135–176. doi:10.1016/j.jep.2015.10.027
231. Al Muqarrabun LM, Ahmat N (2015) Medicinal uses, phytochemistry and pharmacology of family Sterculiaceae: a review. *European Journal of Medicinal Chemistry* **92**, 514–530. doi:10.1016/j.ejmech.2015.01.026
232. Thabet AA, Youssef FS, El-Shazly M, Singab ANB (2018) *Sterculia* and *Brachychiton*: a comprehensive overview on their ethnopharmacology, biological activities, phytochemistry and the role of their gummy exudates in drug delivery. *The Journal of Pharmacy and Pharmacology* **70**, 450–474. doi:10.1111/jphp.12876
233. Nonato MG, Takayama H, Garson MJ (2008) *Pandanus* alkaloids: chemistry and biology. *The Alkaloids* **66**, 215–249.
234. Adkar PP, Bhaskar VH (2014) *Pandanus odoratissimus* (Kewda): a review on ethnopharmacology, phytochemistry, and nutritional aspects. *Advances in Pharmacological and Pharmaceutical Sciences* **2014**, 1–19. doi:10.1155/2014/120895
235. Parmar VS, Jain SC, Bisht KS, Jain R, Taneja P *et al.* (1997) Phytochemistry of the genus *Piper*. *Phytochemistry* **46**, 597–673. doi:10.1016/S0031-9422(97)00328-2
236. Xiang C-P, Shi Y-N, Liu F-F, Li H-Z, Zhang Y-J *et al.* (2016) A survey of the chemical compounds of *Piper* spp. (Piperaceae) and their biological activities. *Natural Product Communications* **11**, 1403–1408. doi:10.1177/1934578X1601100948
237. Peerzada AM, Ali HH, Naeem M, Latif M, Bukhari AH *et al.* (2015) *Cyperus rotundus* L.: traditional uses, phytochemistry, and pharmacological activities. *Journal of Ethnopharmacology* **174**, 540–560. doi:10.1016/j.jep.2015.08.012
238. Pan C, Chen YG, Ma XY, Jiang JH, He F *et al.* (2011) Phytochemical constituents and pharmacological activities of plants from the genus *Adiantum*: a review. *Tropical Journal of Pharmaceutical Research* **10**, 681–692. doi:10.4314/tjpr.v10i5.18
239. Sharma BR, Gautam LN, Adhikari D, Karki R (2017) A comprehensive review on chemical profiling of *Nelumbo nucifera*: potential for drug development. *Phytotherapy Research* **31**, 3–26. doi:10.1002/ptr.5732
240. Semwal DK, Semwal RB, Vermaak I, Viljoen A (2014) From arrow poison to herbal medicine: the ethnobotanical, phytochemical and pharmacological significance of *Cissampelos* (Menispermaceae). *Journal of Ethnopharmacology* **155**, 1011–1028. doi:10.1016/j.jep.2014.06.054
241. Hungerford NL, Sands DPA, Kitching W (1998) Isolation and structure of some constituents of the Australian medicinal plant *Tinospora smilacina* ('snakevine'). *Australian Journal of Chemistry* **51**, 1103–1111. doi:10.1071/C98034
242. Pathak AK, Jain DC, Sharma RP (2008) Chemistry and biological activities of the genera *Tinospora*. *International Journal of Pharmacology* **33**, 277–287.
243. Singh D, Chaudhuri PK (2017) Chemistry and pharmacology of *Tinospora cordifolia*. *Natural Product Communications* **12**, 299–308. doi:10.1177/1934578X1701200240
244. Chawla R, Kumar S, Sharma A (2012) The genus *Clematis* (Ranunculaceae): chemical and pharmacological perspectives. *Journal of Ethnopharmacology* **143**, 116–150. doi:10.1016/j.jep.2012.06.014
245. Cassels BK, Eckhardt G, Kaussmann EU, Tschesche R (1974) Cyclopeptide alkaloids of *Zizyphus oenoplia*. *Tetrahedron* **30**, 2461–2466. doi:10.1016/S0040-4020(01)97117-0
246. Mahajan RT, Chopda MZ (2009) Phytopharmacology of *Ziziphus jujuba* Mill.: a plant review. *Pharmacognosy Reviews* **3**, 320–329.
247. Gao QH, Wu CS, Wang M (2013) The jujube (*Ziziphus jujuba* Mill.) fruit: a review of current knowledge of fruit composition and health benefits. *Journal of Agricultural and Food Chemistry* **61**, 3351–3363. doi:10.1021/jf4007032
248. Prabhavathi S, Vajayalakshmi S (2015) New hydroxycarboxylic acid from the root bark of *Ziziphus oenoplia*. *Journal of Pharmacognosy and Phytochemistry* **3**, 150–152.
249. Banerjee S, Bandyopadhyay A (2015) *Buchanania lanzan* Spreng.: a veritable storehouse of phytomedicines. *Asian Journal of Pharmaceutical and Clinical Research* **8**, 18–22.
250. Zhao L, Huo C-H, Shen L-R, Yang Y, Zhang Q *et al.* (2010) Chemical constituents of plants from the genus *Melia*. *Chemistry & Biodiversity* **7**, 839–859. doi:10.1002/cbdv.200900043
251. Auzi AA, Hartley TG, Waterman PG (1997) Distribution of flavonoids, alkaloids, acetophenones and phloroglucinols in the genus *Bosistoa* (Rutaceae). *Biochemical Systematics and Ecology* **25**, 611–617. doi:10.1016/S0305-1978(97)00062-8
252. Fernandez LS, Buchanan MS, Carroll AR, Feng Y-J, Quinn RJ *et al.* (2009) Flinderoles A-C: antimalarial bis-

indole alkaloids from *Flindersia* species. *Organic Letters* **11**, 329–332. doi:10.1021/ol802506n
253. Simonsen HT, Adsersen A, Bremner P, Heinrich M, Wagner Smitt U *et al.* (2004) Antifungal constituents of *Melicope borbonica*. *Phytotherapy Research* **18**, 542–545. doi:10.1002/ptr.1482
254. Yamamoto M, Akita T, Koyama Y, Sueyoshi E, Matsunami K *et al.* (2008) Euodionosides A-G: megastigmane glucosides from leaves of *Euodia meliaefolia*. *Phytochemistry* **69**, 1586–1596. doi:10.1016/j.phytochem.2008.01.020
255. Zhao N, Li ZL, Li DH, Sun YT, Shan DT *et al.* (2015) Quinolone and indole alkaloids from the fruits of *Euodia rutaecarpa* and their cytotoxicity against two human cancer cell lines. *Phytochemistry* **109**, 133–139. doi:10.1016/j.phytochem.2014.10.020
256. Eliaser EM, Ho JH, Hashim NM, Rukayadi Y, Ee GCL *et al.* (2018) Phytochemical constituents and biological activities of *Melicope lunu-ankenda*. *Molecules (Basel, Switzerland)* **23**, 2708.
257. Fan S, Zhang C, Luo T, Wang J, Tang Y *et al.* (2019) Limonin: a review of its pharmacology, toxicity, and pharmacokinetics. *Molecules (Basel, Switzerland)* **24**, 3679. doi:10.3390/molecules24203679
258. Vu VT, Chen XL, Kong LY, Luo JG (2020) Melipatulinones A-C, three lignan-phloroglucinol hybrids from *Melicope patulinervia*. *Organic Letters* **22**, 1380–1384. doi:10.1021/acs.orglett.9b04680
259. Yao Q, Gao Y, Lai C, Wu C, Zhao CL *et al.* (2020) The phytochemistry, pharmacology and applications of *Melicope pteleifolia*: a review. *Journal of Ethnopharmacology* **251**, 112546. doi:10.1016/j.jep.2020.112546
260. Rashid MA, Armstrong JA, Gray AI, Waterman PG (1991) Pyranocoumarins as chemotaxonomic markers in *Eriostemon coccineus* and *Philotheca citrina*. *Phytochemistry* **30**, 4033–4035. doi:10.1016/0031-9422(91)83459-X
261. Ghisalberti EL (1998) Phytochemistry of the Australian Rutaceae: *Boronia, Eriostemon* and *Phebalium* species. *Phytochemistry* **47**, 163–176. doi:10.1016/S0031-9422(97)00571-2
262. Sultana N, Sarker SD, Armstrong JA, Wilson PG, Waterman PG (2003) The coumarins of *Philotheca* sensu lato: distribution and systematic significance. *Biochemical Systematics and Ecology* **31**, 681–691. doi:10.1016/S0305-1978(02)00226-0
263. Chlouchi A, Muyard F, Girard C, Waterman PG, Bévalot F (2005) Coumarins from the twigs of *Diplolaena mollis* P. G. Wilson (Rutaceae). *Biochemical Systematics and Ecology* **33**, 967–969. doi:10.1016/j.bse.2005.03.002
264. Chlouchi A, Girard C, Tillequin F, Bévalot F, Waterman PG *et al.* (2006) Coumarins and furoquinoline alkaloids from *Philotheca deserti* var. *deserti* (Rutaceae). *Biochemical Systematics and Ecology* **34**, 71–74. doi:10.1016/j.bse.2005.07.002
265. Chlouchi A, Girard C, Bévalot F, Waterman PG, Muyard F (2007) Taxonomically significant coumarins from three *Philotheca* species (Rutaceae). *Biochemical Systematics and Ecology* **35**, 251–254. doi:10.1016/j.bse.2006.09.011
266. Fokialakis N, Mitaku S, Mikros E, Skaltsounis A-L, Tillequin F *et al.* (1999) Megistosarcimine and megistosarconine, two alkaloids from *Sarcomelicope megistophylla*. *Phytochemistry* **52**, 1745–1748. doi:10.1016/S0031-9422(99)00323-4
267. Fokialakis N, Magiatis P, Skaltsounis A-L, Tillequin F, Sevenet T (2000) Megistolactone, a new alkalods from *Sarcomelicope megistophylla*. *Zeitschrift für Naturforschung C* **55**, 874–876. doi:10.1515/znc-2000-11-1204
268. Fokialakis N, Magiatis P, Aligiannis N, Mitaku S, Tillequin F *et al.* (2001) Furomegistines I and II, two furopyridine alkaloids from the bark of *Sarcomelicope megistophylla*. *Phytochemistry* **57**, 593–596. doi:10.1016/S0031-9422(01)00065-6
269. Tillequin F (2002) *Sarcomelicope* alkaloids as leads for the discovery of new antitumor acronycine derivatives. *Phytochemistry Reviews* **1**, 355–368. doi:10.1023/A:1026076825006
270. Tillequin F (2007) Rutaceous alkaloids as models for the design of novel antitumor drugs. *Phytochemistry Reviews* **6**, 65–79. doi:10.1007/s11101-006-9010-8
271. Nguyen HT, Lallemand M-C, Boutefnouchet S, Michel S, Tillequin F (2009) Antitumor *Psoropermum* xanthones and *Sarcomelicope* acridones: privileged structures implied in DNA alkylation. *Journal of Natural Products* **72**, 527–539. doi:10.1021/np800644y
272. kundu P, Laskar S (2010) A brief resume on the genus *Ailanthus*: chemical and pharmacological aspects. *Phytochemistry Reviews* **9**, 379–412. doi:10.1007/s11101-009-9157-1
273. Liu J-H, Jin H-Z, Zhang W-D, Yan S-K, Shen Y-H (2009) Chemical constituents of plants from the genus *Brucea*. *Chemistry & Biodiversity* **6**, 57–70. doi:10.1002/cbdv.200700409
274. Salehi B, Krochmal-Marczak B, Skiba D, Patra JK, Das SK *et al.* (2020) *Convolvulus* plant: a comprehensive review from phytochemical composition to pharmacy. *Phytotherapy Research* **34**, 315–328. doi:10.1002/ptr.6540
275. Austin DF (2008) *Evolvulus alsinoides* (Convolvulaceae): an American herb in the Old World. *Journal of Ethnopharmacology* **117**, 185–198. doi:10.1016/j.jep.2008.01.038
276. Meira M, Da Silva EP, David JM, David JP (2012) Review of the genus *Ipomoea*: traditional uses, chemistry and biological activities. *Revista Brasileira de Farmacognosia-Brazilian Journal of Pharmacognosy* **22**, 682–713. doi:10.1590/S0102-695X2012005000025
277. Schimming T, Tofern B, Mann P, Richter A, Jenett-Siems K *et al.* (1998) Distribution and taxonomic significance of calystegines in the Convolvulaceae. *Phytochemistry* **49**, 1989–1995. doi:10.1016/S0031-9422(98)00430-0
278. Tofern B, Mann P, Kaloga M, Jenett-Siems K, Witte L *et al.* (1999) Aliphatic purrolidine amides from two tropical convolvulaceous species. *Phytochemistry* **52**, 1437–1441. doi:10.1016/S0031-9422(99)00245-9

279. Schimming T, Jenett-Siems K, Mann P, Tofern-Reblin B, Milson J *et al.* (2005) Calystegines as chemotaxonomic markers in the Convolvulaceae. *Phytochemistry* **66**, 469–480. doi:10.1016/j.phytochem.2004.12.024
280. Griffin WJ, Lin GD (2000) Chemotaxonomy and geographical distribution of tropane alkaloids. *Phytochemistry* **53**, 623–637. doi:10.1016/S0031-9422(99)00475-6
281. Arroo R, Woolley J, Oksman-Caldentey K-M (2007) Tropane alkaloid containing plants: Henbane, Belladonna, Datura and Duboisia. In *Transgenic Crops VI.* (Eds E-C Pua and MR Davey) pp. 189–204. Springer, Berlin.
282. Nugroho LH, Verpoorte R (2002) Secondary metabolism in tobacco. *Plant Cell, Tissue and Organ Culture* **68**, 105–125. doi:10.1023/A:1013853909494
283. Dewey RE, Xie J (2013) Molecular genetics of alkaloid biosynthesis in *Nicotiana* tabacum. *Phytochemistry* **94**, 10–27. doi:10.1016/j.phytochem.2013.06.002
284. Jassbi AR, Zare S, Asadollahi M, Schuman MC (2017) Ecological roles and biological activities of secialized metabolites from the genus *Nicotiana. Chemical Reviews* **117**, 12227–12280. doi:10.1021/acs.chemrev.7b00001
285. Kaunda JS, Zhang YJ (2019) The genus *Solanum*: an ethnopharmacological, phytochemical and biological properties review. *Natural Products and Bioprospecting* **9**, 77–137. doi:10.1007/s13659-019-0201-6
286. Achenbach H, Hubner H, Brandt W, Reiter M (1994) Cardioactive steroid saponins and other constituents from the aerial parts of *Tribulus cistoides. Phytochemistry* **35**, 1527–1543. doi:10.1016/S0031-9422(00)86890-9
287. Achenbach H, Hubner H, Reiter M (1996) Cholestane- and pregnane-type glycosides from the roots of *Tribulus cistoides. Phytochemistry* **41**, 907–917. doi:10.1016/0031-9422(95)00622-2
288. Kostova I, Dinchev D (2005) Saponins in *Tribulus terrestris*: chemistry and bioactivity. *Phytochemistry Reviews* **4**, 111–137. doi:10.1007/s11101-005-2833-x
289. Chhatre S, Nesari T, Somani G, Kanchan D, Sathaye S (2014) Phytopharmacological overview of *Tribulus terrestris. Pharmacognosy Reviews* **8**, 45–51. doi:10.4103/0973-7847.125530
290. Neychev V, Mitev V (2016) Pro-sexual and androgen enhancing effects of *Tribulus terrestris* L.: fact or fiction. *Journal of Ethnopharmacology* **179**, 345–355. doi:10.1016/j.jep.2015.12.055
291. Santos HO, Howell S, Teixeira FJ (2019) Beyond tribulus (*Tribulus terrestris* L.): the effects of phytotherapics on testosterone, sperm and prostate parameters. *Journal of Ethnopharmacology* **235**, 392–405. doi:10.1016/j.jep.2019.02.033

5
Miscellaneous-type specialised metabolites

INTRODUCTION

A persistent difficulty encountered throughout this work has been the clear disambiguation and categorisation of the complex and substantial group of organic molecules under assessment. It was inevitable that a 'miscellaneous'-type natural produc (NP) class would result for the compounds that did not neatly fit the previous three categories, but arranging this disorganised group into a coherent system has proven problematic. Many of the textbooks that address this do so on biosynthetic grounds and this strategy has been followed here, with the remaining NP classes grouped together as: a 'phenolic-like' subclass of miscellaneous compounds of the polyketide (tetra-, penta-, hexa-, hepta-, octa- etc.) and quinone infraclasses (benzoquinone, naphthoquinone, anthraquinone and phenanthrenequinone), a 'terpenoid-like' subclass of miscellaneous compounds of the meroterpenoid infraclass (a large and complex group formed through the condensation of isoprenoids and various other NP sub-/infraclasses, primarily terpenophenolics and alkaloterpenoids) and an 'alkaloid-like' subclass of miscellaneous compounds of both simple (aromatic amines) and complex (cyanogenic glycosides, glucosinolates) infraclasses (*Fig. 5.1*). Within these subclasses, certain compounds or infraclasses are situated tightly between plant primary and secondary metabolism and in an attempt to avoid going beyond the intended scope of this book, certain miscellaneous-type specialised metabolites will be omitted from this chapter on primary metabolic grounds; that is, carbohydrates (mono-, di-, oligo- and polysaccharides), lipids (fatty acids, polyacetylenes, waxes, latexes) and polypeptides (proteins). Biosynthesis and biological activities are clearly going to be disparate within this section, so will be addressed on a case-by-case basis. This chapter will proceed through an introduction to the class itself, followed by more detailed breakdowns of each of the terpenoid-type metabolic subclasses. The end of the chapter will present a summary of the representation of these metabolites reported to be present across the customary medicinal flora of Australia, detailed in *Table 5.1*. This information can also be explored using the associated website (www.cmfoa.info).

'PHENOLIC-LIKE' SUBCLASS

Polyketides

The polyketide-type NP infraclass are grouped together based on their biosynthesis, namely the polyketide pathway. As with the mevalonic acid pathway seen in terpenoid biosynthesis and the primary metabolic fatty acid pathway addressed in general biochemistry textbooks, the polyketide pathway functions through the repeated condensation of acetate/malonate units (C_2 extension).[1] The great structural diversity arising from this metabolic pathway results from the cyclic condensations that

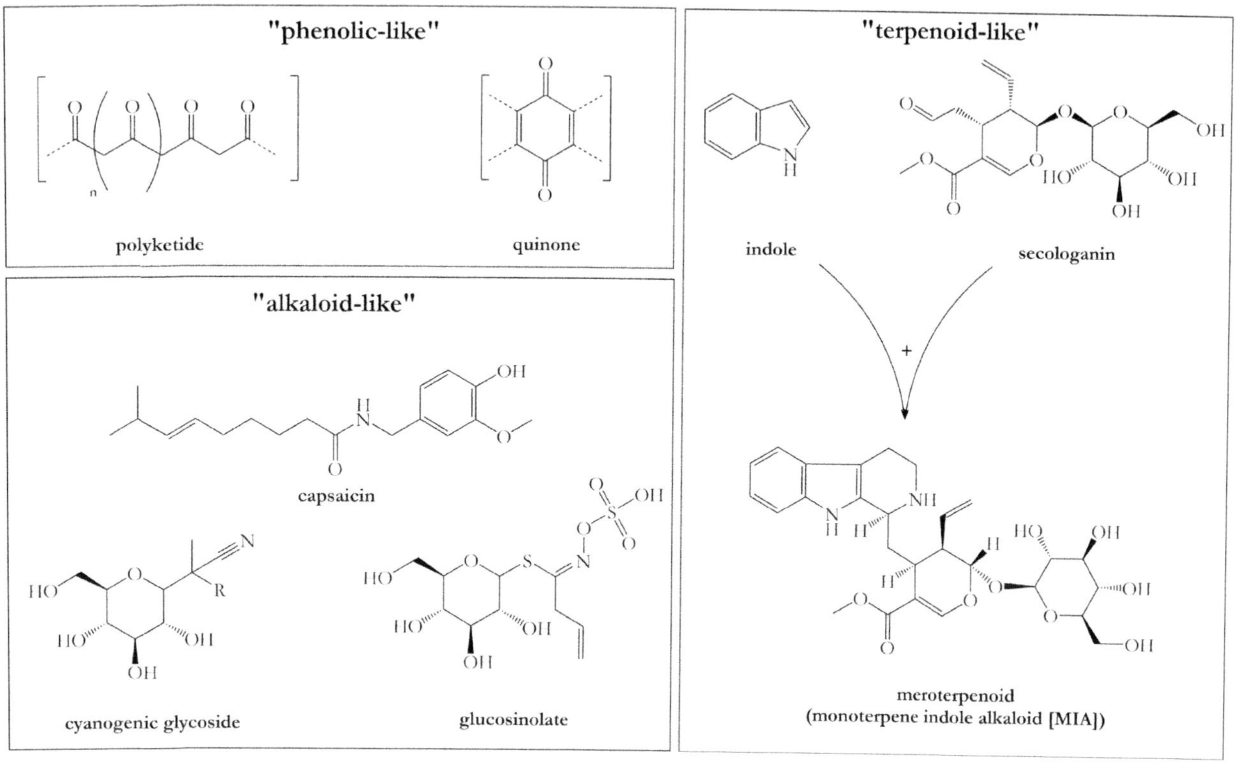

Fig. 5.1. Major subclass groupings of the miscellaneous-type specialised metabolite class.

can occur at various places along the growing polyketide chain, often giving phenolic/aromatic moieties, which is why, although they have 'phenolic-like' moieties, they are being addressed as part of phenolic-type specialised metabolite class. The primitive simplicity of this biosynthetic pathway is evinced by its distribution across the more ancient domains of life: algae, bacteria, fungi and lichen.[1] This being said, certain polyketides can also be found throughout the plant kingdom, such as the ubiquitous quinones (addressed separately below) that are integral to all plant life through their electron accepting role in the electron transport chain of cellular respiration. The intermediates that are produced through the action of enzymes of the polyketide synthase (PKS) grouping show varying degrees of reduction/oxygenation and have been shown to predominantly cyclise through one of two mechanisms facilitated by the enzyme active site (with no intermediary release): intramolecular aldol condensation and intramolecular Claisen condensation (*Fig. 5.2*).[2] For the purposes of this chapter, the polyketides are being treated as a singular group, further classified based on their number of C_2 units: tetra-, penta-, hexa-, hepta-, octa- etc., but certain structural variations are worth noting, such as the macrocylic macrolides (mentioned below, but not elaborated upon due to their microbial origins) and the di-ketone quinones (addressed separately below). The potent biological activities exhibited by these NPs are thought to play several roles for their progenitors: hormonal or developmental regulation, ecological interactions, or defensive as a phytoalexin/phytoanticipin. Of particular note regarding their use by the pathogenic microbial sphere against plants is their phytotoxicity (causing the death and decomposition of their host plant).[1]

Much of the research that has been conducted on the large group of natural polyketides has pertained to microbially derived NPs, which unfortunately lie outside the scope of this book. Some significant examples are the neurotoxic, carcinogenic tetraketide patulin isolated from *Penicillium* spp. and *Aspergillus* spp.; the pentaketide citrinin implicated in the 'yellow rice disease' of the 20th century isolated from *Aspergillus* spp. and *Penicillium citrinum*; emodin, the important octaketide intermediate in many fungal metabolites (also found in the higher plants); the octaketide tetracycline antibiotics (e.g. terramycin, aureomycin, tetracyline) isolated from *Streptomyces* spp.; the decaketide aflatoxin B1 (one of the most toxic and carcinogenic NPs known) isolated from *Aspergillus* spp. (and others); and the macrolide-type erythromycin antibiotics isolated from *Streptomyces erythreus* [now

Table 5.1. Phenolic-type specialised metabolite distribution across the customary medicinal flora of Australia

Order	Family	Genus	Curcuminoid	Tetraketide	Urushiol	Benzoquinone	Naphthoquinone	Antrhaquinone	Phenanthrenequinone	Cannabinoid	Alpha-acid	MIA	Ergot-type	Ipecacuanha-type	Mono-	Sesqui-	Di-	Tri-	Simple	Cyanogenic glycoside	Glucosinolate
Alismatales	Araceae	*Alocasia*[83,84]																		+	
Alismatales	Araceae	*Rhaphidophora*[85,86]										+									
Apiales	Apiaceae	*Centella*[87,88]	+																		
Asparagales	Asphodelaceae	*Dianella*[89,90]					+	+													
Asparagales	Orchidaceae	*Dendrobium*[91]						+								+					
Asterales	Asteraceae	*Acmella*[92]								+									+		
Asterales	Asteraceae	*Chrysocephalum*[93,94]								+	+										
Asterales	Asteraceae	*Eclipta*[95,96]																+			
Asterales	Asteraceae	*Gnaphalium*[97]								+	+										
Asterales	Asteraceae	*Helichrysum*[97]								+	+										
Asterales	Campanulaceae	*Lobelia*[98]																	+		
Boraginales	Boraginaceae	*Ehretia*[99,100]				+														+	
Brassicales	Capparaceae	*Capparis*[101]		+															+		+
Brassicales	Gyrostemonaceae	*Codonocarpus*[102]																	+		
Caryophyllales	Amaranthaceae	*Achyranthes*[103]				+		+													
Caryophyllales	Amaranthaceae	*Atriplex*[104,105]						+													
Caryophyllales	Droseraceae	*Drosera*[106]					+														
Caryophyllales	Plumbaginaceae	*Plumbago*[107]					+														
Caryophyllales	Polygonaceae	*Persicaria*[108–111]				+		+											+		
Caryophyllales	Portulaceae	*Portulaca*[112]																	+		
Cucurbitales	Cucurbitaceae	*Trichosanthes*[113–117]																	+		
Cycadales	Cycadaceae	*Cycas*[118]																	+		
Dioscoreales	Dioscoreaceae	*Dioscorea*[119,120]						+	+												
Ericales	Lecythidaceae	*Barringtonia*[121]																	+		
Ericales	Primulaceae	*Aegiceras*[122]				+	+												+		

Order	Family	Genus	Curcuminoid	Tetraketide	Urushiol	Benzoquinone	Naphthoquinone	Antrhaquinone	Phenanthrenequinone	Cannabinoid	Alpha-acid	MIA	Ergot-type	Ipecacuanha-type	Mono-	Sesqui-	Di-	Tri-	Simple	Cyanogenic glycoside	Glucosinolate
Fabales	Fabaceae	*Acacia*[123]																		+	
Fabales	Fabaceae	*Caesalpinia*[124]					+														
Fabales	Fabaceae	*Canavalia*[125-128]																	+		
Fabales	Fabaceae	*Cassia*[129]						+													
Fabales	Fabaceae	*Chamaecrista*[129]						+													
Fabales	Fabaceae	*Entada*[130,131]																	+		
Fabales	Fabaceae	*Erythrophleum*[132]															+				
Fabales	Fabaceae	*Indigofera*[133]																	+		+
Fabales	Fabaceae	*Mucuna*[134-136]																		+	
Fabales	Fabaceae	*Senna*[129]						+													
Gentianales	Apocynaceae	*Alstonia*[137]										+									
Gentianales	Apocynaceae	*Carissa*[138]																		+	
Gentianales	Apocynaceae	*Cynanchum*[139]																+			
Gentianales	Apocynaceae	*Marsdenia*[140]																	+		
Gentianales	Apocynaceae	*Ochrosia*[141-145]										+									
Gentianales	Apocynaceae	*Tabernaemontana*[146-148]										+									
Gentianales	Loganiaceae	*Strychnos*[149-153]										+									
Gentianales	Rubiaceae	*Coelospermum*[154-157]						+													
Gentianales	Rubiaceae	*Morinda*[158-160]						+													
Gentianales	Rubiaceae	*Nauclea*[161,162]										+									
Gentianales	Rubiaceae	*Timonius*[163,164]										+									
Lamiales	Lamiaceae	*Ajuga*[165]				+															
Lamiales	Lamiaceae	*Clerodendrum*[166]						+												+	
Lamiales	Lamiaceae	*Plectranthus*[167]							+												
Lamiales	Lamiaceae	*Prunella*[168]						+													

Table 5.1. continued

Order	Family	Genus	Curcuminoid	Tetraketide	Urushiol	Benzoquinone	Naphthoquinone	Antrhaquinone	Phenanthrenequinone	Cannabinoid	Alpha-acid	MIA	Ergot-type	Ipecacuanha-type	Mono-	Sesqui-	Di-	Tri-	Simple	Cyanogenic glycoside	Glucosinolate
Lamiales	Lamiaceae	*Vitex*[169–171]															+		+		
Lamiales	Scrophulariaceae	*Eremophila*[172]																		+	
Laurales	Lauraceae	*Litsea*[173]				+															
Malvales	Malvaceae	*Hibiscus*[174–177]								+									+		
Malvales	Malvaceae	*Malva*[178,179]					+														
Malvales	Malvaceae	*Sida*[180]																	+		
Myrtales	Lythraceae	*Ammannia*[181]					+														
Piperales	Piperaceae	*Piper*[182,183]														+			+		
Poales	Cyperaceae	*Cyperus*[184]														+					
Poales	Cyperaceae	*Remirea*[185–188]				+															
Rosales	Rhamnaceae	*Ventilago*[189–195]					+	+													
Rosales	Rhamnaceae	*Ziziphus*[196–199]						+											+		
Rosales	Rosaceae	*Rubus*[200,201]						+													
Sapindales	Anacardiaceae	*Semecarpus*[202–205]			+																
Sapindales	Meliaceae	*Melia*[206]						+													
Sapindales	Rutaceae	*Geijera*[207–209]																	+		
Sapindales	Rutaceae	*Zieria*[210–214]																		+	
Sapindales	Sapindaceae	*Allophylus*[215]					+	+											+		
Solanales	Convolvulaceae	*Evolvulus*[216]											+								
Solanales	Convolvulaceae	*Ipomoea*[217]											+						+		
Solanales	Convolvulaceae	*Operculina*[218–221]																	+		
Solanales	Solanaceae	*Nicotiana*[222–224]																	+		
Solanales	Solanaceae	*Solanum*[225]																+	+		
Zingiberales	Zingiberaceae	*Curcuma*[226]	+																		

Note: grey rows have been taxonomically reclassified since respective publications; genera with no reported phytochemistry have been omitted (see www.cmfoa.info .

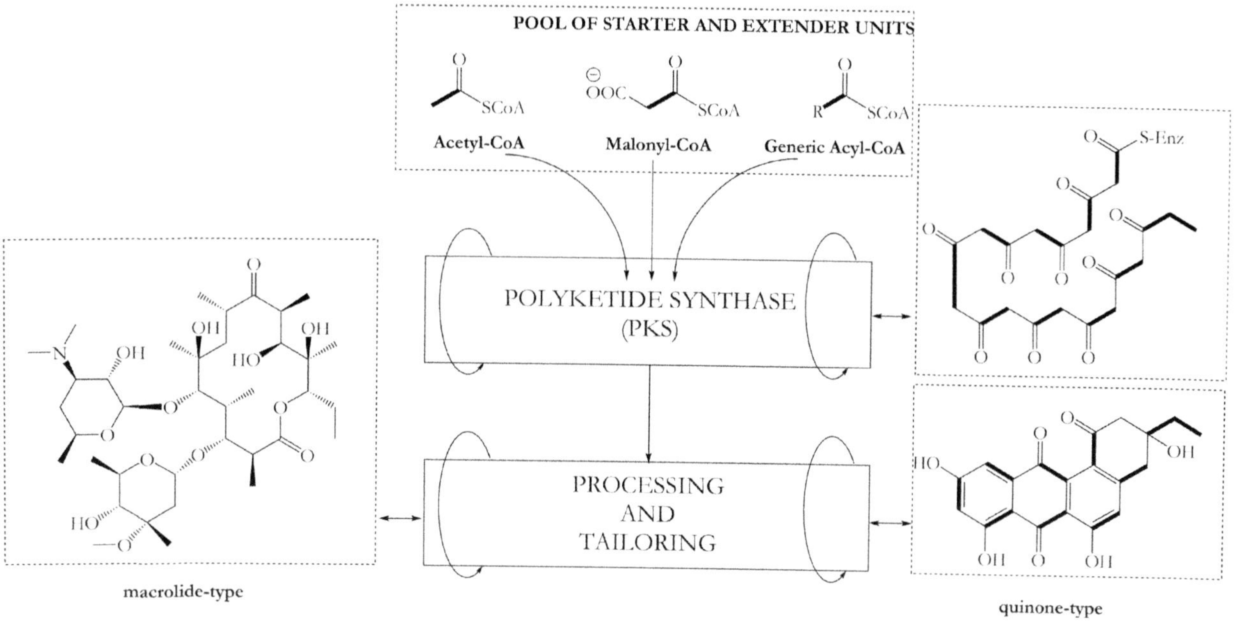

Fig. 5.2. Generic, simplified polyketide metabolic pathway showing two significant potential outcomes (macrolide-type and quinone-type polyketides).

Saccharopolspora erythraea ((Waksman 1923) Labeda 1987, Pseudonocardiaceae) (*Fig. 5.3*).[1] Lichens are also well known for their polyketide NP content, with the fungal member(s) of this symbiotic organism thought to synthesise these compounds from the carbohydrates produced by the algal member(s), with the process thought to be mediated by their bacterial member(s).[1,3] These metabolites, such as depsides, have been critical in the chemotaxonomy of the over 18,000 known lichen species.[1,3]

The polyketides are more restricted in their presence as specialised metabolites in the plant kingdom. The most well-known polyketide-type NPs are the curcuminoids of the Zingiberaceae [*Curcuma* (turmeric)], which arise from the diketide-CoA synthase mediated malonyl-CoA extension of a *p*-coumaroyl-CoA precursor (or its *O*-methylated feruloyl-CoA structural relative) derived from the shikimate/chorismate pathway.[1,4] These two phenolic-diketide-CoA intermediates are then subsequently condensed with another *p*-coumaroyl-CoA (or feruloyl-CoA) unit by the action of curcumin synthase to yield the three major curcuminoid sub-types: bisdemethoxycurcumin, demethoxycurcumin and curcumin (*Fig. 5.4*).[4] The curcuminoids have been implicated in a huge range of promising biological activities that have been studied extensively. Indeed, turmeric extracts (active ingredients: curcumin among others) are one of the most popular nutritional supplements on the market today. Briefly, the biological activities ascribed to curcuminoids (viz. turmeric extract) are: antioxidant, antiprotozoal, antivenom, antimicrobial, antimalarial, anti-inflammatory, antiproliferative, antiangiogenic, antitumour, antiageing and cardioprotective. Hence, the application of these extracts both clinically and traditionally includes, but is not limited to, the treatment of ulcers, parasites, skin diseases, immune disorders, neurological disorders and viral infections.[5]

Other minor examples of the plant-derived polyketide infraclass are the cytotoxic (against KB and P-338 cell lines) tetraketides isolated from the Annonaceae [*Melodorum fruticosum* (Lour. 1790)] and the pentaketide content of the essential oil from clove [*Syzygium aromaticum* ((L.) Merr. & L.M.Perry 1939, Myrtaceae)].[1] Over 400 acetogenins (C_{32-34} compounds with polyketide head groups) have been isolated from the Annonaceae (*Annona, Simina, Goniothalamus, Rollinia, Uvaria*) with potent antimicrobial, antiparasitic, cytotoxic (ED_{50} < 10^{-7} µg/mL) and antineoplastic activity (*Fig. 5.5*).[1,6] Certain unusual polyketide-type compounds arise from the extension of fatty acid precursors, such as the extreme irritants of the urushiol-type, which arise from palmitoleic acid in the Anacardiaceae [*Toxicodendron radicans* ((L.) Kuntze 1891, poison ivy)] in particular and the C_{22-24} anacardic acids (e.g. geranicardic acid) found primarily in pest-resistant members of the Geraniaceae (*Pelargonium*), but similar compounds have been isolated from the Anacardiaceae, Ginkgoaceae, Myrisitaceae, Myrcinaceae, Poaceae and Proteaceae (*Fig. 5.5*).[7]

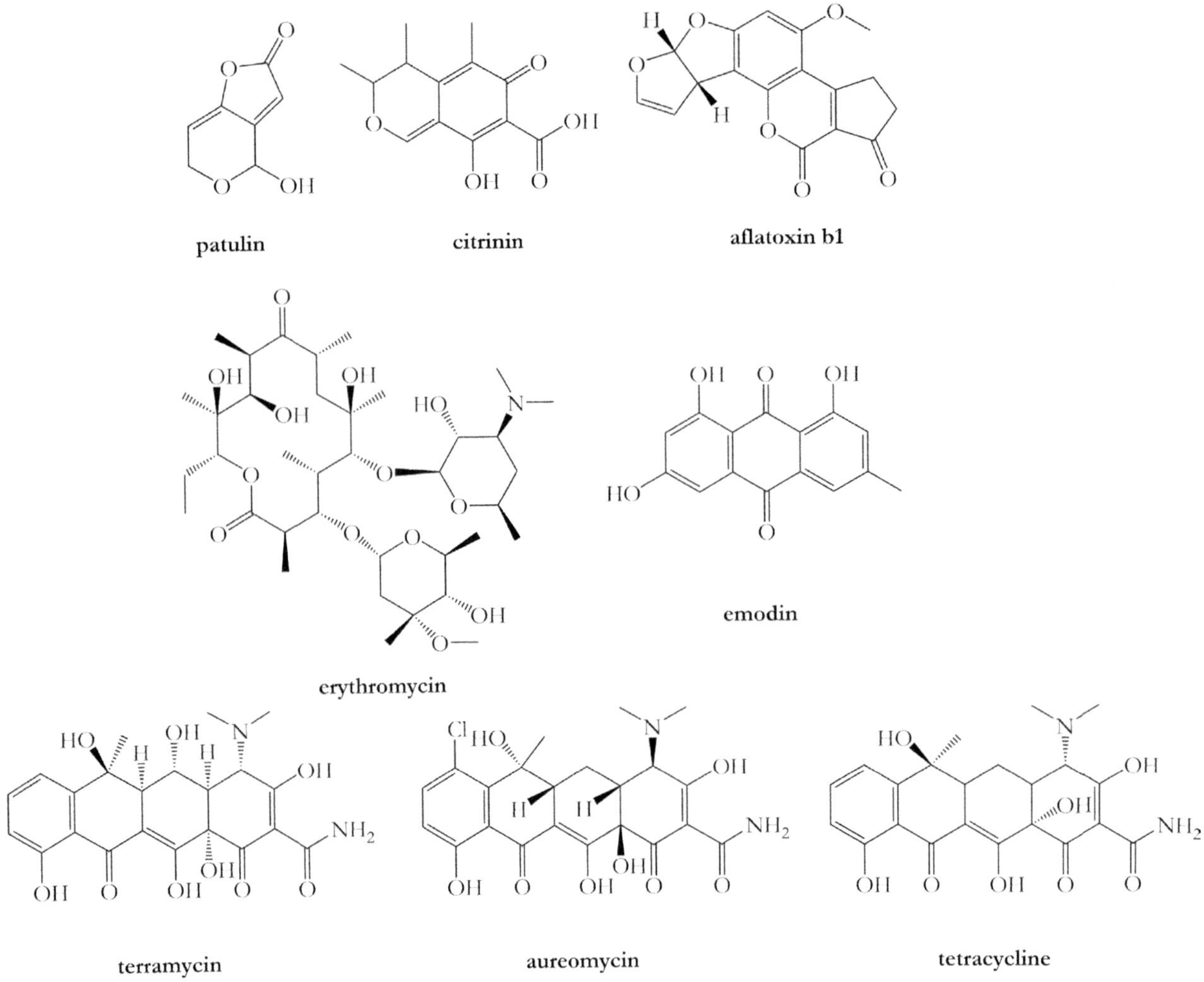

Fig. 5.3. Clinically important microbially derived polyketide drugs.

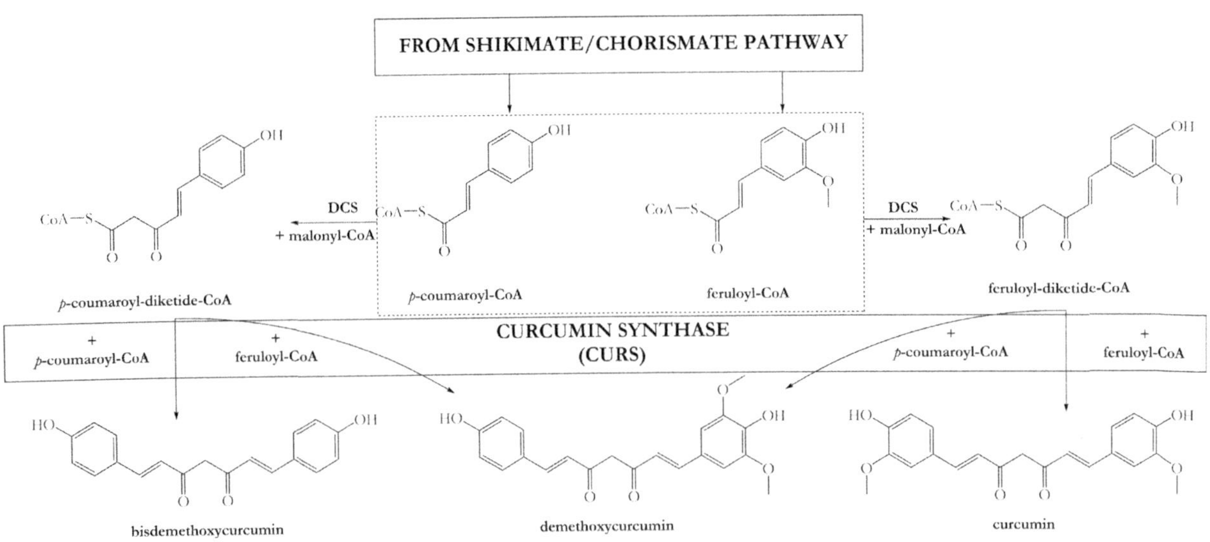

Fig. 5.4. Curcuminoid metabolic pathway, showing how the three major curcuminoid subtypes arise via diketide-CoA-synthase, a type III polyketide synthase.

Bioactive polyketides of the methylenebisphloroglucinol-type [e.g. the anthelmintic, purgative filicins (e.g. filixic acid)], have been isolated primarily from the Dryopteridaceae (*Dryopteris*) and similar compounds have also been isolated from various higher plant families such as the Clusiaceae (*Hypericum*), Euphorbiaceae (*Mallotus*) and Myrtaceae (*Eucalyptus*).[1] The distribution of miscellaneous-type specialised metabolites of the curcuminoid-type polyketide infraclass reported from the customary medicinal flora of Australia is presented in *Table 5.1*.

Quinones

Quinones $[C_6(C_4)_n]$ are an infraclass of 'phenolic-like' miscellaneous-type specialised metabolites that vary in the number and arrangement of their aromatic ring structures (benzoquinone, naphthoquinone, anthraquinone and phenanthrenequinone) and are grouped together due to their important biochemical role arising from the redox properties of their dicarbonyl (quinone) moiety (*Fig. 5.6*).[8] It is difficult to do justice to the primary functionality of quinones in nature and their ubiquity in biological systems has resulted in one *p*-benzoquinone derivative being named 'ubiquinone', also known as coenzyme Q, that exists in three redox states: fully oxidised (ubiquinone), partially reduced (semiquinone) and fully reduced (ubiquinol), all with a variable length isoprenyl tail (where its Q_6-Q_{10} subscript comes from).[8] Coenzyme Q and many other biological quinones are integral members of the electron transport chain in oxidative respiration, as well as constitutive reactive oxygen species (ROS) scavengers (i.e. antioxidants).[8] Their wide ranging biological activity arises from two primary mechanisms: redox cycling and nucleophilic reactions with proteins/enzymes (binding of thiol, amine and hydroxyl moieties).[9,10] This could be applied equally well to the benzoquinones,

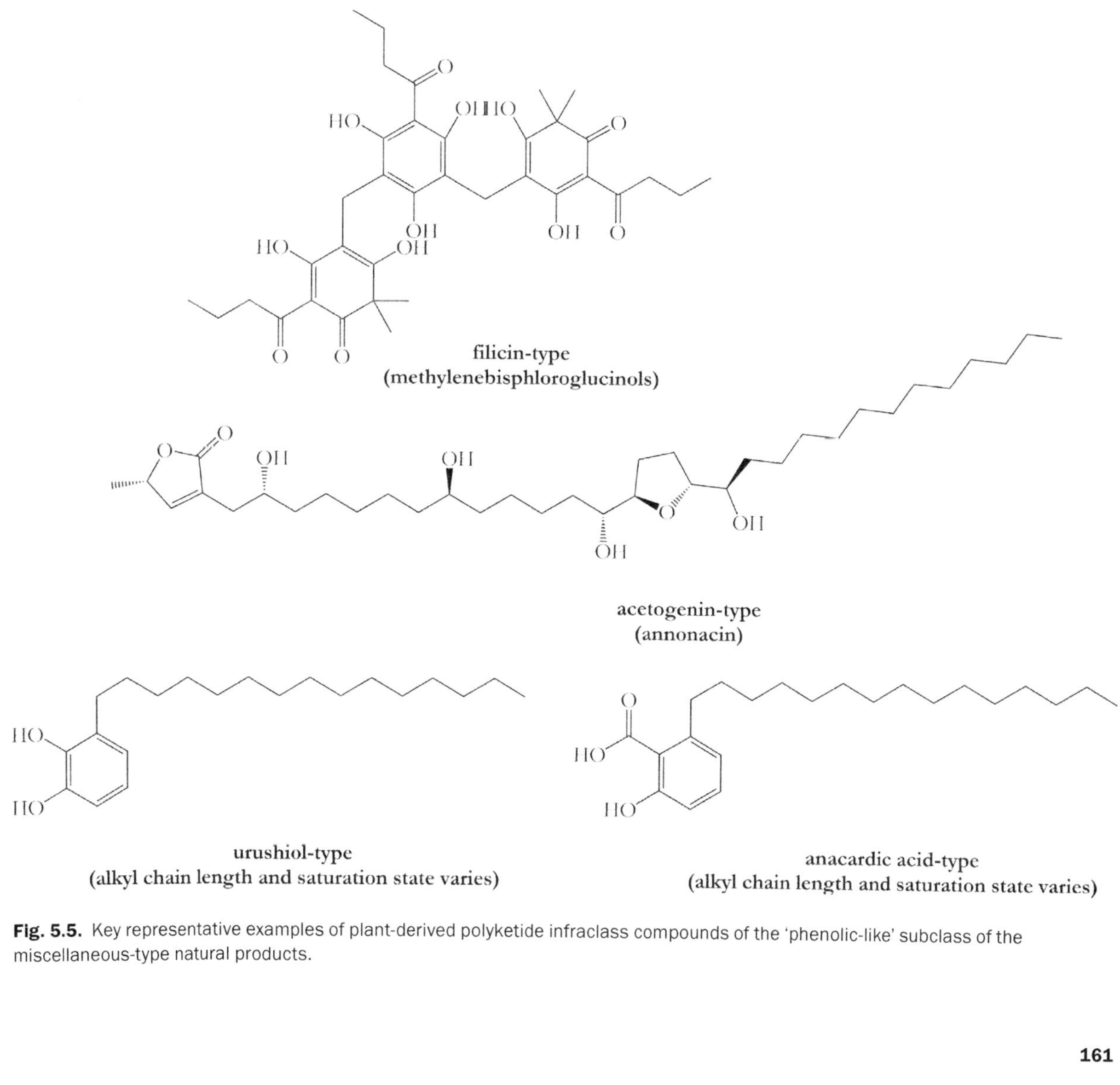

Fig. 5.5. Key representative examples of plant-derived polyketide infraclass compounds of the 'phenolic-like' subclass of the miscellaneous-type natural products.

naphthoquinones, anthraquinones or any other compound that bears a catechol/quinone moiety (e.g. quercetin, catechin, procyanidin, podophyllotoxin etc.).

Benzoquinones have been reported to have antibiotic, antimalarial, antineoplastic, anticoagulant and herbicidal activity.[11] Naphthoquinones have shown antineoplastic, antibacterial, anti-inflammatory, antiparasitic and antifungal activity.[12] It is thought that in addition to the general mechanisms used by all the quinones, naphthoquinones can inhibit electron transport, uncouple oxidative phosphorylation processes and intercalate DNA. Anthraquinones have long been applied for their laxative qualities, but they have also been shown to possess hepatoprotective, antineoplastic, antimicrobial, antifungal, antiviral, antidiabetic and antioxidant activity.[13] Phenanthrenequinones have been investigated much less than the other quinones, but structure–activity relationships (SAR) studies of denbinobin have shown antiproliferative, antineoplastic, anti-inflammatory and antimigratory activities.[14] *Table 5.1* presents the customary medicinal flora of Australia that were reported to possess quinone-type compounds of the 'phenolic-like' subclass of the miscellaneous-type class in the literature reviewed as part of this book.

o-Benzoquinones arise biosynthetically from the oxidation of many phenols, specifically catechols and, likewise, *p*-benzoquinones are derived from phenols, 1,4-dihydroxybenzenes, hydroquinones and dimethoxybenzenes.[15] Polyketide synthesis occurs differently in fungi, lichens, plants and animals, leading to aromatic compounds such as naphthoquinones and

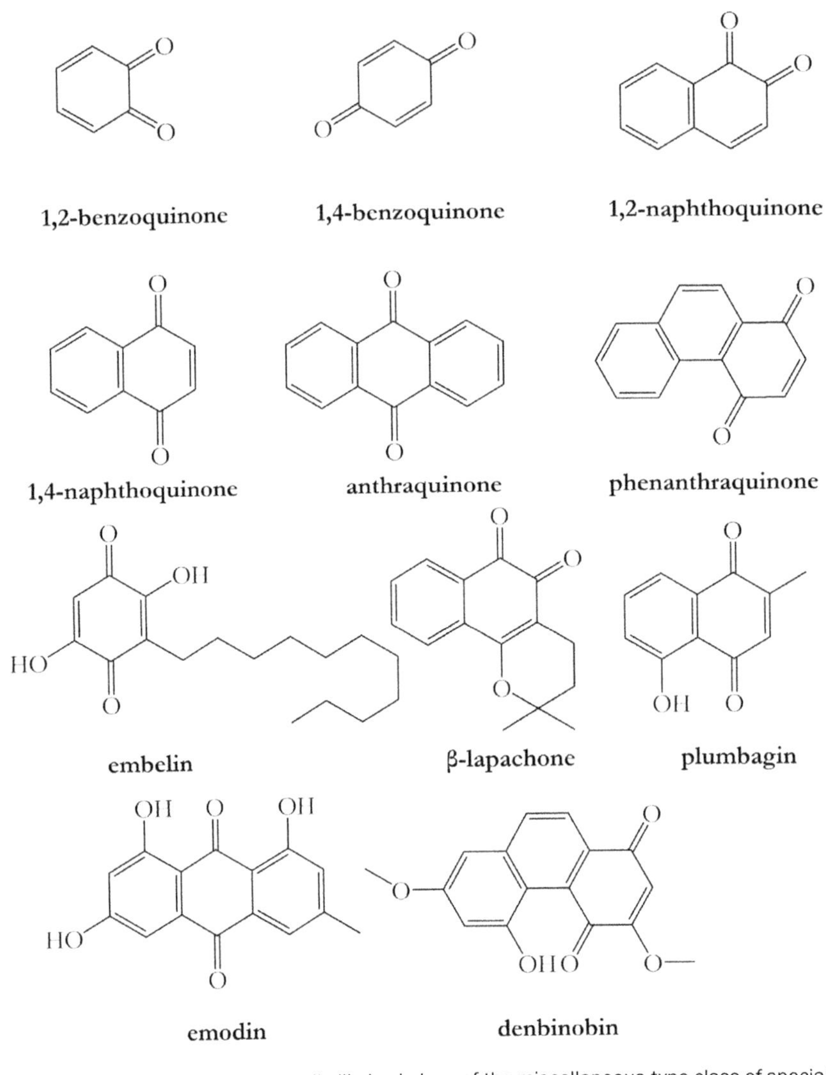

Fig. 5.6. Major quinone-type infraclasses of the 'phenolic-like' subclass of the miscellaneous-type class of specialised metabolites (and representative examples).

anthraquinones, but also to complex compounds such as tetracycline antibiotics and macrolides such as erythromycin.[16] Three types of polyketide synthase enzymes (PKS I, II and III) have been identified in nature and they catalyse the assembly and cyclisation ('folding') of poly-β-ketoacyl intermediates. Two approaches are followed to achieve this folding: eukaryotes (fungi, plants, animals) produce polyketides with two intact C_2 units in the first aromatic ring (called F-mode, for Fungal), whereas prokaryotes yield three intact C_2 units in the first aromatic ring (called S-mode, for *Streptomyces*).[16,17]

Naphthoquinones are synthesised from naphthalene, which has its origin in a polyketide (pentaketide) formed through successive Claisen condensations from one unit of acetyl CoA and four units of malonyl CoA through the action of PKS (*Fig. 5.2*).[18] The first condensation/dehydration occurs with the loss of CoASH to yield 1,3,6,8-tetrahydroxynaphthalene, which can go through a series of dehydrations to yield naphthalene in a variety of hydroxylated forms.[18] Cytochrome P450 generates 1,2-epoxyareneoxide from naphthalene, which can be hydrated via epoxide hydrolase to 1,2-dihydroxydihydronaphthalene. This forms a branching point in which 1,2-naphthoquinones can be formed directly via dihydrodiol dehydrogenase or 1,4-naphthoquinones can be formed through a multistep process from the dehydrated 1-naphthol.[17] As with the initial naphthalene step, 1-naphthol can be oxidised by P450 to yield 1-naphthol-3,4-areneoxide, which through epoxide hydrolase yields 1,3,5-trihydroxydihydronaphthalene that dehydrates to dihydronaphthalene and finally oxidises to yield 1,4-naphthoquinone.[17] Anthraquinones are formed via the polyketide pathway, from one acetyl CoA unit and seven malonyl CoA units. Whether synthesised in eukaryotes or prokaryotes, anthraquinones need to pass through an open-chain octaketide precursor before condensation/folding.[16] A ^{13}C-labelling study demonstrated that eukaryotes (fungi, plants and animals) all follow the F-mode folding pathway to produce anthraquinones.[16] Phenanthrenequinones are thought to arise differently, from the oxidation of phenanthrenes that are biosynthesised by the intramolecular oxidative coupling of various phenolic-type stilbene precursors.[14] However, the details of this biosynthetic pathway are not fully understood and various phenanthrenes and abietane-type diterpenes have been proposed as potential progenitors of the phenanthrenequinones.[14]

Benzoquinones

Benzoquinones are monocyclic aromatic C_6 compounds bearing either a 1,2- or 1,4-dicarbonyl (quinone) moiety and commonly called *o*-benzoquinones and *p*-benzoquinones, respectively. Although hypothetically possible, 1,3-benzoquinones are not observed in nature.[15] *o*-Quinones arise naturally from the oxidation of catechol moieties and are ubiquitous (but also transient due to their electrophilicity) in biological systems.[15] *o*-Quinones are present as a moiety borne by many of the other specialised metabolites discussed in this book and have wide ranging applications in biological systems.[12,15] *p*-Benzoquinones are also ubiquitous in nature and, similar to *o*-benzoquinones, their presence is rarely in isolation and they feature widely as a moiety incorporated into other natural products.[11] Hence, it is difficult to enumerate exactly how many benzoquinones have been isolated, as they occur as moieties across many separate classes and infraclasses of natural products. *o*-Benzoquinones are formed *in planta* and are integral to the prooxidant/antioxidant balance within cells, either quenching ROS to form quinones or proceeding to various biological targets in their active form.[15]

These biological functions can be categorised as detoxifying, chemoprotective or toxic. Glutathione (GSH) is a common cellular antioxidant; *o*-quinones have a coupled relationship with GSH and the depletion of one or the other can cause oxidative stress. The *o*-quinones can alkylate proteins and elicit a chemoprotective activity, through the antioxidant-response mediating protein complex KEAP1/NRF2 or through the inflammation-response mediating protein complex [IκB kinase / nuclear factor kappa B (NF-κB)].[15] Alternatively, *o*-quinones can elicit a stress response through protein alkylation, via heat-shock protein, protein disulfide isomerase, binding immunoglobin protein or glutathione-*S*-transferase (GST). Toxic effects arise from DNA adducts/oxidation leading to mutagenesis, cytochrome P450 oxidative enzyme inhibition or protein alkylation.[15] The *p*-benzoquinones are similarly integral to life; their molecular light absorbance in the UV-visible spectrum has led to them (and most of the quinones) being significant as pigments (viz. dyes).

The benzoquinone subclass exhibit similar biological activity as the phenolic compounds discussed in Chapter 2, primarily in their function as defensive phytoalexin/phytoanticipins.[11] Very few *o*-benzoquinones have been isolated as individual NPs, because they are most often seen as a moiety attached to a larger compound (and grouped in those sub-/infraclasses accordingly). That being said, much of the efficacy of those compounds may be attributable to the redox/nucleophilic properties of this common substituent. Although innumerable compounds have been reported as bearing a 1,4-benzoquinone moiety, the most notable 'true' *p*-benzoquinone is embelin (2,5-dihydroxy-3-undecyl-1,4-benzoquinone; *Fig. 5.6*), which is isolated from the berries of *Embelia ribes* and has been the subject of significant studies regarding its biological activity.[11,12] Embelin is identified as having numerous beneficial pharmacological properties, the two most substantial of which are its role in human cancer prevention/treatment and application to central nervous system (CNS) disorders.[19,20] The oncogenic pathways through which embelin is implicated with cancer are numerous: intrinsic/extrinsic apoptotic cell death, cancer cell autophagy and modulation of the nuclear factor kappa B (NF-κB), tumour protein 53 (p53), phosphoinositide 3-kinase/protein kinase B/mammalian target of rapamycin (PI3K/AKT/mTOR) and signal transducer and activator of transcription 3 (STAT3) signalling pathways.[19] The CNS pathways through which embelin is implicated as acting are also numerous and under active research: epilepsy [chloride/sodium cation (Cl^-/Na^+) conductance], traumatic brain injury (modulation of p53 and NF-κB), anxiety (benzodiazepine receptor mediating GABA A site), depression (inhibition of 5-hydroxytryptamine and noradrenaline reuptake), cerebral ischemia (restoration of antioxidant enzymes and reduced lipid peroxides), Huntington's disease (downregulation of malondialdehyde and upregulation of antioxidants GSH, GST, CAT), multiple sclerosis (modulation of transforming growth factor and STAT3 signalling pathway), among other studies.[20] The distribution of miscellaneous-type specialised metabolites of the benzoquinone infraclass reported from the customary medicinal flora of Australia is presented in *Table 5.1*.

Naphthoquinones

Naphthoquinones are bicyclic aromatic C_6C_4 compounds bearing either a 1,2- or 1,4-dicarbonyl (quinone) moiety. They arise metabolically from naphthalene, are relatively ubiquitous in nature and have been used for millennia by humans as dyes and medicines.[8] The most notable 1,2-naphthoquinone is β-lapachone, isolated from *Tabebuia rosea* [(Bertol.) DC. 1845, Bignoneaceae], which has been extensively studied for its antineoplastic and anti-inflammatory activities (*Fig. 5.7*).[21] The most notable 1,4-naphthoquinones are juglone [primarily from *Juglans nigra* (L. 1753, Juglandaceae)], plumbagin [from the genera *Plumbago* (Plumbaginaceae), *Drosera* (Droseraceae) and *Nepenthes* (Nepenthaceae)], lawsone ['henna' dye, from the genera *Lawsonia* (Lythraceae) and *Eichhornia* (Pontederiaceae)], lapachol [from *Handroanthus impetiginosus* ((Mart. ex DC.) Mattos 1970, Bignoneaceae)] and vitamin K (*Fig. 5.7*).[8,12] Naphthoquinones are widely distributed in the plant kingdom and are present in the Ferns [Schizeales (Schizeaeceae)], the Magnoliids

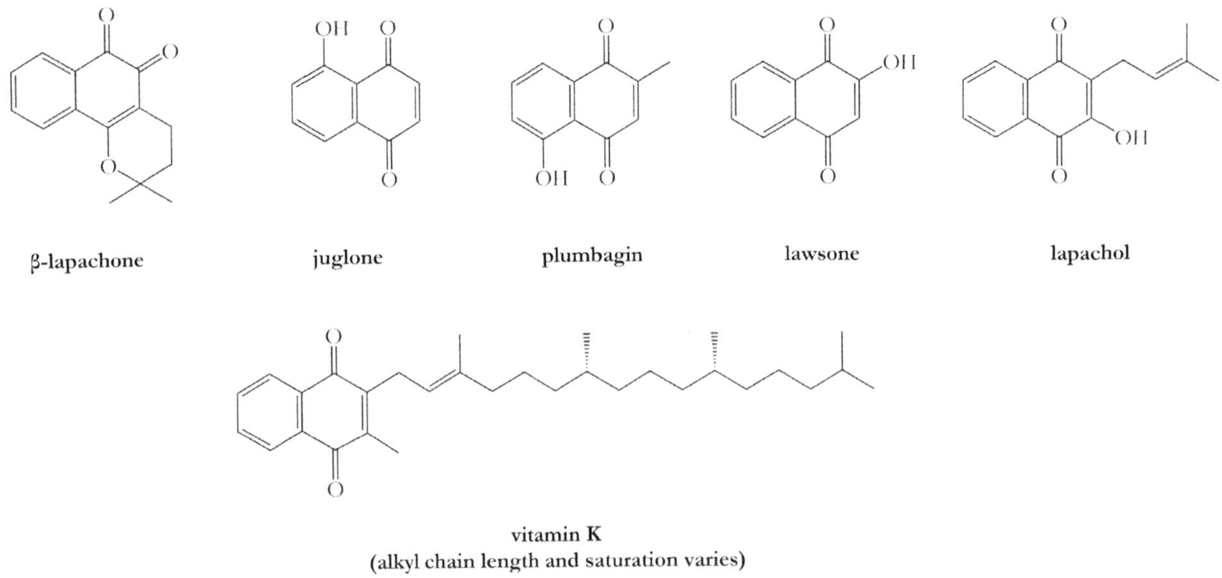

Fig. 5.7. Key representative examples of plant-derived naphthoquinone infraclass compounds of the 'phenolic-like' subclass of the miscellaneous-type natural products.

[Magnoliales (Annonaceae)], the Monocots [Asparagales (Iridaceae, Asparagaceae), Poales (Eriocaulaceae)] and the Eudicots [Proteales (Proteaceae), Zygophyllales (Zygophyllaceae), Malpighiales (Malpighiaceae), Fabales (Leguminosae), Rosales (Ulmaceae), Fagales (Juglandaceae), Myrtales (Lythraceae), Sapindales (Meliaceae, Rutaceae), Malvales (Malvaceae), Caryophyllales (Droseraceae, Nepenthaceae, Drosophyllaceae, Ancistrocladaceae, Dioncophyllaceae, Plumbaginaceae, Polygonaceae), Ericales (Balsaminaceae, Ebenaceae, Ericaceae), Boraginales (Boraginaceae), Gentianales (Gentianaceae, Rubiaceae) and Lamiales (Calceolariaceae, Gesneriaceae, Scrophulariaceae, Lamiaceae, Paulowniaceae, Pedaliaceae, Bignoniaceae, Acanthaceae)].[8,12] The major orders that are considered particularly 'rich' in naphthoquinones are the Caryophyllales, Ericales and Lamiales. Importantly, it should be noted that naphthoquinones have not been isolated from any Gymnosperms (conifers, cycads, *Ginkgo* and *Gnetales*).[8,12]

Naphthoquinones are more specific in their action; for instance, antimicrobial/antifungal naphthoquinones are thought to preserve the prey of carnivorous 'pitcher plants' (Nepenthaceae) from decomposition during glandular digestion.[12] These compounds are often found in plant roots and have a significant allelopathic effect on the surrounding rhizosphere. Further to this, naphthoquinones have been implicated in mycorrhizal, plant-fungal interactions that are beneficial to the plant. It appears that the antimicrobial (phytoalexin/phytoanticipin) and rhizobial aspects of the naphthoquinones is their most universal function, outside of the carnivorous families.[12] The activity of naphthoquinones is widespread and is not particular to any structural skeleton. Cytotoxicity is the primary medicinal function attributed to naphthoquinones, with various representatives (e.g. plumbagin) inhibiting cell proliferation and cell invasion, while inducing apoptosis and cell-cycle arrest.[21,22] These effects have largely been attributed to NF-κB inactivation in cancer cells, but also numerous other mechanisms.[12] Naphthoquinones have shown activity against breast, prostate, ovarian, pancreatic, lung, liver, renal, cervical and skin cancer cell lines, as well as myeloma and leukaemia.[21,22] Naphthoquinones are also powerful broadspectrum antibacterial compounds (semiquinone enhancement of ROS) and are well-documented anti-inflammatory compounds (modulation of various cell-signalling processes) and antiparasitic (generation of ROS, inhibition of endogenous antioxidant enzymes).[23] The distribution of miscellaneous-type specialised metabolites of the naphthoquinone infraclass reported from the customary medicinal flora of Australia is presented in *Table 5.1*.

Anthraquinones

Anthraquinones are symmetrical tricyclic aromatic $C_4C_6C_4$ compounds with a para-dicarbonyl (quinone or 9,10-anthracenedione) moiety on the central ring and are categorised into five groups (simple, hydroxy, anthranones, anthranols and bianthraquinone dimers) (*Fig. 5.8*).[8] Approximately 700 anthraquinones have been isolated, with around 200 of these coming from higher plants and the remainder being isolated from lichens and fungi.[8,13] The most notable compounds are emodin [*Rheum* (Polygonaceae), *Rhamnus* (Rhamnaceae) and *Reynoutria* (Polygonaceae)], parietin/physcion [giving the characteristic yellow colour of lichens from the *Caloplaca* (Teloschistaceae) and *Xanthoria* (Teloschistaceae), as well as being present in the roots of *Rumex* (Polygonaceae)] and rhein [*Rheum* (Polygonaceae), *Cassia* (Fabaceae)] (*Fig. 5.8*).[13] Anthraquinones are widely distributed in the plant kingdom and the following families (and genera) are considered 'rich' in anthraquinones: Fabaceae (*Cassia*), Liliaceae (*Aloe*), Polygonaceae (*Rheum*, *Rumex*), Rhamnaceae (*Rhamnus*), Rubiaceae (*Asperula*, *Coelospermum*, *Coprosma*, *Galium*, *Morinda* and *Rubia*) and Scrophulariaceae (*Digitalis*).[13]

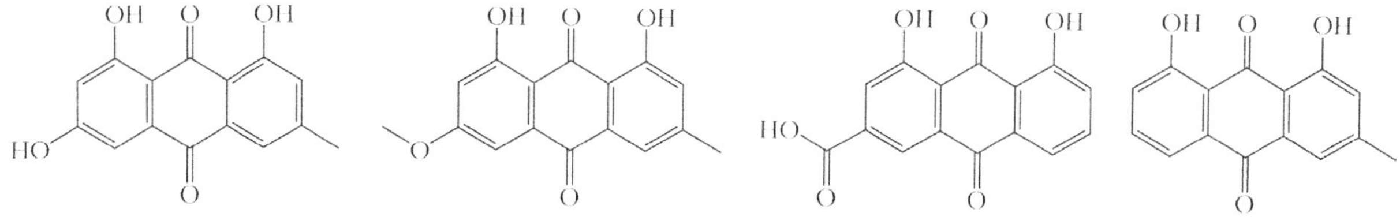

emodin parietin/physcion rhein chrysophanol

Fig. 5.8. Key representative examples of plant-derived anthraquinone infraclass compounds of the 'phenolic-like' subclass of the miscellaneous-type natural products.

Anthraquinones (e.g. chrysophanol) have a similar function to the other quinones, including pigmentation, defence and rhizobial symbiosis.[8] Anthraquinone pigments have been implicated in photosynthesis, absorbing at wavelengths complementary to chlorophyll (analogous to the carotenoids).[13,16,24] The biological activity of anthraquinones follows a similar pattern to the quinones just discussed; antineoplasticity occurs through tumour cell apoptosis, reversal of multidrug resistance, inhibition of enzymes, metabolism and the cell cycle, and inhibition of metastasis.[25] The presence of a C-1, C-5 or C-8 hydroxyl seems key to antineoplastic functionality; emodin, which is one of the most extensively studied medicinal anthraquinones, features C-1, C-3 hydroxylation.[25,26] Efficacy has been demonstrated against human oesophageal (EC-109), cervical (HeLa), hepatoma (J5), breast (MCF-7), ovarian (HO-8910), colon (HCT-116, SW-48), gastric (SNU-1, AGS, MKN-28, MKN-45), melanoma (A375), laryngiocarcinoma (Hep-2) and lung (H1229, HTB179, A549, H520) cancer cell lines.[25] Antimicrobial activity is well substantiated, through inhibition of bacterial respiratory metabolism, destruction of bacterial cell membranes/walls and biofilms and suppression of protein synthesis.[8] Anti-inflammatory activity has been well researched, with anthraquinones inhibiting inflammatory kinases and reducing lipopolysaccharide-induced Ca^{2+} concentrations.[8] The distribution of miscellaneous-type specialised metabolites of the anthraquinone infraclass reported from the customary medicinal flora of Australia is presented in *Table 5.1.*

Phenanthrenequinones

Phenanthrenequinones (or phenanthraquinones) are based on a $C_4C_6C_4$ phenanthrene skeleton and feature dicarbonyl (quinone) moieties at varying locations on the structure. Phenanthrenequinones are the rarest of the quinones, resulting from their origin as phenanthrenes, which are not widely represented in the plant kingdom. For this reason, the phenanthrenequinones have not been subject to a comprehensive review process and the sparse information presented here was inferred from more general reviews on phenanthrenes.[14] The 1,4,5,8-phenanthrenediquinone moniliformin was isolated in 2005 from the stems of *Dendrobium moniliforme* [(L.) Sw. 1799, Orchidaceae]. Phenanthrenes have been isolated from the following families: Annonaceae, Aristolochiaceae, Cannabaceae, Combretaceae, Dioscoreaceae, Euphorbiaceae, Juncaceae, Labiatae, Lauraceae, Malphigiaceae, Orchidaceae and Stemonaceae and are rare among this already sparse representation.[14] The majority of the phenanthrenequinones have been isolated from the Orchidaceae (*Calanthe*, *Cymbidium*, *Dendrobium*, *Bletilla*, *Oncidium* and *Odontioda*), although they have been reported from the Labiatae (*Salvia*) and Dioscoreaceae (*Dioscorea*).[14,27,28]

The limited presence of phenanthrenequinones in both nature and the literature has precluded any conclusive evidence regarding their function within plants and why they are mostly restricted to such a small group of genera in the Orchidaceae, although it would be sensible to purport that the role they play is similar to that of the rest of the natural quinones. Very few phenanthraquinones have been studied for their medicinal properties, with most analyses pertaining to denbinobin, which is notably cytotoxic (*Fig. 5.8*).[14] Denbinobin exhibits apoptosis induction through modulation of kinase (mitogen-activated protein kinase kinase kinase 7 (MAP3K7 or TAK1), NF-κB, AKT), caspase-dependent (caspase-3, cytochrome c, second mitochondria-derived activator of caspases) and caspase-independent (apoptosis-inducing factor) apoptotic mediators, plus ROS generation through activation of the apoptosis signalling kinase cascade (increased levels of pro-apoptotic Bcl-2-like protein 'Bim') causing mitochondrial dysfunction and cell death. The quinone moiety of denbinobin was deemed essential to MAPK-independent ROS production via SAR studies. Denbinobin has also been implicated in enhancing polymerised tubulin levels and stabilising microtubules (inhibiting cell division) and antimigratory effects.[14]

'TERPENOID-LIKE' SUBCLASS (MEROTERPENOIDS)

The MEP/MVA pathways of isoprenoid biosynthesis are common among all plants and the traditional terpenoid subclasses (mono-, sesqui-, di-, tri-, etc.) do not directly encompass the proliferation of secondary derivatives that form from the condensation of a terpenoid subunit and another natural product sub-/infraclass. These 'terpenoid-like' hybrids have been clustered under the title 'meroterpenoid', with many members of this diverse subclass showing significant biological activity. As this grouping is paraphyletic, it is very difficult to delineate what belongs in this group and what belongs to the group of the non-terpenoid substituent of the hybrid (or how to categorise a terpenoid that cyclised to yield a structure that is predominantly phenolic-like); hence, unfortunately, some specific examples may be unavoidably missed herein. The two major branches of the meroterpenoids arise from the condensation of a terpenoid and a phenolic (yielding the 'terpenophenolics') and the condensation of a terpenoid and an alkaloid (yielding the 'alkaloterpenoids'). Finally, it is worth mentioning that, once again, both microbially and marine invertebrate-derived meroterpenoids abound and lie outside the scope of this book.[29,30]

Terpenophenolics

The terpenophenolics are common among the Cannabaceae [*Cannabis* (hemp/marijuana), *Humulus* (hops)], with the most significant and well-studied infraclass being the cannabinoids and the α-acids (α-acids).[31] Smaller and more disparate infra-/subinfraclasses occur in a variety of plant families, grouped here under 'other terpenophenolics', such as the bakuchiol-type monoterpenophenolics [of the Amaranthaceae (*Aerva*), Fabaceae (*Otholobium*, *Psoralea*), Piperaceae (Piper), Ulmaceae (*Ulmus*)], the ferruginol-type diterpenophenolics [of the Cupressaceae (subfamily Cuppresoideae) and Podocarpaceae (*Podocarpus*)], the prenylated coumarins (primarily of the Apiaceae and Rutaceae, but also the Fabaceae and Moraceae) and the terpe-curcumins [of the Zingiberaceae (*Curcuma*)] (*Fig. 5.9*).[32-35]

Cannabinoids

The term 'cannabinoid' originally applied to the psychoactive prenylated phenolic meroterpenoids derived from the globally significant medicinal plants of the genus *Cannabis* (Cannabaceae). Since then, with the diversity of compounds isolated from these plants, their significant pharmacological activity and the discovery of both their receptors and endogenous ligands, this definition has been expanded to include any compound that can bind these receptors (even if they bear no resemblance to their terpenophenolic counterparts).[36] Natural cannabinoids (those derived from *Cannabis*; *Fig. 5.10*) accumulate on the glandular trichomes of the calyces and bracts of female flowers of the genus *Cannabis* (primarily *C. sativa* and *C. indica*), which have been cultivated by humans for at least the past 2700 years, as evinced by the archaeological findings in the Yanghai tombs of the Gobi desert in Western China, as well as the mention of the plant [*Má* (麻)] in the ancient Chinese medical pharmacopeia attributed to the mythical '*Shennong* (神農)'.[37] The cannabinoids appear to play a defensive role against desiccation, pathogenic attack, herbivory and UV light.[38]

Despite their 'phenolic' moiety, the major group of natural cannabinoids [e.g. Δ^9-tetrahydrocannabinol (THC), cannabidiol (CBD), cannabivarin (CBV), tetrahydrocannabivarin(THV); *Fig. 5.10*], arise from the geranylphosphate:olivetolate geranyl transferase (GOT)-mediated hybridisation of a monoterpene from the MEP isoprenoid pathway (C_{10} geranyl diphosphate), combined with a polyketide-pathway derived, alkylated resorcinol ring (either C_{10} divarinic acid or C_{12} olivetolic acid).[36] THC acid synthase (THCAS), CBD acid synthase (CBDAS) and spontaneous decarboxylation facilitates the formation of the four major natural cannabinoids.[36] THC and CBD are the best known and studied compounds of this group, which, besides being responsible for the

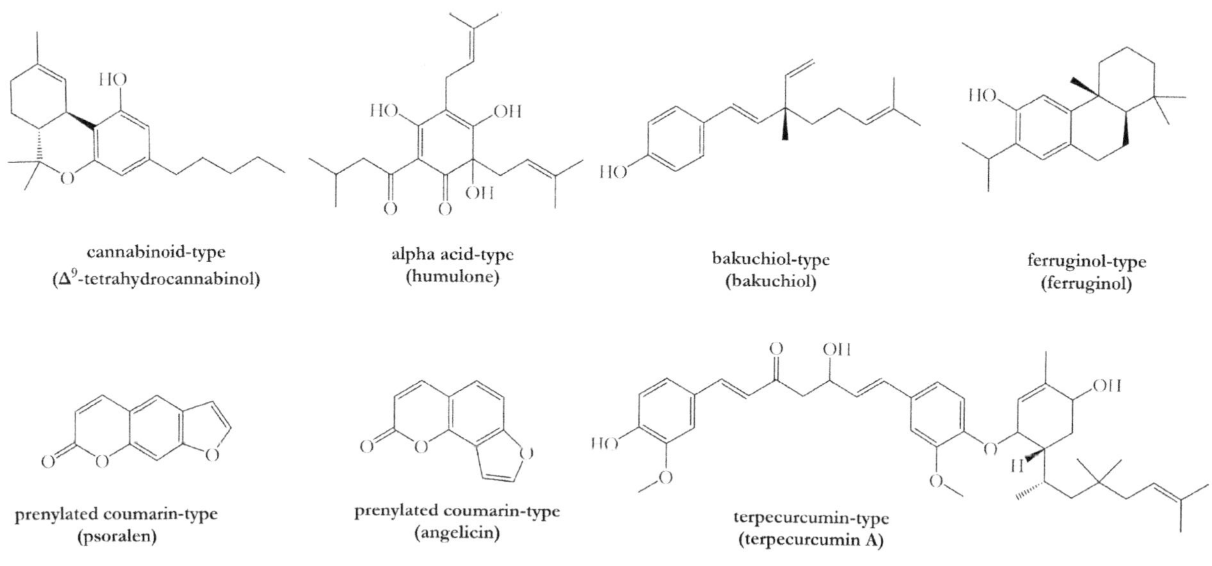

Fig. 5.9. Major terpenophenolic-type infraclasses of 'terpenoid-like' subclass of specialised metabolites (and representative examples).

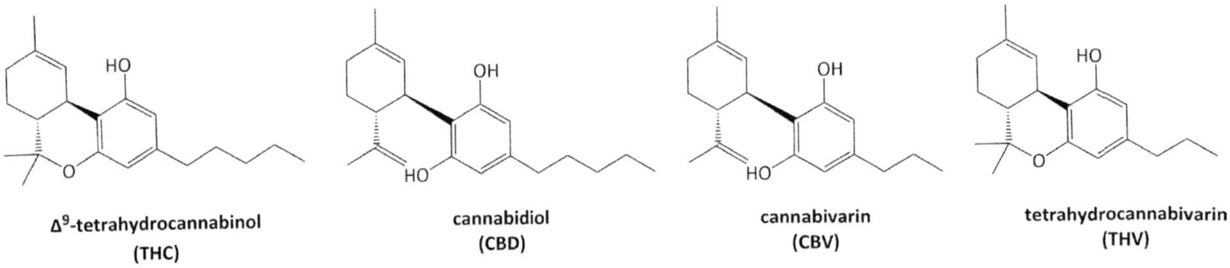

Fig. 5.10. Key representative examples of plant-derived cannabinoid-type compounds of the 'terpenoid-like' subclass of miscellaneous-type specialised metabolites.

intoxicating/recreational effects of *C. sativa* (L. 1753) and *C. indica* (Lam. 1785), are notable for their psychoactive and analgesic effects, respectively. THC principally binds the cannabinoid 1 (CB1) G-coupled protein receptor of the CNS, which inhibits calcium influx to the presynaptic nerve needed for excitatory neurotransmitter release. Although the mechanism of CBD is less well understood, other members of the minor group of natural cannabinoids (e.g. prenylated flavones such as cannflavins A and B, also meroterpenoids), which share structural features with CBD (proximity of alkylation site relative to the hydroxylated aromatic ring), have shown marked analgesic and anti-inflammatory activities due to the inhibition of two enzymes in the prostaglandin E2 inflammatory pathway that could fairly be ascribed to CBD as well, with further substantiation required and likely forthcoming in the near future.[36] The distribution of miscellaneous-type specialised metabolites of the cannabinoid-type terpenophenolic infraclass reported from the customary medicinal flora of Australia is presented in *Table 5.1.*

Alpha-acids

The terpenophenolic α-acids (e.g. humulone) are implicated in the 'bitter' flavour imparted by hops [*Humulus lupulus* (L. 1753, Cannabaceae)] to the alcoholic beverage beer.[39] Although beer has a history dating back 13,000 years, it would have been somewhat unrecognisable in its primitive gruel-like form and certainly did not contain any hops (let alone malted barley). Hops were not commonly added to something modern humans would recognise as beer until roughly the 9th through the 11th centuries in Europe.[40] Similar to the cannabinoids of the *Cannabis* genus of the same family Cannabaceae, *Humulus* spp. produce α-acids in the resinous trichomes of their flowers. The α-acids are considered phloroglucinol-derivatives arising from the phloroisovalerophenone synthase mediated polyketide (acyltransferase) pathway, before prenylation with two units of dimethylallyl pyrophosphate (DMAPP) to yield a phenolic core with three isoprenoid side-chains (*Fig. 5.11*).[41] The α-acids acquire the designation 'acid' due to significant keto-enol tautomerism, which is facilitated by the vinylogous electronic resonance of the structure, and they are thought to play a role akin to the cannabinoids. They have also been considered 'phyto-ionophores', which mimic the antimicrobial action of the ionophore antibiotics (e.g. monensin, lasalocid) through the selective inhibition of Gram-positive bacteria [e.g. *Clostridium perfringens* (Veillon & Zuber 1898)] via the disruption of membrane integrity.[42]

Other terpenophenolics

The complete enumeration of this terpenophenolic infraclass, and the biosynthetic pathways through which they are formed, is too disparate and specific to detail here, although the most notable members are shown in *Fig. 5.9.* A brief overview of the biological activities of this minor subinfraclass of terpenophenolics can be described as follows. Bakuchiol-type monoterpenophenolics share a structural similarity to resveratrol and have demonstrated antioxidant, anti-inflammatory, antiandrogenic and antibacterial properties, as well as showing promise in preclinical antineoplastic models.[32] Ferruginol-type diterpenophenolics are of the abietane diterpenoid structure (see Chapter 3) and have shown in vitro antineoplastic, anti-inflammatory, antibacterial and gastroprotective biological activity, as well as reduction in human colon, breast and lung tumour cells and oncogene-transformed cells.[33] In ovarian cancer cells, ferruginol induced dose-dependent apoptosis, limited migration and caused cell-cycle arrest.[33] Prenylated furanocoumarin terpenophenolics (e.g. psoralen, angelicin) are thought to be phytoalexin-/phytoanticipin-type defence compounds because they are known to be toxic by intercalating DNA (causing cell

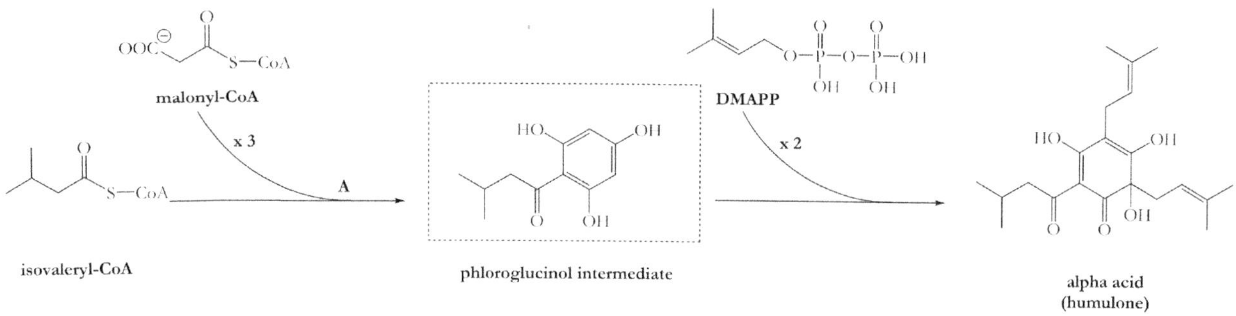

Fig. 5.11. Metabolic pathway resulting in a generic α-acid (humulone). A = phosphoisovalerophenone synthase (a polyketide synthase/ acyltransferase), DMAPP = dimethylallylpyrophosphate

death) and causing inflammation through the activation of the arachidonic acid cascade leading to phytophotodermatitis [most notably in the serious blisters and scars caused by giant hogweed, *Heracleum mantegazzianum* (Sommier & Levier 1895, Apiaceae)].[43,44] Terpecurcuminoids (e.g. terpecurcumin A), formed from the condensation of a curcuminoid subunit with a bisabolane-type sesquiterpene, have been implicated in significant cytotoxicity against A549 (lung), HEPG2 (liver) and MDA-MB-231 (ductal) human cancer cell lines, with lower IC_{50} (10.3–19.4 μM) than curcumin alone (31.3–49.2 μM).[45]

Alkaloterpenoids

The alkaloterpenoids are potentially the largest and most diverse infraclass of all the known NPs (or subclass, for that matter), with a low estimate of 2,000–4,500 members isolated, depending on definition.[46] Of these, the largest subinfraclass are the monoterpenoid indole alkaloids (MIA) that form from the condensation of an indole (tryptamine) unit and a secoiridoid glycoside (secologanin) unit, which can cyclise with substantial diversity.[46] Beyond the MIAs, there are also ergot alkaloterpenes (indole condensed with a hemiterpene), ipecacuanha-type isoquinoline alkaloterpenes (dopamine condensed with secologanin) and a wide range of 'pseudoalkaloterpenoids' whose *N*-source is often vague or unknown, variously condensed with mono-, di- and triterpenoidal units (*Fig. 5.12*).[46] Of all the groups of specialised metabolites that have been addressed in this book, this hybrid group is the most difficult to reconcile. In fact, all of the textbooks and recent literature sources that have been drawn upon still struggle with how best to organise this complex NP group, as basing it on original plant source (genera) only appears to create more confusion due to the significant overlap between families. Since the latter half of the 20th century, an attempt has been made to systematise this group of hybrid compounds into formal divisions, which is admirable, but unfortunately much of the literature relies on the traditional delineation based on genera. An attempt will be made throughout this section to adhere as closely as possible to the systematic grouping system, but some concessions must be made.

The 4100 MIAs [also called terpenoid indole alkaloids (TIAs)] reported in the literature in 1991–1993, have essentially all come from three closely related plant families: Apocynaceae, Loganiaceae and Rubiaceae.[46] Their biological activities and pharmaceutical applications are vast and well substantiated; for example, ajmalicine (hypertension), ajmaline (cardiac arrhythmia), ibogaine (anti-addictive), quinine (antimalarial), strychnine (convulsant), vincristine/vinblastine (antineoplastic) and yohimbine (adrenergic receptor blocker) (*Fig. 5.13*).[46] The reader will notice that within this grouping of compounds there is little structural similarity; in fact, quinine (one of the most notable members) no longer even bears an indole moiety, but is still technically considered an MIA. The structural variations arise from the biosynthetic complexity of the strictosidine synthase-mediated Schiff base (Pictet-Spengler) condensation of the formative tryptamine (amine moiety) and secologanin (carbonyl moiety) subunits, from which a variety of structural rearrangements become possible due to the long, flexible *N*-ethylethanimine bridge between the *N*-heterocyclic indole moiety and the *O*-heterocyclic terpenoid moiety.[47] It is considered that most, if not all, MIAs are derived from the natural intermediate (and common branching point) strictosidine (previously named isovincoside), but then the newly formed piperidine-like *N*-heterocycle of strictosidine can readily break and reform in numerous conformations.

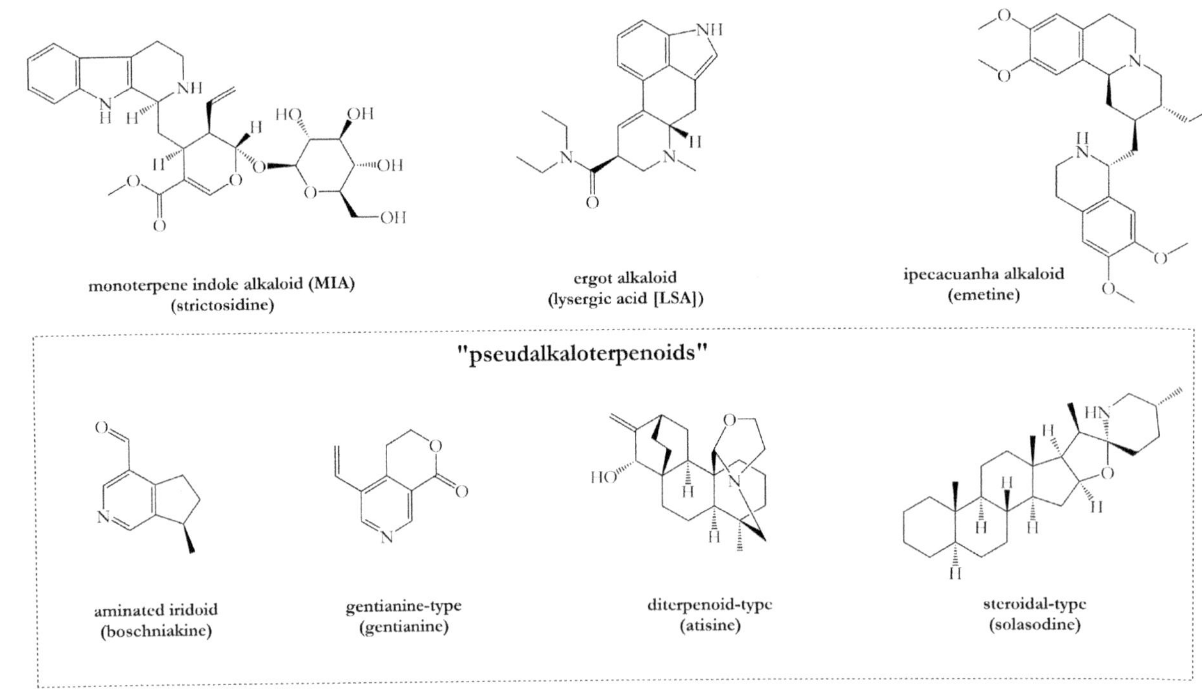

Fig. 5.12. Major subinfraclasses of the alkaloterpenoid-type infraclass of the 'terpenoid-like' subclass of specialised metabolites, (and representative examples).

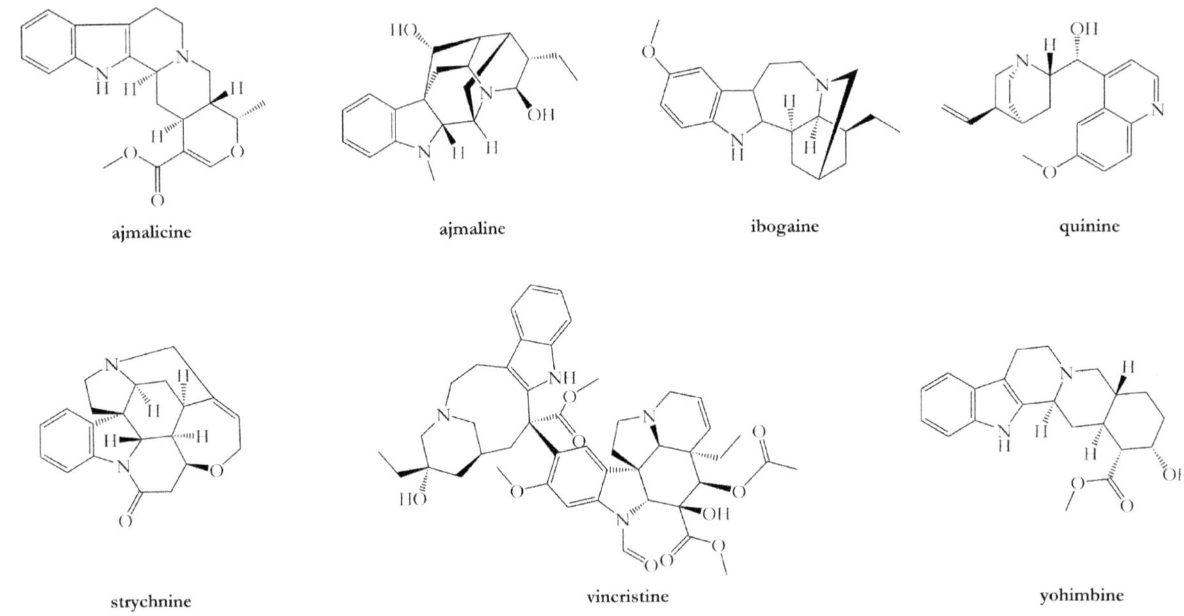

Fig. 5.13. Key representative examples of plant-derived MIA-type compounds of the 'terpenoid-like' subclass of miscellaneous-type specialised metabolites.MIA = monoterpenoid indole alkaloid.

As mentioned above, rather than basing the various MIA groups on the unwieldy number of plant sources or chemical structures [e.g. the strictosidine, corynantheine, vallesiachotamine, adifoline, talbotine, stemmadiene, mavacurine, chinchonamine, sarpagine, perakine, picraline, corynine, ajmaline, peraksine, oxindole, pseudoxindole, condylocarpine, akuammicine, strychnine, secondine, quebranchamine, aspidospermine, schizophylline, vincamine, schizozygine, vindolimine, pleiocarpine, kopsine, fruticosine and catharanthine groups, among others(!)], a classification system based on biosynthetic pathway was proposed in 1965 by Le Men and Taylor and further elaborated upon in 1971 by Kompis *et al.*[46,48,49] This system lists five 'Classes' [1-5, (with a clearly different meaning from the use of class/subclass/infraclass used throughout this book)], with an internally consistent carbon numbering system, each with their own 'subClasses', based on biosynthetic complexity: Class 1 [ajmalicine-type (formerly *Corynanthe-Strychnos*)], Class 2 (secodine-type), Class 3 [3a: vincamine-type, 3b: vindoline-type (formerly *Aspidosperma*)], Class 4 (fruticosine-type) and Class 5 [catharanthine-type (formerly *Iboga*)] (*Fig. 5.14*). It is worth noting that many genera (or species) produce diverging MIA groups and may contain more than one of these classes alongside other more conventional alkaloids [e.g. *Catharanthus roseus* ((L.) G. Don 1837, Apocynaceae)] producing alkaloids from MIA Classes 1, 2, 3 and 5, as well as numerous other specialised metabolites.

The 5-Class structural classification system and synoptic biosynthetic pathway proceeds as shown in *Fig. 5.15*. Strictosidine proceeds through three intermediates before settling as 4,21-dehydrogeissoschizine (the key branching-point for the five Classes). 4,21-dehydrogeissoschizine can branch here directly to Class 1 (ajmalicine-type) MIAs, directly to Class 3a (vincamine-type) MIAs or via preakuammicine, then stemmadenine, onto Class 2 (secodine-type) MIA dehydrosecodine, which is another branching point for MIAs of Class 3b, 4 and 5. Dehydrosecodine can branch directly to Class 3b (vindoline-type via tabersonine) MIAs (taking another six steps to get to vindoline itself) or to Class 5 (catharanthine-type) MIAs. Class 3b (vindoline-type) MIAs are the source of Class 4 (fruticosine-type) MIAs.[46] Certain other MIAs 'break the rules' so to speak, such as the quinine-type quinoline 'pseudo-MIAs' (which proceed from strictosidine through corynantheal and several intermediates to yield the chinchonidinone branching point) and the strychnine-type MIAs, which are actually augmented Class 3b (vindoline-type) MIAs, just to name a few.[46] Unfortunately, this is as much detail as can be afforded, with the numerous enzymes implicated and pathways followed throughout

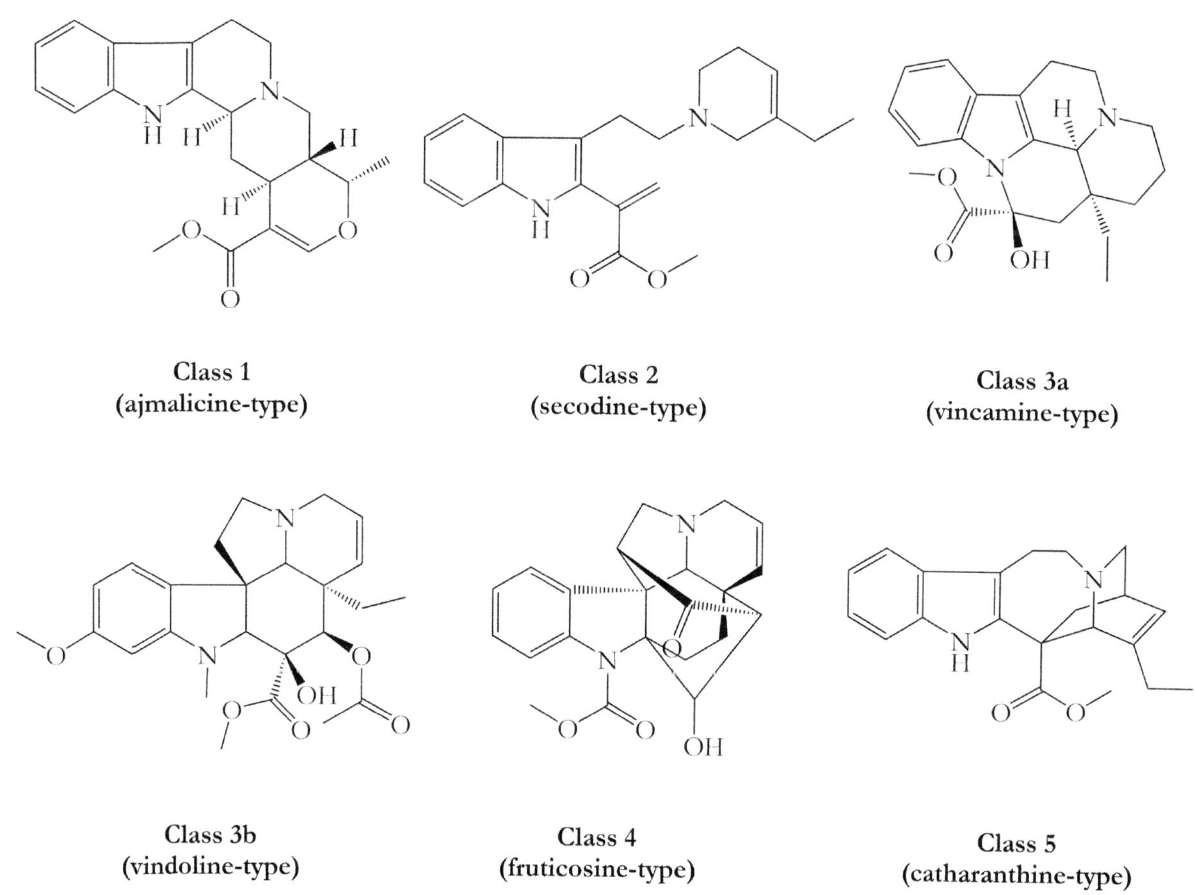

Fig. 5.14. Biogenetic classification of monoterpene indole alkaloids (MIAs).

this metabolic network lying outside the scope of this book, although an interested reader could look to any of the literature sources for a more thorough breakdown of this important and fascinating group of NPs.[50–53]

From the vast range of literature available, the bioactivity of some key MIAs can be briefly outlined here. Ajmalicine, isolated primarily from the Apocynaceae (*Rauvolfia*, *Catharanthus*) and Rubiaceae (*Corynanthe*), has been used for the treatment of circulatory disorders, as it exhibits potent vasodilatory (hypotensive) properties (marketed under various brand names).[46,54,55] Yohimbine is isolated primarily from the Rubiaceae (*Corynanthe*, *Ladenbergia*), but has also been reported from the Apocynaceae (*Amsonia*, *Aspidosperma*, *Catharanthus*, *Rauvolfia*, *Vellesia*), Loganiaceae (*Strycnos*) and even the Euphorbiaceae (*Alchornea*), and is used in the treatment of arteriosclerosis due to it being a pre-/postsynaptic α2-adrenergic antagonist and serotoninergic system antagonist, which depletes catecholamines and leads to hypotensive/cardiostimulatry effects.[46,54–56] Ajmaline, isolated from the Apocynaceae (*Catharanthus*, *Rauvolfia*) and marketed as Gilurytmal, Ritmos and Aritmina, has powerful coronary dilating and 1A antiarrhymic effects (useful in patients with Brugada syndrome), through its ability to block sodium ions and hERG (human 'ether-a-go-go'-related gene) potassium ion channels and lengthen the heart's refractory period, ultimately leading to bradycardia.[46,54,55,57] Quinine,

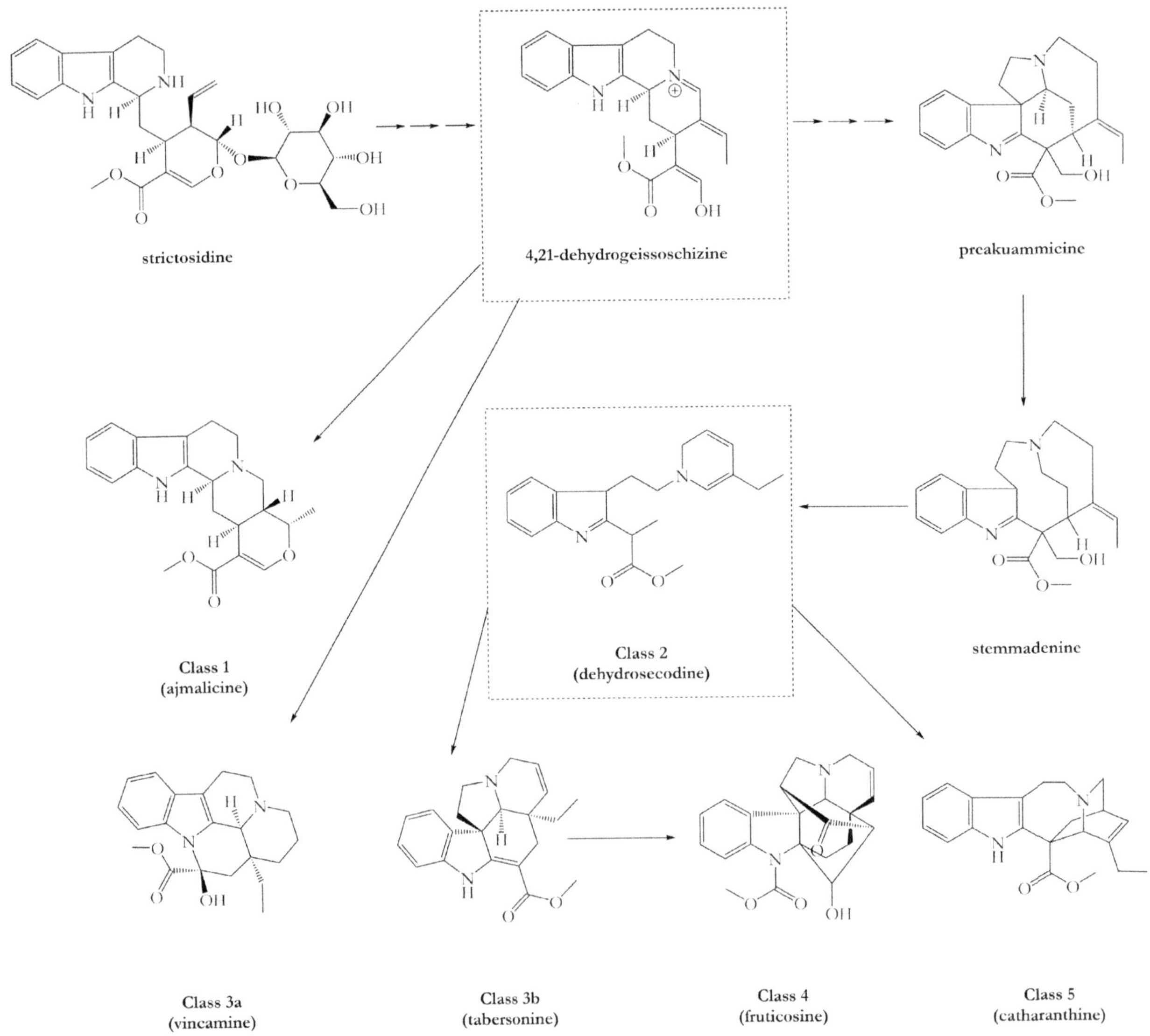

Fig. 5.15. The MIA metabolic pathway resulting in the five biogenetic Classes from the literature.[46,48,49] MIA = monoterpenoid indole alkaloid.

isolated from the Rubiaceae (*Chinchona*), is likely the most well-known member of the MIAs and has been used in the treatment of malaria since at least the 17th century due to its toxicity to the malarial pathogen *Plasmodium falciparum* (Welch 1897) through interfering with the parasite's ability to metabolise haemoglobin, functioning in a fashion similar to the better understood and more widely used synthetic quinoline antimalarial drug, chloroquine.[46,54,55] Vincristine (marketed under the name Oncovin) and vinblastine (marketed under the name Velban) are isolated primarily from the Apocynaceae (*Catharanthus*) and both are used as chemotherapy medications against a variety of cancer types, with their clinical efficacy comes from the binding of tubulin (inhibiting the assembly of microtubules), causing cell-cycle arrest and apoptosis.[46,54,55,58–60] Strychnine, isolated from the Loganiaceae (*Strychnos*), is a notorious poison (minimum lethal oral dose in humans of 30–120 mg) associated with the bisbenzylisoquinoline curares discussed in Chapter 4, which operates through the antagonism of glycine and acetylcholine receptors in the CNS, causing hyperpolarization of nervous membrane potentials that leads to spastic muscle contractions (opisthotonus) and death by asphyxiation.[46,54,55,61,62]

One of the most significant groups of MIAs, with pharmacological peculiarities that warrant their treatment as a separate group, are the iboga alkaloids (primarily ibogaine, alongside noribogaine, voacangine, ibogaline, ibogamine and ibochine). Isolated from the roots of *Tabernanthe iboga* (Baill 1888, Apocynaceae), iboga alkaloids have been used for millenia as a rite of passage by the Pygmies of the Congo basin and more recently in the Bwiti religion of West-Central Africa (Gabon, Cameroon, the Republic of the Congo).[63 64] Marketed as a stimulant from 1900 to 1970 under the name Lambarène, and despite being demoted from its 'world-changing' and 'life-saving' status of the 1980s (marketed by Howard Lotsof), ibogaine is under serious investigation for the treatment of addiction (e.g. alcohol, opiates, cocaine, amphetamines, etc.), in lieu of a better (less toxic) alternative, with claims of a single 4–25 mg/kg dose interrupting addictive drug behaviour for 6–36 months.[63,64] The extremely complex mechanism of action of ibogaine and its chemical relatives can only be glimpsed heren, with the reader being encouraged to pursue the many significant articles/volumes detailing the bioactivity of these compounds. Ibogaine interacts with various neurotransmitter systems simultaneously: μ- and κ-opioid receptors (MOR and KOR), sigma-1 and -2 (σ1 and σ2) receptors and *N*-methyl-D-aspartate (NMDA) receptors of the CNS, as well as various other serotonin and dopamine receptors and reuptake sites, with KOR agonism and NMDA antagonism being suggested as the most likely candidates for narcotic self-administration attenuation.[63,64]

It has been demonstrated that ibogaine pretreatment blocks morphine-induced dopamine release and morphine-induced locomotor hyperactivity, while enhancing similar contra-effects due to stimulants. It has been recently substantiated that ibogaine's mechanism of action differs from other molecules implicated in opiate tolerance and withdrawal, but specific targets are still yet to be discovered.[63,64] The lipophilicity of the iboga alkaloids (as the primary gut human metabolite noribogaine) promotes in vivo fat-sequestration and the duration of effects observed with these compounds differs markedly from the expected pharmacokinetic parameters, with mood elevation occurring generally 24 h after ingestion and some effects experienced well beyond 30 days after ingestion.[63,64] Besides these currently elusive anti-addictive properties, iboga alkaloids have also been implicated in cytotoxic, antiproliferative and anti-inflammatory bioactivities.[63,64] Neurotoxicity and cardiac side-effects have been observed in animal and human studies of these compounds, including a fatal case of a woman who underwent four high-dose treatments (10–30 mg/kg).[63,64]

The distribution of the miscellaneous-type specialised metabolites of the MIA-type alkaloterpenoid infraclass reported from the customary medicinal flora of Australia is presented in *Table 5.1*. Native plant genera bearing compounds of this class have documented customary use among Aboriginal and Torres Strait Islander peoples, with a key example being *Alstonia* (narcotic, tonic, emollient, analgesic, fever, antimicrobial, digestive, diuretic, menstrual). *Alstonia constricta* (F. Muell 1858, Apocynaceae) has been reported to bear a variety of MIA-type alkaloterpenoids (e.g. alstonine, alstonidine, alstonilidine, quebrachidine, rauwolscine, reserpine, vincamajine, vincamedine, etc.).[65,66] These MIA type alkaloterpenoids are of the strictosidine and Class 1 type and thus bear similar biological activities as described for those categories above. A recent study investigated the antiplasmodial efficacy of the Class 1 type MIA alstonine against the malaria-causing parasite *Plasmodium falciparum* (Welch 1897), demonstrating action activity (IC_{50}) of 0.17 μM, with a selectivity index of > 1111 versus two human cell lines and no difference when trialed against multidrug resistant *P. falciparum* (Welch 1897) lines.[67]

Ergot alkaloids

The alkaloterpenes known as the ergot alkaloids arise from the condensation of a tryptamine (indole) subunit with a hemiterpene derived from the MVA pathway discussed in Chapter 3. Ergot alkaloterpenes are divided into three major groups: clavines, lysergic acid (LSA) alkylamides (lysergamides) and LSA peptide-derivatives.[68] They are best known to originate from the fungal kingdom, primarily from the rye ergot fungus *Claviceps purpurea* ((Fr.) Tul. 1853, Clavicipitaceae) and other fungal genera (*Aspergillus, Penicillium, Rhizopus*), although ergot alkaloids have been isolated from the plant kingdom as well [Convolvulaceae (*Argyria, Convolvulus, Cuscuta, Ipomoea, Operculina, Rivea, Strictocarpia, Turbina*)] and thus warrant discussion here.[68] Ergot alkaloids are produced biosynthetically through the prenylation of tryptophan by DMAPP, to yield 4-dimethylallyltryptophan (DMAT) through the action of DMAT synthetase.[68] DMAT proceeds through several intermediates, facilitated by the chanoclavine-I-cyclase enzyme complex, to yield the clavines (e.g. chanoclavine I) and ultimately LSA (*Fig. 5.16*).[68] Ergotamine, ergocryptine, ergocornine and ergometrine are important LSA peptide-derivatives involved in the physiological effects of ergot consumption.[68] The fungus is known to grow on rye [Poaceae (*Secale*)] and the ergot alkaloterpenes it produces are known to cause ergotism through continued consumption, with symptoms beginning with gastrointestinal effects (nausea, vomiting, diarrhoea), proceeding to CNS effects [seizures/spasms and mental effects (headaches then hallucinations/mania/psychosis)] and finally gangrenous effects (due to the vasoconstrictive effects of ergotamine and ergocristine).[68] Ergotism has been implicated throughout history with several notorious outbreaks of 'Saint Anthony's fire' throughout the second millennium CE, even being implicated in the distasteful accusations of 'witchcraft' occurring throughout that period.[69]

The biological activity of the ergot alkaloterpenes varies somewhat, with structural functionalisation leading to both agonism and antagonism of the dopamine and serotonin systems of the CNS.[70] The various semisynthetic lysergamides (e.g. ergine, ergonovine, etc.) are somewhat outshined by their (in)famous 'problem child' lysergic acid diethylamide (LSD), synthesised by the Swiss chemist Albert Hoffmann in 1938 from LSA while working for Sandoz seeking a new respiratory stimulant.[71] After unintentionally ingesting some of the substance in 1943 (possibly through skin contact) and reporting on its intense psychoactivity, it was examined over the following two decades by several differing research (e.g. psychiatric) and governmental (e.g. Project MK Ultra) groups while looking for a marketable use for it.[71] Eventually, the psychedelic 'consciousness-expanding' aspects of its neuropsychopharmacology led to it becoming inextricably bound to the 1960s countercultural movement, becoming criminalised in 1971 and currently having no use in modern medicine, although renewed interest in psychoactive compounds in psychological/therapeutic settings may lead to it being considered as an object of research once again.[71]

Ipecacuanha alkaloids

The curious group of isoquinoline-type alkaloterpenoids known as the ipecacuanha alkaloids are primarily isolated from *Cephaelis ipecacuanha* [(Brot.) Willd. 1804, Rubiaceae], the root of which can be up to 2% alkaloids (approximately two-thirds of which is emetine and the other third is cephaeline).[46] They are also found in other

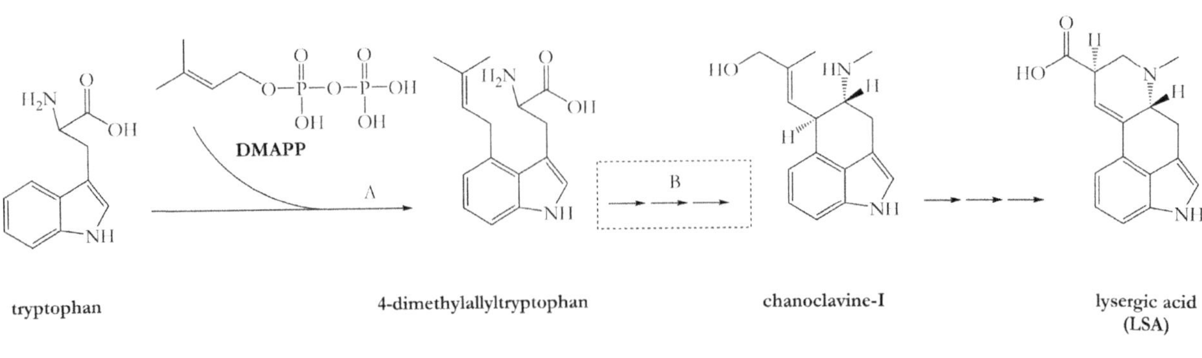

Fig. 5.16. Ergot alkaloterpenoid metabolic pathway. A = 4-dimethylallyltryptophan (DMAT) synthetase, B: chanoclavine-I-cyclase enzyme complex. DMAPP = dimethylallylpyrophosphate.

members of the Rubiaceae (*Borreria, Bothriospora, Capirona, Ferdinandusa, Hillia, Psychotria, Remija, Tocoyena*), as well as the Alangiaceae (*Alangium*) and Araliaceae (*Hedera*).[46] Their biosynthesis takes one of two paths, both of which are similar to the MIA production of their rubiaceous counterparts. In the case of the ipecacuanha alkaloids, instead of tryptamine as the amine source, dopamine provides the Schiff base and secologanin is the carbonyl that again goes through a Pictet-Spengler condensation to form either deacetylipecoside or deacetylisoipecoside.[46] Deacetylipecoside is the basis of the minor pathway and leads directly to ipecoside or cyclises to yield alangiside. Deacetylisoipecoside is the basis of the major pathway and it cyclises to yield protoemetine, then another dopamine unit condenses to yield emetine and cephaeline in the 2:1 ratio just described (*Fig. 5.17*).[46] The medical use of the ipecacuanha alkaloids was initially as an antimicrobial [against *Entamoeba histolytica* (Schaudinn 1903)] and today emetine and tubulosine (a bisalkaloterpene that also features a MIA moiety) are used to treat amoebic dysentery caused by the same pathogen, as well as treatment of schistosomiasis, leishmaniasis and malaria, with global consumption on the order of 10,000–100,000 kg, although the toxicity of these compounds (e.g. for emetine: LD_{50} of 135 mg/kg and 12.1 mg/kg in mice and rats, respectively), is concerning.[46]

'Pseudoalkaloterpenoids'

Apart from the alkaloterpenes just discussed, there is a final large and disparate group in which their nitrogen source is either unknown or does not arise from an amino acid, necessitating the putative description, 'pseudoalkaloterpenoids'. This subinfraclass of compounds, further divided via the usual mono-, sesqui-, di- and triterpenoidal categorisation strategy, generally arises from an ammonia-derived amination of an existing terpenoid from the MVA/MEP pathway.[72] Few

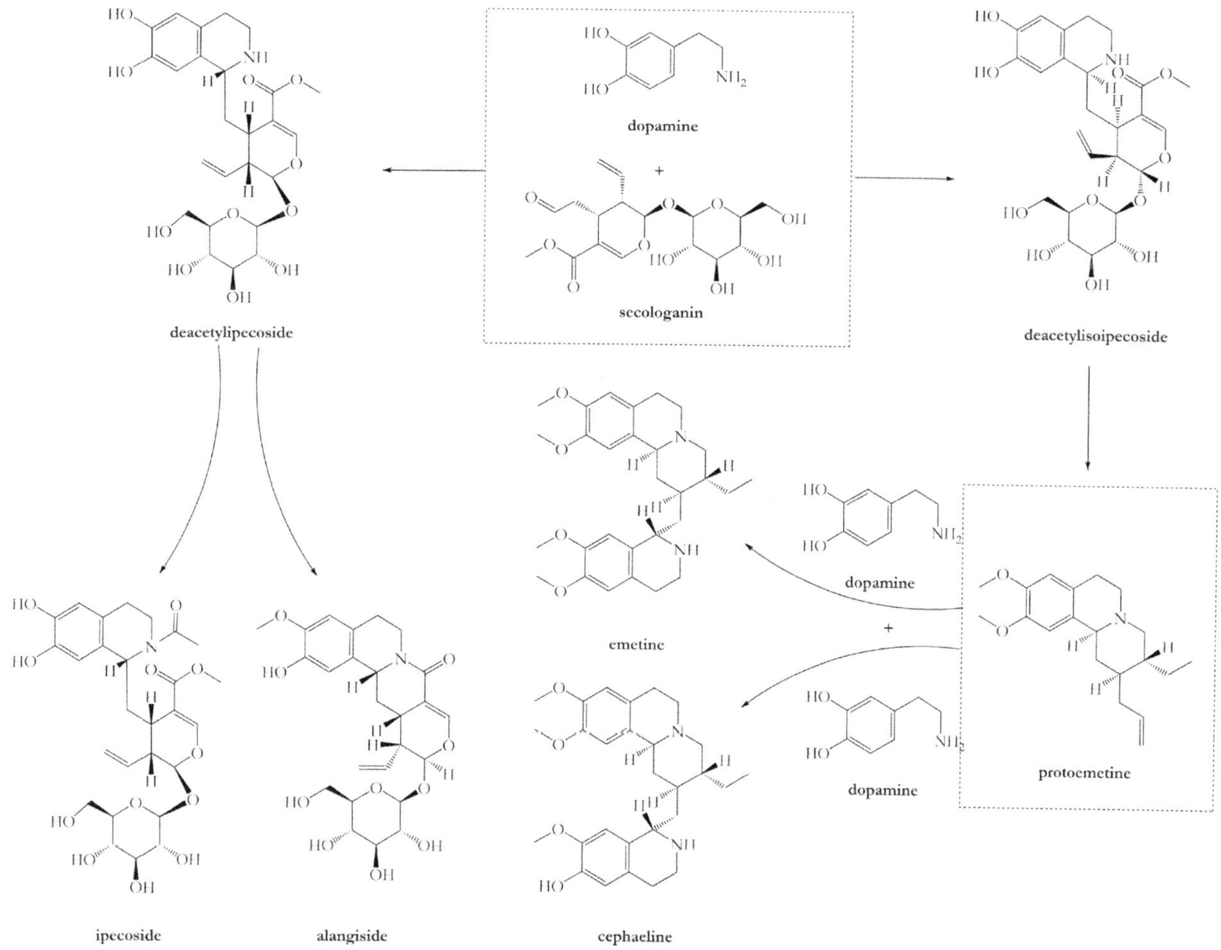

Fig. 5.17. Ipecacuanha alkaloterpenoid metabolic pathway.

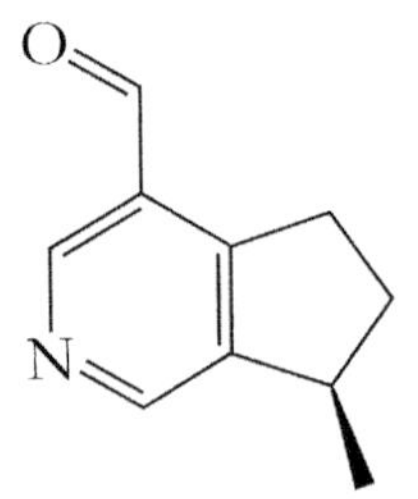

boschniakine

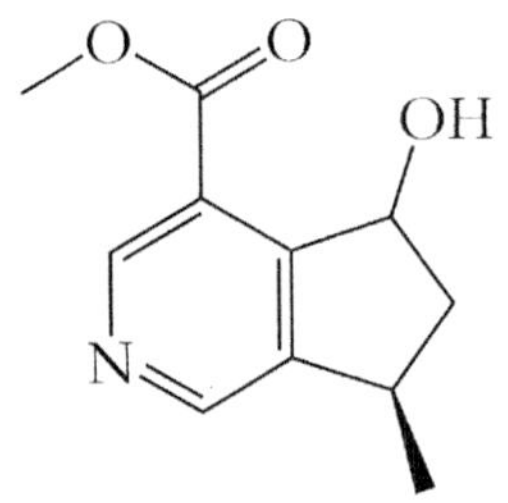

rhexifoline

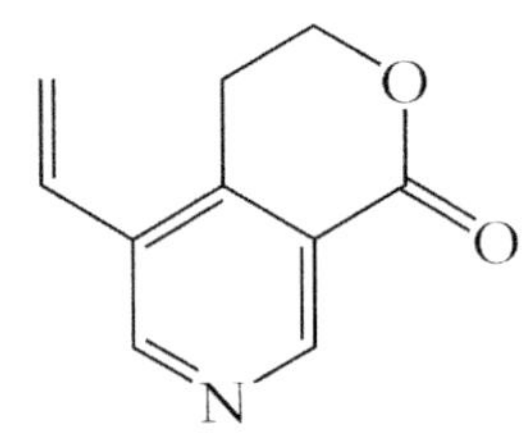

gentianine

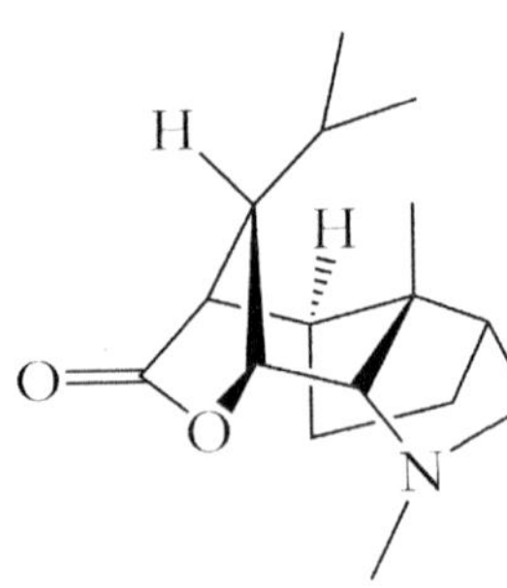

dendrobine

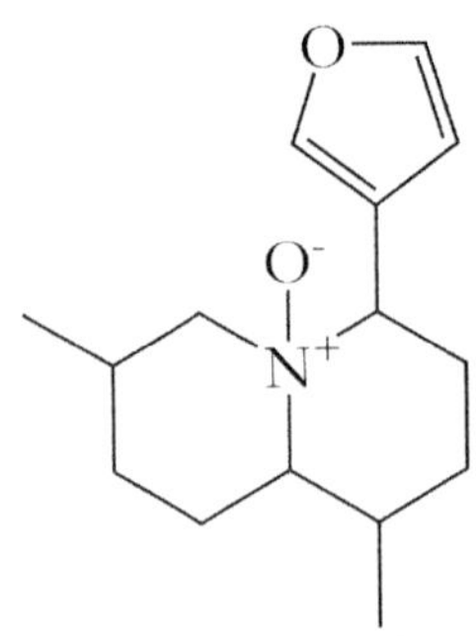

nupharidine

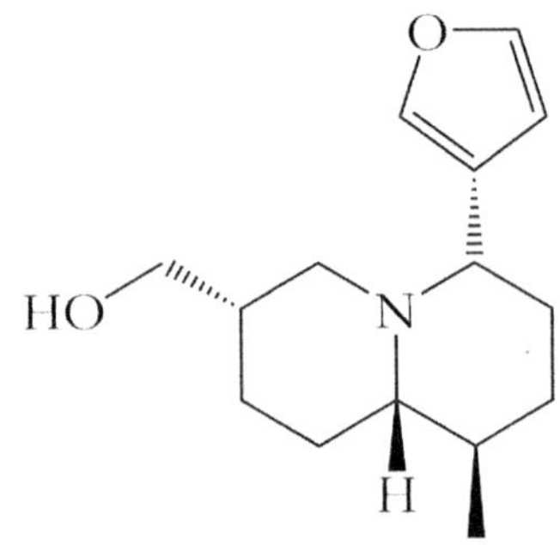

castoramine

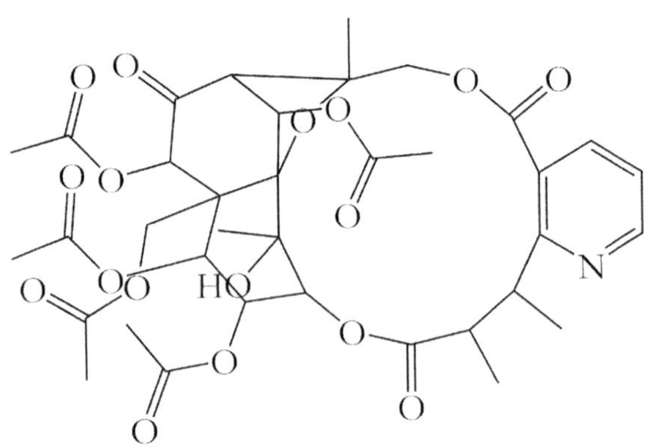

evonine

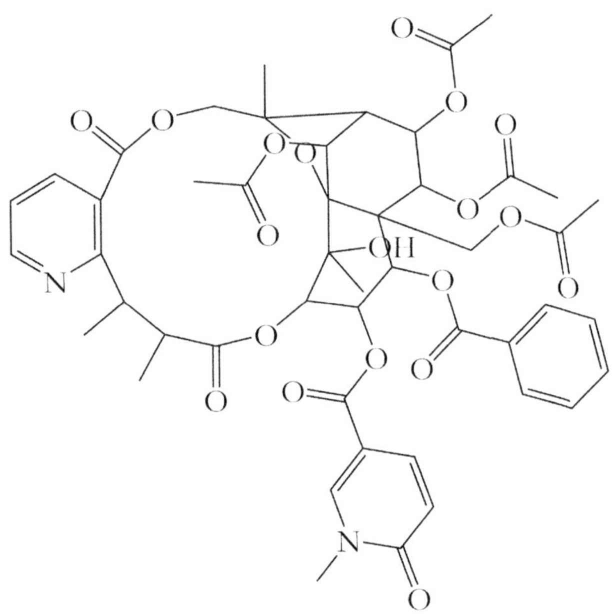

hippocrateine I

Fig. 5.18. Key representative examples of plant-derived 'pseudoalkaloterpenoid'-type compounds of the 'terpenoid-like' subclass of miscellaneous-type specialised metabolites.

monoterpenoid-type pseudoalkaloterpenoids have been reported *sensu stricto*, possibly because aminated monoterpenes would just be reported as a monoterpene. Some important examples of this subinfraclass would be the aminated iridoids (boschniakine, rhexifoline) of the Scrophulariaceae (*Castilleja*, *Penstemon*), the CNS stimulating gentiopicroside-derived gentianine-type pseudoalkaloterpenoids reported from the gentianaceous plants of the Western herbal medicine tradition and some of the aminated iridoids of the sedating herb, valerian or *Valeriana officinalis* (L. 1753, Caprifoliaceae) also of the Western herbal medicine tradition (*Fig. 5.18*).[72] Approximately 130 sesquiterpenoid-type pseudoalkaloterpenoids have been reported, with some of the primary sources being the Orchidaceae (*Dendrobium*) from the Chinese herbal pharmacopoeia (dendrobine, similar to the sesquiterpene copaborneol) and the Nymphaceae (*Nuphar*), which contain members of this subinfraclass featuring quinolizidine moieties (nupharidine, castoramine). The closely related Celastraceae (*Catha*, *Euonymus*, *Tripterygium*) and Hippocrataceae (*Hippocratea*) bear complex evoninoate-type pseudoalkaloterpenoids (e.g. evonine, hippocrateine I) (*Fig. 5.18*).[72]

Over 640 diterpenoid-type pseudoalkaloterpenoids have been reported, of three major types: kaurene-type (often highly toxic), macrocyclic-type and cardiac glycoside-type. The kaurene-type members of this subinfraclass have been isolated from the Ranunculaceae (*Aconitum*, *Delphinium*), Garryaceae (*Garrya*), Asteraceae (*Inula*), Escalloniaceae (*Anopterus*) and Rosaceae (*Spiraea*) and are divided into four structural types: atisine, heteratisine, lycoctonine (of which aconitine is a member, with an LD_{50} in mice of 0.17 mg/kg) and veatchine (*Fig. 5.19*).[72] The macrocyclic-type members of this subinfraclass have been isolated from the Taxaceae [*Taxus* (yew trees)] and have led to a particularly significant compound, paclitaxel (marketed as Taxol and on the WHO's list of essential medicines), which, together with its structural relatives, is extremely cytotoxic with broad antineoplastic activity and possessing a mitotic inhibition mechanism different to any other type of antineoplastic drug on the market.[72] Certain

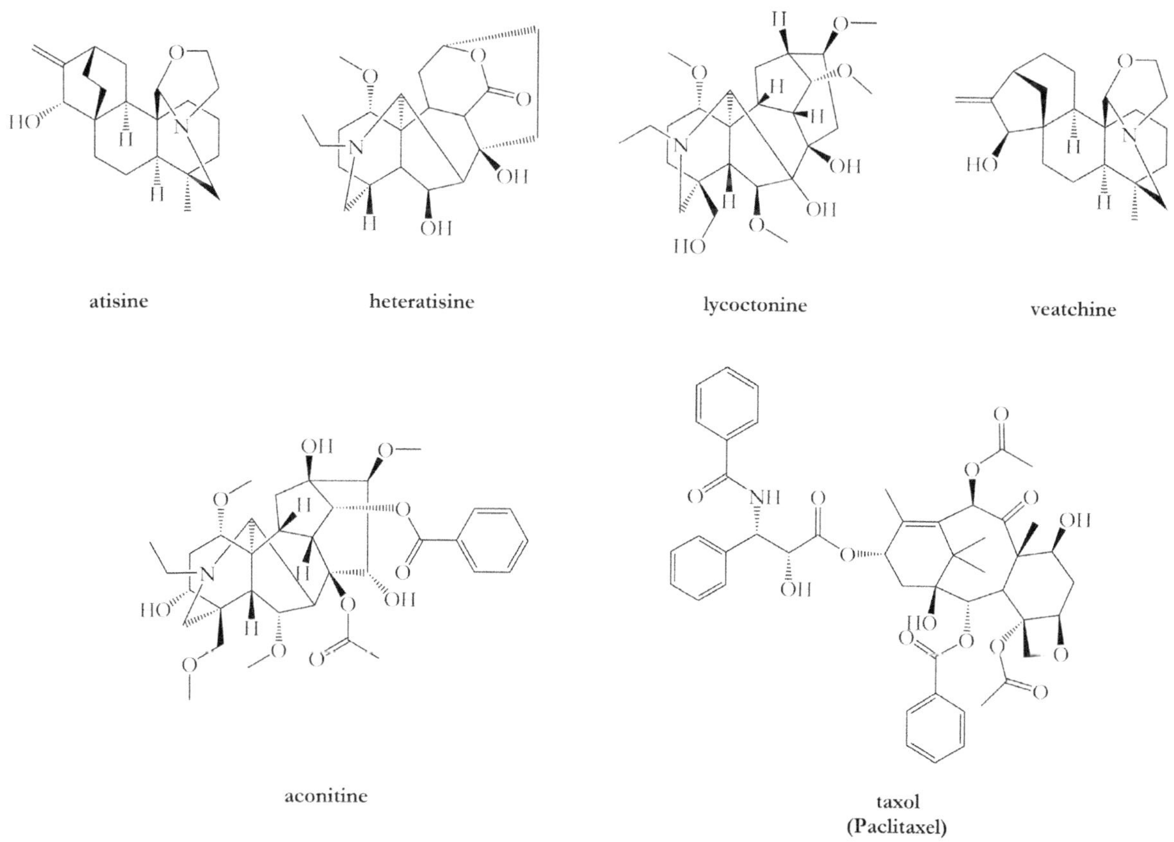

Fig. 5.19. Key representative examples of plant-derived diterpene 'pseudoalkaloterpenoid'-type compounds of the 'terpenoid-like' subclass of miscellaneous-type specialised metabolites.

members of the Fabaceae (*Erythrophleum*) produce toxic compounds of a structural form akin to the cardiac glycosides described in Chapter 3, but with an *N*-alkylaminoethyl moiety instead of a glycoside attached to the C_{19} diterpenoid core, that similarly inhibit (Na^+, K^+) ATPase activity causing cardiac activity and intense local anaesthesia.[72]

Finally there are ~500 steroidal-type (triterpenoid) pseudoalkaloterpenoids, isolated from the Solanaceae (*Cestrum*, *Cyphomandra*, *Lycopersicon*, *Nicotiana*, *Solanum*), Liliaceae (*Fritillaria*, *Veratrum*, *Zygadenus*), Apocynaceae (*Funtumia*, *Holarrhena*, *Malouetia*), Buxaceae (*Buxus*, *Pachysandra*, *Sarcococca*) and Daphniphyllaceae (*Daphniphyllum*).[72] The Solanaceae-derived members of this structural subinfraclass are divided into five types, which are usually derived as glycoalkaloterpenoids: spirosolanes (e.g. solasodine), solanidanes (e.g. solanidine), 22,26-epiminocholestanes (e.g. verazine), α-epiminocyclohemiacetals (e.g. solanocapsine) and 3-aminospirostanes (e.g. jurubidine) (*Fig. 5.20*).[72] These compounds appear to perform a phytoalexic role within this family, as they are potent antimicrobials, but unfortunately this is problematic for humans as they can be toxic and several important, common food crops (the nightshades) produce these compounds when exposed to light or under pathogenic attack: tomato [*Lycopersicon esculentum* (Mill. 1768)], potato [*Solanum tuberosum* (L. 1753)] and eggplant [*Solanum melongena* (L. 1753)]. This fact is likely behind the common anecdotal recommendation not to eat 'greened' potato, which considering they produce α-solanine (LD_{50} of 42 mg/kg in mice and likely teratogenic), is good advice. Members of the Liliaceae, particularly *Veratrum*, also produce steroidal-type pseudoalkaloterpenoids of two basic types: jerveratrum (e.g. jervine) and ceveratrum (e.g. protoveratrine A), which are highly toxic due to their ability to depolarise membranes through binding to sodium channels in the CNS and teratogenic effects best evinced in the bizarre cyclopamine-caused cyclopism seen in lambs born from sheep that graze these plants (*Fig. 5.20*).[72] Both the Apocynaceae and Buxaceae also produce compounds of this subinfraclass, which have been implicated in the traditional and contemporary applications of compounds from these plants (e.g. holaphylline, conessine) in the treatment of amoebic dysentery. Finally, in Japan the Daphniphyllaceae (*Daphniphyllum*) have long been known to cause livestock deaths, which have since been attributed to daphniphylline-type steroidal pseudoalkaloterpenoids.[72]

'ALKALOID-LIKE' SUBCLASS

Simple alkaloid-like compounds

Although of global interest, the ~700 known non-protein amino acids are not widely reported in the literature concerning the specialised metabolic constituents of Australian plants and as such lie outside the scope of this book; a thorough breakdown of this group of compounds that the reader can refer to is included in the cited reference work.[46] What remains to be addressed here are the simple alkaloid-like miscellaneous compounds that contain nitrogen, but are not currently understood to be derived from amino acids and do not belong to any of the alkaloid subtypes *sensu stricto*. Some simple amine compounds that have not found membership in any of the sub-/infraclasses discussed throughout this book are the alkylamides and the di/polyamines (*Fig. 5.21*). The well-known pungent principles of fruits (chilli peppers) from the genus *Capsicum* (Solanaceae) have an *N*-(4-hydroxy-3-methoxybenzyl) alkyl amide structures commonly known as 'capsaicinoids' and arise from the converging phenylpropanoid (Chapter 3) and primary metabolic branched chain fatty acid pathways (acetyl-CoA extension) (*Fig. 5.21*).[73] Capsaicin is the major capsaicinoid (0.5–1.5% of some varieties) that imparts the 'spicy' flavour of hot chilli peppers and it is thought to confer disease resistance to the fruit, as well as being implicated in a particular co-evolutionary ecological role in which certain animals (e.g. birds) appear not to have the same burning/irritation response that other animals (e.g. mammals) experience from ingesting these compounds.[74] The mechanism of the heat-response in mammalian sensory systems is through interaction with the vanilloid receptor subtype 1, which responds to activation with a pain response (as it does with heat, protons (acid) or physical abrasion). What is notable is that capsaicinoids do not actually cause any direct tissue damage (chemical burn), only the semblance of it and the associated pain response; it is the subsequent inflammation that can cause tissue damage in extreme cases (e.g. consuming an extremely hot chilli).[74] Capsaicinoid consumption increases the respiratory rate (bronchoconstriction) through stimulation of C fibres and consequent release of neuropeptides, while also elevating blood pressure (vasoconstriction).[74]

The important alkaloidal synthetic intermediates, putrescine and cadaverine (discussed in Chapter 4), are moderately toxic diamine compounds resulting from decarboxylation of ornithine and arginine, respectively.

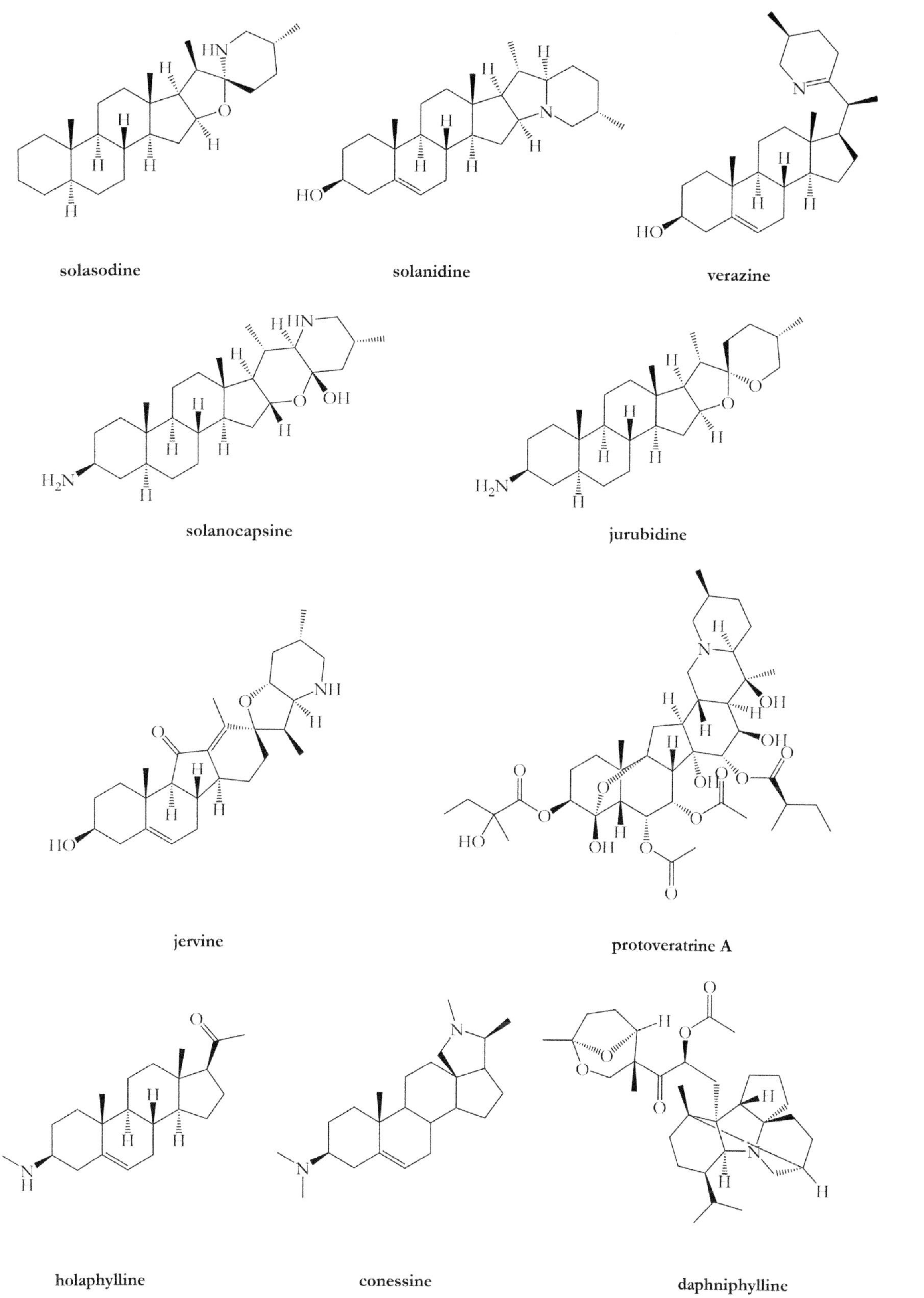

Fig. 5.20. Key representative examples of plant-derived steroidal 'pseudoalkaloterpenoid'-type compounds of the 'terpenoid-like' subclass of miscellaneous-type specialised metabolites.

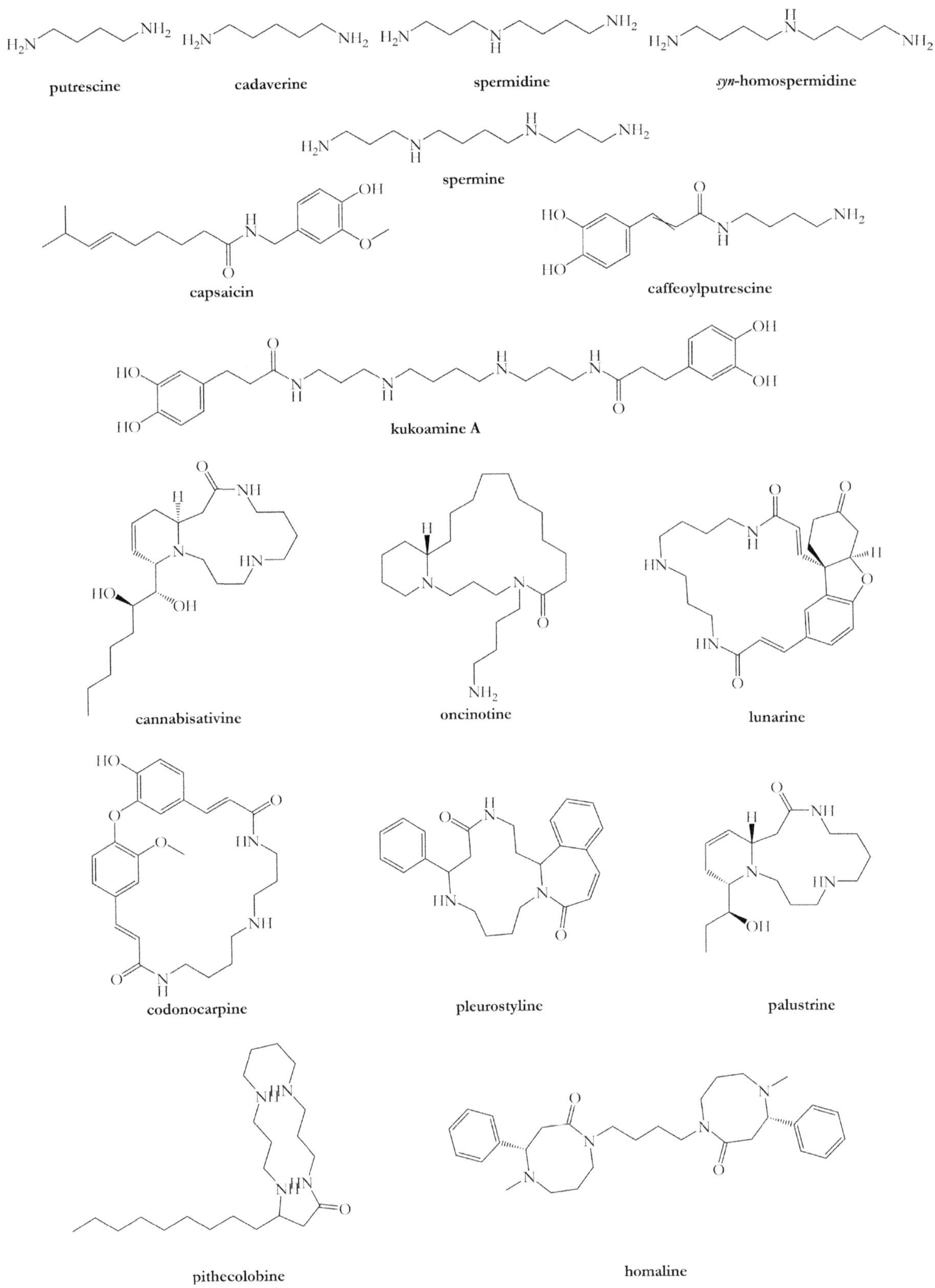

Fig. 5.21. Key representative examples of plant-derived compounds of the simple alkaloid-like infraclass of miscellaneous-type specialised metabolites.

The polyamines encountered in the alkaloidal biosynthetic pathways also discussed in Chapter 4, such as spermine and spermidine, as well as the diamines just mentioned, are commonly found in amidated form in the plant kingdom, being condensed with phenolic precursors to form compounds such as caffeoylputrescine and kukoamine A (*Fig. 5.21*).[46] All of the di-/polyamines just mentioned play important primary metabolic roles *in planta*, such as influencing membrane fluidity and permeability, stabilising chlorophyllic thylakoid membranes, regulating nucleic acid structure, slowing DNA denaturation and showing marked antisensescent activity.[46] At least 125 di-/polyamine-type alkaloids have been reported from the Acanthaceae [*Aphelandra* (e.g. cannabisativine)], Apocynaceae [*Oncinotis* (e.g. oncinotine)], Brassicaceae [*Lunaria* (e.g. lunarine)], Cannabaceae [*Cannabis* (e.g. cannabisativine)], Capparidaceae [*Capparis* (e.g. codonocarpine)], Celastraceae [*Pleurostylia* (e.g. pleurostyline)], Equisetaceae [*Equisetum* (e.g. palustrine)], Fabaceae [*Albizia*, *Pithecelobium* (e.g. pithecolobine)], Gyrostemonaceae [*Codonocarpus* (e.g. codonocarpine)], Homaliaceae [*Homalium* (e.g. homaline)] and Santalaceae [*Santalum* (e.g. *syn*-homospermidine)] (*Fig. 5.21*).[46] The distribution of miscellaneous-type specialised metabolites of the simple alkaloid-like infraclass reported from the customary medicinal flora of Australia is presented in *Table 5.1*.

Complex alkaloid-like compounds

The final group of compounds in this chapter are also difficult to categorise; they contain a nitrogen atom (in the case of the cyanogenic glycosides) or both a nitrogen atom and a sulfur atom (in the case of the glucosinolates), so it seems reasonable to term them 'alkaloid-like', although their biosynthesis, role *in planta* and biological activity is markedly different from the rest of the alkaloids, and thus they were relegated here. *Table 5.1* presents the customary medicinal flora of Australia that were reported to possess complex compounds of the 'alkaloid-like' subclass of the miscellaneous-type class in the literature reviewed as part of this book.

Cyanogenic glycosides

The Guignard and Feigl-Anger cyanogenic colour test methods applied to the plant kingdom have determined that some 2650 species, distributed over 130 families (550 genera), bear cyanide in their metabolome.[75] Actual isolation of these cyanogens has been undertaken on at least 500 species of these plants (~one fifth), yet only approximately 60 unique structures have been identified from all of this work.[46,75] These compounds are ubiquitously isolated as glycosides; thus the term 'cyanogenic glycosides' (CGs) arose (*Fig. 5.22*). CGs are compartmentalised away from the other organelles of the plant, alongside two enzymes required for the release (upon damage to the plant) of hydrogen cyanide, β-glycosidase to cleave the sugar moiety and release the aglycone and a hydroxynitrile lyase to dissociate the cyanohydrin to yield a carbonyl compound and hydrogen cyanide.[75] It is worth mentioning that the Sapindaceae, Hippocastanaceae and Boraginaceae esterify their α-hydroxynitriles with a C_{18-20} long-chain fatty acid and form what would correctly be called cyanogenic lipids (rather than CGs), of which only four have been isolated, but the end result is the same upon release of the aglycone.[46,75] The representation of CGs within the same species is extremely variable, with the expression of cyanogenesis appearing to fluctuate with environment and stress. That said, certain families are known for their expression of these compounds, which is particularly problematic for humans regarding the food crops sorghum [*Sorghum* spp. (Poaceae)] and cassava [*Manihot esculenta* (Crantz 1766, Euphorbiaceae)].[46,75]

CG aglycones are derived primarily from a select few amino acids (L-tyrosine, L-phenylalanine, valine, isoleucine, L-leucine), although nicotinic acid and non-protein amino acids have also been implicated.[46,75] The general pathway proceeds from the amino acid ([e.g. L-tyrosine, to *N*-hydroxytyrosine (via an *N*-hydroxylating cytochrome P450 enzyme)], which is converted to 2-nitro-3-(*p*-hydroxyphenyl)propionic acid and further converted to 1-*aci*-nitro-2-(*p*-hydroxyphenyl)ethane, which is the immediate precursor to (*E*)-*p*-hydroxyphenylacetaldehyde oxime, with both (*E*) and (*Z*) isomeric forms being accepted into the next phase of the biosynthesis that produces *p*-hydroxyphenylacetonitrile, which is subsequently converted to a hydroxynitrile by a *C*-hydroxylating cytochrome P450 enzyme, then glycosylated by a β-glycosidase enzyme (*Fig. 5.22*).[46,75] This same oxime-to-nitrile strategy appears to be followed for all of the amino acid precursors studied. CGs (viz. their breakdown products) play a defensive role *in planta*, as concentrated hydrogen cyanide is extremely toxic to most organisms through its inhibition of cytochrome oxidase and other respiratory enzymes.[75] Most animals have the ability to detoxify low-level cyanide poisoning, particularly the Euarthropoda, so it is suggested that they may be designed as a deterrent to mammalian

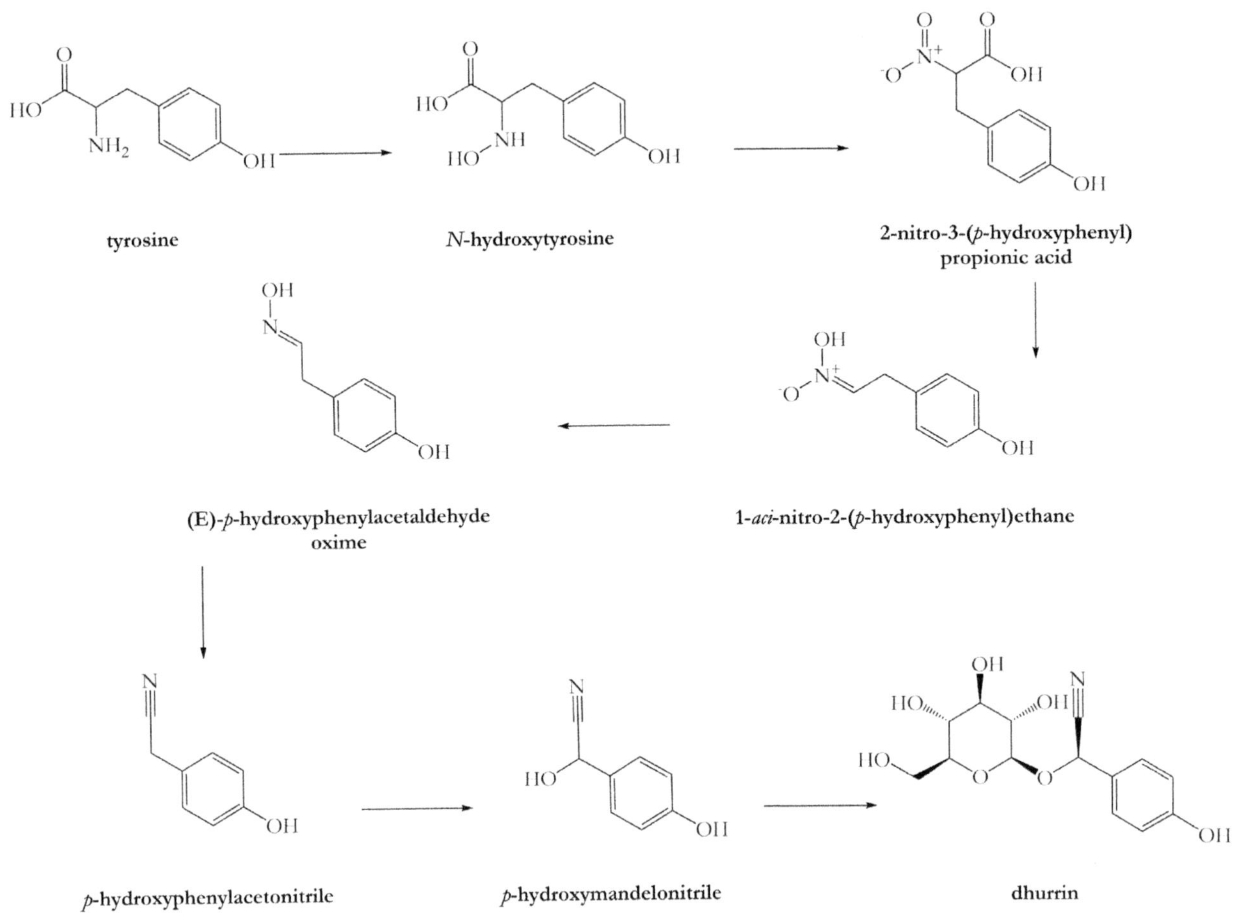

Fig. 5.22. Cyanogenic glycoside (CG) metabolic pathway.

herbivory.[75] Intact CGs are considered to be relatively non-toxic in mammals, being excreted rapidly via the urine.[75]

Some distribution trends are appreciable across the plant kingdom, with overall cyanogenesis being most common among the Araceae, Asteraceae, Euphorbiaceae, Fabaceae, Flacourtiaceae, Malesherbiaceae, Papaveraceae, Passifloraceae, Poaceae, Proteaceae, Ranunculaceae, Rosaceae, Sapindaceae and Turneraceae. Species from the genera *Prunus* (Rosaceae) and *Passiflora* (Passifloraceae) are rarely, if ever, found without the presence of CGs.[75] Some trends can be classified based on the origin of the nitrogen atom, with phenylalanine-derived CGs (e.g. amygdalin) predominating in the Rosidae [e.g. Rosaceae (*Malus, Prunus*)], Asteridae [e.g. Asteraceae (*Anthemis*)] and the Polypodiopsida (ferns), tyrosine-derived CGs (e.g. dhurrin) predominating in the monocotyledonous angiosperms [e.g. Poaceae (*Sorghum*)], the gymnosperms and the Magnoliidae, valine- and isoleucine-derived CGs (e.g. linamarin and lotaustralin, respectively) almost always co-occur and predominate in the Asteraceae, Euphorbiaceae, Fabaceae and Linaceae, leucine-derived CGs (e.g. proacacipetalin) are almost exclusively found in the genus *Acacia* (Fabaceae) (*Fig. 5.23*).[46,75] As mentioned, the cyanogenic lipids are restricted to the Sapindaceae, Hippocastanaceae and Boraginaceae. The nicotinic acid-derived acalyphin is derived from *Acalypha indica* (Euphorbiaceae). The non-protein amino acid-derived cyanogens (e.g. deidaclin) are found in the Caricaceae, Flacourtiaceae, Malesherbiaceae, Passifloraceae and Turneraceae (*Fig. 5.23*).[46,75] The distribution of miscellaneous-type specialised metabolites of the cyanogenic glycoside-type complex alkaloid-like infraclass reported from the customary medicinal flora of Australia is presented in *Table 5.1*.

Glucosinolates

The pungent taste and aroma associated with the cruciferous vegetables [Brassicaceae (e.g. mustard, horseradish, cabbage)] is due to the sulfurous glucosinolates

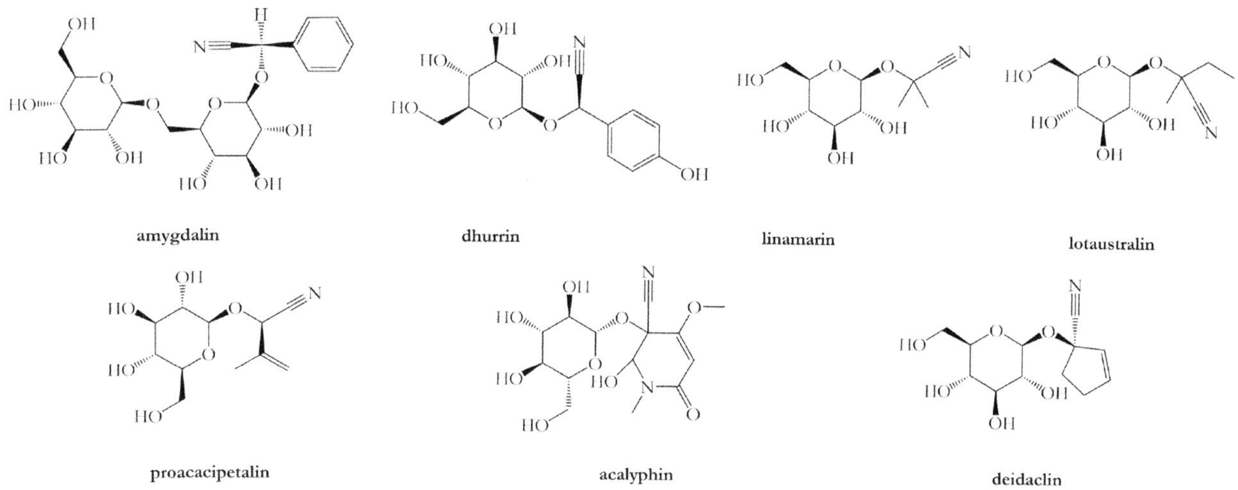

Fig. 5.23. Key representative examples of plant-derived cyanogenic glycoside-type compounds of the 'alkaloid-like' subclass of miscellaneous-type specialised metabolites.

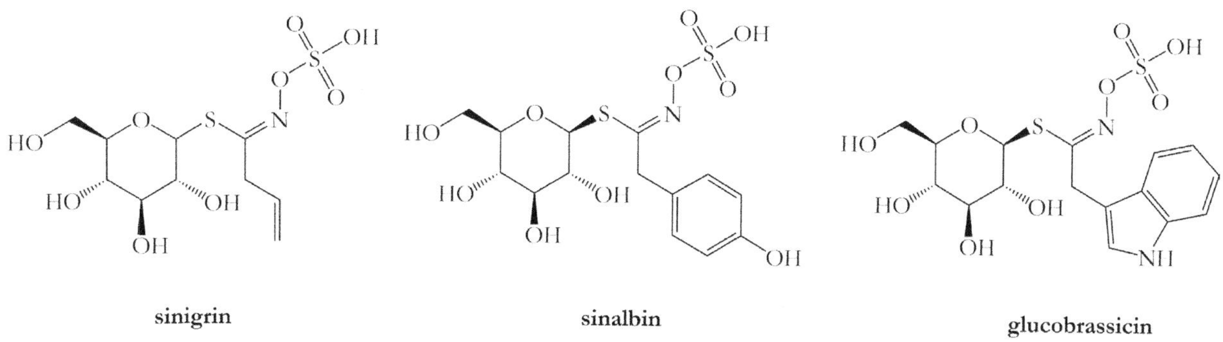

Fig. 5.24. Key representative examples of plant-derived glucosinolate-type compounds of the 'alkaloid-like' subclass of miscellaneous-type specialised metabolites.

they contain (e.g. sinigrin, sinalbin, glucobrassicin).[76] Approximately 100 members of this subinfraclass have been isolated, primarily from the Brassicaceae, all of which essentially produce these compounds (typical content is between 0.1–0.6%, dry weight).[76] Besides predominating in the Brassicaceae, they have been isolated from 13 other families, four of which are closely related to the Brassicaceae (Capparidaceae, Morindaceae, Resedaceae and Tovariaceae) and nine that are not (Bataceae, Bretschneideraceae, Carricaceae, Euphorbiaceae, Gyrostemonaceae, Limnanthaceae, Plantaginaceae, Salvadoraceae and Tropaeolaceae).[77,78] The glucosinolates are biosynthetically related to the CGs, which they resemble in several ways, such as structurally being derived from many of the same select few amino acids (alongside alanine and tryptophan) and being accompanied with specific β-glucosidases (myrosinases/thioglucosidases) required to release their active aglycones (isothiocyanates, thiocyanates, nitriles) (*Fig. 5.24*).[76]

Their biosynthesis is thought to be akin to that of the CGs, arising in a straightforward fashion from their amino acid precursors, but this has yet to be thoroughly established.[76] Historical human consumption of the Brassicaceae suggests glucosinolates do not cause adverse effects in humans, even with prolonged dietary use; in fact, consumption of cruciferous-derived glucosinolates is linked to a decreased incidence of colourectal cancer.[76,78] On the other hand, concentrated intake can lead to adverse effects and, similar to the CGs, it is the volatile aglycone that is responsible. Glucosinolate-derived isothiocyanates have shown potent broadspectrum antimicrobial activity.[76] A high glucosinolate diet has been implicated in goitre formation (goitrogenesis) in livestock, likely due to inhibition of iodine uptake. Glucosinolates are implicated in allomonic/

kairomonic inhibition of certain herbivorous insects, warding off certain species but trapping other species into a permanent nutritional relationship.[76] The distribution of miscellaneous-type specialised metabolites of the glucosinolate-type complex alkaloid-like infraclass reported from the customary medicinal flora of Australia is presented in *Table 5.1*.

REPRESENTATION

This section presents a synopsis of the specialised metabolic trends contained within the flora shown in *Table 5.1*, with a focus on any emergent trends in therapeutic use that correlate with reported biological activity in the sub/infraclass.

'Phenolic-like'

Polyketide-type

Curcuminoid

A summary of the Australian customary medicinal flora reported to bear curcuminoids of the polyketide-type infraclass of the 'phenolic-like' subclass of miscellaneous natural products is as follows: Apiales [Apiaceae (*Centella*)] and Zingiberales [Zingiberaceae (*Curcuma*)]. There was no overlap of the customary uses reported for these two genera.

Tetraketide

A summary of the Australian customary medicinal flora reported to bear tetraketides of the polyketide-type infraclass of the 'phenolic-like' subclass of miscellaneous natural products is as follows: Brassicales [Capparaceae (*Capparis*)]. The genus *Capparis* is used customarily for inflammation, cough, bites/stings and for bacteridical applications.[79] The specific utility of various tetraketide therapeutics is outlined in the relevant section above.

Urushiol

A summary of the Australian customary medicinal flora reported to bear urushiols of the polyketide-type infraclass of the 'phenolic-like' subclass of miscellaneous natural products is as follows: Sapindales [Anacardiaceae (*Semecarpus*)]. The genus *Semecarpus* is used customarily for inflammation and as an antiparasitic.[79]

Quinone-type

Benzoquinone

A summary of the Australian customary medicinal flora reported to bear benzoquinones of the quinone-type infraclass of the 'phenolic-like' subclass of miscellaneous natural products is as follows: Boraginales [Boraginaceae (*Ehretia*)], Caryophyllales [Amaranthaceae (*Achyranthes*), Polygonaceae (*Persicaria*)], Ericales [Primulaceae (*Aegiceras*)], Lamiales [Lamiaceae (*Ajuga*)], Laurales [Lauraceae (*Litsea*)] and Poales [Cyperaceae (*Remirea*)]. There is a moderate trend in the customary use of the plants from these genera, with 57% being used as analgesics (*Aegiceras, Ehretia, Litsea, Persicaria*).[79] A prominent pharmaceutical from this infraclass, embelin, is reported to have analgesic properties, alongside antimalarial, antineoplastic and anti-inflammatory activities.[80]

Naphthoquinone

A summary of the Australian customary medicinal flora reported to bear naphthoquinones of the quinone-type infraclass of the 'phenolic-like' subclass of miscellaneous natural products is as follows: Asparagales [Asphodelaceae (*Dianella*)], Caryophyllales [Droseraceae (*Drosera*), Plumbaginaceae (*Plumbago*)], Ericales [Primulaceae (*Aegiceras*)], Fabales [Fabaceae (*Caesalpinia*)], Malvales [Malvaceae (*Malva*)], Myrtales [Lythraceae (*Ammannia*)], Rosales [Rhamnaceae (*Ventilago*)] and Sapindales [Sapindaceae (*Allophylus*)]. No significant customary therapeutic trends were observed over the plant genera reported to bear naphthoquinones.

Anthraquinone

A summary of the Australian customary medicinal flora reported to bear anthraquinones of the quinone-type infraclass of the 'phenolic-like' subclass of miscellaneous natural products is as follows: Asparagales [Asphodelaceae (*Dianella*), Orchidaceae (*Dendrobium*)], Caryophyllales [Amaranthaceae (*Achyranthes, Atriplex*), Polygonaceae (*Persicaria*)], Dioscoreales [Dioscoreaceae (*Dioscorea*)], Fabales [Fabaceae (*Cassia, Chamaecrista, Senna*)], Gentianales [Rubiaceae (*Coelospermum, Morinda*)], Lamiales [Lamiaceae (*Clerodendrum, Prunella*)], Rosales [Rhamnaceae (*Ventilago, Ziziphus*), Rosaceae (*Rubus*)] and Sapindales [Meliaceae (*Melia*), Sapindaceae (*Allophylus*)]. No significant customary therapeutic trends were observed over the plant genera reported to bear anthraquinones.

Phenanthraquinone

A summary of the Australian customary medicinal flora reported to bear phenanthraquinones of the quinone-type infraclass of the 'phenolic-like' subclass of miscellaneous natural products is as follows: Dioscoreales [Dioscoreaceae (*Dioscorea*)] and Lamiales [Lamiaceae (*Plectranthus*)]. There was no overlap of the customary uses reported for these two genera.

'Terpenoid-like'

Terpenophenolic-type

Cannabinoid

A summary of the Australian customary medicinal flora reported to bear cannabinoids of the terpenophenolic-type infraclass of the 'terpenoid-like' subclass of miscellaneous natural products is as follows: Asterales [Asteraceae (*Acmella, Chrysocephalum, Gnaphalium, Helichrysum*)] and Malvales [Malvaceae (*Hibiscus*)]. No significant customary therapeutic trends were observed over the plant genera reported to bear these compounds.

Alpha-acids

A summary of the Australian customary medicinal flora reported to bear α-acids of the terpenophenolic-type infraclass of the 'terpenoid-like' subclass of miscellaneous natural products is as follows: Asterales [Asteraceae (*Chrysocephalum, Gnaphalium, Helichrysum*)]. Plants from the genera *Chrysocephalum* and *Gnaphalium* are both reported to be used as a customary therapeutic tonic.[79]

Alkaloterpenoid-type

Ergot

A summary of the Australian customary medicinal flora reported to bear ergot-alkaloids of the alkaloterpenoid-type infraclass of the 'terpenoid-like' subclass of miscellaneous natural products is as follows: Solanales [Convolvulaceae (*Evolvulus, Ipomoea*)]. There was no overlap of the customary uses reported for these two genera.

Ipecacuanha

None of the Australian plant genera reported to have customary therapeutic use in the literature have been found to bear compounds of this type.

'Pseudoalkaloterpenoid' (combined)

A summary of the Australian customary medicinal flora reported to bear 'pseudoalkaloterpenoids' of the alkaloterpenoid-type infraclass of the 'terpenoid-like' subclass of miscellaneous natural products is as follows: Asparagales [Orchidaceae (*Dendrobium*)], Asterales [Asteraceae (*Eclipta*)], Fabales [Fabaceae (*Erythrophleum*)], Gentianales [Apocynaceae (*Cynanchum*)], Lamiales [Lamiaceae (*Vitex*)], Piperales [Piperaceae (*Piper*)], Poales [Cyperaceae (*Cyperus*)] and Solanales [Solanaceae (*Solanum*)]. No significant customary therapeutic trends were observed over the plant genera reported to bear these compounds.

'Alkaloid-like'

Simple-type

A summary of the Australian customary medicinal flora reported to bear compounds of the simple-type infraclass of the 'alkaloid-like' subclass of miscellaneous natural products is as follows: Asterales [Asteraceae (*Acmella*), Campanulaceae (*Lobelia*)], Brassicales [Capparaceae (*Capparis*)], Brassicales [Gyrostemonaceae (*Codonocarpus*)], Caryophyllales [Polygonaceae (*Persicaria*), Portulaceae (*Portulaca*)], Cucurbitales [Cucurbitaceae (*Trichosanthes*)], Cycadales [Cycadaceae (*Cycas*)], Ericales [Lecythidaceae (*Barringtonia*), Primulaceae (*Aegiceras*)], Fabales [Fabaceae (*Canavalia, Entada, Indigofera*)], Gentianales [Apocynaceae (*Marsdenia*)], Lamiales [Lamiaceae (*Vitex*)], Malvales [Malvaceae (*Hibiscus, Sida*)], Piperales [Piperaceae (*Piper*)], Rosales [Rhamnaceae (*Ziziphus*)], Sapindales [Rutaceae (*Geijera*), Sapindaceae (*Allophylus*)] and Solanales [Convolvulaceae (*Ipomoea, Operculina*), Solanaceae (*Nicotiana, Solanum*)]. Of these various plant genera, seven were reported to be used as bactericidal agents (*Capparis, Cycas, Hibiscus, Ipomoea, Persicaria*), seven were reportedly used for inflammation (*Barringtonia, Canavalia, Capparis, Codonocarpus, Ipomoea, Sida, Solanum*), and seven are used as analgesics (*Aegiceras, Barringtonia, Canavalia, Entada, Ipomoea, Persicaria, Piper*).[79]

Complex-type

Cyanogenic glycoside

A summary of the Australian customary medicinal flora reported to bear cyanogenic glycosides of the complex-type infraclass of the 'alkaloid-like' subclass of miscellaneous natural products is as follows: Alismatales [Araceae (*Alocasia*)], Boraginales [Boraginaceae (*Ehretia*)], Fabales [Fabaceae (*Acacia, Mucuna*)], Gentianales [Apocynaceae (*Carissa*)], Lamiales [Lamiaceae (*Clerodendrum*), Scrophulariaceae (*Eremophila*)] and Sapindales [Rutaceae (*Zieria*)]. Of these genera, 63% (*Acacia, Alocasia, Carissa, Eremophila, Mucuna*) are reported to be used customarily for inflammation.[79] Despite the toxicity of its cyanide moiety, the notable CG amydalin has been well studied in the literature regarding its role in the inflammatory cyclooxygenase-2 pathway.[81]

Glucosinolate

A summary of the Australian customary medicinal flora reported to bear glucosinolates of the simple-type infraclass of the 'alkaloid-like' subclass of miscellaneous natural products is as follows: Brassicales [Capparaceae (*Capparis*)] and Fabales [Fabaceae (*Indigofera*)]. Plants from both of these genera are reported to be used customarily to treat coughs and colds.[79] Glucosinolates and their degradation products have been shown to have antitussive and anti-asthmatic properties, alongside those discussed above. Plants from the genus *Sinapis* are notable bearers of glucosinolate compounds, and have a 2,000-year history in China for the treatment of coughs and other respiratory problems caused by phlegm retention.[82]

REFERENCES

1. Seigler DS (1998) Polyketides. In *Plant Secondary Metabolism.* (Ed. DS Seigler) pp. 56–75. Springer, New York, NY.
2. Hertweck C (2009) The biosynthetic logic of polyketide biodiversity. *Angewandte Chemie International Edition* **48**, 4688–4716. doi:10.1002/anie.200806121
3. Culberson CF (1969) *Chemical and Botanical Guide to Lichen Products.* University of North Carolina Press, Chapel Hill, NC.
4. Katsuyama Y, Kita T, Funa N, Horinouchi S (2009) Curcuminoid biosynthesis by two type III polyketide synthases in the herb *Curcuma longa. The Journal of Biological Chemistry* **284**, 11160–11170. doi:10.1074/jbc.M900070200
5. Amalraj A, Pius A, Gopi S, Gopi S (2017) Biological activities of curcuminoids, other biomolecules from turmeric and their derivatives: a review. *Journal of Traditional and Complementary Medicine* **7**, 205–233. doi:10.1016/j.jtcme.2016.05.005
6. Bories C, Loiseau P, Cortes D, Myint SH, Hocque-Miller R *et al.* (1991) Antiparasitic activity of *Annona muricata* and *Annona cherimola* seeds. *Planta Medica* **57**, 434–436. doi:10.1055/s-2006-960143
7. Walters DS, Craig R, Mumma RO (1990) Fatty acid incorporation in the biosynthesis of anacardic acids of geraniums. *Phytochemistry* **29**, 1815–1822. doi:10.1016/0031-9422(90)85022-8
8. Seigler DS (1998) Benzoquinones, naphthoquinones, and anthraquinones. In *Plant Secondary Metabolism.* (Ed. DS Seigler) pp. 76–93. Springer, New York, NY.
9. El-Najjar N, Gali-Muhtasib H, Ketola RA, Vuorela P, Urtti A *et al.* (2011) The chemical and biological activities of quinones: overview and implications in analytical detection. *Phytochemistry Reviews* **10**, 353–370. doi:10.1007/s11101-011-9209-1
10. Kutyrev AK (1991) Nucleophilic reactions of quinones. *Tetrahedron* **47**, 8043–8065. doi:10.1016/S0040-4020(01)91002-6
11. Abraham I, Joshi R, Pardasani P, Pardasani RT (2011) Recent advances in 1,4-benzoquinone chemistry. *Journal of the Brazilian Chemical Society* **22**, 385–421. doi:10.1590/S0103-50532011000300002
12. Hook I, Mills C, Sheridan H (2014) Bioactive naphthoquinones from higher plants. In *Studies in Natural Product Chemistry.* (Ed. A-U Rahman) pp. 119–160. Elsevier, Oxford, UK.
13. Duval J, Pecher V, Poujol M, Lesellier E (2016) Research advances for the extraction, analysis and uses of anthraquinones: a review. *Industrial Crops and Products* **94**, 812–833. doi:10.1016/j.indcrop.2016.09.056
14. Toth B, Hohmann J, Vasas A (2018) Phenanthrenes: a promising group of plant secondary metabolites. *Journal of Natural Products* **81**, 661–678. doi:10.1021/acs.jnatprod.7b00619
15. Bolton JL, Dunlap TL, Dietz BM (2018) Formation and biological targets of botanical o-quinones. *Food and Chemical Toxicology* **120**, 700–707. doi:10.1016/j.fct.2018.07.050
16. Bringmann G, Irmer A, Feineis D, Gulder TA, Fiedler HP (2009) Convergence in the biosynthesis of acetogenic natural products from plants, fungi, and bacteria. *Phytochemistry* **70**, 1776–1786. doi:10.1016/j.phytochem.2009.08.019
17. Kumagai Y, Shinkai Y, Miura T, Cho AK (2012) The chemical biology of naphthoquinones and its environmental implications. *Annual Review of Pharmacology and Toxicology* **52**, 221–247. doi:10.1146/annurev-pharmtox-010611-134517
18. Ibrahim SRM, Mohamed GA (2016) Naturally occurring naphthalenes: chemistry, biosynthesis, structural elucidation, and biological activities. *Phytochemistry Reviews* **15**, 279–295. doi:10.1007/s11101-015-9413-5
19. Ko JH, Lee SG, Yang WM, Um JY, Sethi G *et al.* (2018) The application of embelin for cancer prevention and therapy. *Molecules (Basel, Switzerland)* **23**, 621. doi:10.3390/molecules23030621
20. Kundap UP, Bhuvanendran S, Kumari Y, Othman I, Shaikh MF (2017) Plant derived phytocompound, embelin in CNS disorders: a systematic review. *Frontiers in Pharmacology* **8**, 76. doi:10.3389/fphar.2017.00076
21. Qiu H-Y, Wang P-F, Ling H-Y, Tang C-Y, Zhu H-L *et al.* (2018) Naphthoquinones: a continuing source for discovery of therapeutic antineoplastic agents. *Chemical Biology & Drug Design* **91**, 681–690. doi:10.1111/cbdd.13141
22. Wellington KW (2015) Understanding cancer and the anticancer activities of naphthoquinones: a review. *RSC Advances* **5**, 20309–20338. doi:10.1039/C4RA13547D
23. Ravichandiran P, Sheet S, Premnath D, Kim AR, Yoo DJ (2019) 1,4-Naphthoquinone analogues: potent antibacterial agents and mode of action evaluation.

Molecules (Basel, Switzerland) **24**, 1437. doi:10.3390/molecules24071437
24. Strotmann H, Brendel K, Boos KS, Schlimme E (1982) Energy transfer inhibition in photosynthesis by anthraquinone dyes. *FEBS Letters* **145**, 11–15. doi:10.1016/0014-5793(82)81196-4
25. Li Y, Jiang JG (2018) Health functions and structure-activity relationships of natural anthraquinones from plants. *Food & Function* **9**, 6063–6080. doi:10.1039/C8FO01569D
26. Hsu S-C, Chung J-G (2012) Anticancer potential of emodin. *Biomedicine (Taipei)* **2**, 108–116. doi:10.1016/j.biomed.2012.03.003
27. Itharat A, Plubrukarn A, Kongsaeree P, Bui T, Keawpradub N *et al.* (2003) Dioscorealides and dioscoreanone, novel cytotoxic naphthofuranoxepins, and 1,4-phenanthraquinone from *Dioscorea membranacea* Pierre. *Organic Letters* **5**, 2879–2882. doi:10.1021/ol034926y
28. Li M, Zhang J-S, Ye Y-M, Fang J-N (2000) Constituents of the roots of *Salvia prionitis*. *Journal of Natural Products* **63**, 139–141. doi:10.1021/np990357k
29. Matsuda Y, Abe I (2016) Biosynthesis of fungal meroterpenoids. *Natural Product Reports* **33**, 26–53. doi:10.1039/C5NP00090D
30. Menna M, Imperatore C, D'aniello F, Aiello A (2013) Meroterpenes from marine invertebrates: structures, occurrence, and ecological implications. *Marine Drugs* **11**, 1602–1643. doi:10.3390/md11051602
31. Page JE, Nagel J (2006) Biosynthesis of terpenophenolic metabolites in hop and cannabis. In *Recent Advances in Phytochemistry*. (Ed. JT Romeo) pp. 179–210. Elsevier, Oxford, UK.
32. Cho H, Jun J-Y, Song E-K, Kang K-H, Baek H-Y *et al.* (2001) Bakuchiol: a hepatoprotective compound of *Psoralea corylifolia* on tacrine-induced cytotoxicity in Hep G2 cells. *Planta Medica* **67**, 750–751. doi:10.1055/s-2001-18347
33. Zhu X-Y, Zhang C-L, Lin Y, Dang M-Y (2020) Ferruginol alleviates inflammation in dextran sulfate sodium-induced colitis in mice through inhibiting COX-2, MMP-9 and NF-kB signaling. *Asian Pacific Journal of Tropical Biomedicine* **10**, 308–315. doi:10.4103/2221-1691.284945
34. Lin X, Ji S, Li R, Dong Y, Qiao X *et al.* (2012) Terpecurcumins A-I from the rhizomes of *Curcuma longa*: absolute configuration and cytotoxic activity. *Journal of Natural Products* **75**, 2121–2131. doi:10.1021/np300551g
35. Murray RDH, Mendez J, Brown SA (1982) *The Natural Coumarins: Occurrence, Chemistry and Biochemistry*. Wiley, New York, NY.
36. Bergman ME, Davis B, Phillips MA (2019) Medically useful plant terpenoids: biosynthesis, occurrence, and mechanism of action. *Molecules (Basel, Switzerland)* **24**, 3961. doi:10.3390/molecules24213961
37. Hanson GR, Venturelli PJ, Fleckenstein AE (2018) Marijuana. In *Drugs and Society*. 13th edn. (Ed. K Rogers) pp. 416–466. Jones and Bartlett Learning, Burlington, MA.
38. Pate DW (1994) Chemical ecology of *Cannabis*. *Journal of the International Hemp Association* **2**, 32–37.
39. Urban J, Dahlberg C, Carroll B, Kaminsky W (2013) Absolute configuration of beer's bitter compounds. *Angewandte Chemie International Edition* **52**, 1553–1555. doi:10.1002/anie.201208450
40. Nelson M (2005) Germanic Europe and the great beer revival. In *The Barbarian's Beverage: A History of Beer in Ancient Europe*. pp. 78–114. Routledge, New York, NY.
41. De Keukeleire J, Ooms G, Heyerick A, Roldan-Ruiz I, Van Bockstaele E *et al.* (2003) Formation and accumulation of alpha-acids, beta-acids, desmethylxanthohumol, and xanthohumol during flowering of hops. *Journal of Agricultural and Food Chemistry* **51**, 4436–4441. doi:10.1021/jf034263z
42. Flythe MD, Kagan IA, Wang Y, Narvaez N (2017) Hops (Humulus lupulus L.) bitter acids: modulation of rumen fermentation and potential as an alternative growth promoter. *Frontiers in Veterinary Science* **4**, 131. doi:10.3389/fvets.2017.00131
43. Pathak MA, Farrington DJ, Fitzpatrick TB (1962) The presently known distribution of furocoumarins (psoralens) in plants. *The Journal of Investigative Dermatology* **39**, 225–239. doi:10.1038/jid.1962.106
44. Camm EL, Wat CK, Towers GHN (1976) An assessment of the roles of furanocoumarins in *Heracleum lanatum*. *Canadian Journal of Botany* **54**, 2562–2566. doi:10.1139/b76-275
45. Lin X, Ji S, Li R, Dong Y, Qiao X *et al.* (2012) Terpecurcumins A-I from the rhizomes of *Curcuma longa*: absolute configuration and cytotoxic activity. *Journal of Natural Products* **75**, 2121–2131. doi:10.1021/np300551g
46. Seigler DS (1998) *Plant Secondary Metabolism*. Springer, New York, NY.
47. Cao N, Wang C-H (2021) Strictosidine synthase, an indispensible enzyme involved in the biosynthesis of terpenoid indole and β-carboline alkaloids. *Chinese Journal of Natural Medicines* **19**, 591–607. doi:10.1016/S1875-5364(21)60059-6
48. Le Men L, Taylor WI (1965) A uniform numbering system for indole alkaloids. *Experientia* **21**, 508–510. doi:10.1007/BF02138961
49. Kompis I, Hesse M, Schmid H (1971) An approach to the biogenetic classification of indole alkaloids. *Lloydia* **34**, 269–291.
50. Mohammed AE, Abdul-Hameed ZH, Alotaibi MO, Bawakid NO, Sobahi TR *et al.* (2021) Chemical diversity and bioactivities of monoterpene indole alkaloids (MIAs) from six Apocynaceae genera. *Molecules (Basel, Switzerland)* **26**, 488. doi:10.3390/molecules26020488
51. Rosales PF, Bordin GS, Gower AE, Moura S (2020) Indole alkaloids: 2012 until now, highlighting the new chemical structures and biological activities. *Fitoterapia* **143**, 104558. doi:10.1016/j.fitote.2020.104558

52. O'Connor SE, Maresh JJ (2006) Chemistry and biology of monoterpene indole alkaloid biosynthesis. *Natural Product Reports* **23**, 532–547. doi:10.1039/b512615k
53. Somei M, Yamada F (2005) Simple indole alkaloids and those with a non-rearranged monoterpenoid unit. *Natural Product Reports* **22**, 73–103. doi:10.1039/b316241a
54. Wink M (1998) Modes of action of alkaloids. In *Alkaloids: Biochemistry, Ecology, and Medical Applications.* (MF Roberts and M Wink) pp. 301–326. Springer Science+Business, New York, NY.
55. Wink M (2010) Introduction: biochemistry physiology and ecological functions of secondary metabolites. In *Biochemistry of Plant Secondary Metabolism.* (Ed. M Wink) pp. 1–19. Wiley-Blackwell, Sussex, UK.
56. Millan MJ, Newman-Tancredi A, Audinot V, Cussac D, Lejeune F *et al.* (2000) Agonist and antagonist actions of yohimbine as compared to fluparoxan at alpha(2)-adrenergic receptors (AR)s, serotonin (5-HT)(1A), 5-HT(1B), 5-HT(1D) and dopamine D(2) and D(3) receptors: significance for the modulation of frontocortical monoaminergic transmission and depressive states. *Synapse (New York, N.Y.)* **35**, 79–95. doi:10.1002/(SICI)1098-2396(200002)35:2<79::AID-SYN1>3.0.CO;2-X
57. Kiesecker C, Zitron E, Lück S, Bloehs R, Scholz EP *et al.* (2004) Class Ia anti-arrhythmic drug ajmaline blocks HERG potassium channels: mode of action. *Naunyn-Schmiedeberg's Archives of Pharmacology* **370**, 423–435. doi:10.1007/s00210-004-0976-8
58. Jordan MA, Wilson L (2004) Microtubules as a target for anticancer drugs. *Nature Reviews. Cancer* **4**, 253–265. doi:10.1038/nrc1317
59. Gobbi PG, Broglia C, Merli F, Dell'olio M, Stelitano C *et al.* (2003) Vinblastine, bleomycin, and methotrexate chemotherapy plus irradiation for patients with early-stage, favorable Hodgkin lymphoma: the experience of the Gruppo Italiano Studio Linfomi. *Cancer* **98**, 2393–2401. doi:10.1002/cncr.11807
60. Jordan MA (2002) Mechanism of action of antitumor drugs that interact with microtubules and tubulin. *Current Medicinal Chemistry* **2**, 1–17.
61. Smith BA (1990) Strychnine poisoning. *The Journal of Emergency Medicine* **8**, 321–325. doi:10.1016/0736-4679(90)90013-L
62. Palatnick W, Meatherall R, Sitar D, Tenenbein M (2008) Toxicokinetics of acute strychnine poisoning. *Clinical Toxicology* **35**, 617–620.
63. Lavaud C, Massiot G (2017) The *Iboga* alkaloids. In *Progress in the Chemistry of Organic Natural Products 105.* (Eds AD Kinghorn, H Falk, S Gibbons and J Kobayashi) pp. 89–136. Springer, Dordrecht, Germany.
64. Mačiulaitis R, Kontrimavičiūtė V, Bressolle FMM, Briedis V (2008) Ibogaine, an anti-addictive drug: pharmacology and time to go further in development. A narrative review. *Human and Experimental Toxicology* **27**, 181–194. doi:10.1177/0960327107087802
65. Crow WD, Hancox NC, Johns SR, Lamberton JA (1970) New alkaloids of *Alstonia constricta. Australian Journal of Chemistry* **23**, 2489–2501. doi:10.1071/CH9702489
66. Allam K, Beutler JA, Le Quesne PW (1987) 14-Ketoalstonidine and other alkaloidal constitutents of the stem bark of *Alstonia constricta. Journal of Natural Products* **50**, 623–625. doi:10.1021/np50052a007
67. Arnold MSJ, Macdonald JR, Quinn RJ, Skinner-Adams TS, Andrews KT *et al.* (2021) Antiplasmodial activity of the natural product compounds alstonine and himbeline. *International Journal for Parasitology. Drugs and Drug Resistance* **16**, 17–22. doi:10.1016/j.ijpddr.2021.04.003
68. Seigler DS (1998) Ergot and other indole alkaloids. In *Plant Secondary Metabolism.* (Ed. DS Seigler) pp. 665–668. Springer, New York, NY.
69. Caporael LR (1976) Ergotism: the satan loose in Salem. *Science* **192**, 21–26. doi:10.1126/science.769159
70. Mantegani S, Brambilla E, Varasi M (1999) Ergoline derivatives: receptor affinity and selectivity. *Il Farmaco* **54**, 288–296. doi:10.1016/S0014-827X(99)00028-2
71. Hofmann A (2009 (1979)) *LSD, My Problem Child: Reflections on Sacred Drugs, Mysticism, and Science.* Multidisciplinary Association for Psychedelic Studies (MAPS), Santa Cruz, CA.
72. Seigler DS (1998) Alkaloids of terpenoid origin (excepting indole alkaloids and ergot alkaloids). In *Plant Secondary Metabolism.* (Ed. DS Seigler) pp. 668–691. Springer, New York, NY.
73. Fujiwake H, Suzuki T, Oka S, Iwai K (1980) Enzymatic formation of capsaicinoid from vanillylamine and iso-type fatty acids by cell-free extracts of *Capsicum annuum* var. *annuum* cv. Karayatsubusa. *Agricultural and Biological Chemistry* **44**, 2907–2912. doi:10.1271/bbb1961.44.2907
74. Fattori V, Hohmann MSN, Rossaneis AC, Pinho-Ribeiro FA, Verri WA Jr (2016) Capsaicin: current understanding of its mechanisms and therapy of pain and other pre-clinical and clinical uses. *Molecules (Basel, Switzerland)* **21**, 844. doi:10.3390/molecules21070844
75. Islamiyat FB, Oke MO, Olaniyan SA, Ajala AS (2016) A review of cyanogenic glycosides in edible plants. In *Toxicology: New Aspects to this Scientific Conundrum.* (Eds ML Larramendy and S Soloneski) pp. 179–191. IntechOpen. https://www.intechopen.com/books/5362.
76. Blažević I, Montaut S, Burčul F, Olsen CE, Burow M *et al.* (2020) Glucosinolate structural diversity, identification, chemical synthesis and metabolism in plants. *Phytochemistry* **169**, 112100. doi:10.1016/j.phytochem.2019.112100
77. Fahey JW, Zalcmann AT, Talalay P (2001) The chemical diversity and distribution of glucosinolates and isothiocyanates among plants. *Phytochemistry* **56**, 5–51. doi:10.1016/S0031-9422(00)00316-2
78. Nwachukwu ID, Slusarenko AJ, Gruhlke MCH (2012) Sulfur and sulfur compounds in plant defence. *Natural Product Communications* **7**, 395–400. doi:10.1177/1934578X1200700323
79. Lassak EV, Mccarthy T (1983) *Australian Medicinal Plants.* Methuen, Sydney.
80. Basha NJ, Basavarajaiah S, Baskaran S, Kumar P (2022) A comprehensive insight on the biological potential of

embelin and its derivatives. *Natural Product Research* **36**, 3054–3068. doi:10.1080/14786419.2021.1955361
81. Figurová D, Tokárová K, Greifová H, Knížatová N, Kolesárová A *et al.* (2021) Inflammation, it's regulation and antiphlogistic effect of the cyanogenic glycoside amygdalin. *Molecules (Basel, Switzerland)* **26**, 5972. doi:10.3390/molecules26195972
82. Dang R, Guan H, Wang C (2023) Sinapis Semen: a review on phytochemistry, pharmacology, toxicity, analytical methods and pharmacokinetics. *Frontiers in Pharmacology* **14**, 1113583. doi:10.3389/fphar.2023.1113583
83. Huang W, Li C, Wang Y, Yi X, He X (2017) Anti-inflammatory lignanamides and monoindoles from Alocasia macrorrhiza. *Fitoterapia* **117**, 126–132. doi:10.1016/j.fitote.2017.01.014
84. Huang W, Yi X, Feng J, Wang Y, He X (2017) Piperidine alkaloids from Alocasia macrorrhiza. *Phytochemistry* **143**, 81–86. doi:10.1016/j.phytochem.2017.07.012
85. Zhang H-J, Tamez PA, Hoang VD, Tan GT, Hung NV *et al.* (2001) Antimalarial compounds from *Rhaphidophora decursiva*. *Journal of Natural Products* **64**, 772–777. doi:10.1021/np010037c
86. Zhang H, Qiu S, Tamez P, Tan GT, Aydogmus Z *et al.* (2002) Antimalarial agents from plants. II: decursivine, a new antimalarial indole alkaloid from *Rhaphidophora decursiva*. *Pharmaceutical Biology* **40**, 221–224. doi:10.1076/phbi.40.3.221.5832
87. Brinkhaus B, Lindner M, Schuppan D, Hahn EG (2000) Chemical, pharmacological and clinical profile of the East Asian medical plant *Centella aslatica*. *Phytomedicine* **7**, 427–448. doi:10.1016/S0944-7113(00)80065-3
88. James JT, Dubery IA (2009) Pentacyclic triterpenoids from the medicinal herb, *Centella asiatica* (L.) Urban. *Molecules (Basel, Switzerland)* **14**, 3922–3941. doi:10.3390/molecules14103922
89. Bloor SJ (2001) Deep blue anthocyanins from blue *Dianella berries*. *Phytochemistry* **58**, 923–927. doi:10.1016/S0031-9422(01)00343-0
90. Dias DA, Silva CA, Urban S (2009) Naphthalene aglycones and glycosides from the Australian medicinal plant, *Dianella callicarpa*. *Planta Medica* **75**, 1442–1447. doi:10.1055/s-0029-1185724
91. Xu J, Han Q-B, Li S-L, Chen X-J, Wang X-N *et al.* (2013) Chemistry, bioactivity and quality control of *Dendrobium*, a commonly used tonic herb in traditional Chinese medicine. *Phytochemistry Reviews* **12**, 341–367. doi:10.1007/s11101-013-9310-8
92. Paulraj J, Govindarajan R, Palpu P (2013) The genus spilanthes ethnopharmacology, phytochemistry, and pharmacological properties: a review. *Advances in Pharmacological and Pharmaceutical Sciences* **2013**, 510298.
93. Zdero C, Bohlmann F, King RM (1992) *Ent*-labdanes, manoyloxide and helipterol derivatives from *Chrysocephalum ambiguum*. *Phytochemistry* **31**, 1631–1638. doi:10.1016/0031-9422(92)83119-J
94. Lourens AC, Viljoen AM, Van Heerden FR (2008) South African *Helichrysum* species: a review of the traditional uses, biological activity and phytochemistry. *Journal of Ethnopharmacology* **119**, 630–652. doi:10.1016/j.jep.2008.06.011
95. Chung IM, Rajakumar G, Lee JH, Kim SH, Thiruvengadam M (2017) Ethnopharmacological uses, phytochemistry, biological activities, and biotechnological applications of *Eclipta prostrata*. *Applied Microbiology and Biotechnology* **101**, 5247–5257. doi:10.1007/s00253-017-8363-9
96. Feng L, Zhai YY, Xu J, Yao WF, Cao YD *et al.* (2019) A review on traditional uses, phytochemistry and pharmacology of *Eclipta prostrata* (L.) L. *Journal of Ethnopharmacology* **245**, 112109. doi:10.1016/j.jep.2019.112109
97. Zheng X, Wang W, Piao H, Xu W, Shi H *et al.* (2013) The genus *Gnaphalium* L. (Compositae): phytochemical and pharmacological characteristics. *Molecules (Basel, Switzerland)* **18**, 8298–8318. doi:10.3390/molecules18078298
98. Folquitto DG, Swiech JND, Pereira CB, Bobek VB, Halila Possagno GC *et al.* (2019) Biological activity, phytochemistry and traditional uses of genus *Lobelia* (Campanulaceae): a systematic review. *Fitoterapia* **134**, 23–38. doi:10.1016/j.fitote.2018.12.021
99. Li L, Peng Y, Yao X, Xu L-J, Wulan T-N *et al.* (2010) Chemical constituents and biological activities of plants from the genus *Ehretia* Linn. *Chinese Herbal Medicines* **2**, 106–111.
100. Shukla A, Kaur A (2018) A systematic review of traditional uses bioactive phytoconstituents of genus *Ehretia*. *Asian Journal of Pharmaceutical and Clinical Research* **11**, 88–100. doi:10.22159/ajpcr.2018.v11i6.25178
101. Tlili N, Elfalleh W, Saadaoui E, Khaldi A, Triki S *et al.* (2011) The caper (*Capparis* L.): ethnopharmacology, phytochemical and pharmacological properties. *Fitoterapia* **82**, 93–101. doi:10.1016/j.fitote.2010.09.006
102. Doskotch RW, Ray AB, Kubelka W, Fairchild EH, Hufford CD *et al.* (1974) *Codonocarpus* alkaloids - III; the structure of codonocarpine. *Tetrahedron* **30**, 3229–3236. doi:10.1016/S0040-4020(01)97493-9
103. He X, Wang X, Fang J, Chang Y, Ning N *et al.* (2017) The genus Achyranthes: a review on traditional uses, phytochemistry, and pharmacological activities. *Journal of Ethnopharmacology* **203**, 260–278. doi:10.1016/j.jep.2017.03.035
104. Clauser M, Dall'acqua S, Loi MC, Innocenti G (2013) Phytochemical investigation on *Atriplex halimus* L. from Sardinia. *Natural Product Research* **27**, 1940–1944. doi:10.1080/14786419.2013.793684
105. Mroczek A (2015) Phytochemistry and bioactivity of triterpene saponins from Amaranthaceae family. *Phytochemistry Reviews* **14**, 577–605. doi:10.1007/s11101-015-9394-4
106. Egan PA, van der Kooy F (2013) Phytochemistry of the carnivorous sundew genus *Drosera* (Droseraceae): future perspectives and ethnopharmacological relevance. *Chemistry & Biodiversity* **10**, 1774–1790. doi:10.1002/cbdv.201200359

107. Garg V, Jyoti, Dutt R (2018) The genus *Plumbago*: an update. *International Journal of Green and Herbal Chemistry* **7**, 313–326.
108. Peng W, Qin R, Li X, Zhou H (2013) Botany, phytochemistry, pharmacology, and potential application of *Polygonum cuspidatum* Sieb.et Zucc.: a review. *Journal of Ethnopharmacology* **148**, 729–745. doi:10.1016/j.jep.2013.05.007
109. Lin L, Ni B, Lin H, Zhang M, Li X *et al.* (2015) Traditional usages, botany, phytochemistry, pharmacology and toxicology of *Polygonum multiflorum* Thunb.: a review. *Journal of Ethnopharmacology* **159**, 158–183. doi:10.1016/j.jep.2014.11.009
110. Ayaz M, Ahmad I, Sadiq A, Ullah F, Ovais M *et al.* (2020) *Persicaria hydropiper* (L.) Delarbre: a review on traditional uses, bioactive chemical constituents and pharmacological and toxicological activities. *Journal of Ethnopharmacology* **251**, 112516. doi:10.1016/j.jep.2019.112516
111. Gou KJ, Zeng R, Ma Y, Li AN, Yang K *et al.* (2020) Traditional uses, phytochemistry, and pharmacology of *Persicaria orientalis* (L.) Spach: a review. *Journal of Ethnopharmacology* **249**, 112407. doi:10.1016/j.jep.2019.112407
112. Iranshahy M, Javadi B, Iranshahi M, Jahanbakhsh SP, Mahyari S *et al.* (2017) A review of traditional uses, phytochemistry and pharmacology of *Portulaca oleracea* L. *Journal of Ethnopharmacology* **205**, 158–172. doi:10.1016/j.jep.2017.05.004
113. Kasai R, Sasaki A, Hashimoto T, Kaneko T, Ohtani K *et al.* (1999) Cycloartane glycosides from *Trichosanthes tricuspidata*. *Phytochemistry* **51**, 803–808. doi:10.1016/S0031-9422(99)00078-3
114. Kanchanapoom T, Kasai R, Yamasaki K (2002) Cucurbitane, hexanorcucurbitane and octanorcucurbitane glycosides from fruits of *Trichosanthes tricuspidata*. *Phytochemistry* **59**, 215–228. doi:10.1016/S0031-9422(01)00430-7
115. Mai LP, Guenard D, Franck M, Van TM, Gaspard C *et al.* (2002) New cytotoxic cucurbitacins from the pericarps of *Trichosanthes tricuspidata* fruits. *Natural Product Letters* **16**, 15–19. doi:10.1080/1057563029001/4755
116. Shaw PC, Lee KM, Wong KB (2005) Recent advances in trichosanthin, a ribosome-inactivating protein with multiple pharmacological properties. *Toxicon* **45**, 683–689. doi:10.1016/j.toxicon.2004.12.024
117. Yu X, Tang L, Wu H, Zhang X, Luo H *et al.* (2018) Trichosanthis Fructus: botany, traditional uses, phytochemistry and pharmacology. *Journal of Ethnopharmacology* **224**, 177–194. doi:10.1016/j.jep.2018.05.034
118. Schneider D, Wink M, Sporer F, Lounibos P (2002) Cycads: their evolution, toxins, herbivores and insect pollinators. *Naturwissenschaften* **89**, 281–294. doi:10.1007/s00114-002-0330-2
119. Kovacs A, Vasas A, Hohmann J (2008) Natural phenanthrenes and their biological activity. *Phytochemistry* **69**, 1084–1110. doi:10.1016/j.phytochem.2007.12.005
120. Lv H, She G (2010) Naturally occurring diarylheptanoids. *Natural Product Communications* **5**, 1687–1708. doi:10.1177/1934578X1000501035
121. Mohanty A, Das D, Ghosh G, Sahu PK (2016) *Barringtonia acutangula*: ethnobotanical and phytopharmacological review. *International Journal of Pharmaceutical Sciences Review and Research* **38**, 22–29.
122. Rajeswari K, Bhaskara Rao T (2015) *Aegicera corniculatum* Linn (Myrsinaceae). *Journal of Chemical and Pharmaceutical Research* **7**, 305–316.
123. Seigler DS (2003) Phytochemistry of Acacia—sensu lato. *Biochemical Systematics and Ecology* **31**, 845–873. doi:10.1016/S0305-1978(03)00082-6
124. Zanin JL, De Carvalho BA, Martineli PS, Dos Santos MH, Lago JH *et al.* (2012) The genus *Caesalpinia* L. (Caesalpiniaceae): phytochemical and pharmacological characteristics. *Molecules (Basel, Switzerland)* **17**, 7887–7902. doi:10.3390/molecules17077887
125. Matsuzaki S, Hamana K, Okada M, Niitsu M, Samejima K (1990) Aliphatic pentaamines found in *Canavalia gladiata*. *Phytochemistry* **29**, 1311–1312. doi:10.1016/0031-9422(90)85449-P
126. Fujimura S, Terada S, Jayavardhanan KK, Panikkar KR, Kimoto E (1993) Primary structures of concanavalin A-like lectins from seeds of two species of *Canavalia*. *Phytochemistry* **33**, 985–987. doi:10.1016/0031-9422(93)85008-F
127. Pattamadilok D, Pengsuparp T, Phummiratch D, Ongpipattanakul B, Meksuriyen D *et al.* (2008) Canarosine: a new guanidine alkaloid from *Canavalia rosea* with inhibitory activity on dopamine D1 receptors. *Journal of Asian Natural Products Research* **10**, 915–918. doi:10.1080/10286020802181513
128. Huang X, Mu B, Lin W, Qiu Y (2012) Pterocarpin and isoflavan derivatives from *Canavalia maritima* (Aubl.) Thou. *Records of Natural Products* **6**, 166–170.
129. Yadav JP, Arya V, Yadav S, Panghal M, Kumar S *et al.* (2010) *Cassia occidentalis* L.: a review on its ethnobotany, phytochemical and pharmacological profile. *Fitoterapia* **81**, 223–230. doi:10.1016/j.fitote.2009.09.008
130. Singh O, Ali M, Akhtar N (2011) Phenolic acid glucosides from the seeds of *Entada phaseoloides* Merill. *Journal of Asian Natural Products Research* **13**, 682–687. doi:10.1080/10286020.2011.584444
131. Sugimoto S, Matsunami K, Otsuka H (2018) Biological activity of *Entada phaseoloides* and *Entada rheedei*. *Journal of Natural Medicines* **72**, 12–19. doi:10.1007/s11418-017-1146-x
132. Son NT (2019) Genus *Erythrophleum*: botanical description, traditional use, phytochemistry and pharmacology. *Phytochemistry Reviews* **18**, 571–599. doi:10.1007/s11101-019-09640-0
133. Gerometta E, Grondin I, Smadja J, Frederich M, Gauvin-Bialecki A (2020) A review of traditional uses, phytochemistry and pharmacology of the genus *Indigofera*. *Journal of Ethnopharmacology* **253**, 112608. doi:10.1016/j.jep.2020.112608

134. Sathiyanarayanan L, Arulmozhi S (2007) *Mucuna pruriens* Linn.: a comprehensive review. *Pharmacognosy Reviews* **1**, 157-162.
135. Tse GG, Kim BB, Mcmurtray AM, Nakamoto BK (2013) Case of levodopa toxicity from ingestion of *Mucuna gigantea*. *Hawai'i Journal of Medicine & Public Health : a Journal of Asia Pacific Medicine & Public Health* **72**, 157-160.
136. Vora R, Joshi AN, Joshi NC (2018) Comparison of L-Dopa content in three species of genus *Mucuna* by different extraction techniques. *Annals of Plant Sciences* **7**, 1973-1977. doi:10.21746/aps.2018.7.1.10
137. Khyade MS, Kasote DM, Vaikos NP (2014) *Alstonia scholaris* (L.) R. Br. and *Alstonia macrophylla* Wall. ex G. Don: a comparative review on traditional uses, phytochemistry and pharmacology. *Journal of Ethnopharmacology* **153**, 1-18. doi:10.1016/j.jep.2014.01.025
138. Kaunda JS, Zhang YJ (2017) The genus *Carissa*: an ethnopharmacological, phytochemical and pharmacological review. *Natural Products and Bioprospecting* **7**, 181-199. doi:10.1007/s13659-017-0123-0
139. Han L, Zhou X, Yang M, Zhou L, Deng X *et al.* (2018) Ethnobotany, phytochemistry and pharmacological effects of plants in genus *Cynanchum* Linn. (Asclepiadaceae). *Molecules (Basel, Switzerland)* **23**, 1194. doi:10.3390/molecules23051194
140. Wang P, Yang J, Zhu Z, Zhang X (2018) *Marsdenia tenacissima*: a review of traditional uses, phytochemistry and pharmacology. *The American Journal of Chinese Medicine* **46**, 1449-1480. doi:10.1142/S0192415X18500751
141. Salim AA, Garson MJ, Craik DJ (2004) New indole alkaloids from the roots of *Ochrosia acuminata*. *Journal of Natural Products* **67**, 1719-1721. doi:10.1021/np049812p
142. Carroll AR, Addepalli R, Fechner G, Smith J, Guymer GP *et al.* (2008) Alkaloids from the Australian rainforest tree *Ochrosia moorei*. *Journal of Natural Products* **71**, 1063-1065. doi:10.1021/np070655e
143. Miller CM, Mccarthy FO (2012) Isolation, biological activity and synthesis of the natural product ellipticine and related pyridocarbazoles. *RSC Advances* **2**, 8883-8918. doi:10.1039/c2ra20584j
144. Zhang B-J, Yan J-M, Wu Z-K, Liu Y-P, Bao M-F *et al.* (2013) Alkaloids from *Ochrosia borbonica*. *Helvetica Chimica Acta* **96**, 2288-2298. doi:10.1002/hlca.201300136
145. Chen AH, Liu QL, Ma YL, Jiang ZH, Tang JY *et al.* (2017) A new monoterpenoid indole alkaloid from *Ochrosia elliptica*. *Natural Product Research* **31**, 1490-1494. doi:10.1080/14786419.2016.1277349
146. Van Beek TA, Verpoorte R, Svendsen AB, Leeuwenberg AJM, Bisset NG (1984) *Tabernaemontana* L. (Apocynaceae): a review of its taxonomy, phytochemistry, ethnobotany and pharmacology. *Journal of Ethnopharmacology* **10**, 1-156. doi:10.1016/0378-8741(84)90046-1
147. Patchayasakul W, Pongchaidecha A, Chattipakorn N, Chattipakorn S (2008) Ethnobotany and ethnopharmacology of *Tabernaemontana divaricata*. *The Indian Journal of Medical Research* **127**, 317-335.
148. Abubakar IB, Loh HS (2016) A review on ethnobotany, pharmacology and phytochemistry of *Tabernaemontana corymbosa*. *The Journal of Pharmacy and Pharmacology* **68**, 423-432. doi:10.1111/jphp.12523
149. Itoh A, Tanaka Y, Nagakura N, Nishi T, Tanahashi T (2006) A quinic acid ester from *Strychnos lucida*. *Journal of Natural Medicines* **60**, 146-148. doi:10.1007/s11418-005-0021-3
150. Galliford CV, Scheidt KA (2007) Pyrrolidinyl-spirooxindole natural products as inspirations for the development of potential therapeutic agents. *Angewandte Chemie International Edition* **46**, 8748-8758. doi:10.1002/anie.200701342
151. Tchinda AT, Tamze V, Ngono ARN, Ayimele GA, Cao M *et al.* (2012) Alkaloids from the stem bark of *Strychnos icaja*. *Phytochemistry Letters* **5**, 108-113. doi:10.1016/j.phytol.2011.11.004
152. Guo R, Wang T, Zhou G, Xu M, Yu X *et al.* (2018) Botany, phytochemistry, pharmacology and toxicity of *Strychnos nux-vomica* L.: a review. *The American Journal of Chinese Medicine* **46**, 1-23. doi:10.1142/S0192415X18500015
153. Li W, Tang GH, Chen L, Tang YQ, Xu YK *et al.* (2018) New pyridocarbazole alkaloids from *Strychnos nitida*. *Natural Product Research* **32**, 1532-1536. doi:10.1080/14786419.2017.1385016
154. Bowie JH, Cooke RG, Wilkin PE (1962) Colouring matters of Australian plants. X: anthraquinones from *Coelospermum* species. *Australian Journal of Chemistry* **15**, 336-341. doi:10.1071/CH9620336
155. Lopez JL, Pusset J (1988) Plantes de Nouvelle-Caledonie, 115. Alcaloides monoterpeniques de *Coelospermum billardieri*. *Journal of Natural Products* **51**, 829-835. doi:10.1021/np50059a002
156. Benkrief R, Skaltsounis A-L, Tillequin F, Koch M (1991) Plantes de Nouvelle-Caledonie, 137. Iridoides de *Coelospermum billardieri*. *Journal of Natural Products* **54**, 532-541. doi:10.1021/np50074a028
157. Magiatis P, Skaltsounis A-L, Tillequin F, Seguin E, Cosson J-P (2002) Coelobillardin, an iridoid glucoside dimer from *Coelospermum billardieri*. *Phytochemistry* **60**, 415-418. doi:10.1016/S0031-9422(02)00118-8
158. Pawlus AD, Kinghorn AD (2007) Review of the ethnobotany, chemistry, biological activity and safety of the botanical dietary supplement *Morinda citrifolia* (noni)*. *The Journal of Pharmacy and Pharmacology* **59**, 1587-1609. doi:10.1211/jpp.59.12.0001
159. Potterat O, Hamburger M (2007) *Morinda citrifolia* (noni) fruit: phytochemistry, pharmacology, safety. *Planta Medica* **73**, 191-199. doi:10.1055/s-2007-967115
160. Zhang JH, Xin HL, Xu YM, Shen Y, He YQ *et al.* (2018) *Morinda officinalis* How.: a comprehensive review of traditional uses, phytochemistry and pharmacology. *Journal of Ethnopharmacology* **213**, 230-255. doi:10.1016/j.jep.2017.10.028

161. Boucherle B, Haudecoeur R, Queiroz EF, De Waard M, Wolfender JL *et al.* (2016) *Nauclea latifolia*: biological activity and alkaloid phytochemistry of a West African tree. *Natural Product Reports* **33**, 1034–1043. doi:10.1039/C6NP00039H
162. Haudecoeur R, Peuchmaur M, Peres B, Rome M, Taiwe GS *et al.* (2018) Traditional uses, phytochemistry and pharmacological properties of African *Nauclea* species: a review. *Journal of Ethnopharmacology* **212**, 106–136. doi:10.1016/j.jep.2017.10.011
163. Johns SR, Lamberton JA (1970) The identification of a new alkaloid from *Timonius kaniensis* (Rubiaceae) as dihydrocupreine. *Australian Journal of Chemistry* **23**, 211–212. doi:10.1071/CH9700211
164. Khan IA, Sticher O, Rali T (1993) New triterpenes from the leaves of *Timonius timon*. *Journal of Natural Products* **56**, 2163–2165. doi:10.1021/np50102a019
165. Luan F, Han K, Li M, Zhang T, Liu D *et al.* (2019) Ethnomedicinal uses, phytochemistry, pharmacology, and toxicology of species from the genus *Ajuga* L.: a systematic review. *The American Journal of Chinese Medicine* **47**, 959–1003. doi:10.1142/S0192415X19500502
166. Wang JH, Luan F, He XD, Wang Y, Li MX (2018) Traditional uses and pharmacological properties of *Clerodendrum* phytochemicals. *Journal of Traditional and Complementary Medicine* **8**, 24–38. doi:10.1016/j.jtcme.2017.04.001
167. Abdel-Mogib M, Albar HA, Batterjee SM (2002) Chemistry of the genus *Plectranthus*. *Molecules (Basel, Switzerland)* **7**, 271–301. doi:10.3390/70200271
168. Bai Y, Xia B, Xie W, Zhou Y, Xie J *et al.* (2016) Phytochemistry and pharmacological activities of the genus *Prunella*. *Food Chemistry* **204**, 483–496. doi:10.1016/j.foodchem.2016.02.047
169. Sena Filho JG, Duringer J, Maia GLA, Tavares JF, Xavier HS *et al.* (2008) Ecdysteroids from *Vitex* species: distribution and compilation of their 13C-NMR spectral data. *Chemistry & Biodiversity* **5**, 707–713. doi:10.1002/cbdv.200890067
170. Rani A, Sharma A (2013) The genus *Vitex*: a review. *Pharmacognosy Reviews* **7**, 188–198. doi:10.4103/0973-7847.120522
171. Yao JL, Fang SM, Liu R, Oppong MB, Liu EW *et al.* (2016) A review on the terpenes from genus *Vitex*. *Molecules (Basel, Switzerland)* **21**, 1179. doi:10.3390/molecules21091179
172. Singab AN, Youssef FS, Ashour ML, Wink M (2013) The genus *Eremophila* (Scrophulariaceae): an ethnobotanical, biological and phytochemical review. *The Journal of Pharmacy and Pharmacology* **65**, 1239–1279. doi:10.1111/jphp.12092
173. Wang YS, Wen ZQ, Li BT, Zhang HB, Yang JH (2016) Ethnobotany, phytochemistry, and pharmacology of the genus *Litsea*: an update. *Journal of Ethnopharmacology* **181**, 66–107. doi:10.1016/j.jep.2016.01.032
174. Wu P-L, Chuang T-H, He C-X, Wu T-S (2004) Cytotoxicity of phenylpropanoid esters from the stems of *Hibiscus taiwanensis*. *Bioorganic & Medicinal Chemistry* **12**, 2193–2197. doi:10.1016/j.bmc.2004.02.020
175. Wu P-L, Wu T-S, He C-X, Su C-H, Lee K-H (2005) Constituents from the stems of *Hibiscus taiwanensis*. *Chemical & Pharmaceutical Bulletin* **53**, 56–59. doi:10.1248/cpb.53.56
176. Vasudeva N, Sharma SK (2008) Biologically active compounds from the genus *Hibiscus*. *Pharmaceutical Biology* **46**, 145–153. doi:10.1080/13880200701575320
177. Maganha EG, Halmenschlager RD, Rosa RM, Henriques Ja P, Ramos ALLD *et al.* (2010) Pharmacological evidences for the extracts and secondary metabolites from plants of the genus *Hibiscus*. *Food Chemistry* **118**, 1–10. doi:10.1016/j.foodchem.2009.04.005
178. Gasparetto JC, Martins CA, Hayashi SS, Otuky MF, Pontarolo R (2012) Ethnobotanical and scientific aspects of *Malva sylvestris* L.: a millennial herbal medicine. *The Journal of Pharmacy and Pharmacology* **64**, 172–189. doi:10.1111/j.2042-7158.2011.01383.x
179. Sharifi-Rad J, Melgar-Lalanne G, Hernandez-Alvarez AJ, Taheri Y, Shaheen S *et al.* (2020) *Malva* species: insights on its chemical composition towards pharmacological applications. *Phytotherapy Research* **34**, 546–567. doi:10.1002/ptr.6550
180. Dinda B, Das N, Dinda S, Dinda M, Silsarma I (2015) The genus *Sida* L.: a traditional medicine: its ethnopharmacological, phytochemical and pharmacological data for commercial exploitation in herbal drugs industry. *Journal of Ethnopharmacology* **176**, 135–176. doi:10.1016/j.jep.2015.10.027
181. Poornima V, Sharanya M, Jeyam M (2014) An ethnomedical, pharmacological and phytochemical review of *Ammannia baccifera* L. *World Journal of Pharmaceutical Research* **3**, 1771–1789.
182. Parmar VS, Jain SC, Bisht KS, Jain R, Taneja P *et al.* (1997) Phytochemistry of the genus *Piper*. *Phytochemistry* **46**, 597–673. doi:10.1016/S0031-9422(97)00328-2
183. Xiang C-P, Shi Y-N, Liu F-F, Li H-Z, Zhang Y-J *et al.* (2016) A survey of the chemical compounds of *Piper* spp. (Piperaceae) and their biological activities. *Natural Product Communications* **11**, 1403–1408. doi:10.1177/1934578X1601100948
184. Peerzada AM, Ali HH, Naeem M, Latif M, Bukhari AH *et al.* (2015) *Cyperus rotundus* L.: traditional uses, phytochemistry, and pharmacological activities. *Journal of Ethnopharmacology* **174**, 540–560. doi:10.1016/j.jep.2015.08.012
185. Allan RD, Correll RL, Wells RJ (1969) Two new phenolic ketones from *Remirea maritima* (Cyperaceae). *Tetrahedron Letters* **10**, 4673–4674. doi:10.1016/S0040-4039(01)88779-7
186. Allan RD, Correll RL, Wells RJ (1969) A new class of quinones from certain members of the family Cyperaceae. *Tetrahedron Letters* **10**, 4669–4672. doi:10.1016/S0040-4039(01)88778-5
187. Allan RD, Wells RJ, Macleod JK (1970) Further phenolic ketones from *Remirea maritima* Aubl. *Tetrahedron Letters* **11**, 3945–3946. doi:10.1016/S0040-4039(01)98632-0
188. Rabelo AS, Oliveira ID, Guimaraes AG, Quintans JS, Prata AP *et al.* (2013) Antinociceptive, anti-inflammatory and

antioxidant activities of aqueous extract from *Remirea maritima* (Cyperaceae). *Journal of Ethnopharmacology* **145**, 11-17. doi:10.1016/j.jep.2012.10.020

189. Cooke RG, Johnson BL (1963) Colouring matters of Australian plants. XI: quinones from *Ventilago viminalis*. *Australian Journal of Chemistry* **16**, 695-702. doi:10.1071/CH9630695
190. Rao BK, Hanumaiah T, Rao CP, Rao GSR, Rao KVJ *et al.* (1983) Anthraquinones in *Ventilago* species. *Phytochemistry* **22**, 2583-2585. doi:10.1016/0031-9422(83)80170-8
191. Hanumaiah T, Marshall DS, Rao BK, Rao CP, Rao GSR *et al.* (1985) Benzoisochromanquinones in *Ventilago* species. *Phytochemistry* **24**, 2373-2378. doi:10.1016/S0031-9422(00)83044-7
192. Hanumaiah T, Rao BK, Rao CP, Rao GSR, Rao JUM *et al.* (1985) Naphthalenes and naphthoquinones from *Ventilago* species. *Phytochemistry* **24**, 1811-1815. doi:10.1016/S0031-9422(00)82557-1
193. Jammula SR, Pepalla SB, Hanumaiah T, Rao KVJ, Thomson RH (1991) Benzisochromanquinones from *Ventilago goughii*. *Phytochemistry* **30**, 3741-3744. doi:10.1016/0031-9422(91)80100-F
194. Lin L-C, Chou C-J, Kuo Y-C (2001) Cytotoxic principles from *Ventilago leiocarpa*. *Journal of Natural Products* **64**, 674-676. doi:10.1021/np000569d
195. Periyasamy K, Kaliyaperumal S (2016) Ethnobotanical, phytochemical and pharmaceutical studies of medicinal plant, *Ventilago maderaspatana* Gaertn. (red creeper): a review. *International Journal of Current Pharmaceutical Research* **8**, 16-18.
196. Cassels BK, Eckhardt G, Kaussmann EU, Tschesche R (1974) Cyclopeptide alkaloids of *Zizyphus oenoplia*. *Tetrahedron* **30**, 2461-2466. doi:10.1016/S0040-4020(01)97117-0
197. Mahajan RT, Chopda MZ (2009) Phytopharmacology of *Ziziphus jujuba* Mill.: a plant review. *Pharmacognosy Reviews* **3**, 320-329.
198. Gao QH, Wu CS, Wang M (2013) The jujube (*Ziziphus jujuba* Mill.) fruit: a review of current knowledge of fruit composition and health benefits. *Journal of Agricultural and Food Chemistry* **61**, 3351-3363. doi:10.1021/jf4007032
199. Prabhavathi S, Vajayalakshmi S (2015) New hydroxycarboxylic acid from the root bark of *Ziziphus oenoplia*. *Journal of Pharmacognosy and Phytochemistry* **3**, 150-152.
200. Rocabado GO, Bedoya LM, Abad MJ, Bermejo P (2008) *Rubus*: a review of its phytochemical and pharmacological profile. *Natural Product Communications* **3**, 423-436. doi:10.1177/1934578X0800300319
201. Lee J, Dossett M, Finn CE (2012) *Rubus* fruit phenolic research: the good, the bad, and the confusing. *Food Chemistry* **130**, 785-796. doi:10.1016/j.foodchem.2011.08.022
202. Oelrichs PB, Macleod JK, Seawright AA, Ng JC (1997) Isolation and characterisation of urushiol components from the Australian native cashew (*Semecarpus australiensis*). *Nature Toxicology* **5**, 96-98.
203. Premalatha B (2000) *Semecarpus anacardium* Linn. Nuts: a boon in alternative medicine. *Indian Journal of Experimental Biology* **38**, 1177-1182.
204. Nagabhushana KS, Umamaheshwari S, Tocoli FE, Prabhu SK, Green IR *et al.* (2002) Inhibition of soybean and potato lipoxygenases by bhilawanols from bhilawan (*Semecarpus anacardium*) nut shell liquid and some synthetic salicylic acid analogues. *Journal of Enzyme Inhibition and Medicinal Chemistry* **17**, 255-259. doi:10.1080/1475636021000006243
205. Semalty M, Semalty A, Badola A, Joshi GP, Rawat MS (2010) *Semecarpus anacardium* Linn.: a review. *Pharmacognosy Reviews* **4**, 88-94. doi:10.4103/0973-7847.65328
206. Zhao L, Huo C-H, Shen L-R, Yang Y, Zhang Q *et al.* (2010) Chemical constituents of plants from the genus *Melia*. *Chemistry & Biodiversity* **7**, 839-859. doi:10.1002/cbdv.200900043
207. Lahey FN, Macleod JK (1967) The coumarins of *Geijera parviflora* Lindl. *Australian Journal of Chemistry* **20**, 1943-1955. doi:10.1071/CH9671943
208. Dreyer DL, Lee A (1972) Extractives of *Geijera parviflora*. *Phytochemistry* **11**, 763-767. doi:10.1016/0031-9422(72)80045-1
209. Shou Q, Banbury LK, Renshaw DE, Smith JE, He X *et al.* (2013) Parvifloranines A and B, two 11-carbon alkaloids from *Geijera parviflora*. *Journal of Natural Products* **76**, 1384-1387. doi:10.1021/np400376r
210. Southwell IA (1981) Methoxystyrenes from the genus *Zieria*. *Phytochemistry* **20**, 1448-1450. doi:10.1016/0031-9422(81)80068-4
211. Flynn TM, Southwell IA (1987) Cyanogenesis in the genus *Zieria*. *Phytochemistry* **26**, 1669-1673. doi:10.1016/S0031-9422(00)82267-0
212. Flynn TM, Southwell IA (1987) Essential oil constituents of the genus *Zieria*. *Phytochemistry* **26**, 1673-1686. doi:10.1016/S0031-9422(00)82268-2
213. Southwell IA, Armstrong JA (1987) Chemical variation within the genus *Zieria*. *Phytochemistry* **26**, 1687-1692. doi:10.1016/S0031-9422(00)82269-4
214. Griffin SG, Leach DN, Markham JL, Johnstone R (1998) Antimicrobial activity of essential oils from *Zieria*. *The Journal of Essential Oil Research* **10**, 165-174. doi:10.1080/10412905.1998.9700870
215. Chavan RB, Gaikwad DK (2016) The ethnobotany, phytochemistry and biological properties of *Allophylus* species used in traditional medicine: a review. *World Journal of Pharmacy and Pharmaceutical Sciences* **5**, 664-682.
216. Austin DF (2008) *Evolvulus alsinoides* (Convolvulaceae): an American herb in the Old World. *Journal of Ethnopharmacology* **117**, 185-198. doi:10.1016/j.jep.2008.01.038
217. Meira M, Da Silva EP, David JM, David JP (2012) Review of the genus *Ipomoea*: traditional uses, chemistry and biological activities. *Revista Brasileira de Farmacognosia-Brazilian Journal of Pharmacognosy* **22**, 682-713. doi:10.1590/S0102-695X2012005000025

218. Ding W, Zeng F, Xu L, Chen Y, Wang Y *et al.* (2011) Bioactive dammarane-type saponins from *Operculina turpethum*. *Journal of Natural Products* **74**, 1868–1874. doi:10.1021/np200274m
219. Ding W, Jiang ZH, Wu P, Xu L, Wei X (2012) Resin glycosides from the aerial parts of *Operculina turpethum*. *Phytochemistry* **81**, 165–174. doi:10.1016/j.phytochem.2012.05.010
220. Gupta S, Ved A (2017) *Operculina turpethum* (Linn.) Silva Manso as a medicinal plant species: a review on bioactive components and pharmacological properties. *Pharmacognosy Reviews* **11**, 158–166.
221. Lira-Ricardez J, Pereda-Miranda R, Castaneda-Gomez J, Fragoso-Serrano M, Simas RC *et al.* (2019) Resin glycosides from the roots of *Operculina macrocarpa* (Brazilian Jalap) with purgative activity. *Journal of Natural Products* **82**, 1664–1677. doi:10.1021/acs.jnatprod.9b00222
222. Nugroho LH, Verpoorte R (2002) Secondary metabolism in tobacco. *Plant Cell, Tissue and Organ Culture* **68**, 105–125. doi:10.1023/A:1013853909494
223. Dewey RE, Xie J (2013) Molecular genetics of alkaloid biosynthesis in *Nicotiana* tabacum. *Phytochemistry* **94**, 10–27. doi:10.1016/j.phytochem.2013.06.002
224. Jassbi AR, Zare S, Asadollahi M, Schuman MC (2017) Ecological roles and biological activities of secialized metabolites from the genus *Nicotiana*. *Chemical Reviews* **117**, 12227–12280. doi:10.1021/acs.chemrev.7b00001
225. Kaunda JS, Zhang YJ (2019) The genus *Solanum*: an ethnopharmacological, phytochemical and biological properties review. *Natural Products and Bioprospecting* **9**, 77–137. doi:10.1007/s13659-019-0201-6
226. Nahar L, Sarker SD (2007) Phytochemistry of the genus *Curcuma*. In *Turmeric: The Genus Curcuma*. (Eds PN Ravindran, K Nirmal Babu and K Sivaraman) pp. 71–106. CRC Press, Boca Raton, FL.

6 Conclusion

This book stands as a compendium, a reference work that brings a plethora of disparate information into one place, but to what end? As outlined in Chapter 1, we, the authors, sincerely hope that this localised data is useful for people within the field of natural product (NP) chemistry in Australia. We hope that this work contributes to the advancement of Indigenous Cultural Intellectual Property (ICIP) and the integration of customary knowledge with modern scientific approaches. What remains in concluding this book is some grounded speculation as to where the field is heading in the immediate and not-so-distant future.

This final chapter will address seven key aspects of this evolution, all of which are predicated upon an ever-increasing development of power and reach coming from the technological sector in the form of artificial intelligence and machine learning (AI/ML). On the more mechanical end of the spectrum, automation in the drug discovery process combined with high-throughput screening (HTS) technologies have drastically accelerated workflows and expanded the possibility of what can be studied within the scope of a single project. This follows naturally into biotechnology, primarily in the form of synthetic biology and bioengineering, which can develop upon findings emerging from more traditional approaches to NP chemistry, leading to robust outcomes that can augment our supply of bioactive compounds and help us to understand and harness their origins. Much of the work coming out of the biotech sector is only comprehensible through modern -omics-based models, which began with genomics but has now grown into various new fields such as transcriptomics, proteomics, metabolomics and interactomics, collectively known as multi-omics as these fields become increasingly integrated. Of these, metabolomics is of the greatest relevance to NP chemistry, as it is explicitly concerned with biological metabolites, and from this comes one of the great techniques available to modern NP chemists – dereplication – or the integration of metabolomics with analytical spectroscopy/spectrometry to enable a drastic increase in capability of targeted and untargeted screening protocols, particularly when combined with molecular networking (i.e. Global Natural Product Social (GNPS) Molecular Networking). The enormous amounts of data generated by many of the techniques just outlined has spilled into the realm of systems biology, as our understanding of biosynthesis, drug action, disease mechanisms and complex synergistic effects in biosystems has driven the emergence of entirely new fields such as network pharmacology or polypharmacology. All of this work is made possible through AI/ML, which are now indispensable tools for the life sciences (viz. society at large) due to their monumental capacity for data analysis, prediction, identification and modelling. Finally, within, behind and above all of this work are the results that are continually pouring out of the fields of ethnopharmacology and ethnomedicine, a place where the integration of traditional therapeutic knowledge systems and modern science is most actively taking place. Each of the topics to be addressed in this concluding chapter can and already do comprise many superb books and review articles that will be referenced throughout, if the interested reader wanted to further investigate any of these points.

AUTOMATION AND HTS

Automation in NP chemistry is rapidly increasing the efficiency and efficacy of the drug discovery process. HTS has application in several technical fields and involves automated processes combined with robotics, data pipelines and detection software that enable the screening of complex metabolic mixtures and the rapid performance of chemical/genetic/pharmacological tests and assays.[1] HTS facilitates the recognition of key or novel metabolites and helps develop our understanding of the pathways through which these compounds biosynthetically arise and/or have their biomedical effects. Research in 2010 showed that HTS facilitates a 1,000-fold increase in screening rate at a 1,000,000-fold decrease in cost relative to a conventional drop-based microfluidic approach.[2] UCLA's HTS Molecular Screening Shared Resources (MSSR) facility can screen over 100,000 compounds per day, and offers an open-access policy, metrics that were unthinkable to the NP chemist of the 20th century.[3] The task for HTS is the acquisition of chemical starting material for drug development, through biochemical and cell-based hit-identification assays. Some key techniques are biochemical assays of both the fluorescence- [fluorescence polarisation and anisotropy, fluorescence resonance energy transfer (FRET) and time-resolved FRET, fluorescence lifetime analysis] and binding variety (Nuclear Magnetic Resonance (NMR), Mass Spectrometry (MS), surface plasmon resonance (SPR), differential scanning fluorimetry (DSF)]), as well as cell-based assays (viability, reporter gene, second messenger and HT-microscopy assays).[4] Hit discovery leads to target deconvolution, which brings an assortment of interdisciplinary assistance from multi-omics that will be discussed in greater detail later in this chapter.

The lead author can speak from experience to the complexity of the traditional approach to NP-based drug discovery and HTS effects a drastic reduction in the time and operating costs involved in the metabolic identification/elucidation pipeline.[5] Computational approaches that segregate NPs into virtual and physical collections and then leverage these to optimise the discovery process are of significant interest to the field. The movement of a drug from an unknown entity to market takes on average 13–15 years and billions of dollars of investment, so any approach that can offer a reduction in these figures are worthy of attention.[5] NPs are more diverse and complex than those populating typical lead-compound libraries,and despite having associated detriments (often not conforming to Lipinski's 'Rule of Five'), also offer promising positive attributes due to their structural obscurity and the potentially novel effects these motifs may have on a biosystem. NP-based drug discovery has serious limitations when considering the amount of effort that may be expended studying a plant only to discover compounds already known (somewhat circumvented by the process of dereplication). Alongside this, material acquisition (which opens up the dangerous realm of biopiracy), extraction, detection and isolation of bioactive NPs is slow and expensive with a low rate of success. Considering these factors, *in silico* approaches, particularly when leveraging the power of cloud-based supercomputing, enable promising cheminformatic optimisation of the process through data-mining, chemical space analysis, prediction of physicochemical (ADMET) profiles, binding and de novo design.[5]

Cheminformatics, the use of *in silico* techniques in chemical theory (often ported from the more established field of bioinformatics), in NP chemistry utilises chemical databases and libraries to virtually screen for potential leads, structural novelty and drug-targets. Prediction of quantitative structure–activity relationships (QSAR), often in tandem with chemometrics and expert systems, can enable superb chemical knowledge representation.[6] It is through the integration with many of these computational techniques that this book hopes to have a lasting effect on the state of inquiry pertaining to the specialised metabolites identified from the customary medicinal flora of Australia. Before moving on to the next section, it is worth briefly addressing the shortcomings of the HTS approach. The most glaring being that global implementation of HTS strategies has not resulted in an increased number of novel medications coming to market. The reason for this will arise later in this concluding chapter, but to address it briefly here, there appears to be somewhat of a paradigm shift occurring in the understanding of the pharmaceutical field as a whole, whereby the focus of drug discovery/development is shifting from one-drug one-target to a more nuanced concept of efficacy that is predicated upon cellular-/organism-level survival pathways and homeostatic balance, alongside enzymatic modification of non-local complexes (protein displacement).[7] Recent developments in HTS are appearing in an attempt to address these shortcomings, such as HTS flow cytometry, HTS of 3D cultures, virtual HTS consensus docking and AI analysis of HTS primary data, but none of these can as yet substitute for the downstream substantiation of animal models for unknown targets.[7]

BIOTECHNOLOGY

Pioneering fields within the biotech sector, such as synthetic biology and bioengineering, naturally augment the advances just detailed regarding automation and HTS. Once lead-compounds are identified and their efficacy quantified, numerous questions remain unanswered. Development of a sustainable source of these compounds is paramount, often where the complexity of a natural chemical structure is not conducive to chemical synthesis. The natural source of these compounds may well be extremely limited and the industrial extraction of these compounds for the benefit of society would not be ethical. Synthetic biology and bioengineering can provide pathways to understand how these compounds are produced and solutions for how we could afford a ready supply of ideal compounds for development. Some examples of these solutions are the engineering of biosynthetic pathways and genomes, as well as the de novo development of microbial consortia, genomes and even more recently cell-free systems.[8,9]

The chemical diversity produced by the natural world is enabled by the evolutionary development of sophisticated biosynthetic pathways. It is one thing to obtain and extract these molecules after the fact, through traditional NP approaches, but it is quite another to manipulate these pathways for ourselves. On top of this, the concept of a dark metabolome suggests that less than 5% of what an organism can produce is ever present in a single extract, due to a limitation of information flow between the genome and the metabolome at any given point in time.[10] It is suggested that we are only able to annotate approximately 2% of the measurable metabolome of a given organism, when run through LC-MS/MS identification strategies, suggesting a broad spectrum of unknown compounds.[11] In-source fragmentation (may be a plausible explanation for much of this mystery, but the fact still remains that the supressed avenues of a genome at a given point in time may be hiding much of the chemical diversity of an organism. The rational reprogramming and manipulation of the pathways within a given genome presents researchers with previously under-expressed or even completely unnatural NPs under normal circumstances.[11]

Beyond this propagation of the dark metabolome, the acquisition of the metabolites that we do know of is greatly simplified if these systems can be manipulated or recreated in the laboratory. Advances in sequencing and recombinant DNA technologies can improve NP yields, even from difficult to culture strains (this approach works primarily with microbially sourced NPs).[12] Increased public sentiment against chemically synthesised therapeutic agents, based on environmental and/or health concerns, leads to a market for more sustainable or 'green' bioproducts. The production of NPs from renewable biomass not only addresses these concerns, but also the more practical concerns of the low-yield, high-cost of conventional extraction or synthesis. Systems metabolic engineering (i.e. using model micro-organisms such as *Escherichia coli* or *Saccharomyces cerevisiae* as microbial cell factories) can maximise the production of amenable target chemicals. Integration with AI can assist the design of novel biosynthetic pathways and integration with HTS can apply a somewhat more random approach that can isolate strains of merit.[13] The potential of genome editing through the Clustered Regularly Interspaced Short Palindromic Repeats (CRISPR) and CRISPR-associated 9 (Cas9) endonuclease system (CRISPR/Cas9) can bring this microbial approach out to the level of medicinal plant species, with modulation of specialised metabolic pathways yielding greater or optimised quantities/qualities of a specific plant metabolome.[14,15] Finally, cell-free biosynthetic approaches depart from the constraints of living cells, by extracting, optimising and augmenting cellular biochemical production mechanisms to a broad set of use-cases.[8] All these approaches again benefit from AI/ML, both in optimisation and interpretation, and these developments are only just beginning; watch this space for significant advances over the coming years.

OMICS

The biggest driver of conceptual change in the life sciences today are what have become known as the 'omics' (biological fields with the suffix -omics, from -ome: 'all constituents considered collectively'). Omics are all about totality, holism, the quantification of the complete set of molecules and processes that comprise the structure, function and dynamics of an organism.[16] Since the original genomics, the study of the genomes of organisms, a range of other -omics have emerged (194 of them by current wikipedia count (2024): https://en.wikipedia.org/wiki/Omics), with transcriptomics (study of cellular RNA and transcriptome), proteomics (study of the cellular protein and proteome), metabolomics (study of cellular metabolism and metabolome) and interactomics (study of the relationships and interactions between proteins and other molecules) having particular significance for NP chemistry.[16] Multiomics involves the integration of multiple 'omes' as a technical approach to the life sciences, often involving AI/ML and big data. The ramifications of this approach will be outlined further in a later section on network pharmacology, whereas here it will suffice just to outline each of the most pertinent -omics approaches and how they relate to NP chemistry and its advance into the future.

Genomics

The backbone of genomics is gene (DNA) sequencing, which enables the analysis of genomes. Sequencing is used in the pursuit of understanding the collective characterisation of the entire suite of genes possessed by a given organism, as well as their relationships and influence on the organism.[17] Understanding the structure and function of the entire genome is made possible by the field of bioinformatics, from which cheminformatics emerged. Advances in next-generation sequencing technologies have facilitated the rapid analysis of a wide range of genomes across the natural kingdoms. The data obtained from these studies enable the identification of biosynthetic gene clusters (BGCs), which are implicated in the production of NPs. BGC analysis can help researchers develop *in silico* models for the resulting molecular structure and mechanistic function of genomic-level output, forming the groundwork for the remaining -omics to be built upon. Gene sequencing affords the opportunity to assess the biosynthetic potential of an organism (particularly microorganisms), before any testing taking place. The investigation of the dark metabolome, through the identification of the products of silent or cryptic pathways, is also made possible through genomics. A challenge that has arisen in the practical implementation of the genomics strategy in NP research is that it is much easier to discover a promising pathway than the resulting NP.[18,19] The advances detailed in the preceding section on biotechnology, such as manipulation of enzymatic machinery, were made possible through genomics. Despite its already stellar contribution to the life sciences, genomics has much more to give, both independently (in the realm of post-genomics), but more importantly in concert with other -omics, particularly metabolomics.[20] The prospect of linking metabolites to the hard code of their biosynthetic pathways has furthered our understanding of the expression of biosynthetic products and the machinery that produce them, as well as the environmental factors that influence their production.[20]

Transcriptomics

The study of the transcriptome, the sum of an organism's RNA transcript, is concerned with the intermediary process that interfaces between the instructions encoded on DNA and the products it encodes. The transcripts present in an organism at any given time offer a rare window into the temporal health and activity of a biosystem: which processes are active and which are dormant. Transcriptomics gives insight into how genes are expressed, how cells are differentiated and how all this is regulated. RNA sequencing (RNA-seq) technologies enable this analysis and provide the researcher with a more nuanced understanding of an organism's cellular response to environmental factors and developmental stages.[16] The chemical ecology of an organism's metabolome is generally geared towards some form of stress response, with the machinery controlled through regulatory pathways encoded in the transcriptome; this is particularly important for NP chemists studying the differing pharmacological profiles of various species (or phenotypes) in various environmental/developmental contexts.[21] Furthermore, comparative transcriptomics helps uncover BGCs that are transcriptionally silent (i.e. orphan BGCs), with the integration of genomics, transcriptomics and metabolomics opening up substantial advances in our understanding of the dark metabolome.[22] RNA interference (RNAi) methodologies have been implemented in the functional validation of NP interactions, as well as deciphering specifically phenotypic gene function through 'reverse genetics', and identifying gene-target knockdown.[16] Of concern with RNAi is the putative disconnect between mRNA levels and their associated protein expressions, leading to false-positive non-correlated results.[16]

Proteomics

Proteins are integral to the function of all living organisms and proteomics concerns itself with the entire set of these macromolecules produced by a particular organism. Proteomics has, unsurprisingly, broad application in the life sciences; when limited to the field of NP chemistry the main application is the identification of drug-targets. NP proteomics typically involves the development of probes (design and synthesis) and then target-fishing with these probes.[23] NP interaction with protein targets is the foundation of our understanding of their pharmacological action. The discovery and development of new medicines is complicated by the fact that individual drugs interact with multiple targets, with solutions to multitarget efficacy emerging over the recent decade from genomics, transcriptomics and proteomics. Compound-Centric Chemical Proteomics (CCCP) and Activity-Based Protein Profiling (ABPP) are key drivers for modern studies of drug-targets and Mechanisms of Action (MOA).[23] MS is key in the later stages of chemical proteomics, with advances being directed at mitigating earlier issues with low-abundance target loss and non-specific results, yielding a field known as quantitative proteomics, which by often implementing Stable Isotope Labelling by Amino acids in Cell culture (SILAC), has been widely used for target identification of NPs.[23] Implementation of chemical proteomics, in conjunction with HTS and multiomics, should lead to accurate high-throughput non-labelled approaches in the future.

Metabolomics

The most obvious -omics to be relevant to NP chemistry is the one that arose organically from within the study of an organism's metabolic profile, hence metabolomics. As has been outlined throughout this book, chemical extracts from natural sources are composed of thousands of NPs, many of which are present in minute relative quantities and having in vivo bioactivities with synergistic mechanisms. Annotation of these compounds, primarily through MS and NMR, is integral to the process of NP isolation/elucidation/identification.[24] Metabolomics arose as a development in response to shortcomings inherent to the traditional approaches to NP chemistry, such as degradation/modification of NPs during analysis, loss of key support compounds through bioactivity-guided fractionation and the substantiated efficacy of whole-organism extracts in customary medical systems. Metabolomics offers expedited drug discovery routes, comprehensive analyses of a complex biosystem's metabolic fingerprint and a broader understanding of the relationship between biochemistry, gene function and therapeutic efficacy.[24] The immense amount of data produced from these studies necessitates the implementation of AI/ML models through informatics approaches such as chemometrics. Practical opportunities facilitated by metabolomic sophistication are optimised sample preparation and extraction, enhanced analytical methods for analysis (GC-MS, LC-MS, LC-MS/MS, CE-MS and NMR) and data processing/interpretation (in concert with NP databases and molecular networking).[24] Untargeted metabolomics methodologies for unbiased signal processing in MS experiments, particularly tandem MS fragmentation via Data-Dependent Acquisition (DDA) or Data-Independent Acquisition (DIA) modes, provide informative compound identification results through matching signal spectra with database records.[25] This approach, despite its substantial advantages, has its own pitfalls, which are well known in the field, such as lacking standards for natural compounds (causing a reliance on *in silico* predictions) and many members of the community still depending on actually purifying the compound and characterising it in isolation (which still often depends on traditional NP isolation strategies). This being stated, one of the key outcomes of metabolomics in NP chemistry is its role in dereplication, which is addressed separately later in this conclusion.

Interactomics

The final -omics of particular import for NP chemistry to be covered briefly here are the biological networks produced through interactomics, coined in 1999 to describe the whole set of molecular interactions in a cell. Often, interactomics is concerned with protein–protein interactions (PPIs), but it also details the interaction between the NP and the protein and indirect genetic interactions.[26] The molecular interaction network, most often the PPI network and its subsets, studies the dynamic physical interactions occurring within the metabolome of an organism or cell (metabolic networks can be considered to be molecular interaction networks with NPs related to one another by enzymes).[26] The genetic interaction network is concerned with the protein–DNA interactome formed by transcription factors, regulatory proteins and target genes, with the field of chemoproteomics emerging as a particularly relevant field concerned with drug-target identification.[26] This top-down approach from systems biology requires both experimental techniques already discussed (HTS) and computational approaches such as bioinformatics. Researchers in this field are attempting to elucidate the systemic properties (topology) of a complex, interacting biosystem and the role that individual members (particularly highly connected 'hubs') play in the network.[26]

Multiomics

Multiomics, panomics or integrative omics, crosses the divide between the different -omics methodologies and, as the name suggests, utilises multiple sources of data to develop integrated models across differing levels of biological scope, often driven by ML technologies. This field is only 20 years old, with less than 10 citations per year in the early 2000s and nearly 4,000 citations on PubMed in 2023. Applications in NP research are primarily focused on drug-target discovery, where any single -omics approach (i.e. thermal proteomic profiling (TPP), cellular thermal shift assay (CETSA), drug affinity response target stability (DARTS), photoaffinity labelling (PAL), activity-based protein profiling (ABPP), small interfering RNA (siRNA), short hairpin RNA (shRNA), and CRISPR) may only afford an individual perspective on the actual system under investigation, which can have a limited ability to clearly define molecular targets in complex biochemical or physiological networks.[16] A clear advantage of the integration of several -omics approaches in accordance with the structural characteristics of specific NPs is adequately leveraging the power of bioinformatic data mining and its integration with AI. This modern approach allows the researcher to differentiate between desirable results (i.e. drug-target associations) and noise (i.e. off-target effects and non-target associated, non-specific protein binding interactions).[16] Through the appreciation of linked data and large-scale network interference, real-time sharing and large-scale analysis of digitised NP databases, complex meta-analyses can be applied to achieve an unbiased view of biological processes or drug interventions.[27,28] Specific applications in NP research are for drug-target identification, drug repurposing and the determination of active constituents in complex mixtures and polydrug combinations. Utilisation of modern NP databases, such as LOTUS, which harmonises, curates, validates and disseminates over 750,000 referenced structure organism pairs, and the ChEMBL interaction database, enables the exploration of NP chemical features and biotarget spaces.[29] This approach is highly relevant to the purposes of the database produced in this project, as the chemotaxonomic correlation between biosynthetic pathways and species taxonomy is a key step in comprehending the therapeutic novelty of a biological source-set.[28] From research such as this, trends that show the overall potential of differing chemobiological sources, such as alkaloids for neural disease, polyketides for protein phosphorylation processes or HDAC receptors or fatty acids for prostanoid/leukotriene physiological processes, can be uncovered.[28] Many of the issues associated with NPs in the drug development process can be attenuated through the implementation of integrative approaches such as these, where potential therapeutic indices and mechanisms can be determined before development. The investigation of NPs and their multitarget efficacy, and the perturbation effects they may have on the cell-signalling pathways of a complex, networked biosystem, are more effectively characterised through the implementation of multiomics strategies.

DEREPLICATION

Dereplication is an aspect of metabolomics with significant importance to NP chemistry since the 1990s.[30] It is the process of rapidly identifying known NPs from mixtures, avoiding redundancy in the laborious process of isolation/purification/characterisation. Originally defined as 'a process of quickly identifying known chemotypes', dereplication has evolved into several differing forms since its origin, differing in their integration of the technique into the workflow.[30] In its simplest form, dereplication is critical in the identification of novelty within a matrix and relies on integration with databases that catalogue the (mostly MS) spectral data of reported compounds. Dereplication is not a new idea in NP chemistry; it simply gains a great deal of utility when built upon modern computational strategies. The comparison of LC-MS/MS fragmentation spectra with databases such as GNPS through molecular networking methodologies has precipitated a revolution in the analysis of complex NP mixtures. Dereplication of this kind, particularly when integrated with HTS, can expedite the hit identification process by some orders of magnitude. Algorithms built over and above these methodologies, such as DEREPLICATOR+ (covering peptides, polyketides, terpenes, benzenoids, alkaloids, flavanoids, etc.), enable even greater annotation of complex spectra than GNPS alone.[31] Complications in the direct search of chemical structure libraries (i.e. PubChem, ChemSpider,

ChEMBL, ChemBank, ChEBI, Dictionary of Natural Products, etc.) encountered throughout the 21st century have resulted in prohibitively time-consuming implementation of workflows for large spectral datasets, something that projects such as DEREPLICATOR+ are looking to solve over the coming years.[31] Dereplication can be fully integrated into untargeted metabolomic profiling studies or it can even be used in taxonomic identification of microbial strains in gene-sequence analysis studies. A 2014 study outlines the rise of dereplication studies from 1996 to 2014, with citations on the Web of Science literature database increasing from the tens to almost 1,000 per annum in that timeframe.[30] Today (in 2024) that figure is > 5,000 and it seems that this metabolomic-based aspect of NP drug discovery shows no sign of slowing down.

NETWORK PHARMACOLOGY

Few developments in this area have had such an impact on me as the new field emerging from systems biology known as network pharmacology. The seminal work on the topic is the article titled *Network pharmacology: the next paradigm in drug discovery*, published in *Nature: Chemical Biology* by Andrew L. Hopkins in 2008, and currently it has 4067 citations.[32] This article outlines the significant decrease in new drug candidates translating to effective medicines in the clinic due to late-stage attrition in the drug development pipeline. Hopkins cites two primary reasons for this: lack of efficacy and clinical safety/toxicology, each accounting for 30% of failures. This article, alongside the myriad studies that have been published subsequently, outlines the ailing notion of designing maximally selective ligands to act on explicit, individual drug-targets. As has been discussed at various points throughout this concluding chapter, NPs interact with numerous targets (not just one). Additionally, a new appreciation for the role of 'polypharmacology' (chemical agents that act on multiple targets or disease pathways) can help in the resolution of issues that typically arise in drug development, such as side effects and toxicity.[33] Rational drug design under this paradigm can expand the current opportunity space for drug-targets and many of the multitude of NPs that have been overlooked due to their low specificity may well be repurposed to new and unexpected ends. The key challenge of network pharmacology is to design therapeutic agents that have mechanistically validated target combinations and optimised multiple SARs, while maintaining drug-like properties. Large-scale functional genomics studies have demonstrated that single-gene knockouts alone have limited effect on phenotype, with this phenotypic robustness being understood through redundancy and alternative signalling routes.[32] Network biology analysis has revealed the correlation between robustness and network structure; scale-free networks are resilient against the random deletion of any node. Network pharmacology suggests that the ideal strategy should be to identify the perturbations that the foreign environmental factor is causing in the network, where the symptoms are emerging from synthetic network behaviours; synthetic lethality, synthetic sickness and/or synthetic rescue.[32] Through engagement with this literature, one of the most apparent trends in the field of network pharmacology is the interest it is generating in the validation of traditional and customary medicinal systems, particularly Traditional Chinese Medicine (TCM). The combination of multiomics, polypharmacology, ML/AI, computational simulations and TCM theory have caused rapid advancements in our understanding of NPs (whole herbal extracts) as therapeutic agents and the superposition of TCM syndrome elements within a perturbed system.[34] It is only a matter of time until this kind of work expands out into less well-established ethnomedical fields.

ARTIFICIAL INTELLIGENCE

Lying behind essentially all of the advancements in the NP field is AI/ML. The exciting developments emerging from computational drug design such as bioactivity prediction and de novo drug design offer substantial potential, but also key challenges.[35] Current projects involve the prediction of chemical structures of BGC clusters based on DNA sequence alone (NP genome/metabolome mining), advanced filtering/prioritisation in the drug development pipeline, better understanding of SAR and drug-target prediction (structural characterisation), increasingly sophisticated molecular modelling (docking) and time-resolved molecular dynamics models, chemical similarity techniques (molecular networking), drug repurposing through the prediction of unexpected activity and optimisation of drugs for certain off-/on-target effects.[35] Some issues that have arisen are high false-positive rates compared with rule-based annotation approaches, accurate imputation of missing data in LC-MS/MS metabolomics approaches, lack of high-quality standardised open-access data (data error, interoperability, compatibility, scope), time requirements and costs for active learning techniques, literature curation (scrapability) and AI benchmarking in catalysis and bioactivity. Overall, it is limited high-quality data that is restraining progress in this direction, rather than algorithms, and simpler models may afford better bias-variance trade-off outcomes.[35] Integration with broader AI fields and techniques such natural language processing, retrieval-augmented generation (RAG) and active learning will help drive innovation in this space. All of these prospects require funding, particularly regarding database maintenance and interoperability, but above all else the field needs to develop as a global scientific community and to coordinate its efforts in the continuation of producing high-quality research.[35]

INTEGRATION: ETHNOPHARMACOLOGY

Of particular importance for both this book, and the development of contemporary NP chemistry in general, is the integration of traditional ethnopharmacological knowledge systems with modern scientific techniques. Efforts to this end are already flourishing within TCM, yet many of the conceptual leads in the field have come from the knowledge systems of cultures that have historical roots in a particular bioregion and these need to be harmonised with this emerging paradigm. Some of the key outcomes of this process would be documentation and preservation of customary knowledge, validation of ethnomedical approaches, sustainable use of natural resources and ethical bioprospecting, equitable access and benefit sharing, collaboration through community engagement and capacity building, development of cultural sensitivity and the allocation/protection of IP rights.[36–38] Ethnopharmacological research has provided many of the leads utilised in NP drug discovery and a deeper interdisciplinary collaboration with the social sciences, such as anthropology, will enrich our understanding of traditional knowledge systems. The therapeutic synergism that has long been the basis for ethnomedicinal approaches in traditional cultures is key to understanding how a plant is utilised within a cultural context, as well as how ethnopharmacological knowledge is distributed in a society and who benefits from it.[39,40] By studying the social relations of customary therapeutic approaches, the generation, maintenance, spread and devolution of cultural traits can be incorporated into this evolving field in a meaningful way. Appreciation of differing intercultural approaches to illness and therapeutics will expand the horizons of the NP field and the way society thinks about health.[40] Bioprospecting implicitly involves collection, then analysis, but always driven by commercialisation.[41,42] It is imperative that the benefits of commercialisation are shared with the progenitors of local knowledge, as ignoring this point leads directly to biopiracy.[41,42] Ethical, sustainable bioprospecting is now predicated and informed by emerging policies such as the Nagoya Protocol (2010), the Convention on Biological Diversity , ICIP, the creation of the Traditional Knowledge Digital Library and all agents in this field should be well versed in how this sector is evolving and should make a concerted effort to integrate these approaches into their own methodology.

FUTURE WORK

By Edward Owen Norman

This project has comprised much of my adult life as a scientist. I am amazed that it has arrived at this point, as it all began as a simple question: 'Which Australian native plant(s) would be ideal to study during my NP research candidature?' Through continual interaction with the field, the literature and discussions with colleagues, it became increasingly apparent that a lot of work had been done to study the chemical constituents of Australian plants over the past centuries, but it was not at all easy to ascertain exactly the sum total of that work. Through some years of diligent reading, book-keeping, organising and tabulating, the beginnings of a summary was formed. Over the next few years that grew into my thesis at RMIT University, which then grew into a website (www.cmfoa.info) as part of a work-integrated learning (WIL) project and now into this book. The future of the project is multifaceted; in the most concrete terms another WIL project is underway, to build upon the first website and to integrate it with a RAG-AI system to facilitate easier and more conversational interaction with users interested in this space (funded by RMIT's RACE cloud supercomputing hub's merit allocation scheme). In my own career I am back studying; from the content of this chapter I find it evident that computational strategies hold a major role in the future of this field and to that end I am applying myself to master the informatic side of NP chemistry. So much amazing research is being produced by NP chemists around the world, but my own role feels one level of abstraction out, bridging the gap between different fields, researchers and stakeholders to communicate the bigger picture of what is happening in the NPs field. As stated throughout, I genuinely, humbly hope this small contribution to the field will expand our appreciation for the superb, living knowledge of the natural world possessed by Indigenous Peoples. Ideally, this work can serve as a foundation for steady developments in this direction, as there is still so much more work to be done.

BROAD PERSPECTIVE

By Professor Sylvia Urban

This book, together with the CMFoA database, represents an important contribution to Australian NP chemistry. Not only has a database compendium been created as a single package, but has been made available publicly for the first time. This resource of information of the uses of Australian plants by the Indigenous Peoples will be an invaluable resource for scientists and the public. For researchers studying Australian plants wanting to understand their traditional uses or wanting to understand which plants have been understudied in terms of their chemical constituents or not studied at all, this compendium is an invaluable resource to guide future investigations. It will unquestionably be a valuable resource to inform future studies, particularly in the field of chemistry.

REFERENCES

1. Inglese J, Auld DS (2007) High throughput screening (HTS) techniques: applications in chemical biology. In *Wiley Encyclopedia of Chemical Biology*.(ERd. TP Begley) pp. 1-15.Wiley
2. Inglese J, Auld DS, Jadhav A, Johnson RL, Simeonov A *et al.* (2006) Quantitative high-throughput screening: a titration-based approach that efficiently identifies biological activities in large chemical libraries. *Proceedings of the National Academy of Sciences of the United States of America* **103**, 11473–11478. doi:10.1073/pnas.0604348103
3. Dove A (2007) High-throughput screening goes to school. *Nature Methods* **4**, 523–532. doi:10.1038/nmeth0607-523
4. Blay V, Tolani B, Ho SP, Arkin MR (2020) High-throughput screening: today's biochemical and cell-based approaches. *Drug Discovery Today* **25**, 1807-1821. doi:10.1016/j.drudis.2020.07.024
5. Dagur P, Shreya, Ghosh R, Rakshit G, Biswas A *et al.* (2023) Computational methods in natural products-based drug discovery. In *CADD and Informatics in Drug Discovery*. (Eds M Rudrapal and J Khan) pp. 99–121. Springer, Singapore.
6. Sushko Y, Novotarskyi S, Körner R, Vogt J, Abdelaziz A *et al.* (2014) Prediction-driven matched molecular pairs to interpret QSARs and aid the molecular optimization process. *Journal of Cheminformatics* **6**, 48. doi:10.1186/s13321-014-0048-0
7. Hushpulian DM, Gaisina IN, Nikulin SV, Chubar T, Savin SS *et al.* (2024) High throughput screening in drug discovery: problems and solutions. *Moscow University Chemistry Bulletin* **79**, 93–104. doi:10.3103/S0027131424700081
8. Bogart JW, Cabezas MD, Vögeli B, Wong DA, Karim AS *et al.* (2021) Cell-free exploration of the natural product chemical space. *ChemBioChem* **22**, 84–91. doi:10.1002/cbic.202000452
9. Qian X, Chen L, Sui Y, Chen C, Zhang W *et al.* (2020) Biotechnological potential and applications of microbial consortia. *Biotechnology Advances* **40**, 107500. doi:10.1016/j.biotechadv.2019.107500
10. Crow JM (2021) Canada's scientists are elucidating the dark metabolome. *Nature* **599**, S14–S15. doi:10.1038/d41586-021-03062-9
11. Giera M, Aisporna A, Uritboonthai W, Siuzdak G (2024) The hidden impact of in-source fragmentation in metabolic and chemical mass spectrometry data interpretation. *Nature Metabolism* **6**, 1647–1648. doi:10.1038/s42255-024-01076-x
12. Pickens LB, Tang Y, Chooi Y-H (2011) Metabolic engineering for the production of natural products. *Annual Review of Chemical and Biomolecular Engineering* **2**, 211–236. doi:10.1146/annurev-chembioeng-061010-114209
13. Yang D, Eun H, Prabowo CPS, Cho S, Lee SY (2022) Metabolic and cellular engineering for the production of natural products. *Current Opinion in Biotechnology* **77**, 102760. doi:10.1016/j.copbio.2022.102760
14. Biswas P, Anand U, Ghorai M, Pandey DK, Jha NK *et al.* (2022) Unraveling the promise and limitations of CRISPR/Cas system in natural product research: Approaches and challenges. *Biotechnology Journal* **17**, 2100507. doi:10.1002/biot.202100507
15. Dey A, Nandy S (2021) CRISPER/Cas in plant natural product research: therapeutics as anticancer and other drug candidates and recent patents. *Recent Patents on Anti-cancer Drug Discovery* **16**, 460–468. doi:10.2174/1574892816666210706155602
16. Zhang H-W, Lv C, Zhang L-J, Guo X, Shen Y-W *et al.* (2021a) Application of omics-and multi-omics-based techniques for natural product target discovery. *Biomedicine and Pharmacotherapy* **141**, 111833. doi:10.1016/j.biopha.2021.111833
17. Prihoda D, Maritz JM, Klempir O, Dzamba D, Woelk CH *et al.* (2021) The application potential of machine learning and genomics for understanding natural product diversity, chemistry, and therapeutic translatability. *Natural Product Reports* **38**, 1100–1108. doi:10.1039/D0NP00055H
18. Hoskisson PA, Seipke RF (2020) Cryptic or silent? The known unknowns, unknown knowns, and unknown unknowns of secondary metabolism. *mBio* **11**, 02642. doi:10.1128/mBio.02642-20
19. Jensen PR, Chavarria KL, Fenical W, Moore BS, Ziemert N (2014) Challenges and triumphs to genomics-based natural product discovery. *Journal of Industrial*

Microbiology & Biotechnology **41**, 203–209. doi:10.1007/s10295-013-1353-8

20. Caesar LK, Montaser R, Keller NP, Kelleher NL (2021) Metabolomics and genomics in natural products research: complementary tools for targeting new chemical entities. *Natural Product Reports* **38**, 2041–2065. doi:10.1039/D1NP00036E
21. Vitale GA, Geibel C, Minda V, Wang M, Aron AT *et al.* (2024) Connecting metabolome and phenotype: recent advances in functional metabolomics tools for the identification of bioactive natural products. *Natural Product Reports* **41**, 885–904. doi:10.1039/D3NP00050H
22. Amos GC, Awakawa T, Tuttle RN, Letzel A-C, Kim MC *et al.* (2017) Comparative transcriptomics as a guide to natural product discovery and biosynthetic gene cluster functionality. *Proceedings of the National Academy of Sciences* **114**, E11121–E11130.
23. Chen X, Wang Y, Ma N, Tian J, Shao Y *et al.* (2020) Target identification of natural medicine with chemical proteomics approach: probe synthesis, target fishing and protein identification. *Signal Transduction and Targeted Therapy* **5**, 72. doi:10.1038/s41392-020-0186-y
24. Salem MA, Perez De Souza L, Serag A, Fernie AR, Farag MA *et al.* (2020) Metabolomics in the context of plant natural products research: from sample preparation to metabolite analysis. *Metabolites* **10**, 37. doi:10.3390/metabo10010037
25. Guo J, Huan T (2020) Comparison of full-scan, data-dependent, and data-independent acquisition modes in liquid chromatography–mass spectrometry based untargeted metabolomics. *Analytical Chemistry* **92**, 8072–8080. doi:10.1021/acs.analchem.9b05135
26. Bludau I, Aebersold R (2020) Proteomic and interactomic insights into the molecular basis of cell functional diversity. *Nature Reviews. Molecular Cell Biology* **21**, 327–340. doi:10.1038/s41580-020-0231-2
27. Van Santen JA, Kautsar SA, Medema MH, Linington RG (2021) Microbial natural product databases: moving forward in the multi-omics era. *Natural Product Reports* **38**, 264–278. doi:10.1039/D0NP00053A
28. Wang D, Li X, Miao Y, Zhang Q (2023) Profiling chemobiological connection between natural product and target space based on systematic analysis. *International Journal of Molecular Sciences* **24**, 11265. doi:10.3390/ijms241411265
29. Rutz A, Sorokina M, Galgonek J, Mietchen D, Willighagen E *et al.* (2022) The LOTUS initiative for open knowledge management in natural products research.*eLife* **11**, e70780. doi:10.7554/eLife.70780
30. Hubert J, Nuzillard J-M, Renault J-H (2017) Dereplication strategies in natural product research: How many tools and methodologies behind the same concept? *Phytochemistry Reviews* **16**, 55–95. doi:10.1007/s11101-015-9448-7
31. Mohimani H, Gurevich A, Shlemov A, Mikheenko A, Korobeynikov A *et al.* (2018) Dereplication of microbial metabolites through database search of mass spectra. *Nature Communications* **9**, 4035. doi:10.1038/s41467-018-06082-8
32. Hopkins AL (2008) Network pharmacology: the next paradigm in drug discovery. *Nature Chemical Biology* **4**, 682–690. doi:10.1038/nchembio.118
33. Reddy AS, Zhang S (2013) Polypharmacology: drug discovery for the future. *Expert Review of Clinical Pharmacology* **6**, 41–47. doi:10.1586/ecp.12.74
34. Ma S, Liu J, Li W, Liu Y, Hui X *et al.* (2023) Machine learning in TCM with natural products and molecules: current status and future perspectives. *Chinese Medicine* **18**, 43. doi:10.1186/s13020-023-00741-9
35. Mullowney MW, Duncan KR, Elsayed SS, Garg N, Van Der Hooft JJ *et al.* (2023) Artificial intelligence for natural product drug discovery. *Nature Reviews. Drug Discovery* **22**, 895–916. doi:10.1038/s41573-023-00774-7
36. Domingo-Fernández D, Gadiya Y, Mubeen S, Bollerman TJ, Healy MD, Chanana S, Sadovsky RG, Healey D, Colluru V (2023) Modern drug discovery using ethnobotany: a large-scale cross-cultural analysis of traditional medicine reveals common therapeutic uses. *iScience* **26**, 107729. doi:10.1016/j.isci.2023.107729
37. Pirintsos S, Panagiotopoulos A, Bariotakis M, Daskalakis V, Lionis C *et al.* (2022) From traditional ethnopharmacology to modern natural drug discovery: A methodology discussion and specific examples. *Molecules (Basel, Switzerland)* **27**, 4060. doi:10.3390/molecules27134060
38. Süntar I (2020) Importance of ethnopharmacological studies in drug discovery: role of medicinal plants. *Phytochemistry Reviews* **19**, 1199–1209. doi:10.1007/s11101-019-09629-9
39. Reyes-García V (2010) The relevance of traditional knowledge systems for ethnopharmacological research: theoretical and methodological contributions. *Journal of Ethnobiology and Ethnomedicine* **6**, 32. doi:10.1186/1746-4269-6-32
40. Millum J (2010) How should the benefits of bioprospecting be shared? *The Hastings Center Report* **40**, 24–33. doi:10.1353/hcr.0.0227
41. Singh SP, Singh AM (2015) Bioprospecting: enhancing the value of biodiversity and intellectual property. In *Plant Genetic Resources and Traditional Knowledge for Food Security*. (Eds R Salgotra and B Gupta) pp. 231–236. Springer, Singapore.
42. Vuong P, Chong S, Kaur P (2022) The little things that matter: how bioprospecting microbial biodiversity can build towards the realization of United Nations Sustainable Development Goals. *npj. Biodiversity (Nepean)* **1**, 4.

Glossary

^{13}C natural stable isotope of carbon
^{13}C NMR carbon-13 NMR SPECTROSCOPY
^{1}H NMR hydrogen atom (proton) NMR spectroscopy
$^{1}O_2$ highly oxidising singlet oxygen
4OMT 4-*O*-methyltransferase
5-HT 5-hydroxytryptamine
6OMT 6-*O*-methyltransferase
A adenine
ACh acetylcholine
AChR acetylcholine receptor
ADC arginine decarboxylase
ADMET absorption, distribution, metabolism, excretion, toxicology
AI artificial intelligence
AIDS acquired immunodeficiency syndrome
AKT protein kinase B (abbrev. from AKR mouse strain)
amu atomic mass unit
ATP adenosine triphosphate
BBIQ bisbenzylisoquinoline
BCE before Current Era
BIQ Benzylisoquinoline
C cytosine
CAT catalase enzyme
CBD cannabidiol
CBDAS cannabidiolic acid synthase
CBV cannabivarin
CE Current Era
CE-MS capillary electrophoresis mass spectrometry
CG cyanogenic glycoside
CK cytokinin
CMFoA Customary Medicinal Flora of Australia (database)
CNS central nervous system
CoA coenzyme A
CoASH coenzyme-A-SH
CRISPR Clustered Regularly Interspaced Short Palendromic Repeats
CuAO copper amino oxidase
CYP cytochrome p450
CYP76AD1 cytochrome P450 enzyme
CYP80A1 cytochrome P450 ENZYME (berbamunine synthase)
CYT80B1 cytochrome P450 enzyme
DAO diamino oxidase
DCM dichloromethane (CH_2Cl_2)
DNA deoxyribonucleic Acid
DOPA L-3,4-dihydroxyphenylalanine
DXP 1-deoxy-D-xylulose-5-phosphate
ED_{50} effective dose 50%
ETM Eastern Traditional Medicine
FDA Food and Drug Administration [USA]
FPP farnesyl pyrophosphate
G Guanine
GA gibberellic acid
GABA gamma aminobutyric acid
gCOSY gradient correlation spectroscopy
gHMBCAD gradient heteronuclear multiple bond coherence with adiabatic pulses
GNPS Global Natural Product Social (molecular networking)
GOT geranylphosphate:olivetolate geranyl transferase
GPP geranyl pyrophosphate
GSH glutathione
GST glutathione-s-transferase
HIV human immunodeficiency virus
HPLC high-pressure liquid chromatography
HPLC-NMR high-pressure liquid chromatography nuclear magnetic resonance spectroscopy
HRMS high-resolution mass spectrometry
HRESIMS high-resolution electrospray-ionisation mass spectrometry
HSQCAD heteronuclear singly bond coherence with adiabatic pulses
HT hydroxy-tryptamine
HTS high-throughput screening
IAA indole acetic acid
ICIP Indigenous Cultural Intellectual Property
IK Indigenous Knowledge
IP intellectual property
IPP isopentenyl pyrophosphate
IR infra-red
KEAP1 Kelch-like ECH-associated protein 1
KEAP1/NRF2 (antioxidant-response mediating protein complex)
LC-HRESIMS liquid chromatography high-resolution electrospray ionisation mass spectrometry
LC-MS^2 or LC-MS/MS liquid-chromatography tandem mass spectrometry

LC-NMR liquid chromatography NMR
LD_{50} lethal dose 50% (median lethal dose)
LSA lysergic acid
m/z mass-to-charge ratio
MAP3K7 or TAK1 mitogen-activated protein kinase kinase kinase 7
MEP *2C*-methyl-D-erythritol-4-phosphate
MIA monoterpenoid indole alkaloid
miR-92a microRNA 92a (of the pancreatic cancer cell line)
ML machine learning
MVA mevalonic acid
NCE new chemical entity
NF-κB nuclear factor kappa-B
NP natural product
NRF2 nuclear factor erythroid 2-related factor 2
NRP non-ribosomal peptide
OP1 or DOP/δ delta opioid-type G protein coupled receptor
OP2 or KOP/κ kappa opioid-type G protein coupled receptor
OP3 or MOP/μ Mu opioid-type G protein coupled receptor
OR optical rotation
p53 tumour protein 53
PAIN pan-assay interference compound
PDA photodiode array
PhGly phenolic glycoside
PKS (I, II, III) polyketide synthase (families 1, 2 and 3)
PLP pyridoxal phosphate
PNS peripheral nervous system
QSAR quantitative SAR
RAG retrieval-augmented generation (AI)
RiPP ribosomally synthesised and post-translationally modified peptide
RMIT Royal Melbourne Institute of Technology
RNA ribonucleic acid
RNAi RNA interference (methodology)
RNA-seq RNA sequencing
ROS reactive oxygen species
SAR Structure-activity relationship
T thymine
TCM Traditional Chinese Medicine
THBIQ tetrahydrobenzylisoquinoline
THC Δ^9-tetrahydrocannabinol
THCAS tetrahydrocannabinolic acid synthase
THIQ tetrahydroisoquinoline
THV tetrahydrocannabivarin
TIA terpenoid indole alkaloid
UCLA University of California, Los Angeles
UPLC ultra-high-pressure liquid chromatography
UV ultra-violet
WIL work-integrated learning
WTM Western Traditional Medicine

Bibliography

Abdel-Mogib M, Albar HA, Batterjee SM (2002) Chemistry of the genus *Plectranthus*. *Molecules (Basel, Switzerland)* **7**, 271–301. doi:10.3390/70200271

Abdelshaheed MM, Fawzy IM, El-Subbagh HI, Youssef KM (2021) Piperidine nucleus in the field of drug discovery. *Future Journal of Pharmaceutical Sciences* **7**, 188. doi:10.1186/s43094-021-00335-y

Abe F, Iwase Y, Yamauchi T, Honda K, Hayashi N (1995) Phenathroindolizidine alkaloids from *Tylophora tanakae*. *Phytochemistry* **39**, 695–699. doi:10.1016/0031-9422(95)00034-5

Aberham A, Pieri V, Croom EM, Jr, Ellmerer E, Stuppner H (2011) Analysis of iridoids, secoiridoids and xanthones in *Centaurium erythraea*, *Frasera caroliniensis* and *Gentiana lutea* using LC-MS and RP-HPLC. *Journal of Pharmaceutical and Biomedical Analysis* **54**, 517–525. doi:10.1016/j.jpba.2010.09.030

Abinaya R, Srinath S, Soundarya S, Sridhar R, Balasubramanian KK *et al.* (2022) Recent developments on synthesis strategies, SAR studies and biological activities of β-Carboline derivatives: an update. *Journal of Molecular Structure* **1261**, 132750. doi:10.1016/j.molstruc.2022.132750

Abraham I, Joshi R, Pardasani P, Pardasani RT (2011) Recent advances in 1,4-benzoquinone chemistry. *Journal of the Brazilian Chemical Society* **22**, 385–421. doi:10.1590/S0103-50532011000300002

Abubakar IB, Loh HS (2016) A review on ethnobotany, pharmacology and phytochemistry of *Tabernaemontana corymbosa*. *The Journal of Pharmacy and Pharmacology* **68**, 423–432. doi:10.1111/jphp.12523

Achenbach H, Hubner H, Brandt W, Reiter M (1994) Cardioactive steroid saponins and other constituents from the aerial parts of *Tribulus cistoides*. *Phytochemistry* **35**, 1527–1543. doi:10.1016/S0031-9422(00)86890-9

Achenbach H, Hubner H, Reiter M (1996) Cholestane- and pregnane-type glycosides from the roots of *Tribulus cistoides*. *Phytochemistry* **41**, 907–917. doi:10.1016/0031-9422(95)00622-2

Adkar PP, Bhaskar VH (2014) *Pandanus odoratissimus* (Kewda): a review on ethnopharmacology, phytochemistry, and nutritional aspects. *Advances in Pharmacological and Pharmaceutical Sciences* **2014**, 1–19. doi:10.1155/2014/120895

Agnew MP, Lauren DR (1991) Determination of ptaquiloside in bracken fern (*Pteridium esculentum*). *Journal of Chromatography. A* **538**, 462–468. doi:10.1016/S0021-9673(01)88870-1

Agostini-Costa TDS (2018) Bioactive compounds and health benefits of some palm species traditionally used in Africa and the Americas: a review. *Journal of Ethnopharmacology* **224**, 202–229. doi:10.1016/j.jep.2018.05.035

Ahmed F, Urooj A (2010) Traditional uses, medicinal properties, and phytopharmacology of *Ficus racemosa*: a review. *Pharmaceutical Biology* **48**, 672–681. doi:10.3109/13880200903241861

Ahmed AS, Nakamura N, Meselhy MR, Makhboul MA, El-Emary N *et al.* (2000) Phenolic constituents from *Grevillea robusta*. *Phytochemistry* **53**, 149–154. doi:10.1016/S0031-9422(99)00484-7

Al Muqarrabun LM, Ahmat N (2015) Medicinal uses, phytochemistry and pharmacology of family Sterculiaceae: a review. *European Journal of Medicinal Chemistry* **92**, 514–530. doi:10.1016/j.ejmech.2015.01.026

Al Muqarrabun LM, Ahmat N, Ruzaina SA, Ismail NH, Sahidin I (2013) Medicinal uses, phytochemistry and pharmacology of *Pongamia pinnata* (L.) Pierre: a review. *Journal of Ethnopharmacology* **150**, 395–420. doi:10.1016/j.jep.2013.08.041

Al-Snafi AE (2016) The contents and pharmacology of *Crotalaria juncea*: a review. *IOSR Journal of Pharmacy* **6**, 77–86.

Ali M, Bhutani KK (1989) Alkaloids from *Tylophora indica*. *Phytochemistry* **28**, 3513–3517. doi:10.1016/0031-9422(89)80376-0

Allam K, Beutler JA, Le Quesne PW (1987) 14-Ketoalstonidine and other alkaloidal constitutents of the stem bark of *Alstonia constricta*. *Journal of Natural Products* **50**, 623–625. doi:10.1021/np50052a007

Allan RD, Correll RL, Wells RJ (1969a) A new class of quinones from certain members of the family Cyperaceae. *Tetrahedron Letters* **10**, 4669–4672. doi:10.1016/S0040-4039(01)88778-5

Allan RD, Correll RL, Wells RJ (1969b) Two new phenolic ketones from *Remirea maritima* (Cyperaceae). *Tetrahedron Letters* **10**, 4673–4674. doi:10.1016/S0040-4039(01)88779-7

Allan RD, Wells RJ, Macleod JK (1970) Further phenolic ketones from *Remirea maritima* Aubl. *Tetrahedron Letters* **11**, 3945–3946. doi:10.1016/S0040-4039(01)98632-0

Allen D, Hatfield G (2004) *Medicinal Plants in Folk Tradition: An Ethnobotany of Britain and Ireland.* Timber Press, Cambridge, UK.

Allen JG, Colegate SM, Mitchell AA, Mulder RJ, Raisbeck MF (2006) The bioactivity-guided isolation and structural identification of toxic cucurbitacin steroidal glucosides from *Stemodia kingii*. *Phytochemical Analysis* **17**, 226–235. doi:10.1002/pca.914

Almubayedh H, Ahmad R (2020) Ethnopharmacology, phytochemistry, biological activities, and therapeutic applications of *Cedrela serrata* Royle: a mini review. *Journal of Ethnopharmacology* **246**, 112206. doi:10.1016/j.jep.2019.112206

Amalraj A, Pius A, Gopi S, Gopi S (2017) Biological activities of curcuminoids, other biomolecules from turmeric and their derivatives: a review. *Journal of Traditional and Complementary Medicine* **7**, 205–233. doi:10.1016/j.jtcme.2016.05.005

Aminpour M, Montemagno C, Tuszynski JA (2019) An overview of molecular modelling for drug discovery with specific illustrative examples of applications. *Molecules (Basel, Switzerland)* **24**, 1693. doi:10.3390/molecules24091693

Amos GC, Awakawa T, Tuttle RN, Letzel A-C, Kim MC *et al.* (2017) Comparative transcriptomics as a guide to natural product discovery and biosynthetic gene cluster functionality. *Proceedings of the National Academy of Sciences* **114**, E11121–E11130.

Anderson P, Leibler M (2017) *Final Report of the Referendum Council: 30 June 2017.* Department of the Prime Minister and Cabinet, Canberra.

Aniszewski T (2007a) Alkaloid chemistry. In *Alkaloids: Secrets of Life: Alkaloid Chemistry, Biological Significance, Applications and Ecological Role.* pp. 61–140. Elsevier, Oxford, UK.

Aniszewski T (2007b) Biological significance of alkaloids. In *Alkaloids: Secrets of Life: Alkaloid Chemistry, Biological Significance, Applications and Ecological Role.* pp. 141–180. Elsevier, Oxford, UK.

Aniszewski T (2007c) Definition, typology and occurrence of alkaloids. In *Alkaloids: Secrets of Life: Alkaloid Chemistry, Biological Significance, Applications and Ecological Role.* pp. 1–60. Elsevier, Oxford, UK.

Arnaud MJ (2010) Pharmacokinetics and metabolism of natural methylxanthines in animal and man. In *Methylxanthines.* (Ed. BB Fredholm) pp. 33–91. Springer-Verlag, Berlin.

Arnold MSJ, Macdonald JR, Quinn RJ, Skinner-Adams TS, Andrews KT *et al.* (2021) Antiplasmodial activity of the natural product compounds alstonine and himbeline. *International Journal for Parasitology. Drugs and Drug Resistance* **16**, 17–22. doi:10.1016/j.ijpddr.2021.04.003

Arroo R, Woolley J, Oksman-Caldentey K-M (2007) Tropane alkaloid containing plants: Henbane, Belladonna, Datura and Duboisia. In *Transgenic Crops VI* (Eds E-C Pua and MR Davey) pp. 189–204. Springer, Berlin.

Arunotayanun W, Gibbons S (2012) Natural product 'legal highs'. *Natural Product Reports* **29**, 1304–1316.

Asakawa Y, Ludwiczuk A, Harinantenaina L, Toyota M, Nishiki M *et al.* (2012) Distribution of drimane sesquiterpenoids and tocopherols in liverworts, ferns and higher plants: Polygonaceae, Canellaceae, and Winteraceae species. *Natural Product Communications* **7**, 685–692. doi:10.1177/1934578X1200700601

Asano N, Nash RJ, Molyneux RJ, Fleet GWJ (2000) Sugar-mimic glycosidase inhibitors: natural occurrence, biological activity and prospects for therapeutic application. *Tetrahedron* **11**, 1645–1680. doi:10.1016/S0957-4166(00)00113-0

Aslam MS, Choudhary BA, Uzair M, Ijaz AS (2013) Phytochemical and ethno-pharmacological review of the genus *Araucaria*: review. *Tropical Journal of Pharmaceutical Research* **12**, 651–659. doi:10.4314/tjpr.v12i4.31

Atanasov AG, Zotchev SB, Dirsch VM, International Natural Product Sciences Taskforce, Supuran CT (2021) Natural products in drug discovery: advances and opportunities. *Nature Reviews. Drug Discovery* **20**, 200–216. doi:10.1038/s41573-020-00114-z

Austin DF (2008) *Evolvulus alsinoides* (Convolvulaceae): an American herb in the Old World. *Journal of Ethnopharmacology* **117**, 185–198. doi:10.1016/j.jep.2008.01.038

Auzi AA, Hartley TG, Waterman PG (1997) Distribution of flavonoids, alkaloids, acetophenones and phloroglucinols in the genus *Bosistoa* (Rutaceae). *Biochemical Systematics and Ecology* **25**, 611–617. doi:10.1016/S0305-1978(97)00062-8

Avoseh O, Oyedeji O, Rungqu P, Nkeh-Chungag B, Oyedeji A (2015) *Cymbopogon* species; ethnopharmacology, phytochemistry and the pharmacological importance. *Molecules (Basel, Switzerland)* **20**, 7438–7453. doi:10.3390/molecules20057438

Ayaz M, Ahmad I, Sadiq A, Ullah F, Ovais M *et al.* (2020) *Persicaria hydropiper* (L.) Delarbre: a review on traditional uses, bioactive chemical constituents and pharmacological and toxicological activities. *Journal of Ethnopharmacology* **251**, 112516. doi:10.1016/j.jep.2019.112516

Ayyad S-EN, Judd AS, Shier T, Hoye TR (1998) Otteliones A and B: potently cytotoxic 4-methylene-2-cyclohenones from *Ottelia alismoides. The Journal of Organic Chemistry* **63**, 8102–8106. doi:10.1021/jo971870a

Ayyanar M, Subash-Babu P (2012) *Syzygium cumini* (L.) Skeels: a review of its phytochemical constituents and traditional uses. *Asian Pacific Journal of Tropical Biomedicine* **2**, 240–246. doi:10.1016/S2221-1691(12)60050-1

Baecker C, Jenett-Siems K, Siems K, Wurster M, Bodtke A *et al.* (2014) Cytotoxic saponins from the seeds of *Pittosporum angustifolium. Journal of Biosciences* **69**, 191–198.

Baecker C, Jenett-Siems K, Siems K, Niedermayer THJ, Wurster M *et al.* (2015) Taraxastane-type triterpenoid saponins isolated from *Pittosporum angustifolium* Lodd. *Journal of Chemical Sciences* **70**, 403–408.

Bagalkotkar G, Sagineedu SR, Saad MS, Stanslas J (2006) Phytochemicals from *Phyllanthus niruri* Linn. and their pharmacological properties: a review. *The Journal of Pharmacy and Pharmacology* **58**, 1559–1570. doi:10.1211/jpp.58.12.0001

Bahramikia S, Yazdanparast R (2012) Phytochemistry and medicinal properties of *Teucrium polium* L. (Lamiaceae). *Phytotherapy Research* **26**, 1581–1593. doi:10.1002/ptr.4617

Bai Y, Xia B, Xie W, Zhou Y, Xie J *et al.* (2016) Phytochemistry and pharmacological activities of the genus *Prunella. Food Chemistry* **204**, 483–496. doi:10.1016/j.foodchem.2016.02.047

Bajguz A, Tretyn A (2003) The chemical characteristic and distribution of brassinosteroids in plants. *Phytochemistry* **62**, 1027–1046. doi:10.1016/S0031-9422(02)00656-8

Bakkali F, Averbeck S, Averbeck D, Idaomar M (2008) Biological effects of essential oils: a review. *Food and*

Chemical Toxicology **46**, 446–475. doi:10.1016/j.fct.2007.09.106

Bakker HJ, Ghisalberti EL, Jefferies PR (1972) Biosynthesis of diterpenes in *Beyeria leschenaultii*. *Phytochemistry* **11**, 2221–2231. doi:10.1016/S0031-9422(00)88381-8

Baky MH, Elshahed M, Wessjohann J, Farag MA (2022) Interactions between dietary flavonoids and the gut microbiome: a comprehensive review. *British Journal of Nutrition* **128**, 577–591. doi:10.1017/S0007114521003627

Baldovini N, Delasalle C, Joulain D (2011) Phytochemistry of the heartwood from fragrant *Santalum* species: a review. *Flavour and Fragrance Journal* **26**, 7–26. doi:10.1002/ffj.2025

Baliga MS, Bhat HP, Baliga BRV, Wilson R, Palatty PL (2011) Phytochemistry, traditional uses and pharmacology of *Eugenia jambolana* Lam. (black plum): a review. *Food Research International* **44**, 1776–1789. doi:10.1016/j.foodres.2011.02.007

Banerjee S, Bandyopadhyay A (2015) *Buchanania lanzan* Spreng.: a veritable storehouse of phytomedicines. *Asian Journal of Pharmaceutical and Clinical Research* **8**, 18–22.

Basha NJ, Basavarajaiah S, Baskaran S, Kumar P (2022) A comprehensive insight on the biological potential of embelin and its derivatives. *Natural Product Research* **36**, 3054–3068. doi:10.1080/14786419.2021.1955361

Bate-Smith EC, Harborne JB, Davenport SM (1966) Occurrence of azaleatin and caryatin in *Eucryphia*. *Nature* **212**, 1065–1066. doi:10.1038/2121065a0

Bate-Smith EC, Davenport SM, Harborne JB (1967) Comparative biochemistry of flavonoids. III: a correlation between chemistry and plant geography in the genus *Eucryphia*. *Phytochemistry* **6**, 1407–1413. doi:10.1016/S0031-9422(00)82883-6

Batiha GE-S, Alkazmi LM, Nadwa EH, Rashwan EK, Beshbishy AM *et al.* (2020) Physostigmine: a plant alkaloid isolated from *Physostigma venenosum*: a review on pharmacokinetics, pharmacological and toxicological activities. *Journal of Drug Delivery and Therapeutics* **10**, 187–190. doi:10.22270/jddt.v10i1-s.3866

Benkrief R, Skaltsounis A-L, Tillequin F, Koch M (1991) Plantes de Nouvelle-Caledonie, 137. Iridoides de *Coelospermum billardieri*. *Journal of Natural Products* **54**, 532–541. doi:10.1021/np50074a028

Bergman ME, Davis B, Phillips MA (2019) Medically useful plant terpenoids: biosynthesis, occurrence, and mechanism of action. *Molecules (Basel, Switzerland)* **24**, 3961. doi:10.3390/molecules24213961

Bhatla SC (2018) Secondary metabolites. In *Plant Physiology, Development and Metabolism* (Eds SC Bhatla and MA Lal) pp. 1099–1166. Springer Nature Singapore, Singapore.

Bhattacharyya P, Chakraborty DP (1987) Carbazole alkaloids. *Fortschritte der Chemie Organischer Naturstoffe* **52**, 159–209. doi:10.1007/978-3-7091-8906-1_4

Bibi H, Ali I, Sadozai SK, Atta Ur R (2006) Phytochemical studies and antibacterial activity of *Centaurium pulchellum* Druce. *Natural Product Research* **20**, 896–901. doi:10.1080/14786410500162047

Bibi SN, Fawzi MM, Gokhan Z, Rajesh J, Nadeem N *et al.* (2019) Ethnopharmacology, phytochemistry, and global distribution of mangroves: a comprehensive review. *Marine Drugs* **17**, 231.

Bick IRC, Leow H-M (1978a) Alkaloids of *Daphnandra johnsonii*. *Australian Journal of Chemistry* **31**, 2539–2544. doi:10.1071/CH9782539

Bick IRC, Sotheeswaran S (1978b) Alkaloids of *Daphnandra apatela*. *Australian Journal of Chemistry* **31**, 2077–2083. doi:10.1071/CH9782077

Bieleski RL, Briggs BG (2005) Taxonomic patterns in the distribution of polyols within the Proteaceae. *Australian Journal of Botany* **53**, 205–217. doi:10.1071/BT04098

Biswas P, Anand U, Ghorai M, Pandey DK, Jha NK *et al.* (2022) Unraveling the promise and limitations of CRISPR/Cas system in natural product research: Approaches and challenges. *Biotechnology Journal* **17**, 2100507. doi:10.1002/biot.202100507

Blackburne ID, Park RJ, Sutherland MD (1971) Terpenoid chemistry XVIII: myodesmone and isomyodesmone, toxic furanoid ketones from *Myoporum deserti* and *M. acuminatum*. *Australian Journal of Chemistry* **24**, 995–1007. doi:10.1071/CH9710995

Blackburne ID, Park RJ, Sutherland MD (1972) Terpenoid chemistry XX: myoporone and dehydromyoporone, toxic furanoid ketones from *Myoporum* and *Eremophila* species. *Australian Journal of Chemistry* **25**, 1787–1796. doi:10.1071/CH9721787

Blay V, Tolani B, Ho SP, Arkin MR (2020) High-throughput screening: today's biochemical and cell-based approaches. *Drug Discovery Today* **25**, 1807–1821. doi:10.1016/j.drudis.2020.07.024

Blažević I, Montaut S, Burčul F, Olsen CE, Burow M *et al.* (2020) Glucosinolate structural diversity, identification, chemical synthesis and metabolism in plants. *Phytochemistry* **169**, 112100. doi:10.1016/j.phytochem.2019.112100

Bloor SJ (2001) Deep blue anthocyanins from blue *Dianella* berries. *Phytochemistry* **58**, 923–927. doi:10.1016/S0031-9422(01)00343-0

Bludau I, Aebersold R (2020) Proteomic and interactomic insights into the molecular basis of cell functional diversity. *Nature Reviews. Molecular Cell Biology* **21**, 327–340. doi:10.1038/s41580-020-0231-2

Boerjan W, Ralph J, Baucher M (2003) Lignin biosynthesis. *Annual Review of Plant Biology* **54**, 519–546. doi:10.1146/annurev.arplant.54.031902.134938

Bogart JW, Cabezas MD, Vögeli B, Wong DA, Karim AS *et al.* (2021) Cell-free exploration of the natural product chemical space. *ChemBioChem* **22**, 84–91. doi:10.1002/cbic.202000452

Bohlmann J, Meyer-Gauen G, Croteau R (1998) Plant terpenoids synthases: molecular biology and phylogenetic analysis. *Proceedings of the National Academy of Sciences of the United States of America* **95**, 4126–4133. doi:10.1073/pnas.95.8.4126

Bolton JL, Dunlap TL, Dietz BM (2018) Formation and biological targets of botanical o-quinones. *Food and Chemical Toxicology* **120**, 700–707. doi:10.1016/j.fct.2018.07.050

Bolz SN, Adasme MF, Schroeder M (2021) Toward an understanding of Pan-Assay Interference Compounds and promiscuity: a structural perspective on binding modes. *Journal of Chemical Information and Modeling* **61**, 2248–2262. doi:10.1021/acs.jcim.0c01227

Bories C, Loiseau P, Cortes D, Myint SH, Hocque-Miller R *et al.* (1991) Antiparasitic activity of *Annona muricata* and *Annona cherimola* seeds. *Planta Medica* **57**, 434–436. doi:10.1055/s-2006-960143

Boucherle B, Haudecoeur R, Queiroz EF, De Waard M, Wolfender JL *et al.* (2016) *Nauclea latifolia*: biological activity and alkaloid phytochemistry of a West African tree. *Natural Product Reports* **33**, 1034–1043. doi:10.1039/C6NP00039H

Bourgaud F, Hehn A, Larbat R, Doerper S, Gontier E *et al.* (2006) Biosynthesis of coumarins in plants: a major pathway still to be unravelled for cytochrome P450 enzymes. *Phytochemistry Reviews* **5**, 293–308. doi:10.1007/s11101-006-9040-2

Bouvier F, Rahier A, Camara B (2005) Biogenesis, molecular regulation and function of plant isoprenoids. *Progress in Lipid Research* **44**, 357–429. doi:10.1016/j.plipres.2005.09.003

Bowie JH, Cooke RG, Wilkin PE (1962) Colouring matters of Australian plants. X: anthraquinones from *Coelospermum* species. *Australian Journal of Chemistry* **15**, 336–341. doi:10.1071/CH9620336

Brielmann HL, Setzer WN, Kaufman PB, Kirakosyan A, Cseke LJ (2006) Phytochemicals: the chemical components of plants. In *Natural Products from Plants* (Eds LJ Cseke A Kirakosyan, PB Kaufman, SL Warber, JA Duke *et al.*) pp. 1–50. CRC Press, Boca Raton, FL.

Bringmann G, Irmer A, Feineis D, Gulder TA, Fiedler HP (2009) Convergence in the biosynthesis of acetogenic natural products from plants, fungi, and bacteria. *Phytochemistry* **70**, 1776–1786. doi:10.1016/j.phytochem.2009.08.019

Brinkhaus B, Lindner M, Schuppan D, Hahn EG (2000) Chemical, pharmacological and clinical profile of the East Asian medical plant *Centella aslatica*. *Phytomedicine* **7**, 427–448. doi:10.1016/S0944-7113(00)80065-3

Brophy JJ, Craig DC, Goldsack RJ, Fookes CJR (1994) Hillyl acetate, a keto-acetate from the leaf stem volatiles of *Syncarpia hillii*. *Phytochemistry* **37**, 1645–1647. doi:10.1016/S0031-9422(00)89584-9

Buchanan MS, Davis RA, Duffy S, Avery VM, Quinn RJ (2009) Antimalarial benzylisoquinoline alkaloid from the rainforest tree *Doryphora sassafras*. *Journal of Natural Products* **72**, 1541–1543. doi:10.1021/np9002564

Buckingham J (2010) *Dictionary of Alkaloids*. CRC Press, Boca Raton FL.

Buettner D, Skemp S (2016) Blue Zones: lessons from the world's longest lived. *American Journal of Lifestyle Medicine* **10**, 318–321. doi:10.1177/1559827616637066

Bunsupa S, Yamazaki M, Saito K (2012) Quinolizidine alkaloid biosynthesis: recent advances and future prospects. *Frontiers in Plant Science* **3**, 1–7. doi:10.3389/fpls.2012.00239

Caesar LK, Montaser R, Keller NP, Kelleher NL (2021) Metabolomics and genomics in natural products research: complementary tools for targeting new chemical entities. *Natural Product Reports* **38**, 2041–2065. doi:10.1039/D1NP00036E

Calixto JB, Santos ARS, Filho VC, Yunes RA (1998) A review of the plants of the genus *Phyllanthus:* their chemistry pharmacology and therapeutic potential. *Medicinal Research Reviews* **18**, 225–258. doi:10.1002/(SICI)1098-1128(199807)18:4<225::AID-MED2>3.0.CO;2-X

Cambie RC, Ser NA, Kokubun T (1997) Heartwood constituents of *Planchonella vitiensis*. *Biochemical Systematics and Ecology* **25**, 677–678. doi:10.1016/S0305-1978(97)00063-X

Camm EL, Wat CK, Towers GHN (1976) An assessment of the roles of furanocoumarins in *Heracleum lanatum*. *Canadian Journal of Botany* **54**, 2562–2566. doi:10.1139/b76-275

Cannon JR, Metcalf BW (1971) Phenolic constituents of *Persoonia elliptica* (Proteaceae). *Australian Journal of Chemistry* **24**, 1925–1931. doi:10.1071/CH9711925

Cannon JR, Chow PW, Fuller MW, Hamilton BH, Metcalf BW *et al.* (1973) Phenolic constituents of *Grevillea robusta* (Proteaceae): the structure of robustol, a novel macrocyclic phenol. *Australian Journal of Chemistry* **26**, 2257–2275. doi:10.1071/CH9732257

Cannon JR, Williams JR, Arbain D, Brossi A, Blount JF *et al.* (1991) Alkaloids of *Templetonia retusa*: chemistry, structure and absolute configuration of (-)-templetine. *Australian Journal of Chemistry* **44**, 509–523. doi:10.1071/CH9910509

Cao N, Wang C-H (2021) Strictosidine synthase, an indispensible enzyme involved in the biosynthesis of terpenoid indole and β-carboline alkaloids. *Chinese Journal of Natural Medicines* **19**, 591–607. doi:10.1016/S1875-5364(21)60059-6

Capon R (2022) Natural products: mainstream but not always natural. *Chemistry in Australia* (**Sept**), 36–37.

Caporael LR (1976) Ergotism: the Satan loose in Salem. *Science* **192**, 21–26. doi:10.1126/science.769159

Carroll AR, Addepalli R, Fechner G, Smith J, Guymer GP *et al.* (2008) Alkaloids from the Australian rainforest tree *Ochrosia moorei*. *Journal of Natural Products* **71**, 1063–1065. doi:10.1021/np070655e

Cassels BK, Eckhardt G, Kaussmann EU, Tschesche R (1974) Cyclopeptide alkaloids of *Zizyphus oenoplia*. *Tetrahedron* **30**, 2461–2466. doi:10.1016/S0040-4020(01)97117-0

Castillo U, Harper JK, Strobel GA, Sears J, Alesi K *et al.* (2003) Kakadumycins, novel antibiotics from *Streptomyces* sp. NRRL 30566, an endophyte of *Grevillea pteridifolia*. *FEMS Microbiology Letters* **224**, 183–190. doi:10.1016/S0378-1097(03)00426-9

Cavalcanti RN, Forster-Carneiro T, Gomes MTMS, Rostagno MA, Prado JM, *et al.* (2013) Uses and applications of extracts from natural sources. In *Natural Product Extraction: Principles and Applications.* (Eds J Prado and M Rostagno) pp. 1–57. Royal Society of Chemistry, London.

Cechinel Filho V, Meyre-Silva C, Niero R (2009) Chemical and pharmacological aspects of the genus *Calophyllum*. *Chemistry & Biodiversity* **6**, 313–327. doi:10.1002/cbdv.200800082

Cely-Veloza W, Kato MJ, Coy-Barrera E (2023) Quinolizidine-type alkaloids: chemodiversity, occurrence, and bioactivity. *ACS Omega* **8**, 27862–27893. doi:10.1021/acsomega.3c02179

Chamy MC, Piovano M, Barbarino JA, Gambaro V (1991) Stemodane diterpenoids from *Stemodia chilensis*. *Phytochemistry* **30**, 1719–1721. doi:10.1016/0031-9422(91)84245-N

Chang J, Song C (2013) Progress in the syntheses of dibenzocyclooctadiene lignans. In *Studies in Natural Product Chemistry*. (Ed. A-U Rahman) pp. 173–203. Elsevier, Oxford, UK.

Chang F-R, Chao Y-C, Teng C-M, Wu Y-C (1998) Chemical constituents from *Cassytha filiformis* II. *Journal of Natural Products* **61**, 863–866.

Chapman AD (2009) Numbers of living species in Australia and the World: report for the Australian Biological Resources Study. Department of the Environment, Water, Heritage and the Arts, Canberra.

Chaudhury RR, Rafei UM (Eds) (2001) Traditional medicine in Asia. World Health Organisation, New Delhi, India.

Chavan RB, Gaikwad DK (2016) The ethnobotany, phytochemistry and biological properties of *Allophylus* species used in traditional medicine: a review. *World Journal of Pharmacy and Pharmaceutical Sciences* **5**, 664–682.

Chawla R, Kumar S, Sharma A (2012) The genus *Clematis* (Ranunculaceae): chemical and pharmacological perspectives. *Journal of Ethnopharmacology* **143**, 116–150. doi:10.1016/j.jep.2012.06.014

Chen K, Yu B (1999) Certain progress of clinical research on Chinese integrative medicine. *Chinese Medical Journal* **112**, 934–937.

Chen JC, Chiu MH, Nie RL, Cordell GA, Qiu SX (2005) Cucurbitacins and cucurbitane glycosides: structure and biological activities. *Natural Product Reports* **22**, 386–399. doi:10.1039/b418841c

Chen W, Vermaak I, Viljoen A (2013) Camphor: a fumigant during the Black Death and a coveted fragrant wood in ancient Egypt and Babylon: a review. *Molecules (Basel, Switzerland)* **18**, 5434–5454. doi:10.3390/molecules18055434

Chen W, Shou X-A, Chen Y, Qin N, Qiao W *et al.* (2014a) A new aurone from *Smilax riparia*. *Chemistry of Natural Compounds* **50**, 989–993. doi:10.1007/s10600-014-1143-1

Chen Y, Yan T, Gao C, Cao W, Huang R (2014b) Natural products from the genus *Tephrosia*. *Molecules (Basel, Switzerland)* **19**, 1432–1458. doi:10.3390/molecules19021432

Chen H-Y, Guh J-H, Chan S-H, Lee S-S (2015) Cytotoxic protobassic acid glycosides from *Planchonella obovata* leaf. *Phytochemistry Letters* **11**, 229–235. doi:10.1016/j.phytol.2015.01.005

Chen AH, Liu QL, Ma YL, Jiang ZH, Tang JY *et al.* (2017) A new monoterpenoid indole alkaloid from *Ochrosia elliptica*. *Natural Product Research* **31**, 1490–1494. doi:10.1080/14786419.2016.1277349

Chen X, Wang Y, Ma N, Tian J, Shao Y *et al.* (2020) Target identification of natural medicine with chemical proteomics approach: probe synthesis, target fishing and protein identification. *Signal Transduction and Targeted Therapy* **5**, 72. doi:10.1038/s41392-020-0186-y

Cherniavsky Lev M, Karlish SJD, Garty H (2015) Cardiac glycosides induced toxicity in human cells expressing a1-, a2-, or a3-isoforms of Na-K-ATPase. *American Journal of Physiology. Cell Physiology* **309**, C126–C135. doi:10.1152/ajpcell.00089.2015

Chhatre S, Nesari T, Somani G, Kanchan D, Sathaye S (2014) Phytopharmacological overview of *Tribulus terrestris*. *Pharmacognosy Reviews* **8**, 45–51. doi:10.4103/0973-7847.125530

Chinou I (2005) Labdanes of natural origin: biological activities. *Current Medicinal Chemistry* **12**, 1295–1317.

Chlouchi A, Muyard F, Girard C, Waterman PG, Bévalot F (2005) Coumarins from the twigs of *Diplolaena mollis* P. G. Wilson (Rutaceae). *Biochemical Systematics and Ecology* **33**, 967–969. doi:10.1016/j.bse.2005.03.002

Chlouchi A, Girard C, Tillequin F, Bévalot F, Waterman PG *et al.* (2006) Coumarins and furoquinoline alkaloids from *Philotheca deserti* var. *deserti* (Rutaceae). *Biochemical Systematics and Ecology* **34**, 71–74. doi:10.1016/j.bse.2005.07.002

Chlouchi A, Girard C, Bévalot F, Waterman PG, Muyard F (2007) Taxonomically significant coumarins from three *Philotheca* species (Rutaceae). *Biochemical Systematics and Ecology* **35**, 251–254. doi:10.1016/j.bse.2006.09.011

Cho H, Jun J-Y, Song E-K, Kang K-H, Baek H-Y *et al.* (2001) Bakuchiol: a hepatoprotective compound of *Psoralea corylifolia* on tacrine-induced cytotoxicity in Hep G2 cells. *Planta Medica* **67**, 750–751. doi:10.1055/s-2001-18347

Choudhury S, Bahadur B, Pullaiah T (2017) Ethnomedicinal plants of north-east India. In *Ethnobotany of India. Vol. 3: North-East India and the Andaman and Nicobar Islands.* (Eds T Pullaiah, KV Krishnamurthy and B Bahadur). Apple Academic Press, Palm Bay, FL.

Christen P, Bieri S, Veuthey JL (2007) Analysis of tropane alkaloids in biological matrices. In *Modern Alkaloids: Structure, Isolation, Synthesis and Biology.* (Eds E Fatorusso and O Taglialatela-Scafati) pp. 339–367. Wiley-VCH, New York, NY.

Chuang T-H, Wu P-L (2007) Cytotoxic 5-alkylresorcinol metabolites from the leaves of *Grevillea robusta*. *Journal of Natural Products* **70**, 319–323. doi:10.1021/np0605687

Chung KT, Wong TY, Wei CI, Huang YW, Lin Y (1998) Tannins and human health: a review. *Critical Reviews in Food Science and Nutrition* **38**, 421–464. doi:10.1080/10408699891274273

Chung IM, Rajakumar G, Lee JH, Kim SH, Thiruvengadam M (2017) Ethnopharmacological uses, phytochemistry, biological activities, and biotechnological applications of *Eclipta prostrata*. *Applied Microbiology and Biotechnology* **101**, 5247–5257. doi:10.1007/s00253-017-8363-9

Ciociola AA, Cohen LB, Kulkarni P (2014) How drugs are developed and approved by the FDA: current process and future directions. *The American Journal of Gastroenterology* **109**, 620–623. doi:10.1038/ajg.2013.407

Clarke P (2008) Aboriginal healing practices and Australian bush medicine. *Journal of the Anthropological Society of South Australia* **33**, 3–38.

Clarke PA (2011) The power of plants. In Aboriginal *People* and *Their Plants*. pp. 96–101. Rosenburg Publishing, Sydney.

Clauser M, Dall'acqua S, Loi MC, Innocenti G (2013) Phytochemical investigation on *Atriplex halimus* L. from Sardinia. *Natural Product Research* **27**, 1940–1944.

Cock IE (2013) The phytochemistry and chemotherapeutic potential of *Tasmannia lanceolata* (Tasmanian pepper): a review. *Pharmacognosy Communications* **3**, 1–13. doi:10.5530/pc.2013.2.1

Colegate SM, Dorling PR, Huxtable CR (1979) A spectroscopic investigation of swainsonine: an alpha-mannosidase inhibitor isolated from *Swainsona canescens*. *Australian Journal of Chemistry* **32**, 2257–2264. doi:10.1071/CH9792257

Collins TL, Jones GL, Sadgrove NJ (2014) Volatiles from the rare Australian desert plant *Prostanthera centralis* B.J.Conn (Lamiaceae): chemical composition and antimicrobial activity. *Agriculture* **4**, 308–316. doi:10.3390/agriculture4040308

Conrad LI, Neve M, Nutton V, Porter R, Wear A (1995) *The Western Medical Tradition: 800 BC to AD 1800*. Cambridge University Press, Cambridge, UK.

Conserva LM, Pereira Ca B, Barbosa-Filho JM (2005) Alkaloids of the Hernandiaceae: occurrence and a compilation of their biological activities. *The Alkaloids. Chemistry and Biology* **62**, 175–243. doi:10.1016/S1099-4831(05)62003-2

Cooke RG, Johnson BL (1963) Colouring matters of Australian plants. XI: quinones from *Ventilago viminalis*. *Australian Journal of Chemistry* **16**, 695–702. doi:10.1071/CH9630695

Cortés-Rojas DF, De Souza CRF, Oliveira WP (2014) Clove (*Syzygium aromaticum*): a precious spice. *Asian Pacific Journal of Tropical Biomedicine* **4**, 90–96. doi:10.1016/S2221-1691(14)60215-X

Cos P, De Bruyne T, Apers S, Berghe DV, Pieters L *et al.* (2003) Phytoestrogens: recent developments. *Planta Medica* **69**, 589–599. doi:10.1055/s-2003-41122

Costa M, Dias TA, Brito A, Proença F (2016) Biological importance of structurally diversified chromenes. *European Journal of Medicinal Chemistry* **123**, 487–507. doi:10.1016/j.ejmech.2016.07.057

Cox SD, Jayasinghe KC, Markham JL (2005) Antioxidant activity in Australian native sarsaparilla (*Smilax glyciphylla*). *Journal of Ethnopharmacology* **101**, 162–168. doi:10.1016/j.jep.2005.04.005

Cragg GM, Newman DJ (2005) Biodiversity: a continuing source of novel drug leads. *Pure and Applied Chemistry* **77**, 7–24. doi:10.1351/pac200577010007

Cresswell I, Murphy H (2017) Australia State of the Environment 2016: biodiversity. Department of the Environment and Energy, Canberra.

Crombie L, Whiting DA (1998) Biosynthesis in the rotenoid group of natural products: Applications of isotope methodology. *Phytochemistry* **49**, 1479–1507. doi:10.1016/S0031-9422(98)00178-2

Crone H (2004) *Paracelsus: The Man Who Defied Medicine – His Real Contribution to Medicine and Science*. Albarello Press, Melbourne.

Crow JM (2021) Canada's scientists are elucidating the dark metabolome. *Nature* **599**, S14–S15. doi:10.1038/d41586-021-03062-9

Crow WD, Hancox NC, Johns SR, Lamberton JA (1970) New alkaloids of *Alstonia constricta*. *Australian Journal of Chemistry* **23**, 2489–2501. doi:10.1071/CH9702489

Crublet M-L, Long C, Sévenet T, Hadi HA, Lavaud C (2003) Acylated flavonol glycosides from leaves of *Planchonia grandis*. *Phytochemistry* **64**, 589–594. doi:10.1016/S0031-9422(03)00215-2

Cseke LJ, Lu CR, Kornfeld A, Kaufman PB, Kirakosyan A (2006) How and why these compounds are synthesised by plants. In *Natural Products from Plants* (Eds LJ Cseke A Kirakosyan, PB Kaufman, SL Warber, JA Duke *et al.*) pp. 51–100. CRC Press, Boca Raton, FL.

Culberson CF (1969) *Chemical and Botanical Guide to Lichen Products*. University of North Carolina Press, Chapel Hill.

Dagur P, Shreya, Ghosh R, Rakshit G, Biswas A *et al.* (2023) Computational methods in natural products-based drug discovery. In *CADD and Informatics in Drug Discovery*. (Eds M Rudrapal and J Khan) pp. 99–121. Springer Singapore, Singapore.

Dang R, Guan H, Wang C (2023) Sinapis Semen: a review on phytochemistry, pharmacology, toxicity, analytical methods and pharmacokinetics. *Frontiers in Pharmacology* **14**, 1113583.

Davies KM, Schwinn KE (2006) Molecular biology and biotechnology of flavonoid biosynthesis. In *Flavonoids: Chemistry, Biochemistry and Applications*. (Eds OM Anderson and KR Markham) pp. 144–218. CRC Press, Boca Raton, FL.

De Andrade TU, Brasil GA, Endringer DC, Da Nobrega FR, De Sousa DP (2017) Cardiovascular activity of the chemical constituents of essential oils. *Molecules (Basel, Switzerland)* **22**, 1539–1557. doi:10.3390/molecules22091539

De Assis Oliveira F, Andrade LN, De Sousa EBV, De Sousa DP (2014) Anti-ulcer activity of essential oil constituents. *Molecules (Basel, Switzerland)* **19**, 5717–5747. doi:10.3390/molecules19055717

De Keukeleire J, Ooms G, Heyerick A, Roldan-Ruiz I, Van Bockstaele E *et al.* (2003) Formation and accumulation of alpha-acids, beta-acids, desmethylxanthohumol, and xanthohumol during flowering of hops. *Journal of Agricultural and Food Chemistry* **51**, 4436–4441. doi:10.1021/jf034263z

De Leo M, Milella L, Braca A, De Tommasi N (2018) Cedrela and Toona genera: a rich source of bioactive limonoids and triterpenoids. *Phytochemistry Reviews* **17**, 751–783. doi:10.1007/s11101-018-9557-1

De Sousa DP (2011) Analgesic-like activity of essential oils constituents. *Molecules (Basel, Switzerland)* **16**, 2233–2252. doi:10.3390/molecules16032233

De Vos P (2010) European Materia Medica in historical texts: longevity of a tradition and implications for future use. *Journal of Ethnopharmacology* **132**, 28–47. doi:10.1016/j.jep.2010.05.035

Deans BJ, Kilah NL, Jordan GJ, Bissember AC, Smith JA (2018) Arbutin derivatives isolated from ancient Proteaceae: potential phytochemical markers present in *Bellendena, Cenarrhenes,* and *Persoonia* genera. *Journal of Natural Products* **81**, 1241–1251. doi:10.1021/acs.jnatprod.7b01038

Deepak M, Handa SS (1998) 3a,24-dihydroxy-urs-12-en-28-oic acid from *Verbena officinalis. Phytochemistry* **49**, 269–271. doi:10.1016/S0031-9422(97)01004-2

Deepak M, Handa SS (2000) Antiinflammatory activity and chemical composition of extracts of *Verbena officinalis. Phytotherapy Research* **14**, 463–465. doi:10.1002/1099-1573(200009)14:6<463::AID-PTR611>3.0.CO;2-G

Dellar JE, Cole MD, Gray AI, Gibbons S, Waterman PG (1994) Antimicrobial sesquiterpenes from *Prostanthera* aff. *Melissifolia* and *P. rotundifolia. Phytochemistry* **36**, 957–960. doi:10.1016/S0031-9422(00)90471-0

Demmig-Adams B (1990) Carotenoids and photoprotection in plants: a role for the xanthophyll zeaxanthin. *Biochimica et Biophysica Acta* **1020**, 1–24. doi:10.1016/0005-2728(90)90088-L

Denisov ET, Afanas'ev IB (2005) *Oxidation and Antioxidants in Organic Chemistry and Biology.* CRC Press, Boca Raton, FL.

Department of Climate Change, Energy, the Environment and Water (2012) Collaborative Australian Protected Areas Database: National Reserve System. The Department, Canberra.

Dewey RE, Xie J (2013) Molecular genetics of alkaloid biosynthesis in *Nicotiana* tabacum. *Phytochemistry* **94**, 10–27. doi:10.1016/j.phytochem.2013.06.002

Dey A, Nandy S (2021) CRISPER/Cas in plant natural product research: therapeutics as anticancer and other drug candidates and recent patents. *Recent Patents on Anti-cancer Drug Discovery* **16**, 460–468. doi:10.2174/1574892816666210706155602

Dias DA, Silva CA, Urban S (2009) Naphthalene aglycones and glycosides from the Australian medicinal plant, *Dianella callicarpa. Planta Medica* **75**, 1442–1447. doi:10.1055/s-0029-1185724

Dias DA, Urban S, Roessner U (2012) A historical overview of natural products in drug discovery. *Metabolites* **2**, 303–336. doi:10.3390/metabo2020303

Dilpreet K, Amandeep K, Jaswinder K (2018) A review of *Aleurites moluccana. World Journal of Pharmacy and Pharmaceutical Sciences* **7**, 341–358.

Dinda B, Debnath S, Harigaya Y (2007a) Naturally occurring iridoids: a review, part 1. *Chemical & Pharmaceutical Bulletin* **55**, 159–222. doi:10.1248/cpb.55.159

Dinda B, Debnath S, Harigaya Y (2007b) Naturally occurring secoiridoids and bioactivity of naturally occurring iridoids and secoiridoids: a review, part 2. *Chemical & Pharmaceutical Bulletin* **55**, 689–728. doi:10.1248/cpb.55.689

Dinda B, Debnath S, Mohanta BC, Harigaya Y (2010) Naturally occurring triterpenoid saponins. *Chemistry & Biodiversity* **7**, 2327–2580. doi:10.1002/cbdv.200800070

Dinda B, Das N, Dinda S, Dinda M, Silsarma I (2015) The genus *Sida* L.: a traditional medicine: Its ethnopharmacological, phytochemical and pharmacological data for commercial exploitation in herbal drugs industry. *Journal of Ethnopharmacology* **176**, 135–176. doi:10.1016/j.jep.2015.10.027

Ding W, Zeng F, Xu L, Chen Y, Wang Y *et al.* (2011) Bioactive dammarane-type saponins from *Operculina turpethum. Journal of Natural Products* **74**, 1868–1874. doi:10.1021/np200274m

Ding W, Jiang ZH, Wu P, Xu L, Wei X (2012) Resin glycosides from the aerial parts of *Operculina turpethum. Phytochemistry* **81**, 165–174. doi:10.1016/j.phytochem.2012.05.010

Divakar PR (2017) Phytopharmacology of *Toona ciliata*: a review. *International Journal of Pharmaceutics* **8**, 30–35.

Dixon RA (2004) Phytoestrogens. *Annual Review of Plant Biology* **55**, 225–261. doi:10.1146/annurev.arplant.55.031903.141729

Dixon RA, Achnine L, Kota P, Liu C-J, Srinivasa Reddy MS, Wang L *et al.* (2002) The phenylpropanoid pathway and plant defence: a genomics perspective. *Molecular Plant Pathology* **3**, 371–390. doi:10.1046/j.1364-3703.2002.00131.x

Domingo V, Arteaga JF, Quilez Del Moral JF, Barrero AF (2009) Unusually cyclized triterpenes: occurrence, biosynthesis and chemical synthesis. *Natural Product Reports* **26**, 115–134. doi:10.1039/B801470C

Domingo-Fernández D, Gadiya Y, Mubeen S, Bollerman TJ, Healy MD *et al.* (2023) Modern drug discovery using ethnobotany: a large-scale cross-cultural analysis of traditional medicine reveals common therapeutic uses. *iScience* **26**, 107729. doi:10.1016/j.isci.2023.107729

Dorling PR, Huxtable CR, Colegate SM (1980) Inhibition of lysosomal alpha-mannosidase by swainsonine, an indolizidine alkaloid isolated from *Swainsona canescens. The Biochemical Journal* **191**, 649–651. doi:10.1042/bj1910649

Doskotch RW, Ray AB, Kubelka W, Fairchild EH, Hufford CD *et al.* (1974) *Codonocarpus* alkaloids. III: the structure of codonocarpine. *Tetrahedron* **30**, 3229–3236. doi:10.1016/S0040-4020(01)97493-9

Douglas SE, Turner S (1991) Molecular evidence for the origin of plastids from a cyanobacterium-like ancestor. *Journal of Molecular Evolution* **33**, 267–273. doi:10.1007/BF02100678

Douglas C, Ellard M, Hauffe KD, Molitor E, De Sa *et al.* (1992) General phenylpropanoid metabolism: regulation by environmental and developmental signals. In *Phenolic metabolism in plants.* (Eds HA Stafford and RK Ibrahim) pp. 63–89. Springer, New York, NY.

Dove A (2007) High-throughput screening goes to school. *Nature Methods* **4**, 523–532. doi:10.1038/nmeth0607-523

Dragon N (2019) Aboriginal healers treat patients in SA hospitals. *Australian Nursing & Midwifery Journal* 7 May. https://anmj.org.au/aboriginal-healers-treat-patients-in-sa-hospitals/ (accessed 8 June 6/2022).

Dreyer DL, Lee A (1972) Extractives of *Geijera parviflora*. *Phytochemistry* **11**, 763–767. doi:10.1016/0031-9422(72)80045-1

Duh C-Y, Phoebe CHJ, Pezzuto JM, Kinghorn AD, Farnsworth NR (1986) Plant anticancer agents. XLII: cytotoxic constituents from *Wikstroemia elliptica*. *Journal of Natural Products* **49**, 706–709. doi:10.1021/np50046a031

Duta-Bratu C-G, Nitulescu GM, Mihai DP, Olaru OT (2023) Resveratrol and other natural oligomeric stilbenoid compounds and their therapeutic applications. *Plants* **12**, 2935. doi:10.3390/plants12162935

Duval J, Pecher V, Poujol M, Lesellier E (2016) Research advances for the extraction, analysis and uses of anthraquinones: a review. *Industrial Crops and Products* **94**, 812–833. doi:10.1016/j.indcrop.2016.09.056

Eade RA, Hunt K, Simes JJH, Stern W (1969) Extractives of Australian timbers. IX: the saponin of *Planchonella pohlmanianum*. *Australian Journal of Chemistry* **22**, 2703–2707. doi:10.1071/CH9692703

Edgar JA, Molyneux RJ, Colegate SM (2015) Pyrrolizidine alkaloids: potential role in the etiology of cancers, pulmonary hypertension, congenital anomalies, and liver disease. *Chemical Research in Toxicology* **28**, 4–20. doi:10.1021/tx500403t

Efange SMN, Brun R, Wittlin S, Connolly JD, Hoye TR *et al.* (2009) Okundoperoxide, a bicyclic cyclofarnesylsesquiterpene endoperoxide from *Scleria striatinux* with antiplasmodial activity. *Journal of Natural Products* **72**, 280–283. doi:10.1021/np800338p

Egan PA, van der Kooy F (2013) Phytochemistry of the carnivorous sundew genus *Drosera* (Droseraceae): future perspectives and ethnopharmacological relevance. *Chemistry & Biodiversity* **10**, 1774–1790. doi:10.1002/cbdv.201200359

El-Najjar N, Gali-Muhtasib H, Ketola RA, Vuorela P, Urtti A *et al.* (2011) The chemical and biological activities of quinones: overview and implications in analytical detection. *Phytochemistry Reviews* **10**, 353–370. doi:10.1007/s11101-011-9209-1

El-Shanawany MA, Mohamed GA, Nafady AM, Ibrahim SRM, Radwan MM *et al.* (2011) A new xanthone from the roots of *Centaurium spicatum*. *Phytochemistry Letters* **4**, 126–128. doi:10.1016/j.phytol.2010.12.008

Eliaser EM, Ho JH, Hashim NM, Rukayadi Y, Ee GCL *et al.* (2018) Phytochemical constituents and biological activities of *Melicope lunu-ankenda*. *Molecules (Basel, Switzerland)* **23**, 2708.

Elkhateeb A, Yamasaki M, Maede Y, Katakura K, Nabeta K *et al.* (2008) Anti-babesial quassinoids from the fruits of *Brucea javanica*. *Natural Product Communications* **3**, 1934578X0800300207. https://doi.org/10.1177/1934578X0800300207

Essa A, Ali O (2010) *Studies in Islamic Civilization: The Muslim Contribution to the Renaissance.* Internation Institute of Islamic Thought (IIIT), Washington DC.

Fahey JW, Zalcmann AT, Talalay P (2001) The chemical diversity and distribution of glucosinolates and isothiocyanates among plants. *Phytochemistry* **56**, 5–51. doi:10.1016/S0031-9422(00)00316-2

Fan Q, Jiang Y-P, Zhu D-Q, Xu W, Huang W-Q *et al.* (2018) Phenols from the rhizome of *Wikstroemia indica*. *Biochemical Systematics and Ecology* **78**, 59–62. doi:10.1016/j.bse.2018.04.004

Fan S, Zhang C, Luo T, Wang J, Tang Y *et al.* (2019) Limonin: a review of its pharmacology, toxicity, and pharmacokinetics. *Molecules (Basel, Switzerland)* **24**, 3679. doi:10.3390/molecules24203679

Fantegrossi WE, Murnane KS, Reissig CJ (2008) The behavioral pharmacology of hallucinogens. *Biochemical Pharmacology* **75**, 17–33. doi:10.1016/j.bcp.2007.07.018

Farag MA, Sakna ST, El-Fiky NM, Shabana MM, Wessjohann LA (2015) Phytochemical, antioxidant and antidiabetic evaluation of eight *Bauhinia* L. species from Egypt using UHPLC-PDA-qTOF-MS and chemometrics. *Phytochemistry* **119**, 41–50. doi:10.1016/j.phytochem.2015.09.004

Farghaly TA, Masaret GS, Abdulwahab HG (2023) The patent review of the biological activity of tropane containing compounds. *Expert Opinion on Therapeutic Patents* **33**, 875–899. doi:10.1080/13543776.2023.2299349

Farouil L, Sylvestre M, Fournet A, Cebrián-Torrejón G (2022) Review on canthin-6-one alkaloids: distribution, chemical aspects and biological activities. *European Journal of Medicinal Chemistry Reports* **5**, 100049. doi:10.1016/j.ejmcr.2022.100049

Fattori V, Hohmann MSN, Rossaneis AC, Pinho-Ribeiro FA, Verri WA Jr (2016) Capsaicin: current understanding of its mechanisms and therapy of pain and other pre-clinical and clinical uses. *Molecules (Basel, Switzerland)* **21**, 844. doi:10.3390/molecules21070844

Feng L, Zhai YY, Xu J, Yao WF, Cao YD *et al.* (2019) A review on traditional uses, phytochemistry and pharmacology of *Eclipta prostrata* (L.) L. *Journal of Ethnopharmacology* **245**, 112109. doi:10.1016/j.jep.2019.112109

Fernandez LS, Buchanan MS, Carroll AR, Feng Y-J, Quinn RJ *et al.* (2009) Flinderoles A-C: antimalarial bis-indole alkaloids from *Flindersia* species. *Organic Letters* **11**, 329–332. doi:10.1021/ol802506n

Feunang YD, Eisner R, Knox C, Chepelev L, Hastings J *et al.* (2016) ClassyFire: automated chemical classification with a comprehensive, computable taxonomy. *Journal of Cheminformatics* **8**, 61.

Figurová D, Tokárová K, Greifová H, Knížatová N, Kolesárová A *et al.* (2021) Inflammation, it's regulation and antiphlogistic effect of the cyanogenic glycoside amygdalin. *Molecules (Basel, Switzerland)* **26**, 5972. doi:10.3390/molecules26195972

Fletcher MT, Hayes PY, Somerville MJ, De Voss JJ (2010) Ptesculentoside, a novel norsesquiterpene glucoside from the Australian bracken fern *Pteridium esculentum*. *Tetrahedron Letters* **51**, 1997–1999. doi:10.1016/j.tetlet.2010.02.032

Flynn TM, Southwell IA (1987a) Cyanogenesis in the genus *Zieria*. *Phytochemistry* **26**, 1669–1672. doi:10.1016/S0031-9422(00)82267-0

Flynn TM, Southwell IA (1987b) Essential oil constituents of the genus *Zieria*. *Phytochemistry* **26**, 1673–1686. doi:10.1016/S0031-9422(00)82268-2

Flythe MD, Kagan IA, Wang Y, Narvaez N (2017) Hops (Humulus lupulus L.) bitter acids: modulation of rumen fermentation and potential as an alternative growth promoter. *Frontiers in Veterinary Science* **4**, 131. doi:10.3389/fvets.2017.00131

Fogel DB (2018) Factors associated with clinical trials that fail and opportunities for improving the likelihood of success: a review. *Contemporary Clinical Trials* **11**, 156–164. doi:10.1016/j.conctc.2018.08.001

Fokialakis N, Mitaku S, Mikros E, Skaltsounis A-L, Tillequin F *et al.* (1999) Megistosarcimine and megistosarconine, two alkaloids from *Sarcomelicope megistophylla*. *Phytochemistry* **52**, 1745–1748. doi:10.1016/S0031-9422(99)00323-4

Fokialakis N, Magiatis P, Skaltsounis A-L, Tillequin F, Sevenet T (2000) Megistolactone, a new alkalods from *Sarcomelicope megistophylla*. *Zeitschrift für Naturforschung C* **55**, 874–876. doi:10.1515/znc-2000-11-1204

Fokialakis N, Magiatis P, Aligiannis N, Mitaku S, Tillequin F *et al.* (2001) Furomegistines I and II, two furopyridine alkaloids from the bark of *Sarcomelicope megistophylla*. *Phytochemistry* **57**, 593–596. doi:10.1016/S0031-9422(01)00065-6

Folquitto DG, Swiech JND, Pereira CB, Bobek VB, Halila Possagno GC *et al.* (2019) Biological activity, phytochemistry and traditional uses of genus *Lobelia* (Campanulaceae): a systematic review. *Fitoterapia* **134**, 23–38. doi:10.1016/j.fitote.2018.12.021

Foster S, Yue C (1992) *Herbal Emissaries: Bringing Chinese Herbs to the West.* Healing Arts Press, Rochester, VT.

Fraga BM (2013) Natural sesquiterpenoids. *Natural Product Reports* **30**, 1226–1264. doi:10.1039/c3np70047j

Frick KM, Kamphuis LG, Siddique KH, Singh KB, Foley RC (2017) Quinolizidine alkaloid biosynthesis in lupins and prospects for grain quality improvement. *Frontiers in Plant Science* **8**, 1–12. doi:10.3389/fpls.2017.00087

Fu PP (2017) Pyrrolizidine alkaloids: metabolic activation pathways leading to liver tumor initiation. *Chemical Research in Toxicology* **30**, 81–93. doi:10.1021/acs.chemrestox.6b00297

Fu PP, Xia Q, Lin G, Chou MW (2004) Pyrrolizidine alkaloids: genotoxicity, metabolism enzymes, metabolic activation, and mechanisms. *Drug Metabolism Reviews* **36**, 1–55. doi:10.1081/DMR-120028426

Fu G, Pang H, Wong YH (2008) Naturally occuring phenylethanoid glycosides: potential leads for new therapeutics. *Current Medicinal Chemistry* **15**, 2592–2613. doi:10.2174/092986708785908996

Fujimura S, Terada S, Jayavardhanan KK, Panikkar KR, Kimoto E (1993) Primary structures of concanavalin A-like lectins from seeds of two species of *Canavalia*. *Phytochemistry* **33**, 985–987. doi:10.1016/0031-9422(93)85008-F

Fujiwake H, Suzuki T, Oka S, Iwai K (1980) Enzymatic formation of capsaicinoid from vanillylamine and iso-type fatty acids by cell-free extracts of *Capsicum annuum* var. *annuum* cv. Karayatsubusa. *Agricultural and Biological Chemistry* **44**, 2907–2912. doi:10.1271/bbb1961.44.2907

Furlan CM, Motta LB, Cursino Dos Santos DYA (2010) Tannins: what do they represent in plant life. In *Tannins: Types, Foods Containing, and Nutrition.* (Ed. GK Petridis) pp. 1–13. Nova Science Publishers, Hauppauge, NY.

Gadzikowska M, Grynkiewicz G (2002) Tropane alkaloids in pharmaceutical and phytochemical analysis. *Acta Poloniae Pharmaceutica* **59**, 149–160.

Galliford CV, Scheidt KA (2007) Pyrrolidinyl-spirooxindole natural products as inspirations for the development of potential therapeutic agents. *Angewandte Chemie International Edition* **46**, 8748–8758. doi:10.1002/anie.200701342

Gangwar M, Goel RK, Nath G (2014) *Mallotus philippinensis* Muell. Arg (Euphorbiaceae): ethnopharmacology and phytochemistry review. *BioMed Research International* **2014**, 213973.

Gao JM, Wu WJ, Zhang JW, Konishi Y (2007) The dihydro-β-agarofuran sesquiterpenoids. *Natural Product Reports* **24**, 1153–1189. doi:10.1039/b601473a

Gao QH, Wu CS, Wang M (2013) The jujube (*Ziziphus jujuba* Mill.) fruit: a review of current knowledge of fruit composition and health benefits. *Journal of Agricultural and Food Chemistry* **61**, 3351–3363. doi:10.1021/jf4007032

Garaniya N, Bapodra A (2014) Ethno botanical and phytopharmacological potential of Abrus precatorius L.: a review. *Asian Pacific Journal of Tropical Biomedicine* **4**, S27–S34. doi:10.12980/APJTB.4.2014C1069

Garg V, Jyoti, Dutt R (2018) The genus *Plumbago*: an update. *International Journal of Green and Herbal Chemistry* **7**, 313–326.

Garrison FH (1966) *History of Medicine.* W. B. Saunders Co., Philadelphia, PA.

Gasparetto JC, Martins CA, Hayashi SS, Otuky MF, Pontarolo R (2012) Ethnobotanical and scientific aspects of *Malva sylvestris* L.: a millennial herbal medicine. *The Journal of Pharmacy and Pharmacology* **64**, 172–189. doi:10.1111/j.2042-7158.2011.01383.x

Gerometta E, Grondin I, Smadja J, Frederich M, Gauvin-Bialecki A (2020) A review of traditional uses, phytochemistry and pharmacology of the genus *Indigofera*. *Journal of Ethnopharmacology* **253**, 112608. doi:10.1016/j.jep.2020.112608

Gershenzon J, Dudareva N (2007) The function of terpene natural products in the natural world. *Nature Chemical Biology* **3**, 408–414. doi:10.1038/nchembio.2007.5

Ghisalberti EL (1996) Bioactive acylphloroglucinol derivatives from *Eucalyptus* species. *Phytochemistry* **41**, 7–22. doi:10.1016/0031-9422(95)00484-X

Ghisalberti EL (1998a) Biological and pharmacological activity of naturally occurring iridoids and secoiridoids. *Phytomedicine* **5**, 147–163. doi:10.1016/S0944-7113(98)80012-3

Ghisalberti EL (1998b) Ethnopharmacology and phytochemistry of *Dodonaea* species. *Fitoterapia* **69**, 99–113.

Ghisalberti EL (1998c) Phytochemistry of the Australian Rutaceae: *Boronia, Eriostemon* and *Phebalium* species. *Phytochemistry* **47**, 163–176. doi:10.1016/S0031-9422(97)00571-2

Ghisalberti EL (2004) The Goodeniaceae. *Fitoterapia* **75**, 429–446. doi:10.1016/j.fitote.2004.01.018

Giera M, Aisporna A, Uritboonthai W, Siuzdak G (2024) The hidden impact of in-source fragmentation in metabolic and chemical mass spectrometry data interpretation. *Nature Metabolism* **6**, 1647–1648. doi:10.1038/s42255-024-01076-x

Gobbi PG, Broglia C, Merli F, Dell'olio M, Stelitano C *et al.* (2003) Vinblastine, bleomycin, and methotrexate chemotherapy plus irradiation for patients with early-stage, favorable Hodgkin lymphoma: the experience of the Gruppo Italiano Studio Linfomi. *Cancer* **98**, 2393–2401. doi:10.1002/cncr.11807

Dioscorides P, Goodyer J (trans.) (2000) *De Materia Medica: Being an Herbal with Many Other Medicinal Matters. Written in Greek in the First Century of the Common Era* [c. 70]. Ibidis, Johannesburg, South Africa.

Gou KJ, Zeng R, Ma Y, Li AN, Yang K *et al.* (2020) Traditional uses, phytochemistry, and pharmacology of *Persicaria orientalis* (L.) Spach - A review. *Journal of Ethnopharmacology* **249**, 112407. doi:10.1016/j.jep.2019.112407

Gould KS, Lister C (2006) Flavonoid functions in plants. In *Flavonoids: Chemistry, Biochemistry and Applications.* (Eds OM Anderson and KR Markham) pp. 397–441. CRC Press, Boca Raton, FL.

Goyal PK (2012) Phytochemical and pharmacological properties of the genus *Grewia*: a review. *International Journal of Pharmacy and Pharmaceutical Sciences* **4**, 72–78.

Goyal N, Sharma SK (2014) Bioactive phytoconstituents from genus Heliotropium. *International Journal of Green Pharmacy* **8**, 217–225.

Grace MH, Jin Y, Wilson GR, Coates RM (2006) Structures, biogenetic relationships, and cytotoxicity of pimarane-derived diterpenes from *Petalostigma pubescens. Phytochemistry* **67**, 1708–1715. doi:10.1016/j.phytochem.2005.09.026

Grayer-Barkmeijer RJ, Tomás-Barberán FA (1993) 8-hydroxylated flavone *O*-glycosides and other flavonoids in chemotypes of *Gratiola officinalis. Phytochemistry* **34**, 205–210. doi:10.1016/S0031-9422(00)90806-9

Grayson DH (2000) Monoterpenoids. *Natural Product Reports* **17**, 385–419. doi:10.1039/a804437f

Green BT, Lee ST, Panter KE, Brown DR (2012) Piperidine alkaloids: Human and food animal teratogens. *Food and Chemical Toxicology* **50**, 2049–2055. doi:10.1016/j.fct.2012.03.049

Greinwald R, Henrichs C, Veen G, Ross JH, Witte L *et al.* (1995) A survey of alkaloids in *Templetonia biloba* (Benth.) Polhill (Fabaceae: Brongniartieae). *Biochemical Systematics and Ecology* **23**, 649–654. doi:10.1016/0305-1978(95)00035-6

Greinwald R, Ross JH, Witte L, Czygan F-C (1996) Alkaloids of *Templetonia incana. Biochemical Systematics and Ecology* **24**, 423–426. doi:10.1016/0305-1978(96)00024-5

Griffin WJ, Lin GD (2000) Chemotaxonomy and geographical distribution of tropane alkaloids. *Phytochemistry* **53**, 623–637. doi:10.1016/S0031-9422(99)00475-6

Griffin SG, Leach DN, Markham JL, Johnstone R (1998) Antimicrobial activity of essential oils from *Zieria. The Journal of Essential Oil Research* **10**, 165–174. doi:10.1080/10412905.1998.9700870

Groger D (1980) Alkaloids derived from tryptophan and anthranilic acid. In *Secondary Plant Products.* (Eds EA Bell and BV Charlwood) pp. 128–159. Springer-Verlag, Berlin.

Grum DS, Cook D, Baucom D, Mott IW, Gardner DR *et al.* (2013) Production of the alkaloid swainsonine by a fungal endophyte in the host *Swainsona canescens. Journal of Natural Products* **76**, 1984–1988. doi:10.1021/np400274n

Grynkiewicz G, Gadzikowska M (2008) Tropane alkaloids as medicinally useful natural products and their synthetic derivatives as new drugs. *Pharmacological Reports* **60**, 439–463.

Gualtieri F (2000) Cholinergic receptors and neurodegenerative diseases. *Pharmaceutica Acta Helvetiae* **74**, 85–89. doi:10.1016/S0031-6865(99)00021-7

Gunatilaka LA (2012) Plant natural products. In *Natural Products in Chemical Biology.* (Ed. N Civjan) pp. 3–31. John Wiley & Sons, NJ

Guo J, Huan T (2020) Comparison of full-scan, data-dependent, and data-independent acquisition modes in liquid chromatography-mass spectrometry based untargeted metabolomics. *Analytical Chemistry* **92**, 8072–8080. doi:10.1021/acs.analchem.9b05135

Guo Z, Vangapandu S, Sindelar RW, Walker LA, Sindelar RD (2005) Biologically active quassinoids and their chemistry: potential leads for drug design. *Current Medicinal Chemistry* **12**, 173–190. doi:10.2174/0929867053363351

Guo Y, Sakulnarmrat K, Konczak I (2014) Anti-inflammatory potential of native Australian herbs polyphenols. *Toxicology Reports* **1**, 385–390. doi:10.1016/j.toxrep.2014.06.011

Guo R, Wang T, Zhou G, Xu M, Yu X *et al.* (2018) Botany, phytochemistry, pharmacology and toxicity of *Strychnos nux-vomica* L.: a review. *The American Journal of Chinese Medicine* **46**, 1–23. doi:10.1142/S0192415X18500015

Gupta S, Ved A (2017) *Operculina turpethum* (Linn.) Silva Manso as a medicinal plant species: a review on bioactive components and pharmacological properties. *Pharmacognosy Reviews* **11**, 158–166. doi:10.4103/phrev.phrev_6_17

Gupta D, Bleakley B, Gupta RK (2008) Dragon's blood: botany, chemistry and therapeutic uses. *Journal of*

Ethnopharmacology **115**, 361–380. doi:10.1016/j.jep.2007.10.018

Gurav S, Senthilkumar DGP, Tembare R, Mani T (2011) *Tylophora indica*: a review on its ethnobotany, phytochemical and pharmacological profile. *Asian Journal of Biochemical and Pharmaceutical Research* **3**, 405–414.

Hajar R (2013) The air of history: part III. *Heart Views* **14**, 43–46[History of Medicine]. doi:10.4103/1995-705X.107125

Halleux R (1996) The reception of Arabic alchemy in the West. In *Encyclopedia of the History of Arabic Science.* (R Rashed and R Morelon) pp. 840–856. Routledge, New York, NY.

Han L, Zhou X, Yang M, Zhou L, Deng X *et al.* (2018) Ethnobotany, phytochemistry and pharmacological effects of plants in genus *Cynanchum* Linn. (Asclepiadaceae). *Molecules (Basel, Switzerland)* **23**, 1194. doi:10.3390/molecules23051194

Hanson JR (1979) Diterpenoids. In *Terpenoids and Steroids.* (Ed. JR Hanson) pp. 160–185. J. W. Arrowsmith Ltd., Bristol, UK.

Hanson JR (2013) Diterpenoids of terrestrial origin. *Natural Product Reports* **30**, 1346–1356. doi:10.1039/c3np70046a

Hanson GR, Venturelli PJ, Fleckenstein AE (2018) Marijuana. In *Drugs and Society* 13th edn. (Ed. K Rogers) pp. 416–466. Jones and Bartlett Learning, Burlington, MA.

Hanumaiah T, Marshall DS, Rao BK, Rao CP, Rao GSR *et al.* (1985a) Benzoisochromanquinones in *Ventilago* species. *Phytochemistry* **24**, 2373–2378. doi:10.1016/S0031-9422(00)83044-7

Hanumaiah T, Rao BK, Rao CP, Rao GSR, Rao JUM *et al.* (1985b) Naphthalenes and naphthoquinones from *Ventilago* species. *Phytochemistry* **24**, 1811-1815. doi:10.1016/S0031-9422(00)82557-1

Harper D (1998) *Early Chinese Medical Literature: The Mawangdui Medical Manuscripts.* Kegan Paul International, New York, NY.

Harris CM, Schneider MJ, Ungemach FS, Hill JE, Harris TM (1988) Biosynthesis of the toxic indolizidine alkaloids slaframine and swainsonine in *Rhizoctonia leguminicola*: metabolism of 1-hydroxyindolizidines. *Journal of the American Chemical Society* **110**, 940–949. doi:10.1021/ja00211a039

Harrison JJEK, Dankyi E, Kingsford-Adaboh R, Ishida H (2011) In search of new leads: a closer look at the therapeutic potential of the constituents of *Millettia thonningii, Millettia machycarpa* and their structural analogues. *International Journal of Pharmacy and Pharmaceutical Sciences* **3**, 71–81.

Hart NK, Lamberton JA (1966) Pyrrolizidine alkaloids from *Planchonella* species (family Sapotaceae). *Australian Journal of Chemistry* **19**, 1259–1264. doi:10.1071/CH9661259

Hart NK, Johns SR, Lamberton JA (1968) Alkaloids of the Sapotaceae: trans-B-methylthioacrylate and tiglate esters of (-)-isoretronecanol from a *Planchonella* species. *Australian Journal of Chemistry* **21**, 1393–1395. doi:10.1071/CH9681393

Hartmann T (2007) From waste products to ecochemicals: 50 years research on plant secondary metabolism. *Phytochemistry* **68**, 2831–2846. doi:10.1016/j.phytochem.2007.09.017

Hartmann T, Ehmke A, Eilert U, Von Borstel K, Theuring C (1989) Sites of synthesis, translocation and accumulation of pyrrolizidine N-oxides in *Senecio vulgaris* L. *Planta* **177**, 98–107. doi:10.1007/BF00392159

Hasan N, Osman H, Mohamad S, Chong WK, Awang K *et al.* (2012) The chemical components of *Sesbania grandiflora* root and their antituberculosis activity. *Pharmaceuticals* **5**, 882–889. doi:10.3390/ph5080882

Hastings J, Adams N, Ennis M, Hull D, Steinbeck C (2011) Chemical ontologies: what they are, what they are for and what are the challenges. *Journal of Cheminformatics* **3**, 04. doi:10.1186/1758-2946-3-S1-O4

Haudecoeur R, Peuchmaur M, Peres B, Rome M, Taiwe GS *et al.* (2018) Traditional uses, phytochemistry and pharmacological properties of African *Nauclea* species: A review. *Journal of Ethnopharmacology* **212**, 106–136. doi:10.1016/j.jep.2017.10.011

Hayashi T, Kishi MP, Kawasaki M, Arisawa M, Shimizu M *et al.* (1987) Scopadulcic acid -A and -B, new diterpenoids with a novel skeleton, from a Paraguayan crude drug "Typycha Kuratu" (*Scoparia dulcis* L.). *Tetrahedron Letters* **28**, 3693–3696. doi:10.1016/S0040-4039(00)96357-3

Hayashi T, Kawasaki M, Miwa Y, Taga T, Morita N (1990) Antiviral agents of plant origin. III: scopadulin, a novel tetracyclic diterpene from *Scoparia dulcis* L. *Chemical & Pharmaceutical Bulletin* **38**, 945–947. doi:10.1248/cpb.38.945

Hayes PY, Chow S, Somerville MJ, De Voss JJ, Fletcher MT (2009) Pimelotides A and B, diterpenoid ketal-lactone orthoesters with an unprecedented skeleton from *Pimelea elongata. Journal of Natural Products* **72**, 2081–2083. doi:10.1021/np900573k

Hayes PY, Chow S, Somerville MJ, Fletcher MT, De Voss JJ (2010) Daphnane- and tigliane-type diterpenoid esters and orthoesters from *Pimelea elongata. Journal of Natural Products* **73**, 1907–1913. doi:10.1021/np1005746

He X, Fang J, Huang L, Wang J, Huang X (2015) *Sophora flavescens* Ait.: Traditional usage, phytochemistry and pharmacology of an important traditional Chinese medicine. *Journal of Ethnopharmacology* **172**, 10–29. doi:10.1016/j.jep.2015.06.010

He X, Bai Y, Zhao Z, Wang X, Fang J *et al.* (2016) Local and traditional uses, phytochemistry, and pharmacology of *Sophora japonica* L.: a review. *Journal of Ethnopharmacology* **187**, 160–182. doi:10.1016/j.jep.2016.04.014

He X, Wang X, Fang J, Chang Y, Ning N *et al.* (2017) The genus Achyranthes: a review on traditional uses, phytochemistry, and pharmacological activities. *Journal of Ethnopharmacology* **203**, 260–278. doi:10.1016/j.jep.2017.03.035

Hertweck C (2009) The biosynthetic logic of polyketide biodiversity. *Angewandte Chemie International Edition* **48**, 4688–4716. doi:10.1002/anie.200806121

Hill AW (1915) The history and function of botanic gardens. *Annals of the Missouri Botanical Garden* **2**, 185–240. doi:10.2307/2990033

Hill RA, Connolly JD (2012) Triterpenoids. *Natural Product Reports* **29**, 780–818. doi:10.1039/c2np20027a

Hill RA, Connolly JD (2013) Triterpenoids. *Natural Product Reports* **30**, 1028–1065. doi:10.1039/C3NP70032A

Hill RA, Connolly JD (2015) Triterpenoids. *Natural Product Reports* **32**, 273–327. doi:10.1039/C4NP00101J

Hill RA, Connolly JD (2017) Triterpenoids. *Natural Product Reports* **34**, 90–122. doi:10.1039/C6NP00094K

Hill RA, Connolly JD (2018) Triterpenoids. *Natural Product Reports* **35**, 1294–1329. doi:10.1039/C8NP00029H

Hill RA, Connolly JD (2020) Triterpenoids. *Natural Product Reports* **37**, 962–998. doi:10.1039/C9NP00067D

Hill RS, Truswell EM, Mcloughlin S, Dettman ME (1999) The evolution of Australian flora: fossil evidence. *Flora of Australia* **1**, 251–320.

Hinrichs TJ (2005) Healing and medicine in China. In *Healing and Medicine*. (Ed. L Jones) pp. 3859–3864. Macmillan Reference USA, Detroit, MI.

Hiradeve SM, Rangari VD (2014) A review on pharmacology and toxicology of *Elephantopus scaber* Linn. *Natural Product Research* **28**, 819–830. doi:10.1080/14786419.2014.883394

Hiteksha SP, Mamta BS (2017) *Thespesia populnea* Linn.: A review. *International Journal of Pharmaceutics* **4**, 1–5.

Hodgson B (2001) *In the Arms of Morpheus: The Tragic History of Laudanum, Morphine, and Patent Medicines*. Firefly Books, Buffalo, NY.

Hodgson D, Ritchie E, Taylor WC (1960) Chemical studies of the Myrtaceae. II: the constituents of *Syncarpia laurifolia* Tenn. *Australian Journal of Chemistry* **13**, 385–389. doi:10.1071/CH9600385

Hofmann A (1980) *LSD, my problem child: reflections on sacred drugs, mysticism, and science*. Multidisciplinary Association for Psychedelic Studies (MAPS), Santa Cruz, CA.

Holeski LM, Keefover-Ring K, Bowers MD, Harnenz ZT, Lindroth RL (2013) Patterns of phytochemical variation in *Mimulus guttatus* (yellow monkeyflower). *Journal of Chemical Ecology* **39**, 525–536. doi:10.1007/s10886-013-0270-7

Holmes EM (1906) Horticulture in relation to medicine. *Journal of the Royal Horticultural Society* **31**, 50–54.

Hook I, Mills C, Sheridan H (2014) Bioactive naphthoquinones from higher plants. In *Studies in Natural Product Chemistry*. (Ed.A-U Rahman) pp. 119–160. Elsevier, Oxford.

Hopkins AL (2008) Network pharmacology: the next paradigm in drug discovery. *Nature Chemical Biology* **4**, 682–690. doi:10.1038/nchembio.118

Horden P (2005) The earliest hospitals in Byzantium, Western Europe, and Islam. *The Journal of Interdisciplinary History* **35**, 361–389. doi:10.1162/0022195052564243

Hosamani KM (1994) Ricinoleic and cyclopropene acids in *Trichodesma zeylanicum* seed oil. *Phytochemistry* **37**, 1621–1624. doi:10.1016/S0031-9422(00)89578-3

Hoskisson PA, Seipke RF (2020) Cryptic or silent? The known unknowns, unknown knowns, and unknown unknowns of secondary metabolism. *mBio* **11**, 02642. doi:10.1128/mBio.02642-20

Hotti H, Rischer H (2017) The killer of Socrates: coniine and related alkaloids in the plant kingdom. *Molecules (Basel, Switzerland)* **22**, 1962. doi:10.3390/molecules22111962

Hsu S-C, Chung J-G (2012) Anticancer potential of emodin. *Biomedicine (Taipei)* **2**, 108–116. doi:10.1016/j.biomed.2012.03.003

Hua S, Zhang Y, Liu J, Dong L, Huang J *et al.* (2018) Ethnomedicine, phytochemistry and pharmacology of *Smilax glabra*: An important traditional Chinese medicine. *The American Journal of Chinese Medicine* **46**, 261–297. doi:10.1142/S0192415X18500143

Huang Y-L, Chen C-C, Hsu F-L, Chen C-F (1996) A new lignan from *Phyllanthus virgatus*. *Journal of Natural Products* **59**, 520–521. doi:10.1021/np9603380

Huang Y-L, Chen C-C, Hsu F-L, Chen C-F (1998) Tannins, flavonol sulfonates, and a norlignan from *Phyllanthus virgatus*. *Journal of Natural Products* **61**, 1194–1197. doi:10.1021/np970336v

Huang X, Mu B, Lin W, Qiu Y (2012) Pterocarpin and isoflavan derivatives from *Canavalia maritima* (Aubl.) Thou. *Records of Natural Products* **6**, 166–170.

Huang W, Li C, Wang Y, Yi X, He X (2017a) Anti-inflammatory lignanamides and monoindoles from Alocasia macrorrhiza. *Fitoterapia* **117**, 126–132. doi:10.1016/j.fitote.2017.01.014

Huang W, Yi X, Feng J, Wang Y, He X (2017b) Piperidine alkaloids from Alocasia macrorrhiza. *Phytochemistry* **143**, 81–86. doi:10.1016/j.phytochem.2017.07.012

Hubert J, Nuzillard J-M, Renault J-H (2017) Dereplication strategies in natural product research: How many tools and methodologies behind the same concept? *Phytochemistry Reviews* **16**, 55–95. doi:10.1007/s11101-015-9448-7

Hufford CD, Oguntimein BO, Muhammad I (1992) New stemodane diterpenes from *Stemodia maritima*. *Journal of Natural Products* **55**, 48–52. doi:10.1021/np50079a007

Hungerford NL, Sands DPA, Kitching W (1998) Isolation and structure of some constituents of the Australian medicinal plant *Tinospora smilacina* ('snakevine'). *Australian Journal of Chemistry* **51**, 1103–1111. doi:10.1071/C98034

Hushpulian DM, Gaisina IN, Nikulin SV, Chubar T, Savin SS *et al.* (2024) High throughput screening in drug discovery: problems and solutions. *Moscow University Chemistry Bulletin* **79**, 93–104. doi:10.3103/S0027131424700081

Ibrahim SRM, Mohamed GA (2016) Naturally occurring naphthalenes: chemistry, biosynthesis, structural elucidation, and biological activities. *Phytochemistry Reviews* **15**, 279–295. doi:10.1007/s11101-015-9413-5

Ilanko A, Cock IE (2019) The interactive antimicrobial activity of conventional antibiotics and *Petalostigma* spp. extracts against bacterial triggers of some autoimmune inflammatory diseases. *Pharmacognosy Journal* **11**, 292–309. doi:10.5530/pj.2019.11.45

Inglese J, Auld DS (2007) High throughput screening (HTS) techniques: applications in chemical biology. In *Wiley Encyclopedia of Chemical Biology*.(ERd. TP Begley) pp. 1-15.Wiley

Inglese J, Auld DS, Jadhav A, Johnson RL, Simeonov A *et al.* (2006) Quantitative high-throughput screening: a titration-based approach that efficiently identifies biological activities in large chemical libraries. *Proceedings of the National Academy of Sciences of the United States of America* **103**, 11473–11478. doi:10.1073/pnas.0604348103

IP Australia (2019) *Protection of Indigenous Knowledge in the Intellectual Property System: consultation report.* Australian Government, Canberra.

Iranshahy M, Javadi B, Iranshahi M, Jahanbakhsh SP, Mahyari S *et al.* (2017) A review of traditional uses, phytochemistry and pharmacology of *Portulaca oleracea* L. *Journal of Ethnopharmacology* **205**, 158–172. doi:10.1016/j.jep.2017.05.004

Ishikura M, Abe T, Choshi T, Hibino S (2015) Simple indole alkaloids and those with a nonrearranged monoterpenoid unit. *Natural Product Reports* **32**, 1389–1471. doi:10.1039/C5NP00032G

Islam MT, Mubarak MS (2020) Pyrrolidine alkaloids and their promises in pharmacotherapy. *Advances in Traditional Medicine* **20**, 13–22. doi:10.1007/s13596-019-00419-4

Islamiyat FB, Oke MO, Olaniyan SA, Ajala AS (2016) A review of cyanogenic glycosides in edible plants. In *Toxicology: New Aspects to this Scientific Conundrum.* (Eds ML Larramendy and S Soloneski) pp. 179–191. IntechOpen. https://www.intechopen.com/books/5362.

Itharat A, Plubrukarn A, Kongsaeree P, Bui T, Keawpradub N *et al.* (2003) Dioscorealides and dioscoreanone, novel cytotoxic naphthofuranoxepins, and 1,4-phenanthraquinone from *Dioscorea membranacea* Pierre. *Organic Letters* **5**, 2879–2882. doi:10.1021/ol034926y

Itoh A, Tanaka Y, Nagakura N, Nishi T, Tanahashi T (2006) A quinic acid ester from *Strychnos lucida. Journal of Natural Medicines* **60**, 146–148. doi:10.1007/s11418-005-0021-3

James JT, Dubery IA (2009) Pentacyclic triterpenoids from the medicinal herb, *Centella asiatica* (L.) Urban. *Molecules (Basel, Switzerland)* **14**, 3922–3941. doi:10.3390/molecules14103922

Jammula SR, Pepalla SB, Hanumaiah T, Rao KVJ, Thomson RH (1991) Benzisochromanquinones from *Ventilago goughii. Phytochemistry* **30**, 3741–3744. doi:10.1016/0031-9422(91)80100-F

Jasshi AR, Zare S, Asadollahi M, Schuman MC (2017) Ecological roles and biological activities of secialized metabolites from the genus *Nicotiana. Chemical Reviews* **117**, 12227–12280. doi:10.1021/acs.chemrev.7b00001

Jensen PR, Chavarria KL, Fenical W, Moore BS, Ziemert N (2014) Challenges and triumphs to genomics-based natural product discovery. *Journal of Industrial Microbiology & Biotechnology* **41**, 203–209. doi:10.1007/s10295-013-1353-8

Jiménez C, Riguera R (1994) Phenylethanoid glycosides in plants: structure and biological activity. *Natural Product Reports* **11**, 591–606. doi:10.1039/NP9941100591

Jin Z (2006) Imidazole, oxazole and thiazole alkaloids. *Natural Product Reports* **23**, 464–496. doi:10.1039/b502166a

Johne S (1984) The quinazoline alkaloids. In *Progress in the Chemistry of Organic Natural Products.* (Eds W Herz, H Grisebach, GW Kirby and C Tamm,) pp. 159–219. Springer-Verlag, New York, NY.

Johns SR, Lamberton JA (1970) The identification of a new alkaloid from *Timonius kaniensis* (Rubiaceae) as dihydrocupreine. *Australian Journal of Chemistry* **23**, 211–212. doi:10.1071/CH9700211

Johnson JB, Mani JS, Broszczak D, Prasad SS, Ekanayake CP *et al.* (2021) Hitting the sweet spot: a systematic review of the bioactivity and health benefits of phenolic glycosides from medicinally used plants. *Phytotherapy Research* **35**, 3484–3508. doi:10.1002/ptr.7042

Jordan MA (2002) Mechanism of action of antitumor drugs that interact with microtubules and tubulin. *Current Medicinal Chemistry* **2**, 1–17.

Jordan MA, Wilson L (2004) Microtubules as a target for anticancer drugs. *Nature Reviews. Cancer* **4**, 253–265. doi:10.1038/nrc1317

Josephson-Storm J (2017) *The Myth of Disenchantment: Magic, Modernity, and the Birth of the Human Sciences.* University of Chicago Press, Chicago, Ill.

Kalbhen DA (1971) Nutmeg as a narcotic: a contribution to the chemistry and pharmacology of nutmeg (*Myristica fragrans*). *Angewandte Chemie International Edition* **10**, 370–374. doi:10.1002/anie.197103701

Kanchanapoom T, Kasai R, Yamasaki K (2002a) Cucurbitane, hexanorcucurbitane and octanorcucurbitane glycosides from fruits of *Trichosanthes tricuspidata. Phytochemistry* **59**, 215–228. doi:10.1016/S0031-9422(01)00430-7

Kanchanapoom T, Kasai R, Yamasaki K (2002b) Cucurbitane, hexanorcucurbitane and octanorcucurbitane glycosides from fruits of Trichosanthes tricuspidata. *Phytochemistry* **59**, 215–228. doi:10.1016/S0031-9422(01)00430-7

Kanemoto M, Matsunami K, Otsuka H, Shinzato T, Ishigaki C *et al.* (2008) Chlorine-containing iridoid and iridoid glucoside, and other glucosides from leaves of *Myoporum bontioides. Phytochemistry* **69**, 2517–2522. doi:10.1016/j.phytochem.2008.07.002

Karan R, Agarwal P, Sinha M, Mahato N (2021) Recent advances on quinazoline derivatives: a potential bioactive scaffold in medicinal chemistry. *ChemEngineering* **5**, 73. doi:10.3390/chemengineering5040073

Kasai R, Sasaki A, Hashimoto T, Kaneko T, Ohtani K *et al.* (1999) Cycloartane glycosides from *Trichosanthes tricuspidata. Phytochemistry* **51**, 803–808. doi:10.1016/S0031-9422(99)00078-3

Kasim SM, Al-Dabbagh BM, Mustafa YF (2022) A review on the biological potentials of carbazole and its derived products. *Eurasian Chemical Communications* **4**, 495–512.

Katoh T (2013) Enantioselective total synthesis of otteliones A and B, novel and powerful antitumor agents from the freshwater plant *Ottelia alismoides*. *Natural Product Communications* **8**, 973–980. doi:10.1177/1934578X1300800724

Katsuyama Y, Kita T, Funa N, Horinouchi S (2009) Curcuminoid biosynthesis by two type III polyketide synthases in the herb *Curcuma longa*. *The Journal of Biological Chemistry* **284**, 11160–11170. doi:10.1074/jbc.M900070200

Kaufman PB, Kirakosyan A, Mckenzie M, Dayanandan P, Hoyt JE, Li C (2006) The uses of plant natural products by humans and risks associated with their use. In *Natural Products from Plants* (Eds LJ Cseke A Kirakosyan, PB Kaufman, SL Warber, JA Duke *et al.*) pp. 442–474 CRC Press, Boca Raton, FL.

Kaunda JS, Zhang YJ (2017) The genus *Carissa*: an ethnopharmacological, phytochemical and pharmacological review. *Natural Products and Bioprospecting* **7**, 181–199. doi:10.1007/s13659-017-0123-0

Kaunda JS, Zhang YJ (2019) The genus *Solanum*: an ethnopharmacological, phytochemical and biological properties review. *Natural Products and Bioprospecting* **9**, 77–137. doi:10.1007/s13659-019-0201-6

Kawasaki M, Hayashi T, Arisawa M, Morita N, Berganza LH (1988) 8-hydroxytricetin 7-glucuronide, a B-glucuronidase inhibitor from *Scoparia dulcis*. *Phytochemistry* **27**, 3709–3711. doi:10.1016/0031-9422(88)80811-2

Keylor MH, Matsuura BS, Stephenson CR (2015) Chemistry and biology of resveratrol-derived natural products. *Chemical Reviews* **115**, 8976–9027. doi:10.1021/cr500689b

Khan IA, Sticher O, Rali T (1993) New triterpenes from the leaves of *Timonius timon*. *Journal of Natural Products* **56**, 2163–2165. doi:10.1021/np50102a019

Khong PW, Lewis KG (1977) New chemical constituents of *Planchonia careya*. *Australian Journal of Chemistry* **30**, 1311–1322. doi:10.1071/CH9771311

Khong PW, Lewis KG (1979) New chemical constituents of *Planchonia careya*. II: bark constituents soluble in light petroleum. *Australian Journal of Chemistry* **32**, 1621–1626. doi:10.1071/CH9791621

Khyade MS, Kasote DM, Vaikos NP (2014) *Alstonia scholaris* (L.) R. Br. and *Alstonia macrophylla* Wall. ex G. Don: a comparative review on traditional uses, phytochemistry and pharmacology. *Journal of Ethnopharmacology* **153**, 1–18. doi:10.1016/j.jep.2014.01.025

Kiesecker C, Zitron E, Lück S, Bloehs R, Scholz EP *et al.* (2004) Class Ia anti-arrhythmic drug ajmaline blocks HERG potassium channels: mode of action. *Naunyn-Schmiedeberg's Archives of Pharmacology* **370**, 423–435. doi:10.1007/s00210-004-0976-8

Kinghorn AD, Balandrin MF (1984) Quinolizidine alkaloids of the Leguminosae: structural types, analysis, chemotaxonomy and biological activities. In *Alkaloids: Chemical and Biological Perspectives*. (Ed. SW Pelletier) pp. 105–148. Wiley, New York, NY.

Kisiel W, Piozzi F (1999) Phenylethanoid glycosides from *Prostanthera melissifolia*. *Phytochemistry* **51**, 1083–1085. doi:10.1016/S0031-9422(99)00167-3

Ko JH, Lee SG, Yang WM, Um JY, Sethi G *et al.* (2018) The application of embelin for cancer prevention and therapy. *Molecules (Basel, Switzerland)* **23**, 621. doi:10.3390/molecules23030621

Kokanova-Nedialkova Z, Neialkov PT, Nikolov SD (2009) The genus *Chenopodium*: phytochemistry, ethnopharmacology and pharmacology. *Pharmacognosy Reviews* **3**, 280–306.

Kompis I, Hesse M, Schmid H (1971) An approach to the biogenetic classification of indole alkaloids. *Lloydia* **34**, 269–291.

Koo KA, Lorent K, Gong W, Windsor P, Whittaker SJ *et al.* (2015) Biliatresone, a reactive natural toxin from *Dysphania glomulifera* and *D. littoralis*: discovery of the toxic moiety 1,2-diaryl-2-propenone. *Chemical Research in Toxicology* **28**, 1519–1521. doi:10.1021/acs.chemrestox.5b00227

Kostova I, Dinchev D (2005) Saponins in *Tribulus terrestris*: chemistry and bioactivity. *Phytochemistry Reviews* **4**, 111–137. doi:10.1007/s11101-005-2833-x

Koua FHM (2011) *Striga hermonthica* (Del.) Benth: phytochemistry and pharmacological properties outline. *Journal of Applied Pharmaceutical Science* **1**, 1–5.

Koua FHM, Babiker HA, Halfawi A, Ibrahim RO, Abbas FM *et al.* (2011) Phytochemical and biological study of *Striga hermonthica* (Del.) Betnh callus and intact plant. *Research in Pharmaceutical Biotechnology* **3**, 85–92.

Kovacs A, Vasas A, Hohmann J (2008) Natural phenanthrenes and their biological activity. *Phytochemistry* **69**, 1084–1110. doi:10.1016/j.phytochem.2007.12.005

Kreek MJ, Bart G, Lilly C, Laforge KS, Nielsen DA (2005) Pharmacogenetics and human molecular genetics of opiate and cocaine addictions and their treatment. *Pharmacological Reviews* **57**, 1–26. doi:10.1124/pr.57.1.1

Krinsky NI, Johnson EJ (2005) Carotenoid actions and their relation to health and disease. *Molecular Aspects of Medicine* **26**, 459–516. doi:10.1016/j.mam.2005.10.001

Krishna PM, Rao KNV, Sandhya S, Banji D (2012) A review on phytochemical, ethnomedical and pharmacological studies on genus *Sophora*, Fabaceae. *Revista Brasileira de Farmacognosia-Brazilian Journal of Pharmacognosy* **22**, 1145–1154. doi:10.1590/S0102-695X2012005000043

Kumagai Y, Shinkai Y, Miura T, Cho AK (2012) The chemical biology of naphthoquinones and its environmental implications. *Annual Review of Pharmacology and Toxicology* **52**, 221–247. doi:10.1146/annurev-pharmtox-010611-134517

Kumar S, Ritika (2020) A brief review of the biological potential of indole derivatives. *Future Journal of Pharmaceutical Sciences* **6**, 121. doi:10.1186/s43094-020-00141-y

Kumar D, Kumar S, Gupta J, Arya R, Gupta A (2011) A review on chemical and biological properties of *Cayratia trifolia* Linn. (Vitaceae). *Pharmacognosy Reviews* **5**, 184–188. doi:10.4103/0973-7847.91117

Kumar R, Anjum N, Tripathi YC (2015) Phytochemistry and pharmacology of *Santalum album* L.: a review. *World Journal of Pharmaceutical Research* **4**, 1842–1876.

Kundap UP, Bhuvanendran S, Kumari Y, Othman I, Shaikh MF (2017) Plant derived phytocompound, embelin in CNS disorders: a systematic review. *Frontiers in Pharmacology* **8**, 76. doi:10.3389/fphar.2017.00076

kundu P, Laskar S (2010) A brief resume on the genus *Ailanthus*: chemical and pharmacological aspects. *Phytochemistry Reviews* **9**, 379–412. doi:10.1007/s11101-009-9157-1

Kurkin VA (2003) Phenylpropanoids from medicinal plants: distribution, classification, structural analysis, and biological activity. *Chemistry of Natural Compounds* **39**, 123–153. doi:10.1023/A:1024876810579

Kutyrev AK (1991) Nucleophilic reactions of quinones. *Tetrahedron* **47**, 8043–8065. doi:10.1016/S0040-4020(01)91002-6

Lahey FN, Macleod JK (1967) The coumarins of *Geijera parviflora* Lindl. *Australian Journal of Chemistry* **20**, 1943–1955. doi:10.1071/CH9671943

Lahiri R, Ansari AA, Vankar YD (2013) Recent developments in design and synthesis of bicyclic azasugars, carbasugars and related molecules as glycosidase inhibitors. *Chemical Society Reviews* **42**, 5102–5118. doi:10.1039/c3cs35525j

Lai YW, Wang SW, Hu YY, Hwang TL, Cheng MJ *et al.* (2020) Anti-inflammatory alkaloids from the root bark of *Hernandia nymphaeifolia*. *Phytochemistry* **173**, 112326. doi:10.1016/j.phytochem.2020.112326

Lakshmi V, Pandey K, Mishra SK, Srivastava S, Mishra M *et al.* (2009) An overview of family Hernandiaceae. *Records of Natural Products* **3**, 1–22.

Lange BM (2015) Biosynthesis and biotechnology of high-value *p*-menthane monoterpenes, including menthol, carvone, and limonene. *Advances in Biochemical Engineering/Biotechnology* **148**, 319–353. doi:10.1007/10_2014_289

Lange BM (2016) Online resources for gene discovery and biochemical research with aromatic and medicinal plants. *Phytochemistry Reviews* **15**, 489–510. doi:10.1007/s11101-015-9450-0

Lassak EV, Mccarthy T (1983) *Australian Medicinal Plants.* Methuen, Sydney.

Lavaud C, Massiot G (2017) The *Iboga* alkaloids. In *Progress in the Chemistry of Organic Natural Products.* (Eds AD Kinghorn, H Falk, S Gibbons and J Kobayashi) pp. 89–136. Springer, Dordrecht.

Le Men L, Taylor WI (1965) A uniform numbering system for indole alkaloids. *Experientia* **21**, 508–510. doi:10.1007/BF02138961

Le NH, Malterud KE, Diallo D, Paulsen BS, Nergard CS *et al.* (2012) Bioactive polyphenols in *Ximenia americana* and the traditional use among Malian healers. *Journal of Ethnopharmacology* **139**, 858–862. doi:10.1016/j.jep.2011.12.031

Lee T-H, Juang S-H, Hsu F-L, Wu C-Y (2005) Triterpene acids from the leaves of *Planchonella duclitan* (Blanco) Bakhuizan. *Journal of the Chinese Chemical Society (Taipei)* **52**, 1275–1280. doi:10.1002/jccs.200500184

Lee S-S, Tseng C-C, Chen C-K (2010) Three new benzophenone glucosides from the leaves of *Planchonella obovata*. *Helvetica Chimica Acta* **93**, 522–529. doi:10.1002/hlca.200900264

Lee J, Dossett M, Finn CE (2012) *Rubus* fruit phenolic research: the good, the bad, and the confusing. *Food Chemistry* **130**, 785–796. doi:10.1016/j.foodchem.2011.08.022

Leete E (1990) Recent developments in the biosynthesis of the tropane alkaloids. *Planta Medica* **56**, 339–352. doi:10.1055/s-2006-960979

Leete E, Muir A, Neil Towers GH (1982) Biosynthesis of psilotin from [2',3'-13C2,1'-14C,4-3H]phenylalanine studied with 13C-NMR. *Tetrahedron Letters* **23**, 2635–2638. doi:10.1016/S0040-4039(00)87416-X

Leitereg TJ, Guadagni DG, Harris J, Mon TR, Teranishi R (1971) Chemical and sensory data supporting the difference between the odors of the enantiomeric carvones. *Journal of Agricultural and Food Chemistry* **19**, 785–787. doi:10.1021/jf60176a035

Leonelli F, Mostarda A, De Angelis L, Lamba D, Demitri N *et al.* (2016) Proof of the structure of the *Stemodia chilensis* tetracyclic diterpenoid (+)-19-acetoxystemodan-12-ol by synthesis from (+)-podocarpic acid: X-ray structure determination of a key intermediate. *Journal of Natural Products* **79**, 1155–1159. doi:10.1021/acs.jnatprod.5b00834

Leu YL, Hwang TL, Kuo PC, Liou KP, Huang BS *et al.* (2012) Constituents from *Vigna vexillata* and their anti-inflammatory activity. *International Journal of Molecular Sciences* **13**, 9754–9768. doi:10.3390/ijms13089754

Leung AY (2006) Traditional toxicity documentation of Chinese Materia Medica: an overview. *Toxicologic Pathology* **34**, 319–326. doi:10.1080/01926230600773958

Li Y, Jiang JG (2018) Health functions and structure-activity relationships of natural anthraquinones from plants. *Food & Function* **9**, 6063–6080. doi:10.1039/C8FO01569D

Li M, Zhang J-S, Ye Y-M, Fang J-N (2000) Constituents of the roots of *Salvia prionitis*. *Journal of Natural Products* **63**, 139–141. doi:10.1021/np990357k

Li X, Dong M, Liu Y, Shi Q-W, Kiyota H (2007) Structures and biological properties of the chemical constituents from the genus *Wedelia*. *Chemistry & Biodiversity* **4**, 823–836. doi:10.1002/cbdv.200790070

Li L, Peng Y, Yao X, Xu L-J, Wulan T-N *et al.* (2010) Chemical constituents and biological activities of plants from the genus *Ehretia* Linn. *Chinese Herbal Medicines* **2**, 106–111.

Li N, Xia Q, Ruan J, Fu PP, Lin G (2011) Hepatotoxicity and tumorigenicity induced by metabolic activation of pyrrolizidine alkaloids in herbs. *Current Drug Metabolism* **12**, 823–834. doi:10.2174/138920011797470119

Li SF, Jiao YY, Zhang ZQ, Chao JB, Jia J *et al.* (2018a) Diterpenes from buds of *Wikstroemia chamaedaphne* showing anti-hepatitis B virus activities. *Phytochemistry* **151**, 17–25. doi:10.1016/j.phytochem.2018.01.021

Li W, Tang GH, Chen L, Tang YQ, Xu YK *et al.* (2018b) New pyridocarbazole alkaloids from *Strychnos nitida*. *Natural Product Research* **32**, 1532–1536. doi:10.1080/14786419.2017.1385016

Li Petri G, Raimondi MV, Spanò V, Holl R, Barraja P *et al.* (2021) Pyrrolidine in drug discovery: a versatile scaffold for novel biologically active compounds. *Topics in Current Chemistry* **379**, 34. doi:10.1007/s41061-021-00347-5

Liao S-G, Wu Y, Yue J-M (2006) Lignans from *Wikstroemia hainanensis*. *Helvetica Chimica Acta* **89**, 73–80.

Liao G, Wu P, Liu Z, Xue J, Li H (2020) 2H-pyranone and isocoumarin derivatives from the endophytic fungus *Pestalotiopsis microspora* SC3082 derived from *Scaevola taccada* (Gaertn.) Roxb. *Natural Product Research* **35**, 3644–3651.

Lietava J (1992) Medicinal plants in a Middle Paleolithic grave Shanidar IV? *Journal of Ethnopharmacology* **35**, 263–266. doi:10.1016/0378-8741(92)90023-K

Lin L-C, Chou C-J, Kuo Y-C (2001) Cytotoxic principles from *Ventilago leiocarpa*. *Journal of Natural Products* **64**, 674–676. doi:10.1021/np000569d

Lin X, Ji S, Li R, Dong Y, Qiao X *et al.* (2012a) Terpecurcumins A-I from the rhizomes of *Curcuma longa*: absolute configuration and cytotoxic activity. *Journal of Natural Products* **75**, 2121–2131. doi:10.1021/np300551g

Lin X, Ji S, Li R, Dong Y, Qiao X *et al.* (2012b) Terpecurcumins A-I from the rhizomes of *Curcuma longa*: Absolute configuration and cytotoxic activity. *Journal of Natural Products* **75**, 2121–2131. doi:10.1021/np300551g

Lin L, Ni B, Lin H, Zhang M, Li X *et al.* (2015) Traditional usages, botany, phytochemistry, pharmacology and toxicology of *Polygonum multiflorum* Thunb.: a review. *Journal of Ethnopharmacology* **159**, 158–183. doi:10.1016/j.jep.2014.11.009

Lin SX, Curtis MA, Sperry J (2020) Pyridine alkaloids with activity in the central nervous system. *Bioorganic & Medicinal Chemistry* **28**, 115820. doi:10.1016/j.bmc.2020.115820

Lira-Ricardez J, Pereda-Miranda R, Castaneda-Gomez J, Fragoso-Serrano M, Simas RC *et al.* (2019) Resin glycosides from the roots of *Operculina macrocarpa* (Brazilian Jalap) with purgative activity. *Journal of Natural Products* **82**, 1664–1677. doi:10.1021/acs.jnatprod.9b00222

Liu J-H, Jin H-Z, Zhang W-D, Yan S-K, Shen Y-H (2009) Chemical constituents of plants from the genus *Brucea*. *Chemistry & Biodiversity* **6**, 57–70. doi:10.1002/cbdv.200700409

Liu J, Ng T, Rui Z, Ad O, Zhang W (2014) Unusual acetylation-dependent reaction cascade in the biosynthesis of the pyrroloindole drug physostigmine. *Angewandte Chemie International Edition* **53**, 136–139. doi:10.1002/anie.201308069

Lloyd O (2019) IP rights and Indigenous Knowledge: the case of gumby gumby. https://www.allens.com.au/insights-news/insights/2019/12/ip-rights-and-indigenous-knowledge-the-case-of-gumby-gumby/ [accessed 9 June 2022].

Lopez JL, Pusset J (1988) Plantes de Nouvelle-Caledonie, 115. Alcaloides monoterpeniques de *Coelospermum billardieri*. *Journal of Natural Products* **51**, 829–835. doi:10.1021/np50059a002

Lourens AC, Viljoen AM, Van Heerden FR (2008) South African *Helichrysum* species: a review of the traditional uses, biological activity and phytochemistry. *Journal of Ethnopharmacology* **119**, 630–652. doi:10.1016/j.jep.2008.06.011

Luan F, Han K, Li M, Zhang T, Liu D *et al.* (2019) Ethnomedicinal uses, phytochemistry, pharmacology, and toxicology of species from the genus *Ajuga* L.: a systematic review. *The American Journal of Chinese Medicine* **47**, 959–1003. doi:10.1142/S0192415X19500502

Lv H, She G (2010) Naturally occurring diarylheptanoids. *Natural Product Communications* **5**, 1687–1708. doi:10.1177/1934578X1000501035

Lytollis W, Scannell RT, An H, Murty VS, Sambi Reddy K *et al.* (1995) 5-Alkylresorcinols from *Hakea trifurcata* that cleave DNA. *Journal of the American Chemical Society* **117**, 12683–12690. doi:10.1021/ja00156a004

Ma X, Gang DR (2004) The Lycopodium alkaloids. *Natural Product Reports* **21**, 752–772. doi:10.1039/b409720n

Ma S, Liu J, Li W, Liu Y, Hui X *et al.* (2023) Machine learning in TCM with natural products and molecules: current status and future perspectives. *Chinese Medicine* **18**, 43. doi:10.1186/s13020-023-00741-9

Maas PA (2018) Indian medicine and Ayurveda. In *The Cambridge History of Science: Vol. 1, Ancient Science - Part IV, India*. (Eds A Jones and L Taub) pp. 532–550. Cambridge University Press, Cambridge, UK.

Maat L, Beyerman HC (1983) The imidazole alkaloids. In *The Alkaloids*. (Ed, A Brossi) pp. 282–334. Academic Press, New York, NY.

Mačiulaitis R, Kontrimavičiūtė V, Bressolle FMM, Briedis V (2008) Ibogaine, an anti-addictive drug: pharmacology and time to go further in development: a narrative review. *Human and Experimental Toxicology* **27**, 181–194. doi:10.1177/0960327107087802

Macleod JK, Rasmussen HB (1999) A hydroxy-B-caryophyllene from *Pterocaulon serrulatum*. *Phytochemistry* **50**, 105–108. doi:10.1016/S0031-9422(98)00460-9

Macleod JK, Rasmussen HB, Willis AC (1997) A new glycoside antimicrobial agent from *Persoonia linearis* x *pinifolia*. *Journal of Natural Products* **60**, 620–622. doi:10.1021/np970006a

Madikizela B, Mcgaw LJ (2017) *Pittosporum viridiflorum* Sims (Pittosporaceae): a review on a useful medicinal plant native to South Africa and tropical Africa. *Journal of Ethnopharmacology* **205**, 217–230. doi:10.1016/j.jep.2017.05.005

Mafu S, Zerbe P (2018) Plant diterpenoid metabolism for manufacturing the biopharmaceuticals of tomorrow: prospects and challenges. *Phytochemistry Reviews* **17**, 113–130. doi:10.1007/s11101-017-9513-5

Maganha EG, Halmenschlager RD, Rosa RM, Henriques Ja P, Ramos ALLD *et al.* (2010) Pharmacological evidences for the extracts and secondary metabolites from plants of the genus *Hibiscus*. *Food Chemistry* **118**, 1–10. doi:10.1016/j.foodchem.2009.04.005

Magiatis P, Skaltsounis A-L, Tillequin F, Seguin E, Cosson J-P (2002) Coelobillardin, an iridoid glucoside dimer from *Coelospermum billardieri*. *Phytochemistry* **60**, 415–418. doi:10.1016/S0031-9422(02)00118-8

Mahajan RT, Chopda MZ (2009) Phytopharmacology of *Ziziphus jujuba* Mill.: a plant review. *Pharmacognosy Reviews* **3**, 320–329.

Mahmoud SS, Croteau RB (2002) Strategies for transgenic manipulation of monoterpene biosynthesis in plants. *Trends in Plant Science* **7**, 366–373. doi:10.1016/S1360-1385(02)02303-8

Mai LP, Guenard D, Franck M, Van TM, Gaspard C *et al.* (2002) New cytotoxic cucurbitacins from the pericarps of *Trichosanthes tricuspidata* fruits. *Natural Product Letters* **16**, 15–19. doi:10.1080/1057563029001/4755

Maia MF, Moore SJ (2011) Plant-based insect repellents: a review of their efficacy, development and testing. *Malaria Journal* **10**, S11. doi:10.1186/1475-2875-10-S1-S11

Majinda RRT, Wanjala CCW, Juma BF (2005) Bioactive non-alkaloidal constituents from the genus *Erythrina*. *Studies in Natural Products Chemistry* **32**, 821–853. doi:10.1016/S1572-5995(05)80070-5

Mallik J, Das J, Banik RK (2018) Pharmacognostic profile and pharmacological activity of different parts of *Amorphophallus campanulatus* (Roxb.) Blume: a complete overview. *Asian Journal of Pharmaceutical Research & Development* **6**, 4–8. doi:10.22270/ajprd.v6i1.354

Manase MJ, Mitaine-Offer AC, Miyamoto T, Tanaka C, Delemasure S *et al.* (2013) New triterpenoid estersaponins from the root barks of *Pittosporum verticillatum* subsp. *verticillatum* and evaluation of cytotoxicities. *Fitoterapia* **91**, 231–235. doi:10.1016/j.fitote.2013.09.002

Manchand PS, White JD, Wright H, Clardy J (1973) Structures of stemodin and stemodinone. *Journal of the American Chemical Society* **95**, 2705–2706. doi:10.1021/ja00789a060

Mangai AS (2018) Pharmacological efficacy of *Trichodesma indicum* (Linn.) R. Br., in folk medicine: an updated review. *Asian Journal of Pharmaceutical and Clinical Research* **11**, 24–27. doi:10.22159/ajpcr.2018.v11i3.24412

Manju M, Varma RS, Parthasarathy MR (1977) New arbutin derivatives from leaves of *Grevillea robusta* and *Hakea saligna*. *Phytochemistry* **16**, 793–794. doi:10.1016/S0031-9422(00)89268-7

Mantegani S, Brambilla E, Varasi M (1999) Ergoline derivatives: receptor affinity and selectivity. *Il Farmaco* **54**, 288–296. doi:10.1016/S0014-827X(99)00028-2

Markham KR (1984) The structures of amentoflavone glycosides isolated from *Psilotum nudum*. *Phytochemistry* **23**, 2053–2056. doi:10.1016/S0031-9422(00)84969-9

Maroyi A (2016) *Ximenia caffra* Sond. (Ximeniaceae) in sub-Saharan Africa: a synthesis and review of its medicinal potential. *Journal of Ethnopharmacology* **184**, 81–100. doi:10.1016/j.jep.2016.02.052

Marsh KJ, Kulheim C, Blomberg SP, Thornhill AH, Miller JT *et al.* (2017) Genus-wide variation in foliar polyphenolics in eucalypts. *Phytochemistry* **144**, 197–207. doi:10.1016/j.phytochem.2017.09.014

Martinez J, Silva LPS (2013) Scale-up of extraction process. In *Natural Product Extraction: Principles and Applications*. (Eds J Prado and M Rostagno) pp. 363–398. Royal Society of Chemistry, London.

Massiot G, Delaude C (1986) Pyrrolidine alkaloids. In *The Alkaloids*. (Ed. A Brossi) pp. 270–322. Academic Press, New York, NY.

Mathew NS, Negi PS (2017) Traditional uses, phytochemistry and pharmacology of wild banana (*Musa acuminata* Colla): a review. *Journal of Ethnopharmacology* **196**, 124–140. doi:10.1016/j.jep.2016.12.009

Matsuda Y, Abe I (2016) Biosynthesis of fungal meroterpenoids. *Natural Product Reports* **33**, 26–53. doi:10.1039/C5NP00090D

Matsumoto M, Inoue K, Kaiji E (1999) Integrating traditional medicine in Japan: the case of Kampo medicines. *Complementary Therapies in Medicine* **7**, 254–255. doi:10.1016/S0965-2299(99)80012-0

Matsuzaki S, Hamana K, Okada M, Niitsu M, Samejima K (1990) Aliphatic pentaamines found in *Canavalia gladiata*. *Phytochemistry* **29**, 1311–1312. doi:10.1016/0031-9422(90)85449-P

Mazur W, Adlercreutz H (1998) Naturally occurring oestrogens in food. *Pure and Applied Chemistry* **70**, 1759–1776. doi:10.1351/pac199870091759

McCormick AC, Unsicker SB, Gershenzon J (2012) The specificity of herbivore-induced plant volatiles in attracting herbivore enemies. *Trends in Plant Science* **17**, 303–310. doi:10.1016/j.tplants.2012.03.012

Mckitterick R (2005) The Carolingian renaissance of culture and learning. In *Charlemagne: Empire and Society*. (Ed. J Story) pp. 151–167. Manchester University Press, Manchester.

Mclennan W (1994) 1994 National Aboriginal and Torres Strait Islander Survey [archived]. Australian Bureau of Statistics, Canberra, Australia. https://www.abs.gov.au/AUSSTATS/abs@.nsf/Lookup/4190.0Main+Features11994?OpenDocument (accessed September 2024).

McRae JM, Yang Q, Crawford RJ, Palombo EA (2008a) Acylated flavonoid tetraglycoside from *Planchonia careya* leaves. *Phytochemistry Letters* **1**, 99–102. doi:10.1016/j.phytol.2008.04.003

McRae JM, Yang Q, Crawford RJ, Palombo EA (2008b) Antibacterial compounds from *Planchonia careya* leaf extracts. *Journal of Ethnopharmacology* **116**, 554–560. doi:10.1016/j.jep.2008.01.007

Medeiros-Neves B, Teixeira HF, Von Poser GL (2018) The genus *Pterocaulon* (Asteraceae): a review on traditional medicinal uses, chemical constituents and biological properties. *Journal of Ethnopharmacology* **224**, 451–464. doi:10.1016/j.jep.2018.06.012

Mehta M, Adem A, Sabbagh M (2012) New acetylcholinesterase inhibitors for Alzheimer's disease. *International Journal of Alzheimer's Disease* **2012**, 728983. doi:10.1155/2012/728983

Meira M, Da Silva EP, David JM, David JP (2012) Review of the genus *Ipomoea*: traditional uses, chemistry and biological activities. *Revista Brasileira de Farmacognosia-Brazilian Journal of Pharmacognosy* **22**, 682–713. doi:10.1590/S0102-695X2012005000025

Meng D, Chen W, Zhao W (2007) Sulfur-containing spiroketal glycosides from *Breynia fructicosa*. *Journal of Natural Products* **70**, 824–829. doi:10.1021/np0606300

Menna M, Imperatore C, D'aniello F, Aiello A (2013) Meroterpenes from marine invertebrates: structures, occurrence, and ecological implications. *Marine Drugs* **11**, 1602–1643. doi:10.3390/md11051602

Menut C, Cabalion P, Hnawia E, Agnaniet H, Waikedre J *et al.* (2005) Two new furanosesquiterpenes from *Myoporum crassifolium* from New Caledonia. *Flavour and Fragrance Journal* **20**, 621–625. doi:10.1002/ffj.1509

Meshnick SR (2002) Artemisinin: mechanisms of action, resistance and toxicity. *International Journal for Parasitology* **32**, 1655–1660. doi:10.1016/S0020-7519(02)00194-7

Metcalf RL (1987) Plant volatiles as insect attractants. *Critical Reviews in Plant Sciences* **5**, 251–301. doi:10.1080/07352688709382242

Metra PL, Sutherland MD (1983) Further skeletal variety in the toxic furanosesquiterpene ketones in the *Myoporum* genus. *Tetrahedron Letters* **24**, 1749–1752. doi:10.1016/S0040-4039(00)81761-X

Miller CM, Mccarthy FO (2012) Isolation, biological activity and synthesis of the natural product ellipticine and related pyridocarbazoles. *RSC Advances* **2**, 8883–8918. doi:10.1039/c2ra20584j

Millan MJ, Newman-Tancredi A, Audinot V, Cussac D, Lejeune F *et al.* (2000) Agonist and antagonist actions of yohimbine as compared to fluparoxan at alpha(2)-adrenergic receptors (AR)s, serotonin (5-HT)(1A), 5-HT(1B), 5-HT(1D) and dopamine D(2) and D(3) receptors. Significance for the modulation of frontocortical monoaminergic transmission and depressive states. *Synapse (New York, N.Y.)* **35**, 79–95. doi:10.1002/(SICI)1098-2396(200002)35:2<79::AID-SYN1>3.0.CO;2-X

Millum J (2010) How should the benefits of bioprospecting be shared? *The Hastings Center Report* **40**, 24–33. doi:10.1353/hcr.0.0227

Mimica-Dukic N, Bozin B (2008) *Mentha* L. species (Lamiaceae) as promising sources of bioactive secondary metabolites. *Current Pharmaceutical Design* **14**, 3141–3150. doi:10.2174/138161208786404245

Mishra L-C, Singh BB, Dagenais S (2001) Ayurveda: a historical perspective and principles of the traditional healthcare system in India. *Alternative Therapies in Health and Medicine* **42**, 36–42.

Mishra MR, Behera RK, Jha S, Panda AK, Mishra A *et al.* (2011) A brief review on phytoconstituents and ethnopharmacology of *Scoparia dulcis* Linn. (Scrophulariaceae). *International Journal of Phytotherapy and Phytopharmacology* **3**, 422–438.

Mizokami SS, Arakawa NS, Ambrosio SR, Zarpelon AC, Casagrande R *et al.* (2012) Kaurenoic acid from *Sphagneticola trilobata* inhibits inflammatory pain: effect on cytokine production and activation of the NO-cyclic GMP-protein kinase G-ATP-sensitive potassium channel signaling pathway. *Journal of Natural Products* **75**, 896–904. doi:10.1021/np200989t

Mogana R, Wiart C (2011) *Canarium* L.: a phytochemical and pharmacological review. *Journal of Pharmacy Research* **4**, 2482–2489.

Mohammed AE, Abdul-Hameed ZH, Alotaibi MO, Bawakid NO, Sobahi TR *et al.* (2021) Chemical diversity and bioactivities of monoterpene indole alkaloids (MIAs) from six Apocynaceae genera. *Molecules (Basel, Switzerland)* **26**, 488. doi:10.3390/molecules26020488

Mohanty A, Das D, Ghosh G, Sahu PK (2016) *Barringtonia acutangula*: ethnobotanical and phytopharmacological review. *International Journal of Pharmaceutical Sciences Review and Research* **38**, 22–29.

Mohimani H, Gurevich A, Shlemov A, Mikheenko A, Korobeynikov A *et al.* (2018) Dereplication of microbial metabolites through database search of mass spectra. *Nature Communications* **9**, 4035. doi:10.1038/s41467-018-06082-8

Mohs RC, Greig NH (2017) Drug discovery and development: role of basic biological research. *Alzheimer's & Dementia* **3**, 651–657. doi:10.1016/j.trci.2017.10.005

Molimau-Samasoni S, Woolner VH, Foliga ST, Munkacsi AB (2021) Functional genomics and metabolomics advance the ethnobotany of the Samoan traditional medicine "matalafi". *Proceedings of the National Acadamy of Sciencesof the United States of America* **118**, e2100880118.

Money T (1979) Sesquiterpenoids. In *Terpenoids and Steroids*. (Ed. JR Hanson) pp. 81–159. J. W. Arrowsmith Ltd., Bristol, UK.

Morikawa H, Kasai R, Otsuka H, Hirata E, Shinzato T *et al.* (2004) Terpenic and phenolic glycosides from the leaves of *Breynia officinalis* Hemsl. *Chemical & Pharmaceutical Bulletin* **52**, 1086–1090. doi:10.1248/cpb.52.1086

Mroczek A (2015) Phytochemistry and bioactivity of triterpene saponins from Amaranthaceae family. *Phytochemistry Reviews* **14**, 577–605. doi:10.1007/s11101-015-9394-4

Muhammad D, Lalun N, Bobichon H, Le Magrex Debar E, Gangloff SC *et al.* (2016) Triterpenoids from the leaves of Alphitonia xerocarpus Baill and their biological activity. *Phytochemistry* **129**, 45–57. doi:10.1016/j.phytochem.2016.07.005

Mulholland DA, Taylor DAH (1992) Limonoids from Australian members of the Meliaceae. *Phytochemistry* **31**, 4163–4166. doi:10.1016/0031-9422(92)80434-G

Mulholland DA, Kotsos M, Mahomed HA, Taylor DAH (1998) Triterpenoids from *Owenia cepiodora*. *Phytochemistry* **49**, 2457–2460. doi:10.1016/S0031-9422(98)00307-0

Mullowney MW, Duncan KR, Elsayed SS, Garg N, Van Der Hooft JJ *et al.* (2023) Artificial intelligence for natural product drug discovery. *Nature Reviews. Drug Discovery* **22**, 895–916. doi:10.1038/s41573-023-00774-7

Multhauf R (1954) John of Rupescissa and the origin of medical chemistry. *Isis* **45**, 359–367. doi:10.1086/348357

Mulyaningsih S, Sporer F, Reichling J, Wink M (2011) Antibacterial activity of essential oils from *Eucalyptus* and of selected components against multidrug-resistant bacterial pathogens. *Pharmaceutical Biology* **49**, 893–899. doi:10.3109/13880209.2011.553625

Murray RDH, Mendez J, Brown SA (1982) *The Natural Coumarins: Occurrence, Chemistry and Biochemistry*. Wiley, New York, NY.

Mutharaj S, Muthusamy P, Ilango K (2019) Pharmacognostical, phytochemical and pharmacological review of *Casuarina equisetifolia* Linn. *World Journal of Pharmaceutical Research* **8**, 328–339.

Nagabhushana KS, Umamaheshwari S, Tocoli FE, Prabhu SK, Green IR *et al.* (2002) Inhibition of soybean and potato lipoxygenases by bhilawanols from bhilawan (*Semecarpus anacardium*) nut shell liquid and some synthetic salicylic acid analogues. *Journal of Enzyme Inhibition and Medicinal Chemistry* **17**, 255–259. doi:10.1080/1475636021000006243

Nagja T, Vimal K, Sanjeev A (2016) *Myristica fragrans*: a comprehensive review. *International Journal of Pharmacy and Pharmaceutical Sciences* **8**, 27–30.

Nahar L, Sarker SD (2007) Phytochemistry of the genus *Curcuma*. In *Turmeric: The Genus Curcuma*. (Eds PN Ravindran, K Nirmal Babu and K Sivaraman) pp. 71–106. CRC Press, Boca Raton, FL.

National Aboriginal Health Strategy Working Party (1989) *A national Aboriginal health strategy*. NAHSWP, Canberra, Australia.

Nebula M, Harisankar HS, Chandramohanakumar N (2013) Metabolites and bioactivities of Rhizophoraceae mangroves. *Natural Products and Bioprospecting* **3**, 207–232. doi:10.1007/s13659-013-0012-0

Negi JS, Singh P, Joshi GP, Rawat MS, Bisht VK (2010) Chemical constituents of *Asparagus*. *Pharmacognosy Reviews* **4**, 215–220. doi:10.4103/0973-7847.70921

Negi JS, Bisht VK, Bandari AK, Bharti MK, Sundriyal RC (2011) Chemical and pharmacological aspects of *Toona* (Meliaceae). *Research Journal of Phytochemistry* **5**, 14–21. doi:10.3923/rjphyto.2011.14.21

Nehlig A, Daval J-L, Debry G (1992) Caffeine and the central nervous system: mechanisms of action, biochemical, metabolic and psychostimulant effects. *Brain Research. Brain Research Reviews* **17**, 139–170. doi:10.1016/0165-0173(92)90012-B

Nelson DL (2005a) *Lehninger Principles of Biochemistry*, 5th edn. W. H. Freeman, New York, NY.

Nelson M (2005b) Germanic Europe and the great beer revival. In *The Barbarian's Beverage: A History of Beer in Ancient Europe*. pp. 78–114 Routledge, New York, NY.

Nerio LS, Olivero-Verbel J, Stashenko E (2010) Repellent activity of essential oils: a review. *Bioresource Technology* **101**, 372–378. doi:10.1016/j.biortech.2009.07.048

Newman DJ, Cragg GM (2020) Natural products as sources of new drugs over nearly four decades from 01/1981 to 09/2019. *Journal of Natural Products* **83**, 770–803. doi:10.1021/acs.jnatprod.9b01285

Neychev V, Mitev V (2016) Pro-sexual and androgen enhancing effects of *Tribulus terrestris* L.: fact or fiction. *Journal of Ethnopharmacology* **179**, 345–355. doi:10.1016/j.jep.2015.12.055

Nguyen HT, Lallemand M-C, Boutefnouchet S, Michel S, Tillequin F (2009) Antitumor *Psoropermum* xanthones and *Sarcomelicope* acridones: Privileged structures implied in DNA alkylation. *Journal of Natural Products* **72**, 527–539. doi:10.1021/np800644y

Nonato MG, Takayama H, Garson MJ (2008) *Pandanus* alkaloids: chemistry and biology. *The Alkaloids* **66**, 215–249.

Norman EO, Lever J, Brkljaca R, Urban S (2019) Distribution, biosynthesis, and biological activity of phenylphenalenone-type compounds derived from the family of plants, Haemodoraceae. *Natural Product Reports* **36**, 753–768. doi:10.1039/C8NP00067K

Norris J (2006) The mineral exhalation theory of metallogenesis in pre-modern mineral science. *Ambix* **53**, 43–65. doi:10.1179/174582306X93183

Nothias L-F, Petras D, Schmid R, Dührkop K, Rainer J *et al.* (2020) Feature-based molecular networking in the GNPS analysis environment. *Nature Methods* **17**, 905–908. doi:10.1038/s41592-020-0933-6

Noviany N, Samadi A, Yuliyan N, Hadi S, Aziz M *et al.* (2020) Structure characterization and biological activity of 2-arylbenzofurans from an Indonesian plant, *Sesbania grandiflora* (L.) Pers. *Phytochemistry Letters* **35**, 211–215. doi:10.1016/j.phytol.2019.12.008

Nugroho LH, Verpoorte R (2002) Secondary metabolism in tobacco. *Plant Cell, Tissue and Organ Culture* **68**, 105–125. doi:10.1023/A:1013853909494

Nutton V (2022) The rediscovery of ancient medicine. In *Renaissance Medicine: A Short History of European Medicine in the Sixteenth Century*. pp. 94–120. Routledge, New York, NY.

Nwachukwu ID, Slusarenko AJ, Gruhlke MCH (2012) Sulfur and sulfur compounds in plant defence. *Natural Product Communications* **7**, 395–400. doi:10.1177/1934578X1200700323

Nyongbela KD, Ndjoko KI, Brun R, Wittlin S, Mbah JA *et al.* (2009) Occurrence of sesquiterpene derivatives in *Scleria striatonux* De Wild (Cyperaceae). *Natural Product Communications* **4**, 5–8. doi:10.1177/1934578X0900400103

Nyongbela KD, Makolo FL, Hoye TR, Efange SMN (2016) Isolation and characterisation of sclerienone C from *Scleria striatinux*. *Natural Product Communications* **11**, 5–6. doi:10.1177/1934578X1601100102

O'Brien MS, Anthony JC (2005) Risk of becoming cocaine dependent: epidemiological estimates for the United States, 2000–2001. *Neuropsychopharmacology* **30**, 1006–1018. doi:10.1038/sj.npp.1300681

O'Connor SE, Maresh JJ (2006) Chemistry and biology of monoterpene indole alkaloid biosynthesis. *Natural Product Reports* **23**, 532–547. doi:10.1039/b512615k

O'Donnell K, Sharrock S (2018) Botanic gardens complement agricultural gene bank in collecting and preserving plant genetic diversity. *Biopreservation and Biobanking* **16**, 384–390. doi:10.1089/bio.2018.0028

Oelrichs PB, Macleod JK, Seawright AA, Ng JC (1997) Isolation and characterisation of urushiol components from the Australian native cashew (*Semecarpus australiensis*). *Nature Toxicology* **5**, 96–98.

Ogawa S, Yazaki Y (2018) Tannins from *Acacia mearnsii* de Wild. bark: tannin determination and biological activities. *Molecules (Basel, Switzerland)* **23**, 837. doi:10.3390/molecules23040837

Oliver SJ (2013) The role of traditional medicine practice in primary healthcare within Aboriginal Australia: a review of the literature. *Journal of Ethnobiology and Ethnomedicine* **9**, 46. doi:10.1186/1746-4269-9-46

Omar M, Defeo J, Youngken HWJ (1983) Chemical and toxicity studies of *Trichodesma africanum* L. *Journal of Natural Products* **46**, 153–156. doi:10.1021/np50026a001

Onyilagha JC, Islam S, Ntamatungiro S (2009) Comparative phytochemistry of eleven species of *Vigna* (Fabaceae). *Biochemical Systematics and Ecology* **37**, 16–19. doi:10.1016/j.bse.2008.11.013

Ovenden SPB, Cobbe M, Kissell R, Birrell GW, Chavchich M *et al.* (2011) Phenolic glycosides with antimalarial activity from *Grevillea* "Poorinda Queen". *Journal of Natural Products* **74**, 74–78. doi:10.1021/np100737q

Packer J, Gaikwad J, Harrington D, Ranganathan S, Jamie J *et al.* (2011) Medicinal plants of New South Wales, Australia. In *Genetic Resources, Chromosome Engineering, and Crop Improvement.* (Ed. RJ Singh) pp. 259–296. CRC Press, Boca Raton, FL.

Padeepika C, Selvakumar R, Krishnakumar T, Sajad UNN, Sajeev MS (2018) Pharmacology and phytochemistry of underexploited tuber crops: a review. *Journal of Pharmacognosy and Phytochemistry* **7**, 1007–1019.

Page JE, Nagel J (2006) Biosynthesis of terpenophenolic metabolites in hop and cannabis. In *Recent Advances in Phytochemistry.* (Ed. JT Romeo) pp. 179–210. Elsevier, Oxford.

Palatnick W, Meatherall R, Sitar D, Tenenbein M (2008) Toxicokinetics of acute strychnine poisoning. *Clinical Toxicology* **35**, 617–620.

Palma M, Barbero GF, Pinero Z, Liazid A, Barroso CG *et al.* (2013) Extraction of natural products: principles and fundamental aspects. In *Natural Product Extraction: Principles and Applications.* (Eds J Prado and M Rostagno) pp. 58–88. Royal Society of Chemistry, London.

Pan JY, Chen SL, Yang MH, Wu J, Sinkkonen J *et al.* (2009) An update on lignans: natural products and synthesis. *Natural Product Reports* **26**, 1251–1292. doi:10.1039/b910940d

Pan C, Chen YG, Ma XY, Jiang JH, He F *et al.* (2011) Phytochemical constituents and pharmacological activities of plants from the genus *Adiantum*: a review. *Tropical Journal of Pharmaceutical Research* **10**, 681–692. doi:10.4314/tjpr.v10i5.18

Parmar VS, Jain SC, Bisht KS, Jain R, Taneja P *et al.* (1997) Phytochemistry of the genus *Piper. Phytochemistry* **46**, 597–673. doi:10.1016/S0031-9422(97)00328-2

Parsons AF, Palframan MJ (2010) *Erythrina* and related alkaloids. *The Alkaloids. Chemistry and Biology* **68**, 39–81. doi:10.1016/S1099-4831(10)06802-1

Patchayasakul W, Pongchaidecha A, Chattipakorn N, Chattipakorn S (2008) Ethnobotany and ethnopharmacology of *Tabernaemontana divaricata. The Indian Journal of Medical Research* **127**, 317–335.

Pate DW (1994) Chemical ecology of *Cannabis. Journal of the International Hemp Association* **2**, 32–37.

Patel S (2016) Plant-derived cardiac glycosides: role in heart ailments and cancer management. *Biomedicine and Pharmacotherapy* **84**, 1036–1041. doi:10.1016/j.biopha.2016.10.030

Patel JR, Tripathi P, Sharma V, Chauhan NS, Dixit VK (2011) *Phyllanthus amarus*: ethnomedicinal uses, phytochemistry and pharmacology: a review. *Journal of Ethnopharmacology* **138**, 286–313. doi:10.1016/j.jep.2011.09.040

Patel K, Gadewar M, Tripathi R, Prasad SK, Patel DK (2012) A review on medical importance, pharmacological activity and biological aspects of beta-carboline alkaloid "Harmine". *Asian Pacific Journal of Tropical Biomedicine* **2**, 660–664. doi:10.1016/S2221-1691(12)60116-6

Patel DK, Kumar R, Sairam K, Hemalatha S (2013) Hybanthus enneaspermus (L.) F. Muell: a concise report on its phytopharmacological aspects. *Chinese Journal of Natural Medicines* **11**, 199–206. doi:10.1016/S1875-5364(13)60017-5

Pathak MA, Farrington DJ, Fitzpatrick TB (1962) The presently known distribution of furocoumarins (psoralens) in plants. *The Journal of Investigative Dermatology* **39**, 225–239. doi:10.1038/jid.1962.106

Pathak AK, Jain DC, Sharma RP (2008) Chemistry and biological activities of the genera *Tinospora. International Journal of Pharmacology* **33**, 277–287.

Pathan H, Williams J (2012) Basic opioid pharmacology: an update. *British Journal of Pain* **6**, 11–16. doi:10.1177/2049463712438493

Patil KS, Bhalsing SR (2016) Ethnomedicinal uses, phytochemistry and pharmacological properties of the genus *Boerhavia. Journal of Ethnopharmacology* **182**, 200–220. doi:10.1016/j.jep.2016.01.042

Pattamadilok D, Pengsuparp T, Phummiratch D, Ongpipattanakul B, Meksuriyen D *et al.* (2008) Canarosine: a new guanidine alkaloid from *Canavalia rosea* with inhibitory activity on dopamine D1 receptors. *Journal of Asian Natural Products Research* **10**, 915–918. doi:10.1080/10286020802181513

Paulraj J, Govindarajan R, Palpu P (2013) The genus spilanthes ethnopharmacology, phytochemistry, and pharmacological properties: a review. *Advances in Pharmacological and Pharmaceutical Sciences* **2013**, 510298.

Pawlus AD, Kinghorn AD (2007) Review of the ethnobotany, chemistry, biological activity and safety of the botanical dietary supplement *Morinda citrifolia* (noni)*. *The Journal of Pharmacy and Pharmacology* **59**, 1587–1609. doi:10.1211/jpp.59.12.0001

Peerzada AM, Ali HH, Naeem M, Latif M, Bukhari AH *et al.* (2015) *Cyperus rotundus* L.: traditional uses, phytochemistry, and pharmacological activities. *Journal of Ethnopharmacology* **174**, 540–560. doi:10.1016/j.jep.2015.08.012

Peng W, Qin R, Li X, Zhou H (2013) Botany, phytochemistry, pharmacology, and potential application of *Polygonum cuspidatum* Sieb.et Zucc.: a review. *Journal of Ethnopharmacology* **148**, 729–745. doi:10.1016/j.jep.2013.05.007

Peng D, Lin X, Jiang L, Huang J, Zeng G *et al.* (2019a) Five macrocyclic glycosides from *Schoenoplectus tabernaemontani. Natural Product Research* **33**, 427–434. doi:10.1080/14786419.2018.1455195

Peng W, Liu Y, Hu M, Zhang M, Yang J *et al.* (2019b) *Toona sinensis*: a comprehensive review on its traditional usages, phytochemisty, pharmacology and toxicology. *Revista Brasileira de Farmacognosia-Brazilian Journal of Pharmacognosy* **29**, 111–124. doi:10.1016/j.bjp.2018.07.009

Pereira A, Maraschin M (2015) Banana (*Musa* spp) from peel to pulp: ethnopharmacology, source of bioactive compounds and its relevance for human health. *Journal of Ethnopharmacology* **160**, 149–163. doi:10.1016/j.jep.2014.11.008

Pereira TN, Webb RI, Reilly PE, Seawright AA, Prakash AS (1998) Dehydromonocrotaline generates sequence-selective N-7 guanine alkylation and heat and alkali stable multiple fragment DNA crosslinks. *Nucleic Acids Research* **26**, 5441–5447. doi:10.1093/nar/26.23.5441

Peres V, Nagem TJ, De Oliveira FF (2000) Tetraoxygenated naturally occurring xanthones. *Phytochemistry* **55**, 683–710. doi:10.1016/S0031-9422(00)00303-4

Periyasamy K, Kaliyaperumal S (2016) Ethnobotanical, phytochemical and pharmaceutical studies of medicinal plant, *Ventilago maderaspatana* Gaertn. (red creeper): a review. *International Journal of Current Pharmaceutical Research* **8**, 16–18.

Pertwee RG (2008) The diverse CB1 and CB2 receptor pharmacology of three plant cannabinoids: Δ9-tetrahydrocannabinol, cannabidiol and Δ9-tetrahydrocannabivarin. *British Journal of Pharmacology* **153**, 199–215. doi:10.1038/sj.bjp.0707442

Pickens LB, Tang Y, Chooi Y-H (2011) Metabolic engineering for the production of natural products. *Annual Review of Chemical and Biomolecular Engineering* **2**, 211–236. doi:10.1146/annurev-chembioeng-061010-114209

Pinder AR (1993) Pyrrole, pyrrolidine, pyridine, piperidine, and related alkaloids. In *Alkaloids and Sulphur Compounds.* (Ed. PG Waterman) pp. 241–269. Academic Press, London, UK.

Pinto MM, Palmeira A, Fernandes C, Resende DI, Sousa E *et al.* (2021) From natural products to new synthetic small molecules: a journey through the world of xanthones. *Molecules (Basel, Switzerland)* **26**, 431. doi:10.3390/molecules26020431

Piovano M, Garbarino J, Tomassini L, Nicoletti M (2009) Cyclohexanones from *Mimulus glabratus* and *M. luteus. Natural Product Communications* **4**, 1637–1638. doi:10.1177/1934578X0900401204

Pirintsos S, Panagiotopoulos A, Bariotakis M, Daskalakis V, Lionis C *et al.* (2022) From traditional ethnopharmacology to modern natural drug discovery: A methodology discussion and specific examples. *Molecules (Basel, Switzerland)* **27**, 4060. doi:10.3390/molecules27134060

Pitanga BP, Nascimento RP, Silva VD, Costa SL (2012) The role of astrocytes in metabolism and neurotoxicity of the pyrrolizidine alkaloid monocrotaline, the main toxin of *Crotalaria retusa. Frontiers in Pharmacology* **3**, 144. doi:10.3389/fphar.2012.00144

Piwowarski JP, Granica S, Kiss AK (2015) *Lythrum salicaria* L. : underestimated medicinal plant from European traditional medicine: a review. *Journal of Ethnopharmacology* **170**, 226–250. doi:10.1016/j.jep.2015.05.017

Pole S (2006) *Ayurvedic Medicine: The Principles of Traditional Practice.* Churchill Livingstone, Edinburgh.

Poornima V, Sharanya M, Jeyam M (2014) An ethnomedical, pharmacological and phytochemical review of *Ammannia baccifera* L. *World Journal of Pharmaceutical Research* **3**, 1771–1789.

Porter R (1997) *The Greatest Benefit to Mankind: Medical History of Humanity.* Harper Collins, London.

Potterat O, Hamburger M (2007) *Morinda citrifolia* (noni) fruit: phytochemistry, pharmacology, safety. *Planta Medica* **73**, 191–199. doi:10.1055/s-2007-967115

Prabhavathi S, Vajayalakshmi S (2015) New hydroxycarboxylic acid from the root bark of *Ziziphus oenoplia. Journal of Pharmacognosy and Phytochemistry* **3**, 150–152.

Prakash BN, Unnikrishnan PM, Hariramamurthi G (2017) Medicinal flora and related traditional knowledge of Western Ghats: a potential source for community-based malaria management through endogenous approach. In *Ethnobotany of India. Vol. 2: Western Ghats and West Coast of Peninsular India.* (Eds T Pullaiah, KV Krishnamurthy and B Bahadur). Apple Academic Press, Palm Bay FL.

Pratap R, Ram VJ (2014) Natural and synthetic chromenes, fused chromenes, and versatility of dihydrobenzo[*h*] chromenes in organic synthesis. *Chemical Reviews* **114**, 10476–10526. doi:10.1021/cr500075s

Premalatha B (2000) *Semecarpus anacardium* Linn. Nuts: a boon in alternative medicine. *Indian Journal of Experimental Biology* **38**, 1177–1182.

Prihoda D, Maritz JM, Klempir O, Dzamba D, Woelk CH *et al.* (2021) The application potential of machine learning and genomics for understanding natural product diversity, chemistry, and therapeutic translatability. *Natural Product Reports* **38**, 1100–1108. doi:10.1039/D0NP00055H

Proudfoot A (2006) The early toxicology of physostigmine: a tale of beans, great men and proud egos. *Toxicological Reviews* **25**, 99–138. doi:10.2165/00139709-200625020-00004

Pullaiah T, Karuppusamy S (2017a) Ethnomedicinal plants of Eastern Ghats and adjacent Deccan region. In *Ethnobotany of India. Vol. 1: Eastern Ghats and Deccan.* (Eds T Pullaiah, KV Krishnamurthy and B Bahadur). Apple Academic Press, Palm Bay, FL.

Pullaiah T, Krishnamurthy KV, Bahadur B (2017b) Introduction. In *Ethnobotany of India. Vol. 4: Western and Central Himalayas.* (Eds T Pullaiah, KV Krishnamurthy and B Bahadur). Apple Academic Press, Palm Bay, FL.

Qian X, Chen L, Sui Y, Chen C, Zhang W *et al.* (2020) Biotechnological potential and applications of microbial consortia. *Biotechnology Advances* **40**, 107500. doi:10.1016/j.biotechadv.2019.107500

Qiu H-Y, Wang P-F, Ling H-Y, Tang C-Y, Zhu H-L *et al.* (2018) Naphthoquinones: a continuing source for discovery of therapeutic antineoplastic agents. *Chemical Biology & Drug Design* **91**, 681–690. doi:10.1111/cbdd.13141

Quinn RA, Nothias L-F, Vining O, Meehan M, Esquenazi E *et al.* (2017) Molecular networking as a drug discovery, drug metabolism, and precision medicine strategy. *Trends in Pharmacological Sciences* **38**, 143–154. doi:10.1016/j.tips.2016.10.011

Rabelo AS, Oliveira ID, Guimaraes AG, Quintans JS, Prata AP *et al.* (2013) Antinociceptive, anti-inflammatory and antioxidant activities of aqueous extract from *Remirea maritima* (Cyperaceae). *Journal of Ethnopharmacology* **145**, 11–17. doi:10.1016/j.jep.2012.10.020

Radwanski ER, Last RL (1995) Tryptophan biosynthesis and metabolism: BIOCHEMICAL and molecular genetics. *The Plant Cell* **7**, 921–934.

Rahaman CH, Mandal SK, Pullaiah T (2017) Ethnomedicinal plants of the Indo-Gangetic region and Central India. In *Ethnobotany of India. Vol. 5: The Indo-Gangetic Region and Central India.* (Eds T Pullaiah, KV Krishnamurthy and B Bahadur). Apple Academic Press, Palm Bay, FL.

Rahman SZ (2001) Unani medicine in India: its origin and fundamental concepts. In *History of Science, Philosophy and Culture in Indian Civilisation, Vol. IV, Part 2 (Medicine and Life Sciences in India).* pp. 298–325. Centre for Studies in Civilisations (Project of History of Indian Science, Philosophy and Culture), New Delhi, India.

Raj D, Luczkiewicz M (2008) Securinega suffruticosa. *Fitoterapia* **79**, 419–427. doi:10.1016/j.fitote.2008.02.011

Rajeswari K, Bhaskara Rao T (2015) *Aegicera corniculatum* Linn (Myrsinaceae). *Journal of Chemical and Pharmaceutical Research* **7**, 305–316.

Rajsekhar PB, Arvind Bharani RS, Angel J, Ramachandran M, Rajsekhar SPV (2016) *Hybanthus enneaspermus* (L) F. Muell: A phytopharmacological review on herbal medicine. *Journal of Chemical and Pharmaceutical Research* **8**, 351–355.

Raju R, Gunawardena D, Ahktar MA, Low M, Reddell P *et al.* (2016) Anti-inflammatory chemical profiling of the Australian rainforest tree *Alphitonia petriei* (Rhamnaceae). *Molecules (Basel, Switzerland)* **21**, 1521. doi:10.3390/molecules21111521

Rakba N, Melhaoui A, Loyer P, Delcros JG, Morel I *et al.* (1999) Bgugaine, a pyrrolidine alkaloid from *Arisarum vulgare*, is a strong hepatotoxin in rat and human liver cell cultures. *Toxicology Letters* **104**, 239–248. doi:10.1016/S0378-4274(98)00375-0

Ralston L, Subramanian S, Matsuno M, Yu O (2005) Partial reconstruction of flavonoid and isoflavonoid biosynthesis in yeast using soybean type I and II chalcone isomerases. *Plant Physiology* **137**, 1375–1388. doi:10.1104/pp.104.054502

Rammohan A, Reddy GM, Bhaskar BV, Gunasekar D, Zyryanov GV (2020) Phytochemistry and pharmacological activities of the genus *Rhynchosia*: a comprehensive review. *Planta* **251**, 9. doi:10.1007/s00425-019-03311-2

Rani A, Sharma A (2013) The genus *Vitex*: a review. *Pharmacognosy Reviews* **7**, 188–198. doi:10.4103/0973-7847.120522

Rao AS (2017) Isolation, absolute configuration and bioactivities of megastigmanes or C13 isonorterpenoids. *Angewandte Chemie International Edition* **3**, 69–91.

Rao BK, Hanumaiah T, Rao CP, Rao GSR, Rao KVJ *et al.* (1983) Anthraquinones in *Ventilago* species. *Phytochemistry* **22**, 2583–2585. doi:10.1016/0031-9422(83)80170-8

Rashid MA, Armstrong JA, Gray AI, Waterman PG (1991) Pyranocoumarins as chemotaxonomic markers in *Eriostemon coccineus* and *Philotheca citrina*. *Phytochemistry* **30**, 4033–4035. doi:10.1016/0031-9422(91)83459-X

Rasmann S, Kollner TG, Degenhardt J, Hiltpold I, Toepfer S *et al.* (2005) Recruitment of entomopathogenic nematodes by insect-damaged maize roots. *Nature* **434**, 732–737. doi:10.1038/nature03451

Rasool N, Ahmad VU, Malik A (1991) Terpenoids from *Pentatropis spiralis*. *Phytochemistry* **30**, 1331–1332. doi:10.1016/S0031-9422(00)95230-0

Ratsch A, Steadman KJ, Bogossian F (2010) The pituri story: a review of the historical literature surrounding traditional Australian Aboriginal use of nicotine in Central Australia. *Journal of Ethnobiology and Ethnomedicine* **6**, 26. doi:10.1186/1746-4269-6-26

Ravichandiran P, Sheet S, Premnath D, Kim AR, Yoo DJ (2019) 1,4-Naphthoquinone analogues: potent antibacterial agents and mode of action evaluation. *Molecules (Basel, Switzerland)* **24**, 1437. doi:10.3390/molecules24071437

Reddy MVB (2017) Review on chemistry, biological activity and biosynthesis of naturally occurring compounds of cadalene type sesquiterpenoids from Malvaceae. *Der Pharma Chemica* **9**, 40–49.

Reddy AS, Zhang S (2013) Polypharmacology: drug discovery for the future. *Expert Review of Clinical Pharmacology* **6**, 41–47. doi:10.1586/ecp.12.74

Refaat J, Kamel MS, Ramadan MA, Ali AA (2012a) *Crinum*: an endless source of bioactive principles: a review. Part II: *Crinum* alkaloids: crinine-type alkaloids. *International Journal of Pharmaceutical Sciences and Research* **3**, 3091–3100.

Refaat J, Kamel MS, Ramadan MA, Ali AA (2012b) *Crinum*: an endless source of bioactive principles: a review. Part I: *Crinum* alkaloids: lycorine-type alkaloids. *International Journal of Pharmaceutical Sciences and Research* **3**, 1883–1890.

Refaat J, Kamel MS, Ramadan MA, Ali AA (2012c) *Crinum*; an endless source of bioactive principles: a review. Part III: *Crinum* alkaloids: belladine-, galanthamine-, lycorenine-, tazettine-type alkaloids and other minor types. *International Journal of Pharmaceutical Sciences and Research* **3**, 3630–3638.

Refaat J, Kamel MS, Ramadan MA, Ali AA (2013) *Crinum*: an endless source of bioactive principles: a review. Part IV: non alkaloidal constituents. *International Journal of Pharmaceutical Sciences and Research* **4**, 941–948.

Reuben S, Cseke LJ, Bhinu VS, Narasimhan K, Jeyakumar M *et al.* (2006) Molecular biology of plant natural products. In *Natural Products from Plants* (Eds LJ Cseke A Kirakosyan, PB Kaufman, SL Warber, JA Duke *et al.*) pp. 166–202 CRC Press, Boca Raton, FL.

Reyes-García V (2010) The relevance of traditional knowledge systems for ethnopharmacological research: theoretical and methodological contributions. *Journal of Ethnobiology and Ethnomedicine* **6**, 32. doi:10.1186/1746-4269-6-32

Ridley DD, Ritchie E, Taylor WC (1970) Chemical studies of the Proteaceae. IV: the structures of the major phenols of *Grevillea striata;* a group of novel cyclophanes. *Australian Journal of Chemistry* **23**, 147–183. doi:10.1071/CH9700147

Rivière C, Nguyen Thi Hong V, Tran Hong Q, Chataigné G, Nguyen Hoai N *et al.* (2010) *Mallotus* species from Vietnamese mountainous areas: phytochemistry and pharmacological activities. *Phytochemistry Reviews* **9**, 217–253. doi:10.1007/s11101-009-9152-6

Roberts MF (1998) Enzymology of alkaloid biosynthesis. In *Alkaloids: Biochemistry, Ecology, and Medical Applications.* (Eds MF Roberts and M Wink) pp. 109–146. Springer Science+Business, New York, NY.

Roberts MF, Wink M (1998) Introduction. In *Alkaloids: Biochemistry, Ecology, and Medical Applications.* (Eds MF Roberts and M Wink) pp. 1–10. Springer Science+Business, New York, NY.

Roberts MF, Strack D, Wink M (2010) Biosynthesis of alkaloids and betalains. In *Biochemistry of Plant Secondary Metabolism.* (Ed. M Wink) pp. 20–91. Wiley-Blackwell, Sussex, UK.

Robinson T (1974) Metabolism and function of alkaloids in plants. *Science* **184**, 430–435. doi:10.1126/science.184.4135.430

Rocabado GO, Bedoya LM, Abad MJ, Bermejo P (2008) *Rubus*: a review of its phytochemical and pharmacological profile. *Natural Product Communications* **3**, 423–436. doi:10.1177/1934578X0800300319

Rodrigues FE, Lima JQ, de Oliveira MDCFD, Vasconcelos JN, Santiago GM *et al.* (2010) Diterpene and other constituents from *Stemodia maritima* (Scrophulariaceae). *Journal of the Brazilian Chemical Society* **21**, 1581–1586. doi:10.1590/S0103-50532010000800025

Rodriguez-Garcia C, Sanchez-Quesada C, Toledo E, Delgado-Rodriguez M, Gaforio JJ (2019) Naturally lignan-rich foods: a dietary tool for health promotion? *Molecules (Basel, Switzerland)* **24**, 917. doi:10.3390/molecules24050917

Romano JD, Tatonetti NP (2019) Informatics and computational methods in natural product drug discovery: a review and perspectives. *Frontiers in Genetics* **10**, 368. doi:10.3389/fgene.2019.00368

Rooney A (2012) *The History of Medicine.* The Rosen Publishing Group, New York, NY.

Rosales PF, Bordin GS, Gower AE, Moura S (2020) Indole alkaloids: 2012 until now, highlighting the new chemical structures and biological activities. *Fitoterapia* **143**, 104558. doi:10.1016/j.fitote.2020.104558

Roy A, Saraf S (2006) Limonoids: overview of significant bioactive triterpenes distributed in plants kingdom. *Biological & Pharmaceutical Bulletin* **29**, 191–201. doi:10.1248/bpb.29.191

Ruddick JP, Evans AK, Nutt DJ, Lightman SL, Rook GA *et al.* (2006) Tryptophan metabolism in the central nervous system: medical implications. *Expert Reviews in Molecular Medicine* **8**, 1–27. doi:10.1017/S1462399406000068

Russell M, Southwell IA, Waterman PG (2001) α-Phellandren-8-ol, a rare monoterpenene alcohol from the oil of *Prostanthera staurophylla* F. Muell. *The Journal of Essential Oil Research* **13**, 446–447. doi:10.1080/10412905.2001.9699723

Russo A, Borrelli F (2005) Bacopa monniera, a reputed nootropic plant: an overview. *Phytomedicine* **12**, 305–317. doi:10.1016/j.phymed.2003.12.008

Rutz A, Sorokina M, Galgonek J, Mietchen D, Willighagen E *et al.* (2022) The LOTUS initiative for open knowledge management in natural products research. *eLife* **11**, e70780. doi:10.7554/eLife.70780

Sadgrove NJ, Jones GL (2014) Medicinal compounds, chemically and biologically characterised from extracts of Australian *Callitris endlicheri* and *C. glaucophylla* (Cupressaceae): used traditionally in Aboriginal and colonial pharmacopoeia. *Journal of Ethnopharmacology* **153**, 872–883. doi:10.1016/j.jep.2014.03.054

Sadowska-Bartosz I, Bartosz G (2021a) Biological properties and applications of betalains. *Molecules (Basel, Switzerland)* **26**, 2520. doi:10.3390/molecules26092520

Sadowska-Bartosz I, Bartosz G (2021b) Biological properties and applications of betalains. *Molecules (Basel, Switzerland)* **26**, 2520. doi:10.3390/molecules26092520

Sagara Y, Fernandez-Belda F, De Meis L, Inesi G (1992) Characterization of the inhibition of intracellular Ca2+ transport ATPases by thapsigargin. *The Journal of Biological Chemistry* **267**, 12606–12613. doi:10.1016/S0021-9258(18)42320-4

Salatino A, Faria Salatino ML, Negri G (2007) Traditional uses, chemistry and pharmacology of *Croton* species (Euphorbiaceae). *Journal of the Brazilian Chemical Society* **18**, 11–33. doi:10.1590/S0103-50532007000100002

Saleem M, Kim HJ, Ali MS, Lee YS (2005) An update on bioactive plant lignans. *Natural Product Reports* **22**, 696–716. doi:10.1039/b514045p

Salehi B, Krochmal-Marczak B, Skiba D, Patra JK, Das SK *et al.* (2020) *Convolvulus* plant: a comprehensive review from phytochemical composition to pharmacy. *Phytotherapy Research* **34**, 315–328. doi:10.1002/ptr.6540

Salem MA, Perez De Souza L, Serag A, Fernie AR, Farag MA *et al.* (2020) Metabolomics in the context of plant natural products research: from sample preparation to metabolite analysis. *Metabolites* **10**, 37. doi:10.3390/metabo10010037

Salim AA, Garson MJ, Craik DJ (2004) New indole alkaloids from the roots of *Ochrosia acuminata*. *Journal of Natural Products* **67**, 1719–1721. doi:10.1021/np049812p

Salminen A, Lehtonen M, Suuronen T, Kaarniranta K, Huuskonen J (2008) Terpenoids: natural inhibitors of NF-κB signalling with anti-inflammatory and anticancer potential. *Cellular and Molecular Life Sciences* **65**, 2979–2999. doi:10.1007/s00018-008-8103-5

Samec D, Pierz V, Srividya N, Wust M, Lange BM (2019) Assessing chemical diversity in *Psilotum nudum* (L.) Beauv., a pantropical whisk fern that has lost many of its fern-like characters. *Frontiers in Plant Science* **10**, 868. doi:10.3389/fpls.2019.00868

Sanger GJ, Andrews PLR (2018) A history of drug discovery for treatment of nausea and vomiting and the implications for future research. *Frontiers in Pharmacology* **9**, 913. doi:10.3389/fphar.2018.00913

Santos HO, Howell S, Teixeira FJ (2019) Beyond tribulus (*Tribulus terrestris* L.): the effects of phytotherapics on testosterone, sperm and prostate parameters. *Journal of Ethnopharmacology* **235**, 392–405. doi:10.1016/j.jep.2019.02.033

Sarker SD, Nahar L (2012) An introduction to natural products isolation. *Methods in Molecular Biology (Clifton, N.J.)* **864**, 1–25. doi:10.1007/978-1-61779-624-1_1

Sati OP, Bahuguna S, Uniyal S, Sakakibara J, Kaiya T *et al.* (1990) A new saponin from *Deeringia amaranthoides*. *Journal of Natural Products* **53**, 466–469. doi:10.1021/np50068a029

Sathiyanarayanan L, Arulmozhi S (2007) *Mucuna pruriens* Linn.: a comprehensive review. *Pharmacognosy Reviews* **1**, 157–162.

Scarborough J, Nutton V (1982) The Preface of Dioscorides' Materia Medica: introduction, translation, and commentary. *Transactions & Studies of the College of Physicians of Philadelphia* **4**, 187–227.

Schiff PB, Fant J, Horwitz SB (1979) Promotion of microtubule assembly *in vitro* by taxol. *Nature* **277**, 665–667. doi:10.1038/277665a0

Schimming T, Tofern B, Mann P, Richter A, Jenett-Siems K *et al.* (1998) Distribution and taxonomic significance of calystegines in the Convolvulaceae. *Phytochemistry* **49**, 1989–1995. doi:10.1016/S0031-9422(98)00430-0

Schimming T, Jenett-Siems K, Mann P, Tofern-Reblin B, Milson J *et al.* (2005) Calystegines as chemotaxonomic markers in the Convolvulaceae. *Phytochemistry* **66**, 469–480. doi:10.1016/j.phytochem.2004.12.024

Schipmann U, Leaman DJ, Cunningham AB, Walter S (2005) Impact of cultivation and collection on the conservation of medicinal plants: global trends and issues. *Acta Horticulturae* (676), 31–44. doi:10.17660/ActaHortic.2005.676.3

Schmeller T, Wink M (1998) Utilisation of alkaloids in modern medicine. In *Alkaloids: Biochemistry, Ecology, and Medical Applications.* (Eds MF Roberts and M Wink) pp. 435–460. Springer Science+Business, New York, NY.

Schneider D, Wink M, Sporer F, Lounibos P (2002) Cycads: their evolution, toxins, herbivores and insect pollinators. *Naturwissenschaften* **89**, 281–294. doi:10.1007/s00114-002-0330-2

Schramm S, Kohler N, Rozhon W (2019) Pyrrolizidine alkaloids: biosynthesis, biological activities and occurrence in crop plants. *Molecules (Basel, Switzerland)* **24**, 498. doi:10.3390/molecules24030498

Seebaluck R, Gurib-Fakim A, Mahomoodally F (2015) Medicinal plants from the genus Acalypha (Euphorbiaceae): a review of their ethnopharmacology and phytochemistry. *Journal of Ethnopharmacology* **159**, 137–157. doi:10.1016/j.jep.2014.10.040

Seigler DS (1998a) Alkaloids of terpenoid origin (excepting indole alkaloids and ergot alkaloids). In *Plant Secondary Metabolism.* pp. 668–691. Springer, New York, NY.

Seigler DS (1998b) Benzoquinones, naphthoquinones, and anthraquinones. In *Plant Secondary Metabolism.* pp. 76–93. Springer, New York, NY.

Seigler DS (1998c) Ergot and other indole alkaloids. In *Plant Secondary Metabolism.* (Ed. DS Seigler) pp. 665–668. Springer, New York, NY.

Seigler DS (1998d) *Plant Secondary Metabolism.* Springer, New York, NY.

Seigler DS (1998e) Polyketides. In *Plant Secondary Metabolism.* pp. 56–75. Springer, New York, NY.

Seigler DS (2003) Phytochemistry of Acacia—sensu lato. *Biochemical Systematics and Ecology* **31**, 845–873. doi:10.1016/S0305-1978(03)00082-6

Semalty M, Semalty A, Badola A, Joshi GP, Rawat MS (2010) *Semecarpus anacardium* Linn.: a review. *Pharmacognosy Reviews* **4**, 88–94. doi:10.4103/0973-7847.65328

Semwal DK, Semwal RB, Vermaak I, Viljoen A (2014) From arrow poison to herbal medicine: the ethnobotanical, phytochemical and pharmacological significance of *Cissampelos* (Menispermaceae). *Journal of Ethnopharmacology* **155**, 1011–1028. doi:10.1016/j.jep.2014.06.054

Sen S, Chakraborty R (2017) Revival, modernization and integration of Indian traditional herbal medicine in clinical practice: importance, challenges and future. *Journal of Traditional and Complementary Medicine* **7**, 234–244. doi:10.1016/j.jtcme.2016.05.006

Sena Filho JG, Duringer J, Maia GLA, Tavares JF, Xavier HS *et al.* (2008) Ecdysteroids from *Vitex* species: distribution and compilation of their 13C-NMR spectral data. *Chemistry & Biodiversity* **5**, 707–713. doi:10.1002/cbdv.200890067

Senadeera SPD, Duffy S, Avery VM, Carroll AR (2017) Antiplasmodial beta-triketones from the flowers of the Australian tree *Angophora woodsiana*. *Bioorganic & Medicinal Chemistry Letters* **27**, 2602–2607. doi:10.1016/j.bmcl.2017.03.065

Sepulveda-Boza S, Delhvi S, Cassels BK (1993) Flavonoids from the twigs of *Eucryphia glutinosa*. *Phytochemistry* **32**, 1301–1303. doi:10.1016/S0031-9422(00)95110-0

Shahat AA, Apers S, Van Miert S, Claeys M, Pieters L *et al.* (2001) Structure elucidation of three new acetylated flavonoid glycosides from *Centaurium spicatum*. *Magnetic Resonance in Chemistry* **39**, 625–629. doi:10.1002/mrc.900

Sharifi-Rad J, Salehi B, Varoni EM, Sharopov F, Yousaf Z *et al.* (2017) Plants of the *Melaleuca* genus as antimicrobial agents: from farm to pharmacy. *Phytotherapy Research* **31**, 1475–1494. doi:10.1002/ptr.5880

Sharifi-Rad J, Melgar-Lalanne G, Hernandez-Alvarez AJ, Taheri Y, Shaheen S *et al.* (2020) *Malva* species: insights on its chemical composition towards pharmacological applications. *Phytotherapy Research* **34**, 546–567. doi:10.1002/ptr.6550

Sharma BR, Gautam LN, Adhikari D, Karki R (2017) A comprehensive review on chemical profiling of *Nelumbo nucifera*: potential for drug development. *Phytotherapy Research* **31**, 3–26. doi:10.1002/ptr.5732

Sharpe WD (1964) Isidore of Seville: the medical writings. An English translation with an introduction and commentary. *Transactions of the American Philosophical Society* **54**, 1–75. doi:10.2307/1005938

Shaw PC, Lee KM, Wong KB (2005) Recent advances in trichosanthin, a ribosome-inactivating protein with multiple pharmacological properties. *Toxicon* **45**, 683–689. doi:10.1016/j.toxicon.2004.12.024

Sheehan DJ, Hitchcock CA, Sibley CM (1999) Current and emerging azole antifungal agents. *Clinical Microbiology Reviews* **12**, 40–79. doi:10.1128/CMR.12.1.40

Shen L-R, Jin S-M, Yin B-W, Du X-F, Wang Y-L *et al.* (2007) Chemical constituents of plants from the genus *Cerbera*. *Chemistry & Biodiversity* **4**, 1438–1449. doi:10.1002/cbdv.200790122

Shi Q-W, Su X-H, Kiyota H (2008) Chemical and pharmacological research of the plants in genus *Euphorbia*. *Chemical Reviews* **108**, 4295–4327. doi:10.1021/cr078350s

Shimada S, Otsuki H, Sukuta M (2007) Transcriptional control of anthocyanin biosynthetic genes in the Caryophyllales. *Journal of Experimental Botany* **58**, 957–967. doi:10.1093/jxb/erl256

Shou Q, Banbury LK, Renshaw DE, Smith JE, He X *et al.* (2013) Parvifloranines A and B, two 11-carbon alkaloids from *Geijera parviflora*. *Journal of Natural Products* **76**, 1384–1387. doi:10.1021/np400376r

Shu J, Liang F, Zhu G, Liu X, Yu J *et al.* (2017) Lignan glycosides from the rhizomes of *Smilax trinervula* and their biological activities. *Phytochemistry Letters* **20**, 1–8. doi:10.1016/j.phytol.2017.03.002

Shukla A, Kaur A (2018) A systematic review of traditional uses bioactive phytoconstituents of genus *Ehretia*. *Asian Journal of Pharmaceutical and Clinical Research* **11**, 88–100. doi:10.22159/ajpcr.2018.v11i6.25178

Siddiqua A, Mittapally S (2017) A review of *Cissus quadrangularis*. *Journal of Pharmaceutical Innovation* **6**, 329–334.

Simoneit BRT, Cox RE, Oros DR, Otto A (2018) Terpenoid compositions of resins from *Callitris* species (Cupressaceae). *Molecules (Basel, Switzerland)* **23**, 3384. doi:10.3390/molecules23123384

Simonsen HT, Adsersen A, Bremner P, Heinrich M, Wagner Smitt U *et al.* (2004) Antifungal constituents of *Melicope borbonica*. *Phytotherapy Research* **18**, 542–545. doi:10.1002/ptr.1482

Simpson K, Spencer CM, Mcclellan KJ (2000) Tropisetron; an update of its use in the prevention of chemotherapy-induced nausea and vomiting. *Drugs* **59**, 1297–1315. doi:10.2165/00003495-200059060-00008

Singab AN, Youssef FS, Ashour ML, Wink M (2013) The genus *Eremophila* (Scrophulariaceae): an ethnobotanical, biological and phytochemical review. *The Journal of Pharmacy and Pharmacology* **65**, 1239–1279. doi:10.1111/jphp.12092

Singh D, Chaudhuri PK (2017) Chemistry and pharmacology of *Tinospora cordifolia*. *Natural Product Communications* **12**, 299–308. doi:10.1177/1934578X1701200240

Singh D, Chaudhuri PK (2018) A review on phytochemical and pharmacological properties of Holy basil (*Ocimum sanctum* L.). *Industrial Crops and Products* **118**, 367–382. doi:10.1016/j.indcrop.2018.03.048

Singh SP, Singh AM (2015) Bioprospecting: enhancing the value of biodiversity and intellectual property. In *Plant Genetic Resources and Traditional Knowledge for Food Security* (Eds RK Salgotra and BB Gupta) pp. 231–236. Springer Singapore

Singh D, Singh B, Goel RK (2011a) Traditional uses, phytochemistry and pharmacology of *Ficus religiosa*: a review. *Journal of Ethnopharmacology* **134**, 565–583. doi:10.1016/j.jep.2011.01.046

Singh O, Ali M, Akhtar N (2011b) Phenolic acid glucosides from the seeds of *Entada phaseoloides* Merill. *Journal of Asian Natural Products Research* **13**, 682–687. doi:10.1080/10286020.2011.584444

Singh H, Mishra A, Mishra AK (2018) The chemistry and pharmacology of *Cleome* genus: a review. *Biomedicine and Pharmacotherapy* **101**, 37–48. doi:10.1016/j.biopha.2018.02.053

Siraisi NG (2012) Medicine, 1450–1620, and the history of science. *Isis* **103**, 491–514. doi:10.1086/667970

Sivin N (1988) Science and medicine in imperial China: the state of the field. *The Journal of Asian Studies* **47**, 41–90. doi:10.2307/2056359

Smith TA (1977a) Phenethylamine and related compounds in plants. *Phytochemistry* **16**, 9–18. doi:10.1016/0031-9422(77)83004-5

Smith TA (1977b) Tryptamine and related compounds in plants. *Phytochemistry* **16**, 171–175. doi:10.1016/S0031-9422(00)86778-3

Smith BA (1990) Strychnine poisoning. *The Journal of Emergency Medicine* **8**, 321–325. doi:10.1016/0736-4679(90)90013-L

Smith LW, Culvenor CCJ (1981) Plant sources of hepatotoxic pyrrolizidine alkaloids. *Journal of Natural Products* **44**, 129–152. doi:10.1021/np50014a001

Sneader W (2005) Plant products analogues and compounds derived from them. In *Drug Discovery: A History*. (Ed. W Sneader) pp. 115–150. Wiley & Sons, Chichester, UK.

Somei M, Yamada F (2005) Simple indole alkaloids and those with a non-rearranged monoterpenoid unit. *Natural Product Reports* **22**, 73–103. doi:10.1039/b316241a

Son NT (2019) Genus *Erythrophleum*: botanical description, traditional use, phytochemistry and pharmacology. *Phytochemistry Reviews* **18**, 571–599. doi:10.1007/s11101-019-09640-0

Sora I, Hall FS, Andrews AM, Itokawa M, Li XF *et al.* (2001) Molecular mechanism of cocaine reward: combined dopamine and serotonin transporter knockouts eliminate cocaine place preference. *Proceedings of the National Academy of Sciences of the United States of America* **98**, 5300–5305. doi:10.1073/pnas.091039298

Southwell IA (1981) Methoxystyrenes from the genus *Zieria*. *Phytochemistry* **20**, 1448–1450. doi:10.1016/0031-9422(81)80068-4

Southwell IA, Armstrong JA (1987) Chemical variation within the genus *Zieria*. *Phytochemistry* **26**, 1687–1692. doi:10.1016/S0031-9422(00)82269-4

Southwell IA, Tucker DJ (1993) Cis-dihydroagarofuran from *Prostanthera* sp. aff. *ovalifolia*. *Phytochemistry* **33**, 857–862. doi:10.1016/0031-9422(93)85290-8

Spenser ID (1985) Stereochemical aspects of the biosynthetic routes leading to the pyrrolizidine and quinolizidine alkaloids. *Pure and Applied Chemistry* **57**, 453–470. doi:10.1351/pac198557030453

Sriphana U, Thongsri Y, Ardwichai P, Poopasit K, Prariyachatigul C *et al.* (2013) New lignan esters from *Alyxia schlechteri* and antifungal activity against *Pythium insidiosum*. *Fitoterapia* **91**, 39–43. doi:10.1016/j.fitote.2013.08.005

Stohs SJ, Ray SD (2013) A review and evaluation of the efficacy and safety of *Cissus quadrangularis* extracts. *Phytotherapy Research* **27**, 1107–1114. doi:10.1002/ptr.4846

Strotmann H, Brendel K, Boos KS, Schlimme E (1982) Energy transfer inhibition in photosynthesis by anthraquinone dyes. *FEBS Letters* **145**, 11–15. doi:10.1016/0014-5793(82)81196-4

Su X-H, Zhang M-L, Li L-G, Huo C-H, Gu Y-C *et al.* (2008) Chemical constituents of the plants of the genus *Calophyllum*. *Chemistry & Biodiversity* **5**, 2579–2608. doi:10.1002/cbdv.200890215

Su Q, Dalal S, Goetz M, Cassera MB, Kingston DG (2016) Antiplasmodial phloroglucinol derivatives from *Syncarpia glomulifera*. *Bioorganic & Medicinal Chemistry* **24**, 2544–2548. doi:10.1016/j.bmc.2016.04.020

Sugimoto S, Matsunami K, Otsuka H (2018) Biological activity of *Entada phaseoloides* and *Entada rheedei*. *Journal of Natural Medicines* **72**, 12–19. doi:10.1007/s11418-017-1146-x

Sultana N, Sarker SD, Armstrong JA, Wilson PG, Waterman PG (2003) The coumarins of *Philotheca* sensu lato: distribution and systematic significance. *Biochemical Systematics and Ecology* **31**, 681–691. doi:10.1016/S0305-1978(02)00226-0

Sun L, Wang Z, Wang Y, Xu J, He X (2020) Anti-proliferative and anti-neuroinflammatory eudesmanolides from *Wedelia (Sphagneticola trilobata* (L.) Pruski). *Fitoterapia* **142**, 104452. doi:10.1016/j.fitote.2019.104452

Süntar I (2020) Importance of ethnopharmacological studies in drug discovery: role of medicinal plants. *Phytochemistry Reviews* **19**, 1199–1209. doi:10.1007/s11101-019-09629-9

Sushko Y, Novotarskyi S, Körner R, Vogt J, Abdelaziz A *et al.* (2014) Prediction-driven matched molecular pairs to interpret QSARs and aid the molecular optimization process. *Journal of Cheminformatics* **6**, 48. doi:10.1186/s13321-014-0048-0

Sutherland MD, Rodwell JL (1989) Terpenoid chemistry. XXVIII: furanosesquiterpene B-ketols from *Myoporum betcheanum*, *M. deserti*, *M. montanum* and other Myoporaceae. *Australian Journal of Chemistry* **42**, 1995–2019. doi:10.1071/CH9891995

Suthiwong J, Thongsri Y, Yenjai C (2017) A new furanocoumarin from the fruits of *Scaevola taccada* and antifungal activity against *Pythium insidiosum*. *Natural Product Research* **31**, 453–459. doi:10.1080/14786419.2016.1188100

Suvarna CHM, Sriya P, Pavan KA (2018) A review on phytochemical and pharmacological properties of *Thespesia populnea*. *Journal of Drug Delivery and Therapeutics* **8**, 1–4. doi:10.22270/jddt.v8i4.1718

Suzuki S, Umezawa T (2007) Biosynthesis of lignans and norlignans. *Journal of Wood Science* **53**, 273–284. doi:10.1007/s10086-007-0892-x

Syros V (2013) Galenic medicine and social stability in early modern Florence and the Islamic empires. *Journal of Early Modern History* **17**, 161–213. doi:10.1163/15700658-12342361

Tan CY, Wang JH, Zhang H, Li X (2002) A new phytosterone from *Cyanotis arachnoidea*. *Journal of Asian Natural Products Research* **4**, 7–11. doi:10.1080/10286020290019631

Tan CY, Wang JH, Li X (2003) Phytoecdysteroid constituents from *Cyanotis arachnoidea*. *Journal of Asian Natural Products Research* **5**, 237–240. doi:10.1080/1028602031000105830

Tan YP, Houston SD, Modhiran N, Savchenko AI, Boyle GM *et al.* (2019a) Stachyonic Acid: A Dengue Virus Inhibitor from Basilicum polystachyon. *Chemistry (Basel, Switzerland)* **25**, 5664–5667.

Tan YP, Xue Y, Savchenko AI, Houston SD, Modhiran N *et al.* (2019b) Basimarols A, B, and C, Highly Oxygenated Pimarane Diterpenoids from Basilicum polystachyon. *Journal of Natural Products* **82**, 2828–2834. doi:10.1021/acs.jnatprod.9b00522

Tanaka N, Yuhara H, Wada H, Murakami T, Cambie RC *et al.* (1993) Phenolic constituents of *Pteridium esculentum*. *Phytochemistry* **32**, 1037–1039. doi:10.1016/0031-9422(93)85251-L

Tang D, Dong Y, Ren H, Li L, He C (2014) A review of phytochemistry, metabolite changes, and medicinal uses of the common food mung bean and its sprouts (*Vigna radiata*). *Chemistry Central Journal* **8**, 4. doi:10.1186/1752-153X-8-4

Tang KSC, Konczak I, Zhao J (2017) Phenolic compounds of the Australian native herb *Prostanthera rotundifolia* and their biological activities. *Food Chemistry* **233**, 530–539. doi:10.1016/j.foodchem.2017.04.137

Tchinda AT, Tamze V, Ngono ARN, Ayimele GA, Cao M *et al.* (2012) Alkaloids from the stem bark of *Strychnos icaja*. *Phytochemistry Letters* **5**, 108–113. doi:10.1016/j.phytol.2011.11.004

Teponno RB, Kusari S, Spiteller M (2016) Recent advances in research on lignans and neolignans. *Natural Product Reports* **33**, 1044–1092. doi:10.1039/C6NP00021E

Thabet AA, Youssef FS, El-Shazly M, Singab ANB (2018) *Sterculia* and *Brachychiton*: a comprehensive overview on their ethnopharmacology, biological activities, phytochemistry and the role of their gummy exudates in drug delivery. *The Journal of Pharmacy and Pharmacology* **70**, 450–474. doi:10.1111/jphp.12876

That QT, Jossang J, Jossang A, Kim PPN, Jaureguiberry G (2007) Wedelolides A and B: novel sesquiterpene sigma-lactones, (9R)-eudesman-9,12-olides, from *Wedelia trilobata. The Journal of Organic Chemistry* **72**, 7102–7105. doi:10.1021/jo070771m

Thimmappa R, Geisler K, Louveau T, O'maille P, Osbourn A (2014) Triterpene biosynthesis in plants. *Annual Review of Plant Biology* **65**, 225–257. doi:10.1146/annurev-arplant-050312-120229

Tian LW, Zhang Z, Long HL, Zhang YJ (2017) Steroidal saponins from the genus *Smilax* and their biological activities. *Natural Products and Bioprospecting* **7**, 283–298. doi:10.1007/s13659-017-0139-5

Tillequin F (2002) *Sarcomelicope* alkaloids as leads for the discovery of new antitumor acronycine derivatives. *Phytochemistry Reviews* **1**, 355–368. doi:10.1023/A:1026076825006

Tillequin F (2007) Rutaceous alkaloids as models for the design of novel antitumor drugs. *Phytochemistry Reviews* **6**, 65–79. doi:10.1007/s11101-006-9010-8

Tiziano D (1999) Chronology. In *The Cambridge history of Hellenistic philosophy*. (Ed. A Keimpe) pp. 52–53. Cambridge University Press, Cambridge, UK.

Tlili N, Elfalleh W, Saadaoui E, Khaldi A, Triki S *et al.* (2011) The caper (*Capparis* L.): ethnopharmacology, phytochemical and pharmacological properties. *Fitoterapia* **82**, 93–101. doi:10.1016/j.fitote.2010.09.006

Tofern B, Mann P, Kaloga M, Jenett-Siems K, Witte L *et al.* (1999) Aliphatic purrolidine amides from two tropical convolvulaceous species. *Phytochemistry* **52**, 1437–1441. doi:10.1016/S0031-9422(99)00245-9

Tomas-Barberan FA, Grayer-Barkmeijer RJ, Gil MI, Harborne JB (1988) Distribution of 6-hydroxy-, 6-methoxy- and 8-hydroxyflavone glycosides in the Labiatae, the Scrophulariaceae and related families. *Phytochemistry* **27**, 2631–2645. doi:10.1016/0031-9422(88)87034-1

Torrsell KBG (1983) *Natural Product Chemistry. A Mechanistic and Biosynthetic Approach to Secondary Metabolism*. Wiley & Sons, Chichester, UK.

Toth B, Hohmann J, Vasas A (2018) Phenanthrenes: a promising group of plant secondary metabolites. *Journal of Natural Products* **81**, 661–678. doi:10.1021/acs.jnatprod.7b00619

Touqeer S, Saeed MA, Ajaib M (2013) A review on the phytochemistry and pharmacology of genus *Tephrosia*. *Phytopharmacology* **4**, 598–637.

Tsai T-H, Wang G-J, Lin L-C (2008) Vasorelaxing alkaloids and flavonoids from *Cassytha filiformis*. *Journal of Natural Products* **71**, 289–291. doi:10.1021/np070564h

Tschanz DW (2003) Arab roots of European medicine. *Heart Views* **4**, 9–10[History of Medicine].

Tschesche R, Delhvi S, Sepulveda S, Breitmaier E (1979) Eucryphin, a new chromone rhamnoside from the bark of *Eucryphia cordifolia*. *Phytochemistry* **18**, 867–869. doi:10.1016/0031-9422(79)80032-1

Tse GG, Kim BB, Mcmurtray AM, Nakamoto BK (2013) Case of levodopa toxicity from ingestion of *Mucuna gigantea*. *Hawai'i Journal of Medicine & Public Health : a Journal of Asia Pacific Medicine & Public Health* **72**, 157–160.

Tu Y, Sun L, Guo M, Chen W (2013) The medicinal uses of *Callicarpa* L. in traditional Chinese medicine: an ethnopharmacological, phytochemical and pharmacological review. *Journal of Ethnopharmacology* **146**, 465–481. doi:10.1016/j.jep.2012.12.051

Ullah W, Uddin G, Siddiqui BS (2012) Ethnic uses, pharmacological and phytochemical profile of genus *Grewia*. *Journal of Asian Natural Products Research* **14**, 186–195. doi:10.1080/10286020.2011.639764

Ulubelen A, Topcu G, Sonmez U (2000) Chemical and biological evaluation of genus *Teucrium*. *Studies in Natural Products Chemistry* **23**, 591–648. doi:10.1016/S1572-5995(00)80139-8

Umscheid CA, Margolis DJ, Grossman CE (2011) Key concepts of clinical trials: a narrative review. *Postgraduate Medical Journal* **123**, 194–204. doi:10.3810/pgm.2011.09.2475

Urban J, Dahlberg C, Carroll B, Kaminsky W (2013) Absolute configuration of beer's bitter compounds. *Angewandte Chemie International Edition* **52**, 1553–1555. doi:10.1002/anie.201208450

Urzua AM, Cassels BK (1978) Alkaloid chemosystematics, chemotaxonomy and biogenesis in the Atherospermataceae. *Lloydia* **41**, 98–113.

Vadivel V, Sriram S, Brindha P (2016) Distribution of flavonoids among Malvaceae family members: a review. *International Journal of Green Pharmacy* **10**, s33–s45.

Valderrama JCM (2000) Distribution of flavonoids in the Myristicaceae. *Phytochemistry* **55**, 505–511. doi:10.1016/S0031-9422(00)00114-X

Van Beek TA, Verpoorte R, Svendsen AB, Leeuwenberg AJM, Bisset NG (1984) *Tabernaemontana* L. (Apocynaceae): a review of its taxonomy, phytochemistry, ethnobotany and pharmacology. *Journal of Ethnopharmacology* **10**, 1–156. doi:10.1016/0378-8741(84)90046-1

Van Santen JA, Kautsar SA, Medema MH, Linington RG (2021) Microbial natural product databases: moving forward in the multi-omics era. *Natural Product Reports* **38**, 264–278. doi:10.1039/D0NP00053A

Vasudeva N, Sharma SK (2008) Biologically active compounds from the genus *Hibiscus*. *Pharmaceutical Biology* **46**, 145–153. doi:10.1080/13880200701575320

Veitch NC, Grayer RJ (2011) Flavonoids and their glycosides, including anthocyanins. *Natural Product Reports* **28**, 1626–1695. doi:10.1039/c1np00044f

Ventola CL (2015) The antibiotic resistance crisis. Part 1: causes and threats. *Pharmacology & Therapeutics* **40**, 277–283.

Verpoorte R, Van Der Heuden R, Van Gulik WM, Ten Hoopen HJG (1991) Plant biotechnology for the production of alkaloids. In *The Alkaloids*. (Ed. A Brossi, A) pp. 1–188. Academic Press, New York, NY.

Ververidis F, Trantas E, Douglas C, Vollmer G, Kretzschmar G *et al.* (2007a) Biotechnology of flavonoids and other phenylpropanoid-derived natural products. Part I:

Chemical diversity, impacts on plant biology and human health. *Biotechnology Journal* **2**, 1214–1234. doi:10.1002/biot.200700084

Ververidis F, Trantas E, Douglas C, Vollmer G, Kretzschmar G *et al.* (2007b) Biotechnology of flavonoids and other phenylpropanoid-derived natural products. Part II: Reconstruction of multienzyme pathways in plants and microbes. *Biotechnology Journal* **2**, 1235–1249. doi:10.1002/biot.200700184

Vieira LMM, Kijjoa A (2005) Naturally-occuring xanthones: recent developments. *Current Medicinal Chemistry* **12**, 2413–2446. doi:10.2174/092986705774370682

Vitaku E, Smith DT, Njardarson JT (2014) Analysis of the structural diversity, substitution patterns, and frequency of nitrogen heterocycles among U.S. FDA approved pharmaceuticals. *Journal of Medicinal Chemistry* **57**, 10257–10274. doi:10.1021/jm501100b

Vitale GA, Geibel C, Minda V, Wang M, Aron AT *et al.* (2024) Connecting metabolome and phenotype: recent advances in functional metabolomics tools for the identification of bioactive natural products. *Natural Product Reports* **41**, 885–904. doi:10.1039/D3NP00050H

Vogt T (2010) Phenylpropanoid biosynthesis. *Molecular Plant* **3**, 2–20. doi:10.1093/mp/ssp106

Voigts L (1979) Anglo-Saxon plant remedies and the Anglo-Saxons. *Isis* **70**, 250–268. doi:10.1086/352199

Vora R, Joshi AN, Joshi NC (2018) Comparison of L-Dopa content in three species of genus Mucuna by different extraction techniques. *Annals of Plant Sciences* **7**, 1973–1977. doi:10.21746/aps.2018.7.1.10

Vu VT, Chen XL, Kong LY, Luo JG (2020) Melipatulinones A-C, three lignan-phloroglucinol hybrids from *Melicope patulinervia*. *Organic Letters* **22**, 1380–1384. doi:10.1021/acs.orglett.9b04680

Vuong QV, Chalmers AC, Bhuyan DJ, Bowyer MC, Scarlett CJ (2015) Botanical, phytochemical, and anticancer properties of the *Eucalyptus* species. *Chemistry & Biodiversity* **12**, 907–924. doi:10.1002/cbdv.201400327

Vuong P, Chong S, Kaur P (2022) The little things that matter: how bioprospecting microbial biodiversity can build towards the realization of United Nations Sustainable Development Goals. *npj. Biodiversity (Nepean)* **1**, 4.

Vuorela P, Leinonenb M, Saikkuc P, Tammelaa P, Rauhad J-P *et al.* (2004) Natural products in the process of finding new drug candidates. *Current Medicinal Chemistry* **11**, 1375–1389. doi:10.2174/0929867043365116

Wagner U, Kratky C (2015) Structure elucidation of natural compounds by X-ray crystallography. In *Progress in the Chemistry of Organic Natural Products*. (Eds AD Kinghorn, H Falk and J Kobayashi) pp. 1–75. Springer, Cham.

Wallace JW, Markham KR (1978) Apigenin and amentoflavone glycosides in the Psilotaceae and their phylogenetic significance. *Phytochemistry* **17**, 1313–1317. doi:10.1016/S0031-9422(00)94580-1

Walters DS, Craig R, Mumma RO (1990) Fatty acid incorporation in the biosynthesis of anacardic acids of geraniums. *Phytochemistry* **29**, 1815–1822. doi:10.1016/0031-9422(90)85022-8

Wang Y (2012) Extracting knowledge from failed development programmes. *Pharmaceutical Medicine* **26**, 91–96. doi:10.1007/BF03256897

Wang G-C, Zhang X-L, Wang Y-F, Li G-Q, Ye W-C *et al.* (2012) Four new dilignans from the roots of *Wikstroemia indica*. *Chemical & Pharmaceutical Bulletin* **60**, 920–923. doi:10.1248/cpb.c12-00132

Wang WX, Li TX, Ma H, Zhang JF, Jia AQ (2013) Tumoral cytotoxic and antioxidative phenylpropanoid glycosides in *Smilax riparia* A. DC. *Journal of Ethnopharmacology* **149**, 527–532. doi:10.1016/j.jep.2013.07.011

Wang YS, Wen ZQ, Li BT, Zhang HB, Yang JH (2016) Ethnobotany, phytochemistry, and pharmacology of the genus *Litsea*: An update. *Journal of Ethnopharmacology* **181**, 66–107. doi:10.1016/j.jep.2016.01.032

Wang JH, Luan F, He XD, Wang Y, Li MX (2018a) Traditional uses and pharmacological properties of *Clerodendrum* phytochemicals. *Journal of Traditional and Complementary Medicine* **8**, 24–38. doi:10.1016/j.jtcme.2017.04.001

Wang P, Yang J, Zhu Z, Zhang X (2018b) *Marsdenia tenacissima*: a review of traditional uses, phytochemistry and pharmacology. *The American Journal of Chinese Medicine* **46**, 1449–1480. doi:10.1142/S0192415X18500751

Wang XF, Liu FF, Zhu Z, Fang QQ, Qu SJ *et al.* (2019) Flueggenoids A - E, new dinorditerpenoids from *Flueggea virosa*. *Fitoterapia* **133**, 96–101. doi:10.1016/j.fitote.2018.12.025

Wang Z-F, You Y-L, Li F-F, Kong W-R, Wang S-Q (2021) Research progress of NMR in natural product quantification. *Molecules (Basel, Switzerland)* **26**, 6308. doi:10.3390/molecules26206308

Wang D, Li X, Miao Y, Zhang Q (2023) Profiling chemobiological connection between natural product and target space based on systematic analysis. *International Journal of Molecular Sciences* **24**, 11265. doi:10.3390/ijms241411265

Watanabe K, Plotnikoff GA, Sakiyama T, Reissenweber-Hewel H (2014) Collaboration of Japanese Kampo medicine and Western biomedicine. *Evidence-Based Complementary and Alternative Medicine* **2014**, 646947. doi:10.1155/2014/646947

Waterman PG (1998) Chemical taxonomy of alkaloids. In *Alkaloids: Biochemistry, Ecology, and Medical Applications*. (MF Roberts and M Wink) pp. 87–108. Springer Science+Business, New York, NY.

Waterman PG, Faulkner DF (1981) Imidazole alkaloids from *Cynometra hankei*. *Phytochemistry* **20**, 2765–2767. doi:10.1016/0031-9422(81)85283-1

Waterman PG, Grundon MF (1983) *Chemistry and Chemical Taxonomy of the Rutales*. Academic Press, London, UK.

Webster C (2008) *Paracelsus: Medicine, Magic and Mission at the End of Time*. Yale University Press, New Haven, CT.

Wehlauch R, Gademann K (2017) Securinega alkaloids: complex structures, potent bioactivities, and efficient total syntheses. *Asian Journal of Organic Chemistry* **6**, 1146–1159. doi:10.1002/ajoc.201700142

Wei X, Ruan W, Vrieling K (2021) Current knowledge and perspectives of pyrrolizidine alkaloids in pharmacological applications: A mini-review. *Molecules (Basel, Switzerland)* **26**, 1970. doi:10.3390/molecules26071970

Wellington KW (2015) Understanding cancer and the anticancer activities of naphthoquinones - a review. *RSC Advances* **5**, 20309–20338. doi:10.1039/C4RA13547D

Wesseling A-M, Demetrowitsch TJ, Schwarz K, Ober D (2017) Variability of pyrrolizidine alkaloid occurrence in species of the grass subfamily Pooideae (Poaceae). *Frontiers in Plant Science* **8**, 2046. doi:10.3389/fpls.2017.02046

Wezeman T, Brase S, Masters KS (2015) Xanthone dimers: a compound family which is both common and privileged. *Natural Product Reports* **32**, 6–28. doi:10.1039/C4NP00050A

Wikipedia (2022). *Category: Medicinal plants of Australia* [Online]. https://en.wikipedia.org/wiki/Category:Medicinal_plants_of_Australia (accessed 9 June 2022).

Williams CA, Harborne JB, Clifford HT (1971) Flavonoid patterns in the monocotyledons: flavanols and flavones in some families associated with the Poaceae. *Phytochemistry* **10**, 1059–1063. doi:10.1016/S0031-9422(00)89939-2

Wills A (1999) Herophilus, Erasistratus, and the birth of neuroscience. *Lancet* **354**, 1719–1720. doi:10.1016/S0140-6736(99)02081-4

Winchester B, Fleet GWJ (1992) Amino-sugar glycosidase inhibitors: versatile tools for glycobiologists. *Glycobiology* **2**, 199–210. doi:10.1093/glycob/2.3.199

Wink M (1987) Quinolizidine alkaloids: biochemistry, metabolism, and function in plants and cell suspension cultures. *Planta Medica* **53**, 509–514. doi:10.1055/s-2006-962797

Wink M (1993) Quinolizidine alkaloids. In *Alkaloids and Sulphur Compounds.* (Ed. PG Waterman) Academic Press, London, UK.

Wink M (1998a) Antimicrobially active alkaloids. In *Alkaloids: Biochemistry, Ecology, and Medical Applications.* (MF Roberts and M Wink) pp. 397–434. Springer Science+Business, New York, NY.

Wink M (1998b) Chemical ecology of alkaloids. In *Alkaloids: Biochemistry, Ecology, and Medical Applications.* (MF Roberts and M Wink) pp. 265–300. Springer Science+Business, New York, NY.

Wink M (1998c) Modes of action of alkaloids. In *Alkaloids: Biochemistry, Ecology, and Medical Applications.* (MF Roberts and M Wink) pp. 301-326. Springer Science+Business, New York, NY.

Wink M (1998d) A short history of alkaloids. In *Alkaloids: Biochemistry, Ecology, and Medical Applications.* (MF Roberts and M Wink) pp. 11–44. Springer Science+Business, New York, NY.

Wink M (2010) Introduction: biochemistry physiology and ecological functions of secondary metabolites. In *Biochemistry of Plant Secondary Metabolism.* (Ed. M Wink) pp. 1–19 Wiley-Blackwell, Sussex.

Wink M (2013) Evolution of secondary metabolites in legumes (Fabaceae). *South African Journal of Botany* **89**, 164–175. doi:10.1016/j.sajb.2013.06.006

Wink M (2019) Quinolizidine and pyrrolizidine alkaloid chemical ecology: a mini review on their similarities and differences. *Journal of Chemical Ecology* **45**, 109–115. doi:10.1007/s10886-018-1005-6

Wink M, Botschen F, Gosmann C, Schafer H, Waterman PG (2010) Chemotaxonomy seen from a phylogenetic perspective and evolution of secondary metabolism. *Annual Review of Plant Biology* **40**, 364–433.

Wiseman H (2006) Isoflavonoids and human health. In *Flavonoids: Chemistry, Biochemistry and Applications.* (Eds OM Anderson and KR Markham) pp. 371-396. CRC PRess, Boca Raton, FL.

Wollenweber E, Dorr M, Rozefelds AC, Minchin P, Forster PI (2000) Variation in flavonoid exudates in *Eucryphia* species from Australia and South America. *Biochemical Systematics and Ecology* **28**, 111–118. doi:10.1016/S0305-1978(99)00045-9

Wong M (1976) *La Médecine Chinoise par les Plantes.* Tchou, Paris, France.

Wong CH, Siah KW (2019) Estimation of clinical trial success rates and related parameters. *Biostatistics (Oxford, England)* **20**, 273–286. doi:10.1093/biostatistics/kxx069

Wooley JG (1993) Tropane alkaloids. In *Alkaloids and Sulphur Compounds.* (Ed. PG Waterman) pp. 133–173. Academic Press, New York, NY.

World Health Organization (2013) WHO Traditional Medicine Strategy: 2014–2023. https://www.who.int/publications/i/item/9789241506096 (accessed September 2024).

Wu P-L, Chuang T-H, He C-X, Wu T-S (2004) Cytotoxicity of phenylpropanoid esters from the stems of *Hibiscus taiwanensis*. *Bioorganic & Medicinal Chemistry* **12**, 2193–2197. doi:10.1016/j.bmc.2004.02.020

Wu P-L, Wu T-S, He C-X, Su C-H, Lee K-H (2005) Constituents from the stems of *Hibiscus taiwanensis*. *Chemical & Pharmaceutical Bulletin* **53**, 56–59. doi:10.1248/cpb.53.56

Wu P, Su MX, Wang Y, Wang GC, Ye WC *et al.* (2012) Supercritical fluid extraction assisted isolation of sesquiterpene lactones with antiproliferative effects from *Centipeda minima*. *Phytochemistry* **76**, 133–140. doi:10.1016/j.phytochem.2012.01.003

Xia X (2017) Bioinformatics and drug discovery. *Current Topics in Medicinal Chemistry* **17**, 1709–1726. doi:10.2174/1568026617666161116143440

Xiang C-P, Shi Y-N, Liu F-F, Li H-Z, Zhang Y-J *et al.* (2016) A survey of the chemical compounds of *Piper* spp. (Piperaceae) and their biological activities. *Natural Product Communications* **11**, 1403–1408. doi:10.1177/1934578X1601100948

Xiao J (2017) Dietary flavonoid aglycones and their glycosides: which show better biological significance? *Critical Reviews in Food Science and Nutrition* **57**, 1874–1905.

Xiao K, Zhang H J, Xuan L J, Zhang J, Xu Y-M *et al.* (2008) Stilbenoids: chemistry and bioactivities. *Studies in Natural Product Chemistry* **34**.

Xiao W, Lei F, Hengqiang Z, Xiaojing L (2013) Isolation and purification of natural products. In *Natural Product Extraction: Principles and Applications.* (Eds J Prado and

M Rostagno) pp. 314–362. Royal Society of Chemistry, London.

Xie X, Yoneyama K, Yoneyama K (2010) The strigolactone story. *Annual Review of Phytopathology* **48**, 93–117. doi:10.1146/annurev-phyto-073009-114453

Xu J, Han Q-B, Li S-L, Chen X-J, Wang X-N *et al.* (2013) Chemistry, bioactivity and quality control of *Dendrobium*, a commonly used tonic herb in traditional Chinese medicine. *Phytochemistry Reviews* **12**, 341–367. doi:10.1007/s11101-013-9310-8

Yadav JP, Arya V, Yadav S, Panghal M, Kumar S *et al.* (2010) *Cassia occidentalis* L.: a review on its ethnobotany, phytochemical and pharmacological profile. *Fitoterapia* **81**, 223–230. doi:10.1016/j.fitote.2009.09.008

Yadav N, Ganie SA, Singh B, Chhillar AK, Yadav SS (2019) Phytochemical constituents and ethnopharmacological properties of *Ageratum conyzoides* L. *Phytotherapy Research* **33**, 2163–2178. doi:10.1002/ptr.6405

Yamada K, Ojika M, Kigoshi H (2007) Ptaquiloside, the major toxin of bracken, and related terpene glycosides: chemistry, biology and ecology. *Natural Product Reports* **24**, 798–813. doi:10.1039/b614160a

Yamamoto M, Akita T, Koyama Y, Sueyoshi E, Matsunami K *et al.* (2008) Euodionosides A-G: megastigmane glucosides from leaves of *Euodia meliaefolia*. *Phytochemistry* **69**, 1586–1596. doi:10.1016/j.phytochem.2008.01.020

Yamazaki Y, Suh DY, Sitthithaworn W, Ishiguro K, Kobayashi Y *et al.* (2001) Diverse chalcone synthase superfamily enzymes from the most primitive vascular plant, *Psilotum nudum*. *Planta* **214**, 75–84. doi:10.1007/s004250100586

Yang S-Z (2008) *A Translation of the Shen Nong Ben Cao Jing*. Blue Poppy Press, Boulder, CO. https://archive.org/stream/ShenNongBenCaoLingTheDivineFarmersMateriaMedica/ShenNong%20Ben%20Cao%20ling%28The%20Divine%20Farmers%20Materia%20Medica%29_djvu.txt (accessed 9 June 2022).

Yang D, Eun H, Prabowo CPS, Cho S, Lee SY (2022) Metabolic and cellular engineering for the production of natural products. *Current Opinion in Biotechnology* **77**, 102760. doi:10.1016/j.copbio.2022.102760

Yang S-Y, Guo L-F, Liu Y, Zou J-B, Zhu H-Y *et al.* (2023) Trichosanates A–G and cucurbitacins W–Y, anticomplement monoterpenoids and cucurbitane-type triterpenoids from the pericarps of Trichosanthes kirilowii. *Bioorganic Chemistry* **139**, 106710. doi:10.1016/j.bioorg.2023.106710

Yao JL, Fang SM, Liu R, Oppong MB, Liu EW *et al.* (2016) A review on the terpenes from genus *Vitex*. *Molecules (Basel, Switzerland)* **21**, 1179. doi:10.3390/molecules21091179

Yao Q, Gao Y, Lai C, Wu C, Zhao CL *et al.* (2020) The phytochemistry, pharmacology and applications of *Melicope pteleifolia*: a review. *Journal of Ethnopharmacology* **251**, 112546. doi:10.1016/j.jep.2020.112546

Yeats RB (1979) Monoterpenoids. In *Terpenoids and Steroids*. (Ed. JR Hanson) pp. 3–80. J. W. Arrowsmith Ltd., Bristol, UK.

Yeung AWK, Tzvetkov NT, Balacheva AA, Georgieva MG, Gan R-Y *et al.* (2020) Lignans: quantitative analysis of the research literature. *Frontiers in Pharmacology* **11**, 37. doi:10.3389/fphar.2020.00037

Yin B-W, Shen L-R, Zhang M-L, Zhao L, Wang Y-L *et al.* (2008) Chemical constituents of plants from the genus *Excoecaria*. *Chemistry & Biodiversity* **5**, 2356–2371. doi:10.1002/cbdv.200890201

Yonekura-Sakakibara K, Higashi Y, Nakabayashi R (2019) The origin and evolution of plant flavonoid metabolism. *Frontiers in Plant Science* **10**, 943. doi:10.3389/fpls.2019.00943

Yu O, Jez JM (2008) Nature's assembly line: biosynthesis of simple phenylpropanoids and polyketides. *The Plant Journal* **54**, 750–762. doi:10.1111/j.1365-313X.2008.03436.x

Yu F, Takahashi T, Moriya J, Kawaura K, Yamakawa J *et al.* (2006) Traditional Chinese medicine and Kampo: a review from the distant past for the future. *The Journal of International Medical Research* **34**, 231–239. doi:10.1177/147323000603400301

Yu X, Tang L, Wu H, Zhang X, Luo H *et al.* (2018) Trichosanthis fructus: botany, traditional uses, phytochemistry and pharmacology. *Journal of Ethnopharmacology* **224**, 177–194. doi:10.1016/j.jep.2018.05.034

Yuan H, Ma Q, Ye L, Piao G (2016) The traditional medicine and modern medicine from natural products. *Molecules (Basel, Switzerland)* **21**, 559. doi:10.3390/molecules21050559

Zanin JL, De Carvalho BA, Martineli PS, Dos Santos MH, Lago JH *et al.* (2012) The genus *Caesalpinia* L. (Caesalpiniaceae): phytochemical and pharmacological characteristics. *Molecules (Basel, Switzerland)* **17**, 7887–7902. doi:10.3390/molecules17077887

Zdero C, Bohlmann F, King RM (1992) *Ent*-labdanes, manoyloxide and helipterol derivatives from *Chrysocephalum ambiguum*. *Phytochemistry* **31**, 1631–1638. doi:10.1016/0031-9422(92)83119-J

Zhan ZJ, Ying YM, Ma LF, Shan WG (2011) Natural disesquiterpenoids. *Natural Product Reports* **28**, 594–629. doi:10.1039/c0np00050g

Zhang H-J, Tamez PA, Hoang VD, Tan GT, Hung NV *et al.* (2001) Antimalarial compounds from *Rhaphidophora decursiva*. *Journal of Natural Products* **64**, 772–777. doi:10.1021/np010037c

Zhang H, Qiu S, Tamez P, Tan GT, Aydogmus Z *et al.* (2002) Antimalarial Agents from Plants II. Decursivine, A New Antimalarial Indole Alkaloid from *Rhaphidophora decursiva*. *Pharmaceutical Biology* **40**, 221–224. doi:10.1076/phbi.40.3.221.5832

Zhang B-J, Yan J-M, Wu Z-K, Liu Y-P, Bao M-F *et al.* (2013) Alkaloids from *Ochrosia borbonica*. *Helvetica Chimica Acta* **96**, 2288–2298. doi:10.1002/hlca.201300136

Zhang QL, Fu BM, Zhang ZJ (2017) Borneol, a novel agent that improves central nervous system drug delivery by enhancing blood-brain barrier permeability. *Drug Delivery* **24**, 1037–1044. doi:10.1080/10717544.2017.1346002

Zhang JH, Xin HL, Xu YM, Shen Y, He YQ *et al.* (2018) *Morinda officinalis* How.: a comprehensive review of traditional uses, phytochemistry and pharmacology. *Journal of Ethnopharmacology* **213**, 230–255. doi:10.1016/j.jep.2017.10.028

Zhang H-W, Lv C, Zhang L-J, Guo X, Shen Y-W *et al.* (2021a) Application of omics-and multi-omics-based techniques for natural product target discovery. *Biomedicine and Pharmacotherapy* **141**, 111833. doi:10.1016/j.biopha.2021.111833

Zhang J, Morris-Natschke SL, Ma D, Shang X-F, Yang C-J *et al.* (2021b) Biologically active indolizidine alkaloids. *Medicinal Research Reviews* **41**, 928–960. doi:10.1002/med.21747

Zhang J, Morris-Natschke SL, Ma D, Shang XF, Yang CJ *et al.* (2021c) Biologically active indolizidine alkaloids. *Medicinal Research Reviews* **41**, 928–960. doi:10.1002/med.21747

Zhao J, Ma J-S (2016) Phytochemicals and biological activities of the genus *Cinnamomum. Journal of Pharmacognosy and Phytochemistry* **4**, 27–34.

Zhao L, Huo C-H, Shen L-R, Yang Y, Zhang Q *et al.* (2010) Chemical constituents of plants from the genus *Melia. Chemistry & Biodiversity* **7**, 839–859. doi:10.1002/cbdv.200900043

Zhao L, Chang W-C, Xiao Y, Liu H-W, Liu P (2013) Methylerythritol phosphate pathway of isoprenoid biosynthesis. *Annual Review of Biochemistry* **82**, 497–530. doi:10.1146/annurev-biochem-052010-100934

Zhao N, Li ZL, Li DH, Sun YT, Shan *et al.* (2015) Quinolone and indole alkaloids from the fruits of *Euodia rutaecarpa* and their cytotoxicity against two human cancer cell lines. *Phytochemistry* **109**, 133–139. doi:10.1016/j.phytochem.2014.10.020

Zhao JY, Wang XJ, Liu Z, Meng FX, Sun SF *et al.* (2019) Nonadride and spirocyclic anhydride derivatives from the plant endophytic fungus *Talaromyces purpurogenus. Journal of Natural Products* **82**, 2953–2962. doi:10.1021/acs.jnatprod.9b00210

Zheng X, Wang W, Piao H, Xu W, Shi H *et al.* (2013) The genus *Gnaphalium* L. (Compositae): phytochemical and pharmacological characteristics. *Molecules (Basel, Switzerland)* **18**, 8298–8318. doi:10.3390/molecules18078298

Zhu X-Y, Zhang C-L, Lin Y, Dang M-Y (2020) Ferruginol alleviates inflammation in dextran sulfate sodium-induced colitis in mice through inhibiting COX-2, MMP-9 and NF-kB signaling. *Asian Pacific Journal of Tropical Biomedicine* **10**, 308–315. doi:10.4103/2221-1691.284945

Zi J, Mafu S, Peters RJ (2014) To gibberellins and beyond! Surveying the evolution of (di)terpenoid metabolism. *Annual Review of Plant Biology* **65**, 259–286. doi:10.1146/annurev-arplant-050213-035705

Zia-Ul-Haq M, Stankovic MS, Rizwan K, Feo VD (2013) *Grewia asiatica* L., a food plant with multiple uses. *Molecules (Basel, Switzerland)* **18**, 2663–2682. doi:10.3390/molecules18032663

Index by chemical entity

Note: page numbers in **bold** refer to chemical structures in figures.

Index by customary use

Index by plant genus

Note: page numbers in **bold** refer to entries in chemical composition tables.

Printed and bound by CPI Group (UK) Ltd, Croydon, CR0 4YY
13/09/2026
14961090-0001